Essential Topics in Applied Mathematics

Essential Topics in Applied Mathematics

Edited by
Gregory Rago

Willford Press
www.willfordpress.com

Published by Willford Press,
118-35 Queens Blvd., Suite 400,
Forest Hills, NY 11375, USA

ISBN: 978-1-64728-348-3

Cataloging-in-Publication Data

Essential topics in applied mathematics / edited by Gregory Rago.
p. cm.
Includes bibliographical references and index.
ISBN 978-1-64728-348-3
1. Mathematics. 2. Mathematical analysis. I. Rago, Gregory.
QA36 .E87 2022
510--dc23

For information on all Willford Press publications
visit our website at www.willfordpress.com

WILLFORD PRESS

Contents

Preface

This book has been a concerted effort by a group of academicians, researchers and scientists, who have contributed their research works for the realization of the book. This book has materialized in the wake of emerging advancements and innovations in this field. Therefore, the need of the hour was to compile all the required researches and disseminate the knowledge to a broad spectrum of people comprising of students, researchers and specialists of the field.

The application of mathematical methods by different fields such as engineering, science, business, industry and computer science is referred to as applied mathematics. It is a combination of mathematical science and specialized knowledge. Applied mathematics is also a professional speciality where the mathematical models are studied and formulated to work on practical problems. Some of the mathematical sciences which are closely related to applied mathematics are scientific computing, statistics, actuarial science, operations research and management science, mathematical economics, and applicable mathematics. The key theories and concepts used within this field are differential equations, approximation theory and applied probability. This book contains some path-breaking studies in the field of applied mathematics. It traces the progress of this field and highlights some of its key concepts and applications. The book is appropriate for students seeking detailed information in this area as well as for experts.

At the end of the preface, I would like to thank the authors for their brilliant chapters and the publisher for guiding us all-through the making of the book till its final stage. Also, I would like to thank my family for providing the support and encouragement throughout my academic career and research projects.

Editor

1

A Note on Caputo's Derivative Operator Interpretation in Economy

Hameed Ur Rehman,[1,2] Maslina Darus,[1] and Jamal Salah[3]

[1]School of Mathematical Sciences, Faculty of Science and Technology, Universiti Kebangsaan Malaysia, Bangi 43600 Selangor Darul Ehsan, Malaysia
[2]Center for Language and Foundation Studies, A' Sharqiyah University, Post Box No. 42, Post Code No. 400 Ibra, Oman
[3]College of Applied and Health Sciences, A' Sharqiyah University, Post Box No. 42, Post Code No. 400 Ibra, Oman

Correspondence should be addressed to Maslina Darus; maslina@ukm.edu.my

Academic Editor: Igor Andrianov

We propound the economic idea in terms of fractional derivatives, which involves the modified Caputo's fractional derivative operator. The suggested economic interpretation is based on a generalization of average count and marginal value of economic indicators. We use the concepts of $T-$ *indicators* which analyses the economic performance with the presence of memory. The reaction of economic agents due to recurrence identical alteration is minimized by using the modified Caputo's derivative operator of order λ instead of integer order derivative n. The two sides of Caputo's derivative are expressed by a brief time-line. The degree of attenuation is further depressed by involving the modified Caputo's operator.

1. Introduction

Fractional calculus is the field of Mathematical study that grows out of the traditional definition of calculus integral and derivative operators as the same way of fractional exponents is an outgrowth of exponents of integer value. The birth of fractional calculus occurred in a letter from G. F. A. de L'Hospital to G. W Leibniz in 1695 posing a possible question "what if the order of derivative $(d^n f/dt^n)$ such that $n = 1/2$", in his reply he wrote "it will lead to a paradox, from which one day useful consequences will be drawn"; for more details, refer to a book by M. Sen [1, p. 1] and an article by Machado [2]. The quoted conversation of the two famous mathematicians evoked interest among the researchers and the theory of noninteger orders was further developed, generalized, and formulated by famous mathematicians such as Riemann, Liouville, L Euler, Letnikov, Grunwald, Marchuad, Weyl, Riesz, Caputo, Abel, and others; see, for example, [3–6]. Fractional derivatives and fractional integrals have vast applications in the field of physics, mechanics, engineering, and biology [7, 8]. In recent progress the fractional derivatives and fractional integrals are applied to narrate the financial process [9–11] and economic process with memory [12]. Fractional derivatives and fractional integration have been interpreted in many ways such as geometric interpretation [13–16], informatic interpretation [17], and economic interpretation [18]. Some of the modern definitions of fractional derivative and fractional integrals are enlisted in the next section.

2. Modern Definitions of Fractional Calculus

Definition 1. Riemann defined the fractional integral of order $\alpha > 0$ as

$$I_a^\alpha f(x) := \frac{1}{\Gamma(\alpha)} \int_a^x (x-t)^{\alpha-1} f(t)\, dt. \tag{1}$$

Definition 2. Riemann-Liouville defined the fractional derivative of order $n-1 < \alpha \le n (n \in N)$ as

$$\begin{aligned} D_a^\alpha f(x) &:= \frac{d^n}{dx^n} I_a^{n-\alpha} f(x) \\ &= \frac{1}{\Gamma(n-\alpha)} \frac{d^n}{dx^n} \int_a^x (x-t)^{n-\alpha-1} f(t)\, dt. \end{aligned} \tag{2}$$

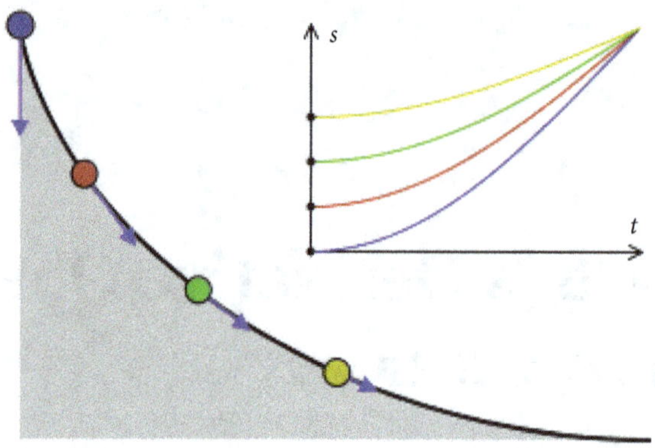

FIGURE 1: Four balls slide down a cycloid curve from different positions, but they arrive at the bottom at the same time, taken from [19].

Definition 3 (see [20]). Abel's integral equation of order α is given by

$$I_0^\alpha f(x) := \frac{1}{\Gamma(\alpha)} \int_a^x (x-t)^{\alpha-1} y(x)\, dx. \quad (3)$$

Remark 4. It is notable that Abel was the first one to use fractional calculus for $\alpha = 1/2$ and solved the famous tautochrone problem [20, 21]. For more details about the solution of problem and animated cycloid curve [19] see Figure 1. The total time of fall for $(f(y) = (2/\sqrt{2g})(ds/dy))$ is

$$T = \int_0^{y_o} (y_o - y)^{-1/2} f(y)\, dy, \quad (4)$$

which is in fact the application of fractional integral of order α, where the equation of cycloid is $ds/dy = T(\sqrt{2g}/\pi)(1/\sqrt{y})$.

3. Economic Meaning of Derivatives

Before elaborating the economic meaning of fractional derivatives, we put up the economic concept of the standard derivative of the first order. In terms of economy the first-order derivatives show the rapidness in alteration of economic indicators with respect to the investigated factor by assuming that other factors remain constant. The first-order derivative of the function of an indicator defines the marginal value of this indicator. The marginal value is the incremental value through the corresponding indicator per unit increase of the determining factor. In economic study, the principal marginal values of indicators are marginal product, margin utility, marginal profit, marginal cost, marginal revenue, marginal tendency to save and consume, marginal tax rate, marginal demand, and some others.

$$\text{Marginal Value} = \text{Cost of Incremental Output} - \text{Cost of Incremental Input}. \quad (5)$$

The economic processes is carried out by calculation of marginal and average value of indicators which are considered as a function of finding factors. According to the standard definition of average values and marginal values of indicators, a single value function as is supposed to be

$$Y = Y(X), \quad (6)$$

where Y is the dependence economic indicator Y by a factor X. The average value AY_X of the indicator Y is given by

$$AY_X := \frac{Y(X)}{X} \quad (7)$$

The economic definition of marginal value MY_X of this indicator is given by the first derivative of (6) and is supposed to be a single-valued function.

$$MY_X := \frac{dY(X)}{dX}, \quad (8)$$

where Y is the dependence economic indicator Y by a factor X.

Mathematically if the dependence $Y = Y(X)$ is directly proportional to $(Y(X) = C.X)$ then marginal and average values are identical such that $MY_X = AY_X$.

Remark 5. Graphical analysis from a sample graph given in Figure 2 chosen from [22] shows the relation between marginal cost and average cost; also see [23].

(1) When the average curve is rising, then $MY_X > AY_X$ so product should decrease.

(2) When the average curve is at its minimum, then $MY_X = AY_X$, showing the peak of the average product.

(3) When the average curve is falling, then $MY_X < AY_X$ so product should increase.

In [18], Tarasova and Tarasov have stated the condition for the applicability of (8) that it defines the marginal value of an indicator under assumption that an indicator Y must be expressed as a single valued function. In general this assumption is not valid [18]. Below are the two pair of examples in [18] of multivalued dependency of an indicator

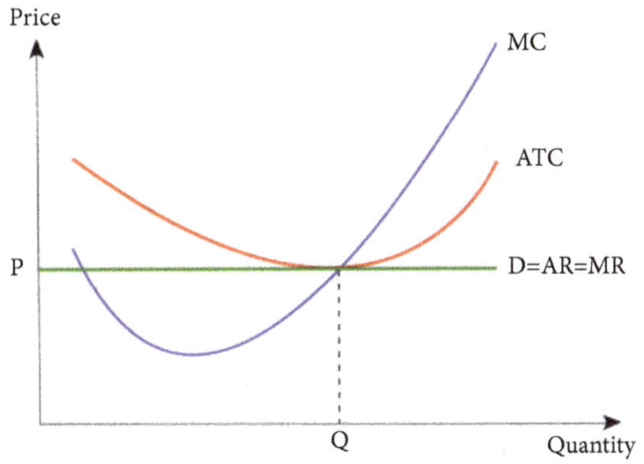

FIGURE 2: Three cost curves (average total cost, marginal cost, and marginal revenue) overlap.

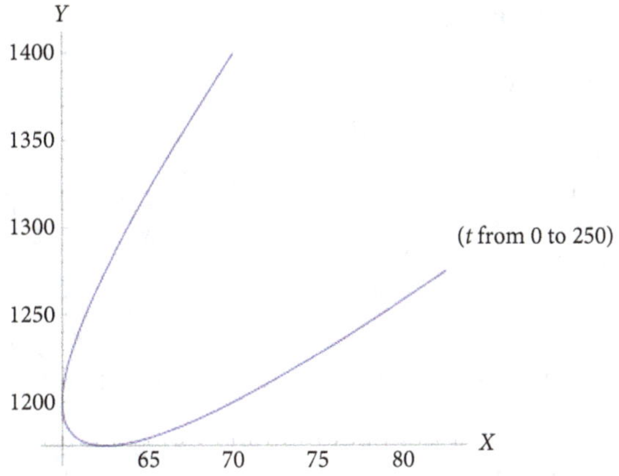

FIGURE 3: The graph in figure defined by equations (9), (10) shows the dependencies of an indicator Y on a factor X.

Y on a factor X, while involving the third variable the time (t) and the indicator and factor are the function of t.

$$X(t) = (0.001)t^2 - (0.2)t + 70, \tag{9}$$

$$Y(t) = (0.01)t^2 - (3.0)t + 1400 \tag{10}$$

and second example

$$X(t) = \left(8.2 \times 10^{-9}\right)t^4 - \left(1.5 \times 10^{-5}\right)t^3 + \left(5.4 \times 10^{-3}\right)t^2 - (0.58)t + 70, \tag{11}$$

$$Y(t) = \left(7.5 \times 10^{-6}\right)t^4 - \left(3.5 \times 10^{-3}\right)t^3 + (0.51)t^2 - (24.0)t + 1700. \tag{12}$$

Note that the above expressions are the simplest appropriate polynomial approximation related to the "real change in the dollar – ruble exchange rate" [24].

Remark 6. Graph in Figure 3 clearly shows that the value of X corresponds to more than one value of Y in many cases, which violates the uniqueness of the function.

So, the formulas enlisted in (7) and (8) are not applicable to determine the average and marginal values of an indicator, as the dependence Y on the factor X is not one to one. So, for this type of economic analysis we need to consider the indicators and factors as a single valued function of time. And hence the indicator $Y = Y(t)$ on a factor $X = X(t)$, then applying to (7) and (8), we obtain

$$AY_X(t) = \frac{Y(t)}{X(t)} \tag{13}$$

$$MY_X(t) = \frac{dY(t)}{dX(t)} \tag{14}$$

Let the average value and marginal value for time $t = T$, so we have

$$AY_X(T) = \frac{Y(T)}{X(T)}, \tag{15}$$

$$MY_X(T) = \left(\frac{dY(t)}{d(t)} \cdot \frac{1}{dX(t)/dt}\right)_{t=T} = \frac{dY(T)}{d(T)} \cdot \frac{1}{dX(T)/dT} \tag{16}$$

provided $X(T) \neq 0$ and $dX(T)/dT \neq 0$.

The expressions (8) and (16) are equivalent and obey the chain rule if the dependence of $Y(t)$ on $X(t)$ can be expressed

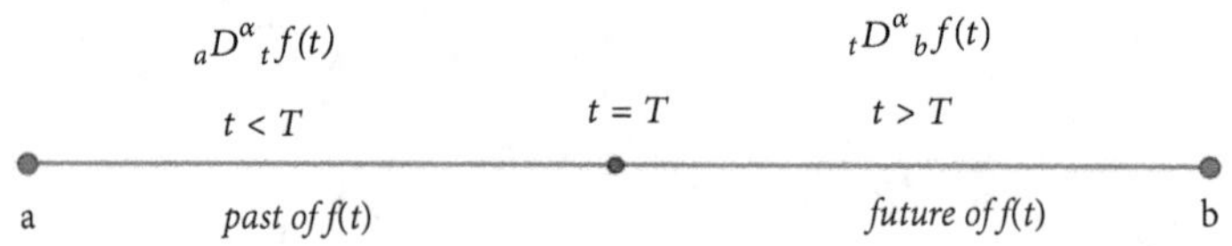

FIGURE 4: Action of derivative on three stages of functions.

as a single valued differentiable function $Y = Y(X)$. It is possible only if the function $X = X(t)$ is reversible, and hence (7) and (15) must be equivalent.

In the study of differential calculus, (16) can be considered as a generalized form of parametric derivative of indicator $Y = Y(t)$ by a factor $X = X(t)$ at time $t = T$, if $dX(T)/dT \neq 0$. It is important to note that (16) is the standard definition of the parametric derivative of first order if the function $X = X(t)$ has an inverse in neighborhood of $t = T$ and first order derivative exists of the functions $X = X(t)$, $Y = Y(t)$. Consequently, expressions (7) and (15) fail to be used for these dependencies; however, (15) and (16) can be used for parametric dependencies given by (9), (10) and/or (11), (12). So, we can say that (16) is associated with marginal indicator which concedes the first-order derivative as a growth of indicator Y per unit increase of the factor X at the given time point $t = T$.

It is worth mentioning that the violation of single valued property of the indicator function $Y(X)$ is due to the presence of memory in economic process in past [12, 18, 25]. Memory of economic agents influences the marginal indicators at the time $t = T$ and can depend on the changes of $Y(t)$ and $X(t)$ on finite time interval $(0, T)$. Average values and marginal values of indicators in (16) depend only on the given time $t = T$ and its diminutive neighborhood. Due to notable repeated similar alteration of the indicator $Y(t)$ and the factor $X(t)$ the economic agents can react on these identical changes. It is avoidable and can be applied only when the economic agents are in state of repression, and hence this approach cannot be always in practice for economic analysis. Mathematically this approach is because of applying integer order derivatives for economic analysis. So, the motivation is to involve a smaller degree to derive modified marginal values of an indicator function $Y(X)$ to put the past memory in a state of repression; see [18] for comparative modeling.

4. Involvement of Modified Caputo's Derivatives Operator in Economy

The Caputo's fractional derivative is one of the most used definitions of a fractional derivative along with other fractional calculus such as $(R-L)$, (*G. Letnikov*) [26] and Caputo-type modification of the Erdély-Kober fractional derivatives. Here it is necessary to mention the Caputo-type modification of the Erdély-Kober fractional derivatives proposed by Luchko and Trujillo in 2007; see [27, 28] defined as follows.

Definition 7 (see[27]). Let $n-1 < \delta \leq n, n \in \mathbb{N}$ and $m-1 < \alpha \leq m, m \in \mathbb{N}$. The operators $({}_*P_{\beta}^{\tau,\alpha} f)(x)$ and $({}_*D_{\beta}^{\gamma,\delta} f)(x)$ are, respectively, called the left-hand and right-hand sided Caputo-type modification of the Erdély-Kober fractional derivatives of orders α and δ, respectively:

$$\left({}_*P_{\beta}^{\tau,\alpha} f\right)(x) := \left(J_{\beta}^{\tau+\alpha,m-\alpha} \prod_{j=0}^{m-1} \left(\tau + j - \frac{1}{\beta} t \frac{d}{dt}\right) f \right)(x) \tag{17}$$

$$\left({}_*D_{\beta}^{\gamma,\delta} f\right)(x) := \left(I_{\beta}^{\gamma+\delta,n-\delta} \prod_{k=0}^{n-1} \left(1 + \gamma + k + \frac{1}{\beta} t \frac{d}{dt}\right) f \right)(x) \tag{18}$$

where $J_{\beta}^{\tau,\alpha}$ and $I_{\beta}^{\gamma,\delta}$ of order α and δ, respectively, called the left-hand and right-hand sided Erdély-Kober fractional integrals defined in [27].

The Caputo approach appears often while modeling applied problems by means of integrodifferential equations. Hence subsequent to our discussion in previous sections, the economic agents can react against identical changes in economic analysis and this is due to the use of integer order derivative. In continuation to this, the concept of noninteger derivative order [3, 6, 29] is used in natural sciences to describe the process with memory. Recently, fractional order derivatives have been used to outline the financial processes [12, 18, 25]. However in this article we investigate the involvement of the modified Caputo's fractional operator [30, 31] for our results. We would like to highlight the two sides of the Caputo's time-fractional derivative by a brief time-line (see Figure 4). We use the concept of left-hand side of Caputo's derivative for our results, concept taken from [1, p. 5].

Definition 8 (see[32]). "The left-hand side Caputo's fractional derivative of order α" is defined as follows:

$$D_0^{\alpha} f(t) = \begin{cases} \dfrac{1}{\Gamma(n-\alpha)} \displaystyle\int_0^t \frac{f^{(n)}(\tau)}{(t-\tau)^{\alpha-n+1}} d\tau, & n-1 < \alpha < n \in \mathcal{N} \\ \dfrac{d^n}{dt^n} f(t), & \alpha = n \in \mathcal{N} \end{cases} \tag{19}$$

Caputo's derivative of order $\alpha > 0$ of the power functions:

$${}_0^C D_T^{\alpha} t^{\beta} = \frac{\Gamma(\beta+1)}{\Gamma(\beta+1-\alpha)} t^{\beta-\alpha}, \quad (\beta > n-1,\ \alpha > n-1,\ t > 0,\ \beta \in \mathcal{R}) \tag{20}$$

Definition 9 (see [31]). Let $f \in A$ be analytic and normalized function. Then the modified Caputo's derivative operator of f is defined by

$$J_{\eta}^{\lambda} f(t) = \frac{\Gamma(2+\eta-\lambda)}{\Gamma(\eta-\lambda)} t^{\lambda-\eta} \int_0^t \frac{\Omega^{\eta} f(\zeta)}{(t-\zeta)^{\lambda-\eta+1}} d\zeta, \qquad (21)$$

where $\eta \in \mathcal{R}$, $(\eta - 1) < \lambda < \eta < 2$ and

$$\Omega^{\eta} f(\zeta) = \Gamma(2-\eta) \zeta^{n} D_{\zeta}^{\eta} f(\zeta) = \frac{\Gamma(2-\eta)\Gamma(n+1)}{\Gamma(n-\eta+1)} \zeta^{n} \qquad (22)$$

Remark 10. Note that $J_0^0 f(t) = f(T)$ and $J_1^1 f(t) = Tf'(T)$.

Definition 11 (see [31]). "Special Case of modified Caputo's derivative operator" is defined as follows: let $\eta = 0$ in Definition 2; we obtain

$$J_0^{\lambda} f(t) = \frac{\Gamma(2-\lambda)}{\Gamma(-\lambda)} t^{\lambda} \int_0^t \frac{f(\tau)}{(t-\tau)^{\lambda+1}} d\tau, \quad -1 < \lambda \le 0. \qquad (23)$$

Denoting the modified Caputo's fractional derivative operator by ${}_0^C J_{\eta}^{\lambda} f(t)$ and omitting the constant term $(\Gamma(\eta - \lambda)/\Gamma(2+\eta-\lambda))t^{\eta-\lambda}$, from (21) by using (22) together with the relation of β–function and γ–function [33], we obtain the power functions of modified Caputo's fractional derivative operator of order $\lambda > 0$ (same as in (20)); in the following form, see for details about below operator [30].

$${}_0^C J_{\eta}^{\lambda} f(t) = \frac{(\Gamma(n+1))^2 \Gamma(2-\eta)\Gamma(\eta-\lambda)}{\Gamma(n+\eta-\lambda+1)\Gamma(n-\eta+1)} t^{n-(\lambda-\eta)}, \qquad (24)$$

where $(\eta - 1 < \lambda < \eta < 2,\ \eta \in \mathcal{R},\ t > 0)$.

Remark 12. Note that the above power form of modified Caputo's fractional derivative corresponds to power function studied by Rubin & Wenzel (1996); see [34], "that is accuracy in a memory task at time t is given by $f(t) = at^{-b} : a, b \in \mathbb{R}$".

For $f(t) = t$, (24) becomes

$${}_0^C J_{\eta,T}^{\lambda} t = \frac{T^{1-(\lambda-\eta)}}{(\eta-\lambda)(\eta-\lambda+1)}. \qquad (25)$$

Considering the effects of memory in economic process the generalized concept of marginal and average values of indicator in terms of Caputo's Left-hand side derivative with $(t \in [t; T])$ is given by

$$MY_X(\alpha, T) = \frac{{}_0^C D_T^{\alpha} Y(t)}{{}_0^C D_T^{\alpha} X(t)}, \qquad (26)$$

where $Y = Y(t)$ is economic indicator and $X = X(t)$ is determining factor. The above equation shows the economic T–indicator at time $t = T$, which confines the economic process with memory, where ${}_0^C D_T^{\alpha}$ is the left-hand side derivative of order $\alpha \ge 0$ given in Definition 8. The parameter $\alpha \ge 0$ demarcates the degree of depression in the memory about the changes of the indicator and factor on the interval $[0, T]$.

Economic research studies show that memory effect leads to abnormality in economic growth. The memory effect with declining order $0 < \alpha < 1$ results in slower growth of output compared with the standard model without memory; on the other hand the descending order of $0 < \alpha < 2$ can give faster growth output, for more details on economic growth model see [35].

For instance to count on changes in results of marginal values, we introduce the modified Caputo's fractional derivative of order λ:

$$MY_X(\lambda, \eta, T) = \frac{{}_0^C J_{\eta,T}^{\lambda} Y(t)}{{}_0^C J_{\eta,T}^{\lambda} X(t)}, \quad (\eta - 1 < \lambda < \eta < 2,\ \eta \in \mathcal{R},\ t > 0), \qquad (27)$$

where ${}_0^C J_{\eta,T}^{\lambda}$ is given by Definition 9 or (24).

Remark 13. Note for $\eta = 0$ expression (27) coincides with the fractional order marginal value of (26) and for $\eta = 1$ it coincides with marginal value of integer order as in (16).

It is also usable to replace $(X(t) = t)$ for economic indicator. So, using (27) the above expression becomes

$$\begin{aligned} MY_X(\lambda, \eta, T) &= \frac{{}_0^C J_{\eta,T}^{\lambda} Y(t)}{{}_0^C J_{\eta,T}^{\lambda} t} \\ &= \frac{(\eta-\lambda)(\eta-\lambda+1)}{t^{1-(\lambda-\eta)}} \left({}_0^C J_{\eta,T}^{\lambda} Y(t)\right) \\ &= (\eta-\lambda)(\eta-\lambda+1) T^{(\lambda-\eta)-1} \left({}_0^C J_{\eta,T}^{\lambda} Y(t)\right), \end{aligned} \qquad (28)$$

where $(\lambda - \eta)$ characterizes the degree of depression of the memory about the changes of indicators and factor on the interval $[0, T]$.

For the sake of numerical analysis of marginal values we apply the integer order derivative, Caputo's fractional order derivative, and modified Caputo's fractional order derivative, respectively, shown in forthcoming equations and tabulation. Manipulating (9), (10), and (16), we obtain the marginal value through integer order:

$$MY_X(1) = \left(\frac{0.02t - 3}{0.002t - 0.2}\right)_{t=1} = 15.0505 \qquad (29)$$

Using (9), (10), and (20), we get the marginal value through Caputo's fractional order:

$$\begin{aligned} MY_X\left(\frac{1}{2}, 1\right) &= \left(\frac{0.0150451t^{3/2} - (6/\sqrt{\pi})t^{1/2} + 1400/\sqrt{\pi}\sqrt{t}}{0.00150451t^{3/2} - 0.225676t^{1/2} + 70/\sqrt{\pi}\sqrt{t}}\right)_{t=1} \\ &= 20.0284 \end{aligned} \qquad (30)$$

TABLE 1: Marginal values for $\eta \in (1/2, 1] \cup \{0\}$.

Depressive value	Marginal Value
η	$MY_X\left(\frac{1}{2}, \eta, 1\right)$
0	20.0284
33/64	20.0288
17/32	20.0292
9/16	20.0305
5/8	20.0336
11/16	20.0382
3/4	20.0455
13/16	20.0579
7/8	20.0832
15/16	20.1616
31/32	20.3265
63/64	20.6906
127/128	21.5932
255/256	24.6707
1	15.0403

Making use of (9), (10), and (27), we get the marginal value through modified Caputo's operator of order η:

$$MY_X\left(\frac{1}{2}, \eta, 1\right) = \frac{{}_0^C J_{\eta,1}^{1/2} Y(t)}{{}_0^C J_{\eta,1}^{1/2} X(t)}, \quad \left(\eta - 1 < \frac{1}{2} < \eta < 2,\ \eta \in \mathscr{R},\ t = 1 > 0\right), \tag{31}$$

where

$$\begin{aligned} {}_0^C J_{\eta,1}^{1/2} Y(t) &= \frac{\Gamma(2-\eta)\Gamma(\eta-0.5)}{(25)\Gamma(\eta+2.5)\Gamma(3-\eta)} t^{\eta+1.5} - \frac{(3)\Gamma(\eta-0.5)}{\Gamma(\eta+1.5)} t^{\eta+0.5} + \frac{(1400)\Gamma(2-\eta)}{\Gamma(\eta+0.5)\Gamma(1-\eta)} t^{\eta-0.5}, \\ {}_0^C J_{\eta,1}^{1/2} X(t) &= \frac{\Gamma(2-\eta)\Gamma(\eta-0.5)}{(250)\Gamma(\eta+2.5)\Gamma(3-\eta)} t^{\eta+1.5} - \frac{\Gamma(\eta-0.5)}{(5)\Gamma(\eta+1.5)} t^{\eta+0.5} + \frac{(70)\Gamma(2-\eta)}{\Gamma(\eta+0.5)\Gamma(1-\eta)} t^{\eta-0.5}. \end{aligned} \tag{32}$$

Now for different choices of $\eta : (\eta - 1 < 1/2 < \eta < 2)$ the marginal values by using the modified Caputo's operator in expression (31) are given in Table 1.

Remark 14. The sharpening behavior of the fractional values can be seen from (Table 1) as $\eta = 1-\epsilon : \epsilon > 0$ (the infinitesimal neighborhood before 1) results in maximum marginal value and on the other hand as $\eta = 1/2+\epsilon : \epsilon > 0$ (the infinitesimal neighborhood after 1/2) results in minimum marginal value. Note that two integer values of $\eta = \{0, 1\}$ where $\{0\}$ (is out of the interval) show anomalous character.

Some of the key notes are as follows:

(1) Equation (15) is the average values of indicator only for the values of the indicator and factor at times 0 and T.

(2) Equation (16) is the marginal values of indicator only for the values of the indicator and factor at $t = T$.

(3) The proposed economic indicator in (27) allows us to describe the dependence of economic process from all state of changes in a finite time interval $[0, T]$.

(4) The supplementary parameter $\eta \in \mathscr{R}$ lowers the rate of the past memory about the changes of indicator and factor on the interval $[0, T]$.

5. Conclusion

We used a modified Caputo's fractional operator towards the rectification of the indicator function $Y(X)$ by involving smaller degree $(\lambda - \eta)$ of forgetting memory. The influences of economic agents have been controlled comparative to the Caputo's fractional derivative of order $\alpha \geq 0$.

Disclosure

The earlier version of this research article was presented in the "First Innovation Conference" on 21 December 2017, At A'Sharqiyah University Ibra, Sultanate of Oman.

Conflicts of Interest

The authors declare that there are no conflicts of interest. All the authors agreed with the content of the manuscript.

Acknowledgments

The work here is supported by UKM grant GUP-2017-064.

References

[1] M. Sen, *Introduction to Fractional-Order Operators and Their Engineering Applications*, 2014.

[2] J. A. T. Machado, A. M. S. F. Galhano, M. S. F. Alexandra, and J. J. Trujillo, "On development of fractional calculus during the last fifty years," *Scientometrics*, vol. 98, no. 1, pp. 577–582, 2014.

[3] S. G. Samko, A. A. Kilbas, and O. I. Marichev, *Fractional Integrals and Derivatives, Theory and Applications*, Gordon and Breach, Pennsylvania, Pa, USA, 1993.

[4] V. S. Kiryakova, *Generalized Fractional Calculus and Applications*, CRC press, 1993.

[5] R. V. Mendes, *Introduction to Fractional Calculus (based on lectures by R. Gorenflo, F. Mainardi and I. Podlubny)*, 2008.

[6] A. A. Kilbas, H. M. Srivastava, and J. J. Trujillo, "Theory and applications of fractional differential equations," *North-Holland Mathematics Studies*, vol. 204, 2006.

[7] T. F. Nonnenmacher and R. Metzler, *Applications of Fractional Calculus Ideas to Biology*, World Scientific, 1998.

[8] R. W. Ibrahim and M. Darus, "Infective disease processes based on fractional differential equation," in *Proceedings of the 3rd International Conference on Mathematical Sciences, ICMS 2013*, pp. 696–703, Malaysia, December 2013.

[9] E. Scalas, R. Gorenflo, and F. Mainardi, "Fractional calculus and continuous-time finance," *Physica A: Statistical Mechanics and its Applications*, vol. 284, no. 1–4, pp. 376–384, 2000.

[10] N. Laskin, "Fractional market dynamics," *Physica A: Statistical Mechanics and its Applications*, vol. 287, no. 3-4, pp. 482–492, 2000.

[11] F. Mainardi, M. Raberto, R. Gorenflo, and E. Scalas, "Fractional calculus and continuous-time finance. II: The waiting-time distribution," *Physica A: Statistical Mechanics and its Applications*, vol. 287, no. 3-4, pp. 468–481, 2000.

[12] V. V. Tarasova and V. E. Tarasov, "Marginal utility for economic processes with memory," *Almanah Sovremennoj Nauki i Obrazovaniya [Almanac of Modern Science and Education]*, vol. 7, 2016.

[13] R. W. Ibrahim and M. Darus, "Differential operator generalized by fractional derivatives," *Miskolc Mathematical Notes*, vol. 12, no. 2, pp. 167–184, 2011.

[14] M. Darus and R. W. Ibrahim, "On classes of analytic functions containing generalization of integral operator," *Journal of the Indonesian Mathematical Society*, vol. 17, no. 1, pp. 29–38, 2011.

[15] J. Salah and M. Darus, "A note on generalized Mittag-Leffler function and applications," *Far East Journal of Mathematical Sciences (FJMS)*, vol. 48, no. 1, pp. 33–46, 2011.

[16] J. Salah, "Fekete-szegö problems involving certain integral operator," *International Journal of Mathematics Trends and Technology*, vol. 7, no. 1, pp. 54–60, 2014.

[17] V. E. Tarasov, "Interpretation of fractional derivatives as reconstruction from sequence of integer derivatives," *Fundamenta Informaticae*, vol. 151, no. 1-4, pp. 431–442, 2017.

[18] V. V. Tarasova and V. E. Tarasov, "Economic Interpretation of Fractional Derivatives," *Progress in Fractional Differentiation and Applications*, vol. 3, no. 1, pp. 1–7, 2017.

[19] Wikipedia, Tautochrone curve – wikipedia, the free encyclopedia, 2017.

[20] E. Hairer and G. Wanner, *L'analyse au fil de l'histoire*, vol. 10, Springer Science & Business Media, 2001.

[21] P. Glaschke, *Tautochrone and brachistochrone shape solutions for rocking rigid bodies*, 2016.

[22] Wikimedia Commons, Costcurve - combined.svg – wikimedia commons, the free media repository, 2014.

[23] AmosWEB Encyclonomic WEB*pedia, Average Product and Marginal Product, LLC, 2000-2018.

[24] V. V. Tarasova and V. E. Tarasov, "On applicability of point price elasticity of demand to exchange trading on us dollar," *Scientific Perspective*, vol. 6, pp. 6–11, 2016.

[25] V. E. Tarasov and V. V. Tarasova, "Long and short memory in economics: fractional-order difference and differentiation," *IRA-International Journal of Management and Social Sciences*, vol. 5, no. 2, pp. 327–334, 2016.

[26] C. Li, D. Qian, and Y. Q. Chen, "On Riemann-Liouville and Caputo derivatives," *Discrete Dynamics in Nature and Society*, vol. 2011, Article ID 562494, 15 pages, 2011.

[27] Y. Luchko and J. J. Trujillo, "Caputo-type modification of the Erdélyi-Kober fractional derivative," *Fractional Calculus and Applied Analysis*, vol. 10, no. 3, pp. 249–267, 2007.

[28] V. Kiryakova and Y. Luchko, "Riemann-Liouville and caputo type multiple Erdélyi-Kober operators," *Open Physics*, vol. 11, no. 10, pp. 1314–1336, 2013.

[29] K. Oldham and J. Spanier, *The Fractional Calculus Theory and Applications of Differentiation and Integration to Arbitrary Order*, vol. 111, Elsevier, 1974.

[30] J. Salah, "A note on the modified caputo's fractional calculus derivative operator," *International Journal of Pure and Applied Mathematics*, vol. 109, no. 3, pp. 665-67, 2016.

[31] J. Salah and M. Darus, "A subclass of uniformly convex functions associated with a fractional calculus operator involving Caputo's fractional differentiation," *Acta Universitatis Apulensis*, no. 24, pp. 295–306, 2010.

[32] M. Caputo, "Linear models of dissipation whose Q is almost frequency independent-II," *The Geophysical Journal of the Royal Astronomical Society*, vol. 13, no. 5, pp. 529–539, 1967.

[33] Wikipedia, Beta function – wikipedia, the free encyclopedia, 2018, https://en.wikipedia.org/wiki/Beta_function.

[34] M. J. Kahana and M. Adler, *Note on the Power Law of Forgetting*, 2017.

[35] V. V. Tarasova and V. E Tarasov, "Fractional Dynamics of Natural Growth And Memory Effect in Economics," *European Research*, vol. 12, no. 23, pp. 30–37, 2016.

2

Multiple Solutions of Mixed Convective MHD Casson Fluid Flow in a Channel

Jawad Raza, Azizah Mohd Rohni, and Zurni Omar

School of Quantitative Sciences, Universiti Utara Malaysia, 06010 Sintok, Kedah, Malaysia

Correspondence should be addressed to Jawad Raza; jawad_6890@yahoo.com

Academic Editor: Waqar A. Khan

A numerical investigation is made to determine the occurrence of the multiple solutions of MHD Casson fluid in a porous channel. Governing partial differential equation of the proposed problem converted into nonlinear ordinary differential equations by using similarity transformation. Numerical technique known as shooting method is used to investigate the existence of the multiple solutions for the variations of different parameters. Effects of physical parameters on velocity profile, temperature, concentration, and skin friction are presented in pictorial and tabulation representation.

1. Introduction

The fluid flow in a channel has abundant practical applications in industry from mathematical and engineering point of view [1–5]. In particular, those fluids in which the shear stresses are not linear proportional to the velocity gradient are characterized as non-Newtonian fluids which are of much interest among the researchers. Among the non-Newtonian fluids, Casson fluid has attracted more attention of researchers due to its applications in the fields of metallurgy, food processing, drilling operations, and bioengineering operations [6, 7]. Some more applications of Casson fluid can be seen in the manufacturing of pharmaceutical products, coal in water, china clay, paints, synthetic lubricants, and biological fluids such as synovial fluids, sewage sludge, jelly, tomato sauce, honey, soup, and blood due to its contents such as plasma, fibrinogen, and protein [8].

Due to the novel application of MHD mixed convection flow in porous medium, in the field of industrial engineering, many researchers are attracted towards it. Design of MHD power generators, nuclear waste processing, and distribution of chemical waste control are some most conspicuous applications among all. The problem of laminar fully developed mixed convection flow in a channel was followed back to 1960 by Tao [9]. Recently, Fersadou et al. [10] investigated the problem of MHD mixed convection flow of nanofluid in a vertical porous channel numerically. Problem of mixed convection MHD flow of an Al_2O_3 water nanofluid in a channel with asymmetric heated walls was examined by Chen et al. [11]. Khan et al. [12] investigated numerically the problem of Blasius and Sakiadis flows of Casson fluid with viscous dissipation and convective boundary conditions. Umavathi and Sultana [13] talked about the mixed convective hydromagnetic flow of micropolar fluid in a vertical channel with related boundary states of third kind. Si et al. [14] utilized homotopy analysis method (HAM) to acquire the arrangement of the problem of micropolar fluid in a channel with heat and mass exchange impacts. The channel is thought to be permeable with extending/contracting dividers. Prakash and Muthtamilselvan [15] executed Crank-Nicolson limited contrast plan to take care of numerical issue of mixed convective MHD stream of micropolar fluid between two vertical permeable dividers with proper limit states of third kind. Recently, Raza et al. [16] examined the rotational effects of channel on the problem of nanofluid with shrinking channel walls.

The Casson constitutive equation was derived by Casson [17] which shows that the rate of strain and stress relationship is nonlinear. Flow of Casson fluid between two rotating cylinders is studied by Eldabe and Salwa [18]. Attia and Sayed-Ahmed [19] considered the Couette flow of electrically conducting Casson fluid between parallel plates. Effect of mass transfer on MHD flow of Casson fluid was discussed analytically by Shehzad et al. [20]. Taylor's series was employed in order to solve nonlinear differential equations. Explicit finite difference method of unsteady Casson fluid flow through parallel plates was investigated by Afikuzzaman et al. [21]. Recently, Reddy et al. [22] discussed the effects of Joule heating and Hall effects on free convection in an electrically conducting Casson fluid in a vertical channel in the presence of viscous dissipation. Analytical solutions were found with the help of homotopy analysis method (HAM) and compared with Adomian Decomposition Method (ADM). Walawander et al. [23], Batra and Jena [24], Sayed Ahmed and Attia [25], Kataria and Patel [26], and Das et al. [27] have reported the flow of Casson fluid under different flow regimes. None of the investigations cited above dealt with multiple solutions of Casson fluid in a channel.

Motivated by the above-cited investigations, the aim of the present study is to investigate the multiple solutions of mixed convection flow of Casson fluid in a porous channel under the influence of the magnetic field. The effects of different parameters such as Reynold number R, magnetic field M, Casson parameter β, and bouncy parameter λ on velocity and temperature profiles are discussed graphically and also in tabulation representation.

2. Formulation of the Problem

Consider the steady, incompressible MHD flow of Casson fluid in a channel. The x-axis is along the centerline of the channel, parallel to the channel surfaces and the y-axis is perpendicular to it. Lower wall of the channel is located at $y = -H$ and upper wall is at $y = H$. The fluid is injected into the channel and extracted out at a uniform velocity V ($V > 0$ suction and $V < 0$ injection) from upper wall and lower wall, respectively. A uniform magnetic field of strength $B_{\circ}$ is applied perpendicular to the velocity field. The induced magnetic field is negligible as compared with the imposed field.

The constitutive equation for Casson fluid can be written as [17]

$$\tau_{ij} = \begin{cases} 2\left(\mu_B + \dfrac{\tau_y}{\sqrt{2\pi}}\right)e_{ij}, & \pi > \pi_c, \\ 2\left(\mu_B + \dfrac{\tau_y}{\sqrt{2\pi_c}}\right)e_{ij}, & \pi < \pi_c, \end{cases} \tag{1}$$

where μ_B is the plastic dynamic viscosity of the non-Newtonian fluid, τ_y is the yield stress of the fluid, π is the product of the component of deformation rate with itself, and π_c is critical value of π based on non-Newtonian model. Under these assumptions the governing equations for MHD boundary layer flow of Casson fluid are expressed as the following equations:

$$\frac{\partial u}{\partial x} + \frac{\partial v}{\partial y} = 0, \tag{2}$$

$$u\frac{\partial u}{\partial x} + v\frac{\partial u}{\partial y} = -\frac{1}{\rho}\frac{\partial p}{\partial x} + \upsilon\left(1 + \frac{1}{\beta}\right)\frac{\partial^2 u}{\partial y^2} - \frac{\sigma B_{\circ}^2 u}{\rho} \pm g\left(\beta_T\left(T - T_2\right) + \beta_C\left(C - C_2\right)\right), \tag{3}$$

$$u\frac{\partial v}{\partial x} + v\frac{\partial v}{\partial y} = -\frac{1}{\rho}\frac{\partial p}{\partial y} + \upsilon\left(1 + \frac{1}{\beta}\right)\frac{\partial^2 v}{\partial x^2}, \tag{4}$$

$$u\frac{\partial T}{\partial x} + v\frac{\partial T}{\partial y} = \frac{k}{\rho C_p}\frac{\partial^2 T}{\partial y^2}, \tag{5}$$

$$u\frac{\partial C}{\partial x} + v\frac{\partial C}{\partial y} = D\frac{\partial^2 C}{\partial y^2} - \kappa_1 C, \tag{6}$$

where ρ is density, μ is dynamic viscosity, ν is kinematic viscosity, σ is electrical conductivity, β is Casson fluid parameter, T is temperature of the fluid, k is thermal conductivity, κ_1 is reaction rate, D is mass diffusion, and C is the concentration field.

Along with boundary conditions,

$$\begin{aligned} u &= 0, \\ v &= \frac{V}{2}, \\ T &= T_2, \\ C &= C_2 \\ &\quad \text{at } y = H, \\ \frac{\partial u}{\partial y} &= 0, \\ v &= 0, \\ T &= T_1, \\ C &= C_1 \\ &\quad \text{at } y = 0. \end{aligned} \tag{7}$$

Introducing stream function such that $\bar{u} = \partial\psi/\partial y$, $\bar{v} = -\partial\psi/\partial x$ and eliminating pressure term from (3) and (4) by introducing vorticity ω, we get

$$u\frac{\partial \omega}{\partial x} + v\frac{\partial \omega}{\partial y} = \upsilon\left(1 + \frac{1}{\beta}\right)\left(\frac{\partial^2 \omega}{\partial x^2} + \frac{\partial^2 \omega}{\partial y^2}\right) - \frac{\sigma B_{\circ}^2 u'}{\rho} \pm \frac{\partial}{\partial y}\left(g\left(\beta_T\left(T - T_2\right) + \beta_C\left(C - C_2\right)\right)\right), \tag{8}$$

where $\omega = (\partial\bar{v}/\partial x - \partial\bar{u}/\partial y)$.

Define

$$\begin{aligned} x^* &= \frac{x}{H}, \\ y^* &= \frac{y}{H}, \\ u &= -Vx^* f'\left(y^*\right), \\ v &= Vf\left(y^*\right), \\ \theta\left(y^*\right) &= \frac{T - T_2}{T_1 - T_2}, \\ \phi\left(y^*\right) &= \frac{C - C_2}{C_1 - C_2}. \end{aligned} \tag{9}$$

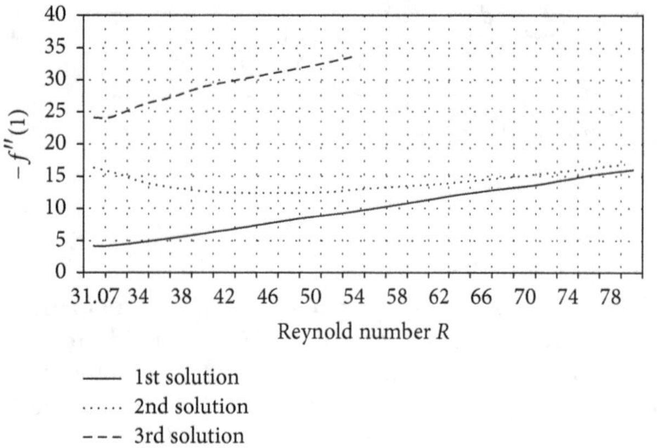

FIGURE 1: Skin friction $-f''(1)$ against the values of Reynold number R.

Then governing nonlinear momentum and energy equations of the proposed problem can be written as

$$\left(1+\frac{1}{\beta}\right) f'''' - M^2 f'' + R\left(f'f'' - ff'''\right) + \lambda\left(\theta' + N\phi'\right) = 0,$$
$$\theta'' - \Pr f\theta' = 0, \tag{10}$$
$$\varphi'' - \mathrm{Sc}\,\gamma\varphi - \mathrm{Sc} f\varphi' = 0.$$

Boundary conditions (7) reduce into

$$f(1) = \frac{1}{2}, \quad f'(1) = 0, \quad \theta(1) = 0, \quad \varphi(1) = 0, \quad f''(0) = 0, \quad f(0) = 0, \quad \theta(0) = 1, \quad \varphi(0) = 1. \tag{11}$$

Here, $R = VH/\nu$ is Reynolds number ($R > 0$ for suction $R < 0$ and for injection), $M^2 = \sigma B_{\circ}^2 H^2/\mu$ is Hartman number, $\lambda = Gr_x/R^2$ is the thermal buoyancy parameter, $Gr_x = VH^4 g\beta_T(T_1 - T_2)/x\nu^3$ is Grashof number, $N = \beta_C(C_1 - C_2)/(T_1 - T_2)$ is concentration buoyancy parameter, $\Pr = \rho C_P HV/\kappa$ is Prantl number, $\mathrm{Sc} = HV/D$ is Smith number, and $\gamma = \kappa_1 H/\nu$ is chemical reaction rate.

3. Numerical Solution of the Problem via Shooting Method

We employ shooting method in order to find the numerical solution of the proposed problem. For this we have to reduce the governing boundary value problem (10) into initial value problem by assuming $x_1 = \eta$, $x_2 = f$, $x_3 = f'$, $x_4 = f''$, $x_5 = f'''$, $x_6 = \theta'$, $x_7 = \varphi'$; then the following system is obtained

$$\begin{pmatrix} x_1' \\ x_2' \\ x_3' \\ x_4' \\ x_5' \\ x_6' \\ x_7' \end{pmatrix} = \begin{pmatrix} 1 \\ x_3 \\ x_4 \\ x_5 \\ \dfrac{\beta}{(1+\beta)}\left(M^2 x_4 - R\left(x_3x_4 - x_2x_5\right) - \lambda\left(x_6 + Nx_7\right)\right) \\ \Pr x_2 x_6 \\ \mathrm{Sc}\,\gamma\varphi + \mathrm{Sc}\,x_2x_7 \end{pmatrix}, \tag{12}$$

with initial condition

$$\begin{pmatrix} x_1(1) \\ x_2(1) \\ x_3(1) \\ x_4(1) \\ x_5(1) \\ x_6(1) \\ x_7(1) \end{pmatrix} = \begin{pmatrix} 1 \\ \frac{1}{2} \\ 0 \\ \alpha_1 \\ \alpha_2 \\ \alpha_3 \\ \alpha_4 \end{pmatrix}. \tag{13}$$

Here, $\alpha_1, \alpha_2, \alpha_3, \alpha_4$ are missing initial conditions. In a shooting method, the missing (unspecified) initial condition at the initial point of the interval is assumed, and the differential equation is then integrated numerically as an initial valued problem to the terminal point. The accuracy of the assumed missing initial condition is then checked by comparing the calculated value of the dependent variable at the terminal point with its given value there. If a difference exists, another value of the missing initial condition must be assumed and the process is repeated. This process is continued until the agreement between the calculated and the given condition at the terminal point is within the specified degree of accuracy. It is important to notice that we have to shoot the values of α_1, α_2, α_3, α_4. Since these values are not given in the boundary condition (13), by trial and error, suitable guess values are made and integration is carried out. The details of shooting method with Maple implementation *shoot* have been described by Meade et al. [28].

4. Results and Discussions

Present section is devoted to discuss the numerical results of our finding both in tabulation and in graphical form. Our main motive is to investigate the multiple solutions of the proposed problem. We have prepared Figures 1–6 in

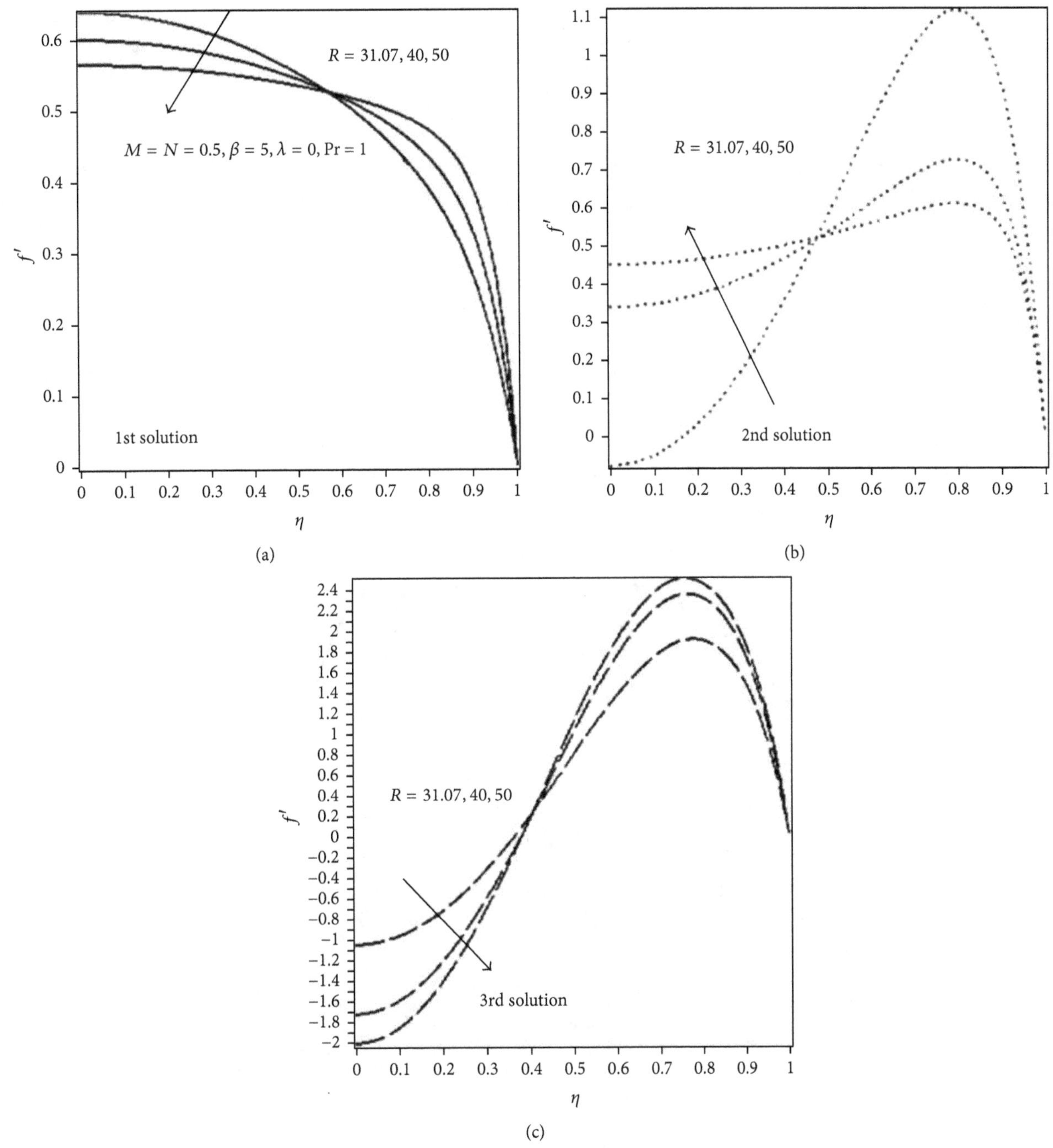

FIGURE 2: Effect of Reynold number R on velocity profile $f'(\eta)$.

order to evaluate the effects of different physical quantities on the velocity, temperature, and mass fraction. The effects include the existence of multiple solutions, Reynold number R, magnetic field M, Casson parameter β, Prandtl number Pr, Smith number Sc, and chemical reaction rate γ.

Based on the finding of multiple solutions it is concluded that there is only one solution in the case of $R < 0$, $\beta \in (0, \infty)$ and $\beta \in (0, 5)$, $R \in [0, \infty)$. However, there exist multiple solutions for $\beta \in [5, \infty)$ and $R \in [31.07, \infty)$ for any value of magnetic number $M \in [0, 2.0]$. So we can say that there is a critical value of Casson number β and suction parameter R such that $(\beta)_{\text{critical}} = 5$ and $(R)_{\text{critical}} = 31.07$. So it can be defended as there is no multiple solutions if $\beta < (\beta)_{\text{critical}}$ and $< (R)_{\text{critical}}$. The said phenomena can be observed from Figure 1. We plot the magnitude of the skin friction $|f''(1)|$ against the values of Reynold number R.

Effects of Reynold number R on velocity profile $f'(\eta)$ for non-buoyant flow case $\lambda = 0$ are presented in Figure 2. It is noticed that with enhanced values of Reynolds number R, velocity profile $f'(\eta)$ decreases near the center of the channel

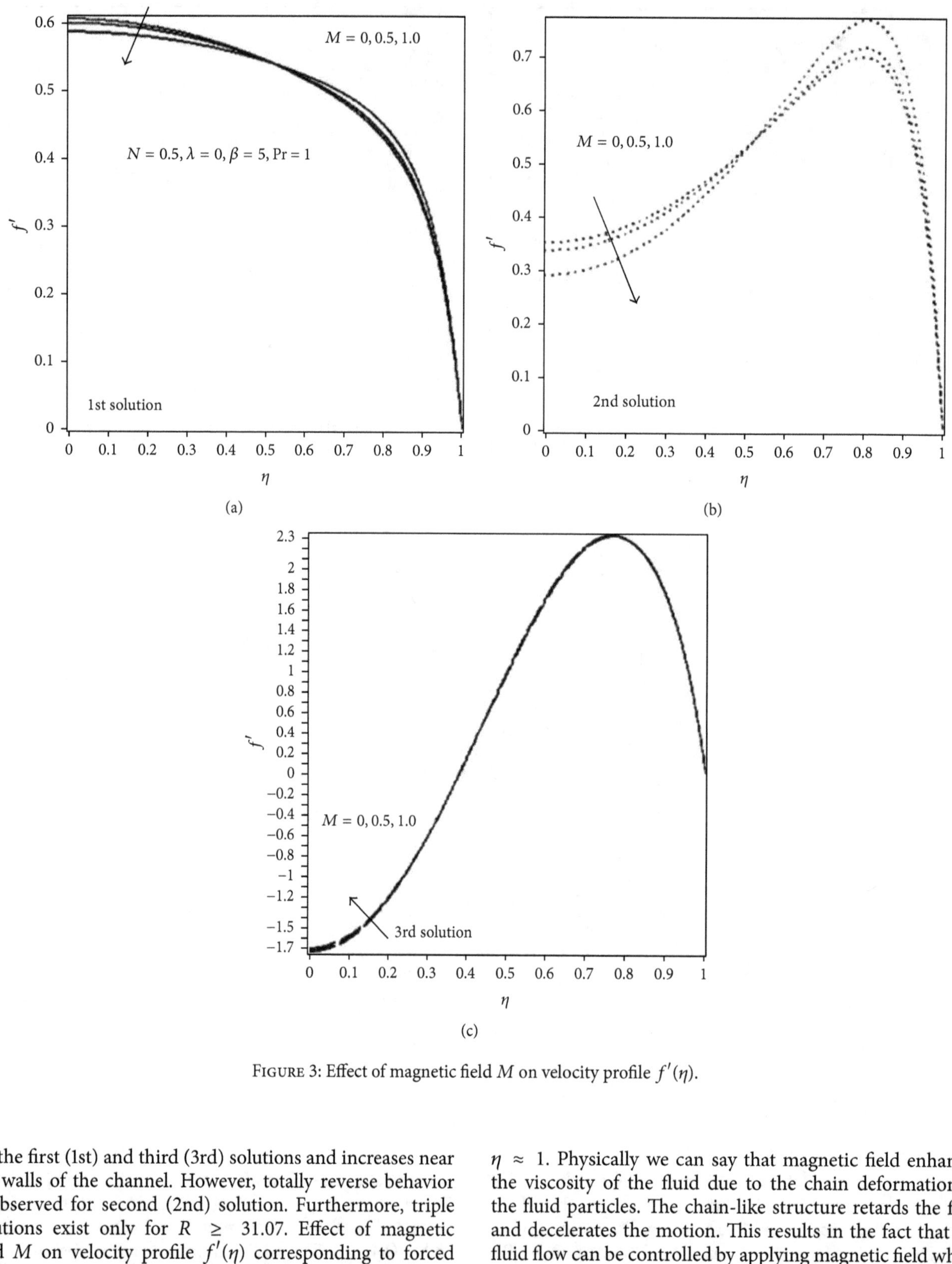

FIGURE 3: Effect of magnetic field M on velocity profile $f'(\eta)$.

for the first (1st) and third (3rd) solutions and increases near the walls of the channel. However, totally reverse behavior is observed for second (2nd) solution. Furthermore, triple solutions exist only for $R \geq 31.07$. Effect of magnetic field M on velocity profile $f'(\eta)$ corresponding to forced convection or non-buoyant flow case $\lambda = 0$ is shown in Figure 3. It is concluded that velocity profile $f'(\eta)$ for 1st and 2nd solutions decreases near the center of the channel $\eta \approx 0$ and increases near the channel walls. Since magnetic field is applied perpendicular to the channel walls, effect of magnetic field can be seen clearly near the channel walls $\eta \approx 1$. Physically we can say that magnetic field enhances the viscosity of the fluid due to the chain deformation of the fluid particles. The chain-like structure retards the flow and decelerates the motion. This results in the fact that the fluid flow can be controlled by applying magnetic field which results in many control based applications including MHD power generation, casting of metals, blood flow in arteries, and many, many more. Effect of Casson parameter β on velocity profile $f'(\eta)$ for the case of forced convection $\lambda = 0$ is depicted in Figure 4. Velocity profile $f'(\eta)$ decreases gradually for the 1st and 3rd solutions near the center of

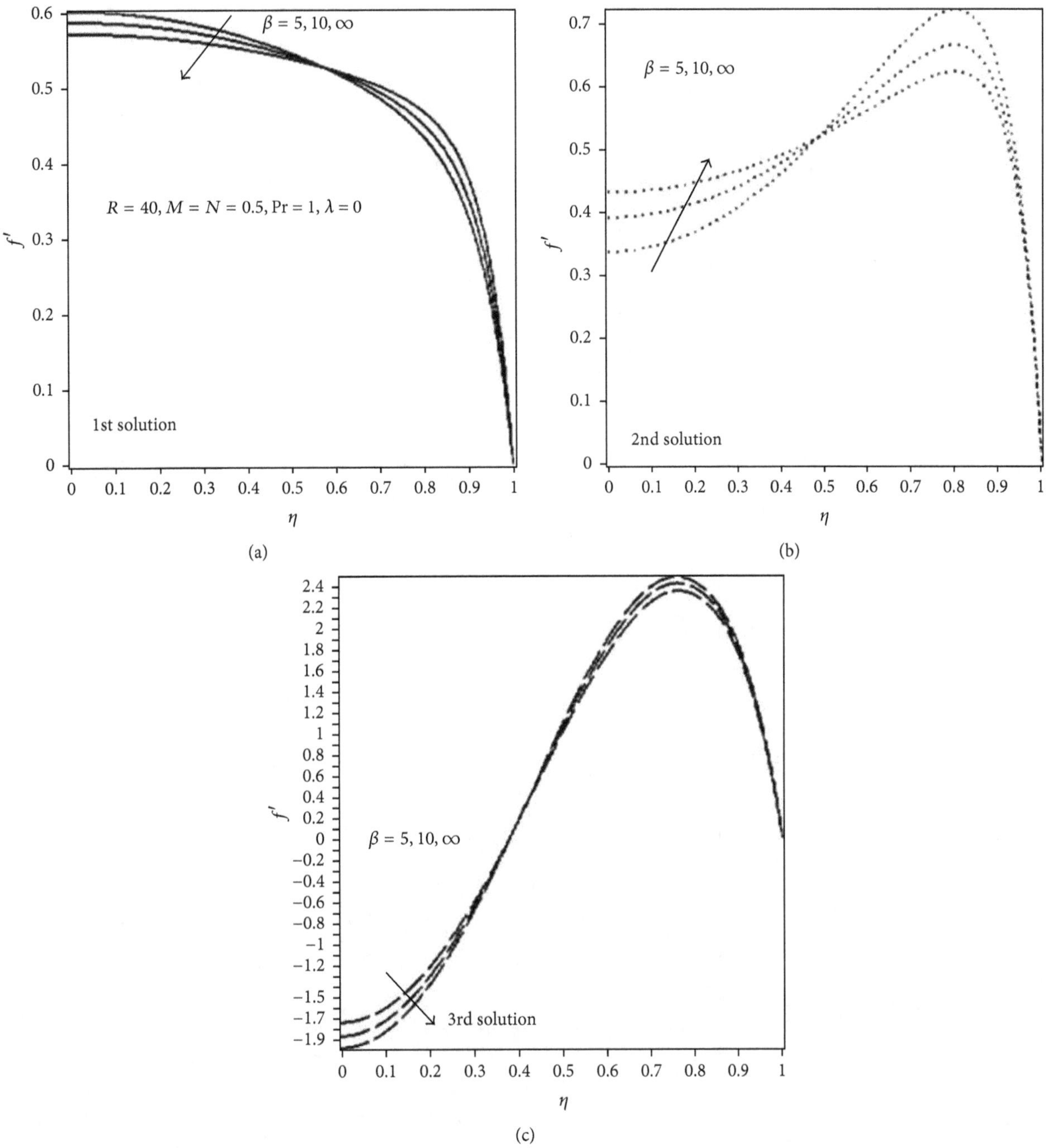

FIGURE 4: Effect of Casson parameter β on velocity profile $f'(\eta)$.

the channel and increases near the channel wall $\eta \approx 1$. However, velocity increases for the 2nd solution as the enhancement of the Casson parameter β is taking place in the neighborhood of $\eta \approx 0$. Furthermore, $\beta \to \infty$ corresponds to the Newtonian fluid. This is because of the fact that fluid is extracted from the walls of the channel with constant velocity that decreases the viscosity of the fluid particles which results in increase in the velocity near the wall of the channel. Figure 5 presented the effect of Prandtl number Pr on temperature profile $\theta(1)$ for non-buoyant $\lambda = 0$ case. It is concluded from these profiles that temperature profile $\theta(1)$ increases strictly monotonically as the Prandtl number Pr increases for 1st and 2nd solutions. On the other hand, temperature profile $\theta(1)$ decreases as the Prandtl number Pr increases for 3rd solution. Effects of Smith number Sc on concentration profile $\phi(\eta)$ for $\lambda = 0$ (non-buoyant case) are plotted in Figure 6. Concentration profile $\phi(\eta)$ decreases by the increase of the strength of Smith number Sc for all triple solutions. Physically it can be argued that an increase in Smith number Sc prompts the increment in the quantity of

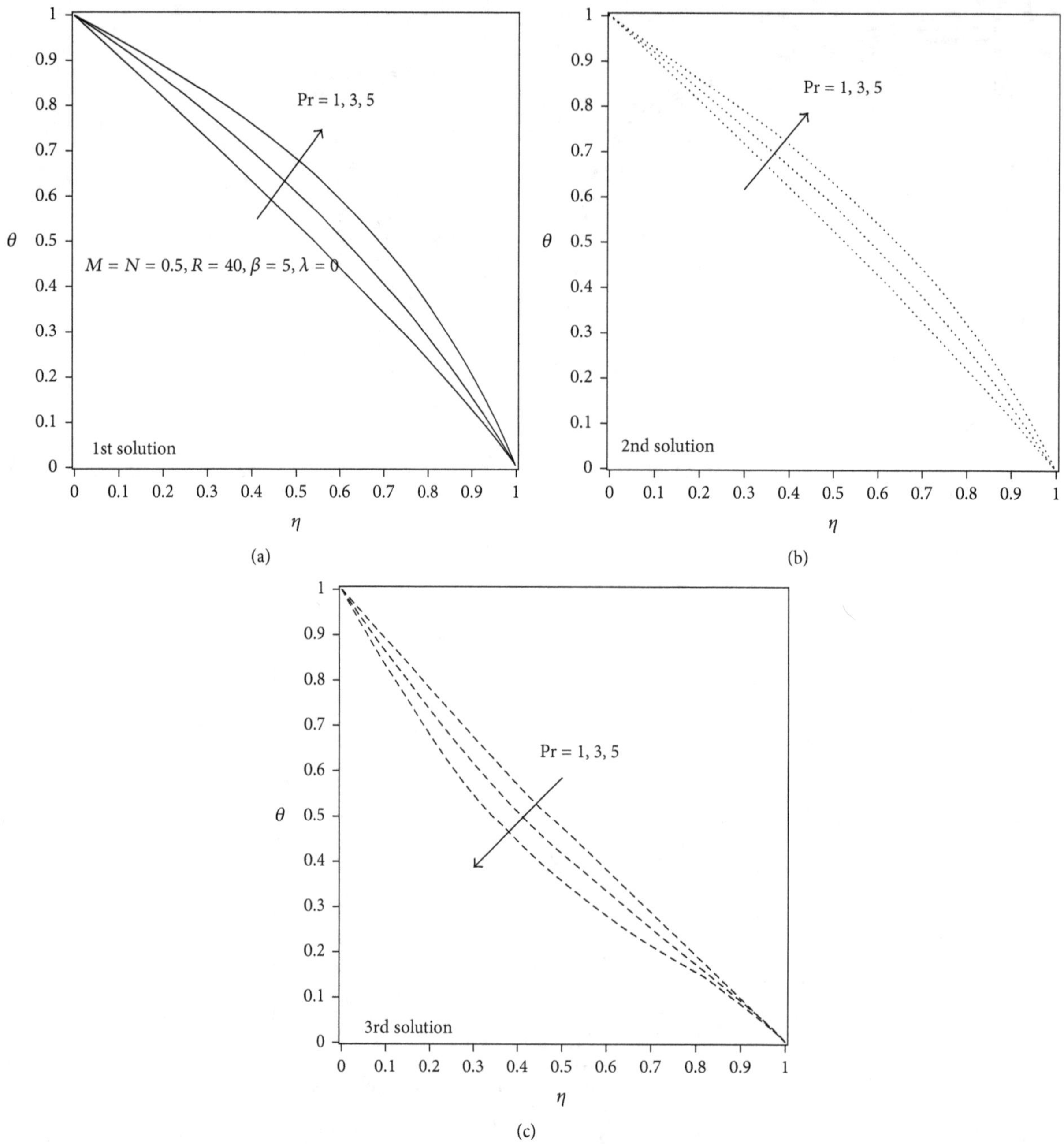

FIGURE 5: Effect of Prandtl number Pr on temperature profile $\theta(\eta)$.

solute atoms experiencing substance reaction coming about the way that concentration of the fluid's molecule diminishes monotonically. In Figure 7, we plot $\theta'(1)$ against the values of Reynold number R for the case of forced convection $\lambda = 0$. It is seen that 1st and 2nd solutions overlap each other as the Reynold number R increases.

In Tables 1 and 2, we represented numerical values of skin friction $f''(1)$ and $\theta'(1)$ for the variations of different physical parameters. Table 1 presented the numerical values of skin friction $f''(1)$ for the variation of buoyancy parameter λ by fixing $R = 36$, $M = N = 0.5$, $\beta = 5$, $\mathrm{Pr} = 1$, $\mathrm{Sc} = 1$, $\gamma = 1.2$. Magnitude of the skin friction increases for

TABLE 1: Skin friction for different values of buoyancy parameters for $R = 36$, $M = N = 0.5$, $\beta = 5$, $\mathrm{Pr} = 1$, $\mathrm{Sc} = 1$, $\gamma = 1.2$.

λ	1st solution $f''(1)$	2nd solution $f''(1)$	3rd solution $f''(1)$
−0.50	−3.29405873	−14.78794525	−27.20927012
−0.25	−4.19831393	−14.22940191	−27.20983823
0	−5.21362887	−13.55733409	−27.21039669
0.25	−6.42985179	−12.68188468	−27.21094599
0.50	−8.22263120	−11.22740570	−27.21148660

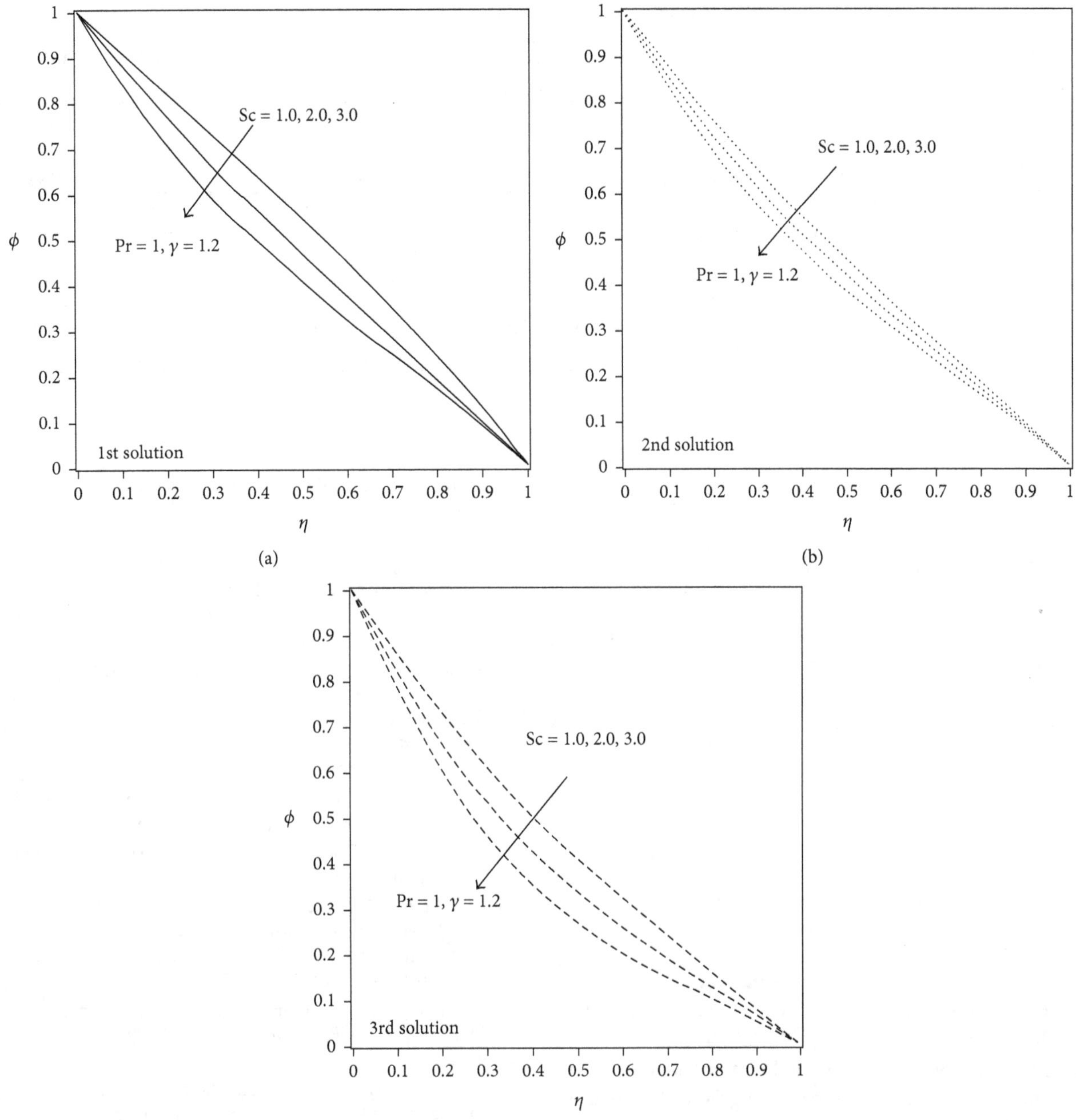

(c)

FIGURE 6: Effect of Smith number Sc on concentration profile $\phi(\eta)$ for $\lambda = 0$.

the 1st and 3rd solutions and decreases for 2nd solution in the case of opposing flow $\lambda < 0$. Furthermore, same behavior is depicted for the case of assisting flow $\lambda > 0$; therefore the fluid velocity near the channel walls $\eta \approx 1$ increases. From Table 2 it is concluded that magnitude of skin friction $f''(1)$ and $\theta'(1)$ increases and decreases, respectively, as Reynold number increases for the case of assisting flow $\lambda > 0$ only for 1st and 3rd solutions by setting $M = N = 0.5$, $\mathrm{Pr} = 1$, $\mathrm{Sc} = 2$, $\gamma = 1.2$. Numerical values of skin friction $f''(1)$ increase for 1st and 3rd solutions and decrease for 2nd solution by the variation of Reynolds number for the case of opposing flow $\lambda < 0$ by setting $M = N = 0.5$, $\mathrm{Pr} = 1$, $\mathrm{Sc} = 2$, $\gamma = 1.2$.

5. Conclusion

Multiple solutions of MHD Casson fluid flow in a channel with heat and mass transfer are analyzed. Here are some important observations have been engendered in the light of numerical investigation:

(i) There is only single solution for the case of injection $R < 0$ for any value of Casson number β or Hartman number M.

(ii) There exist multiple solutions only for $\beta \in [5, \infty)$ and $R \in [31.07, \infty)$ for any value of magnetic number $M \in [0, 2.0]$.

TABLE 2: Skin friction and temperature gradient for different values of Reynold number R.

R	λ	1st solution $f''(1)$	2nd solution $f''(1)$	3rd solution $f''(1)$	1st solution $\theta'(1)$	2nd solution $\theta'(1)$	3rd solution $\theta'(1)$
31.07		−3.56529871	−16.7538394	−23.89501535	−1.20630237	−1.12415316	−1.04645775
35	−0.25	−4.06923819	−14.50163765	−26.71544153	−1.20492033	−1.14909835	−1.01703703
40		−4.70569981	−13.72648291	−28.87613992	−1.20354537	−1.16169799	−0.99886321
31.07		−4.22815985	−16.33685245	−23.92995344	−1.20259758	−1.12723885	−1.045737
35	0	−5.00282319	−13.89608196	−26.71780315	−1.20017552	−1.15257708	−1.01679981
40		−6.10401248	−12.73103850	−28.87353590	−1.19720323	−1.16644810	−0.99874016
31.07		−4.93400210	−15.87676698	−23.96319996	−1.19862360	−1.13057525	−1.04504186
35	0.25	−6.08426152	−13.14036724	−26.72012735	−1.19463415	−1.15685896	−1.01656453
40		−8.28517017	−10.94971752	−28.87095120	−1.18721976	−1.17484905	−0.99861750

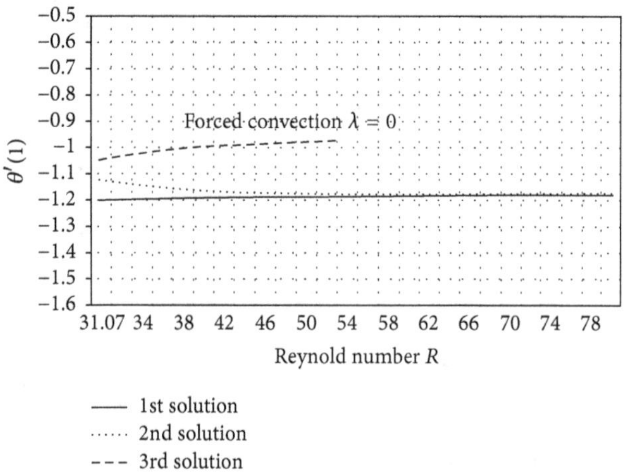

FIGURE 7: Effect of Reynold number R on $\theta'(1)$.

(iii) Effect of Casson number $\beta \geq 5$ and Reynolds number $R > 31.06$ on velocity profile $f'(\eta)$ increases for 1st and 3rd solutions near the channel wall $\eta \approx 1$.

(iv) Effect of Reynold number R on skin friction $f''(1)$ and $\theta'(1)$ is the same for assisting flow $\lambda > 0$ and opposing flow $\lambda < 0$.

Competing Interests

The authors declare that they have no competing interests.

References

[1] B. Li, L. Zheng, X. Zhang, and L. Ma, "The multiple solutions of laminar flow in a uniformly porous channel with suction/injection," *Advanced Studies in Theoretical Physics*, vol. 2, no. 10, pp. 473–478, 2008.

[2] X. Si, L. Zheng, X. Zhang, and Y. Chao, "The flow of a micropolar fluid through a porous channel with expanding or contracting walls," *Central European Journal of Physics*, vol. 9, no. 3, pp. 825–834, 2011.

[3] X.-Y. Si, X.-H. Si, L.-C. Zheng, and X.-X. Zhang, "Homotopy analysis solution for micropolar fluid flow through porous channel with expanding or contracting walls of different permeabilities," *Applied Mathematics and Mechanics-English Edition*, vol. 32, no. 7, pp. 859–874, 2011.

[4] B. J. Gireesha, G. S. Roopa, and C. S. Bagewadi, "Unsteady flow and heat transfer of a dusty fluid through a rectangular channel," *Mathematical Problems in Engineering*, vol. 2010, Article ID 898720, 17 pages, 2010.

[5] Mahesha, B. J. Gireesha, G. K. Ramesha, and C. S. Bagewadi, "Unsteady flow of a dusty fluid through a channel having triangular cross-section in Frenet frame field system," *Acta Universitatis Apulensis*, vol. 25, pp. 53–75, 2011.

[6] K. Ramesh and M. Devakar, "Some analytical solutions for flows of Casson fluid with slip boundary conditions," *Ain Shams Engineering Journal*, vol. 6, no. 3, pp. 967–975, 2015.

[7] D. McDonald, *Blood Flows in Arteries*, Arnold, London, UK, 2nd edition, 1974.

[8] E. W. Merrill, A. M. Benis, E. R. Gilliland, T. K. Sherwood, and E. W. Salzman, "Pressure-flow relations of human blood in hollow fibers at low flow rates," *Journal of Applied Physiology*, vol. 20, no. 5, pp. 954–967, 1965.

[9] L. N. Tao, "On combined free and forced convection in Channels," *Journal of Heat Transfer*, vol. 82, no. 3, pp. 233–238, 1960.

[10] I. Fersadou, H. Kahalerras, and M. El Ganaoui, "MHD mixed convection and entropy generation of a nanofluid in a vertical porous channel," *Computers & Fluids*, vol. 121, pp. 164–179, 2015.

[11] C.-K. Chen, B.-S. Chen, and C.-C. Liu, "Entropy generation in mixed convection magnetohydrodynamic nanofluid flow in vertical channel," *International Journal of Heat and Mass Transfer*, vol. 91, pp. 1026–1033, 2015.

[12] S. I. Khan, U. Khan, N. Ahmed, S. U. Jan, A. Waheed, and S. T. Mohyud-Din, "Effects of viscous dissipation and convective boundary conditions on Blasius and Sakiadis problems for Casson fluid," *National Academy Science Letters*, vol. 38, no. 3, pp. 247–250, 2015.

[13] J. C. Umavathi and J. Sultana, "Mixed convective flow of a micropolar fluid mixture in a vertical channel with boundary conditions of the third kind," *Journal of Engineering Physics and Thermophysics*, vol. 85, no. 4, pp. 895–908, 2012.

[14] X. Si, L. Zheng, P. Lin, X. Zhang, and Y. Zhang, "Flow and heat transfer of a micropolar fluid in a porous channel with expanding or contracting walls," *International Journal of Heat and Mass Transfer*, vol. 67, pp. 885–895, 2013.

[15] D. Prakash and M. Muthtamilselvan, "Effect of radiation on transient MHD flow of micropolar fluid between porous vertical channel with boundary conditions of the third kind," *Ain Shams Engineering Journal*, vol. 5, no. 4, pp. 1277–1286, 2014.

[16] J. Raza, A. M. Rohni, Z. Omar, and M. Awais, "Heat and mass transfer analysis of MHD nanofluid flow in a rotating channel with slip effects," *Journal of Molecular Liquids*, vol. 219, pp. 703–708, 2016.

[17] M. Casson, "A flow equation for pigment-oil suspensions of the printing ink type," in *Rheology of Disperse Systems*, C. C. Mills, Ed., pp. 84–104, Pergamon, Oxford, NY, USA, 1959.

[18] N. T. M. Eldabe and M. G. E. Salwa, "Heat transfer of MHD non-Newtonian Casson fluid flow between two rotating cylinders," *Journal of the Physical Society of Japan*, vol. 64, pp. 41–64, 1995.

[19] H. A. Attia and M. E. Sayed-Ahmed, "Transient MHD couette flow of a Casson fluid between parallel plates with heat transfer," *Italian Journal of Pure and Applied Mathematics*, vol. 27, pp. 19–38, 2010.

[20] S. A. Shehzad, T. Hayat, M. Qasim, and S. Asghar, "Effects of mass transfer on MHD flow of Casson fluid with chemical reaction and suction," *Brazilian Journal of Chemical Engineering*, vol. 30, no. 1, pp. 187–195, 2013.

[21] M. Afikuzzaman, M. Ferdows, and M. M. Alam, "Unsteady MHD casson fluid flow through a parallel plate with hall current," *Procedia Engineering*, vol. 105, pp. 287–293, 2015.

[22] C. Reddy, C. V. Rao, and O. Surender, "Soret, joule heating and hall effects on free convection in a casson fluid saturated porous medium in a vertical channel in the presence of viscous dissipation," *Procedia Engineering*, vol. 127, pp. 1219–1226, 2015.

[23] W. P. Walawender, T. Y. Chen, and D. F. Cala, "An approximate casson fluid model for tube flow of blood," *Biorheology*, vol. 12, no. 2, pp. 111–119, 1975.

[24] R. L. Batra and B. Jena, "Flow of a Casson fluid in a slightly curved tube," *International Journal of Engineering Science*, vol. 29, no. 10, pp. 1245–1258, 1991.

[25] M. E. Sayed Ahmed and H. A. Attia, "Magnetohydrodynamic flow and heat transfer of a non-Newtonian fluid in an eccentric annulus," *Canadian Journal of Physics*, vol. 76, no. 5, pp. 391–401, 1998.

[26] H. R. Kataria and H. R. Patel, "Radiation and chemical reaction effects on MHD Casson fluid flow past an oscillating vertical plate embedded in porous medium," *Alexandria Engineering Journal*, vol. 55, pp. 583–595, 2016.

[27] M. Das, R. Mahato, and R. Nandkeolyar, "Newtonian heating effect on unsteady hydromagnetic Casson fluid flow past a flat plate with heat and mass transfer," *Alexandria Engineering Journal*, vol. 54, no. 4, pp. 871–879, 2015.

[28] D. B. Meade, B. S. Haran, and R. E. White, "The shooting technique for the solution of two-point boundary value problems," *Maple Technical Newsletter*, vol. 3, pp. 1–8, 1996.

3

Noise Folding in Completely Perturbed Compressed Sensing

Limin Zhou, Xinxin Niu, and Jing Yuan

Information Security Center, Beijing University of Posts and Telecommunications, Beijing 100876, China

Correspondence should be addressed to Limin Zhou; zhoulimin.s@163.com

Academic Editor: Jingxin Zhang

This paper first presents a new generally perturbed compressed sensing (CS) model $y = (A + E)(x + u) + e$, which incorporated a general nonzero perturbation E into sensing matrix A and a noise u into signal x simultaneously based on the standard CS model $y = Ax + e$ and is called noise folding in completely perturbed CS model. Our construction mainly will whiten the new proposed CS model and explore in restricted isometry property (RIP) and coherence of the new CS model under some conditions. Finally, we use OMP to give a numerical simulation which shows that our model is feasible although the recovered value of signal is not exact compared with original signal because of measurement noise e, signal noise u, and perturbation E involved.

1. Introduction

Compressed sensing (CS) model, which was proposed by Candes et al. [1] and Donoho [2], had become a hot topic and attracted a lot of researchers to study it over the past years because it can recover a signal as a technique. Thus, it had been widely applied in many areas such as radar systems [3], signal processing [4], and image processing [5]. These applications depended on the main function of CS model to recover the original signal with some related algorithms including convex relaxation [6, 7], greedy pursuit [7], and Bayesian algorithm [8, 9], which were utilized to estimate the best approximation value of the original signal.

The classic and basic CS model in an unperturbed scenario and can be formulated as

$$y = Ax. \tag{1}$$

Here, $y \in R^m$ is the measurement vector or observation value and $A \in R^{m\times n}$ is a full rank measurement matrix with $m \ll n$. Signal $x \in R^n$ is k-sparse if no more than k entries of signal x are nonzero. Thus, x is called a k-sparse signal.

To date, the basic model has mature theories and there are a lot of different algorithms [6, 7] such as match pursuit (BP) [1, 10], orthogonal match pursuit (OMP) [11–15], Compressive Sampling Matching Pursuit (CoSaMP) [16], and Bayesian algorithm [8, 9], which can be recovered exactly as signal value and utilized in many areas [17–20].

But in practice, the measurement vector y in (1) was often contaminated by a noise or an error. More concretely, a noise term $e \in R^m$, called *an additive noise*, was incorporated into $y = Ax$ to result in a *partially perturbed model* [21–23]:

$$y = Ax + e, \tag{2}$$

where a noise or an error e ($e \neq 0$) was uncorrelated with signal x. There were two methods to model noise e in [24]. Here, a noise e was randomly sampled from Gaussian distribution. This model was used in many areas [21–23] and naturally had mature theories in recent years. For example, a number of accuracy algorithms on (2) emerged, for example, BP [1, 21], OMP [21], CoSaMP [16], and Bayesian algorithm [8, 9].

In 2010, Herman and Strohmer [25] first incorporated a randomly nontrivial perturbation E into matrix A in (2) to generate a *general perturbation model* [25–27] as follows:

$$y = (A + E)\,x + e, \tag{3}$$

where $E \in R^{m\times n}$ was called *a general perturbation* or *a multiplicative noise*. They studied influence of E on signal x and indicated that considering this CS model was a must [25–27]. Intuitively, it was harder to analyze the multiplicative noise E compared to the additive noise e because E was related to signal x with Ex.

As for (3), there were two different scenarios from different points of view [25–27]. First, from user's point of view, the sensing process can be formulated as follows:

$$\hat{y} = Ax + e, \quad \hat{A} = A + E. \tag{4}$$

Its recovery process can be expressed by

$$\hat{x} = (\hat{y}, \hat{A}, \ldots). \tag{N_1}$$

Thus, the useful measurement matrix was the perturbed matrix $\hat{A}$ not the original measurement matrix A. The system was researched on the recovery signal with BP in [16, 25] and OMP in [26, 27].

The second model was from designer's perspective [25–27]. The sensing process was just written as

$$\hat{y} = \hat{A}x + e, \quad \hat{A} = A + E \tag{5}$$

and the recovery process was written as

$$\hat{x} = R(\hat{y}, A, \ldots). \tag{N_1'}$$

The useful sensing matrix was A not $\hat{A}$ and the observation value was $\hat{y}$. To the best of our knowledge, no work focused on recovery signal in the context of general perturbation E except for [25–27].

In some practical scenarios, signal itself was often contaminated by noise and such case was applied in sub-Nyquist converter. Though introducing noise to signal was significant, no prolific paper studied such signal noise u except for [24] which first added an unknown random noise $u \in R^n$ to sinal x of $y = Ax + e$ to produce *noise folding CS model* [24]:

$$y = A(x + u) + e. \tag{6}$$

They analyzed the RIP and *coherence* of the equivalent system after whitening and showed that the difference of the RIP and *coherence* between original A and whitened matrix was small [24]. Based on [24–27], we propose a new CS model and study its related properties in Section 3.

2. Preliminaries

In this paper, we will restrict our attention to RIP and *coherence* of our new CS model. By convention, sensing matrix A and perturbation E are assumed to sample independently and identically distributed (i.i.d.) Gaussian random variables since such matrix satisfies RIP and coherence, and so forth [7, 24], with probability one.

Definition 1 (see [7]). A sensing matrix A satisfies the restricted isometry property (RIP) of order k if there exists $\delta_k \in (0, 1)$ s.t.:

$$(1 - \delta_k)\|x\|_2^2 \le \|Ax\|_2^2 \le (1 + \delta_k)\|x\|_2^2 \tag{7}$$

for any k-sparse vector with $k = 1, 2, \ldots$, where δ_k is the smallest nonnegative number called the restricted isometry constant (RIC).

Definition 1′ (see [24]). For (1) and (2), there is another equivalent statement for RIP of A, denoted by RIP′, in some special cases. For any index set $\Lambda \subset \{1, \ldots, N\}$ of size k, let A_Λ denote the submatrix of A consisting of the column vectors indexed by Λ, and the matrix A possesses RIP′ with constants $0 < \alpha_k \le \beta_k$, if

$$\alpha_k \|h\|^2 \le \|A_\Lambda h\|^2 \le \beta_k \|h\| \quad \forall h \in R^k \tag{8}$$

for any index set $\Lambda \subset \{1, \ldots, N\}$ of size k, where N is a positive integer.

For (6), there existed another form of RIP for matrix A which was given by Lemma 2 [24] since matrix A was whitened.

Lemma 2 (see [24]). *As for folding noise model (6), RIP for whitened A can be formulated as*

$$\alpha_k (1 - \rho_1)\|h\|_2^2 \le \|B_\Lambda h\|_2^2 \le \beta_k (1 + \rho_1)\|h\|_2^2, \tag{9}$$

where $\rho_1 = \rho/(1 - \rho)$ with $0 < \rho < 1/2$ and matrix B was obtained after whitening sensing matrix A.

The perturbation E and sensing matrix A in (3) can be quantified below in [25–27]

$$\frac{\|E\|_2}{\|A\|_2} \le \varepsilon_A, \quad \frac{\|E\|_2^{(k)}}{\|A\|_2^{(k)}} \le \varepsilon_A^{(k)}, \quad \|A\|_2^{(k)} = \sigma_{\max}^{(k)}(A), \tag{10}$$

where the symbol $\|A\|_2$ denotes spectral norm of a matrix A, $\|A\|_2^{(k)}$ denotes the largest spectral norm taken over all k-column submatrices of matrix A, and $\sigma_{\max}^{(k)}(A)$ [25] denotes the largest nonzero singular value taken over all k-column submatrices of matrix A. It was appropriate to assume $0 < \varepsilon_A$, $\varepsilon_A^{(k)}$, and $\varepsilon_y \ll 1$.

Lemma 3 (RIP for $\hat{A}$ [25]). *For $k = 1, 2, \ldots$, given the RIC associated with matrix A in (3) and the relative perturbation $\varepsilon_A^{(k)}$, fix the constant*

$$\hat{\delta}_{k,\max} = (1 + \delta_k)\left(1 + \varepsilon_A^{(k)}\right)^2 - 1. \tag{11}$$

Assume that the RIC $\hat{\delta}_k \le \hat{\delta}_{k,\max}$ for matrix $\hat{A} = A + E$ is the smallest nonnegative number, and the RIP for $\hat{A}$ can be written as

$$\left(1 - \hat{\delta}_k\right)\|x\|_2^2 \le \left\|\hat{A}x\right\|_2^2 \le \left(1 + \hat{\delta}_k\right)\|x\|_2^2 \tag{12}$$

for any k-sparse vector x.

From Lemma 3 with (6), there is one equivalent statement for the RIP for $\widehat{A}$ in some special cases given by Lemma 3′ [24].

Lemma 3′ (see [24]). *For any index set $\Lambda \subset \{1,\ldots,N\}$ of size k, let $\widehat{A}_\Lambda$ denote the submatrix of $\widehat{A}$ consisting of the column vectors indexed by Λ. A matrix $\widehat{A}$ possesses RIP with constants $0<\widehat{\alpha}_k \le \widehat{\beta}_k$, if*

$$\widehat{\alpha}_k \|h\|^2 \le \left\|\widehat{A}_\Lambda h\right\|^2 \le \widehat{\beta}_k \|h\|, \quad \forall h \in R^k \tag{13}$$

for any index set $\Lambda \subset \{1,\ldots,N\}$ of size k, where N is a positive integer.

Definition 4 (see [7]). The coherence, $\mu(A)$, of a matrix A is the largest absolute inner product between any two columns A_i, A_j, $i \ne j$, of matrix A as follows:

$$\mu(A) = \max_{1\le i<j\le n} \frac{\left|A_i^T A_j\right|}{\left\|A_i\right\|_2 \left\|A_j\right\|_2}. \tag{14}$$

3. Constructions

3.1. A New Completely Perturbed CS Model. As mentioned above, for (2), (3), and (6), only one noise in (2) or two noises in (3) and (6) affected the CS model. Maybe a noise e, a noise u, and a perturbation E simultaneously affect the CS model although no paper studies this. In terms of the idea, [24] together with [25–27] motivate us to introduce a noise u to *general perturbation model* (3) to generate *noise folding in generally perturbed situation* or to incorporate a nontrivial perturbation E into (6) to produce *a complete perturbation CS model with folding noise,* which for the first time yields so called *noise folding in completely perturbed CS model.* We formulate the CS model as

$$y = (A+E)(x+u)+e, \tag{15}$$

where $e \in R^m$ is a random noise vector with covariance $\sigma^2 I$ and $u \in R^n$ presents a random premeasurement noise vector whose covariance is $\sigma_0^2 I$ independent of e. Here e and u are regarded as *additive noise.* $E \in R^{m\times n}$ is a random perturbation matrix and more details on perturbation E can be seen in [25]. Here we call CS model (15) *noise folding in completely perturbed CS model.* Analogous to (3) in [25–27], (15) can also be considered in two different situations. Similarly, from user's point of view, an incorrect sensing matrix can be obtained via an unknown measurement model:

$$\begin{gathered} \widehat{y} = A(x+u)+e, \\ \widehat{A} = A+E \end{gathered} \tag{16}$$

and the recovery process algorithm can be written as

$$\widehat{x} = R\left(\widehat{y}, \widehat{A}, u, \ldots\right). \tag{N_2}$$

The only difference between (N_1) and (N_2) is noise u in (N_2). From the designer's view, sensing process can be formulated as

$$\begin{gathered} \widehat{y} = \widehat{A}(x+u)+e, \\ \widehat{A} = A+E \end{gathered} \tag{17}$$

and its recovery process is as

$$\widehat{x} = R\left(\widehat{y}, A, u, \ldots\right). \tag{N_2'}$$

Similarly, compared to (N_1'), noise u belongs to (N_2'). In this paper, we only study simply its properties: RIP and *coherence* after whitening. Obviously, (15) can be extended to general multiperturbation CS model:

$$y = \left(A + \sum_{i=1}^{n} E_i\right)(x+u)+e, \quad i = 1,2,\ldots,n, \tag{18}$$

where E_i is perturbation. System (18) can be viewed as a generalization of our proposed CS (15), which implies that the general conclusion of (18) can be obtained from the special conclusion of (15). The concrete results can be seen in the next section. Simultaneously, other general CS systems can be conjectured naturally as follows:

$$\begin{gathered} y = \left(\sum_{i=1}^{s} A_i + E\right)(x+u)+e, \\ y = \left(\sum_{i=1}^{s} A_i + E\right)\left(x+\sum_{i=1}^{s} u_i\right)+e, \\ y = \left(\sum_{i=1}^{s} A_i + \sum_{i=1}^{s} E_i\right)\left(x+\sum_{i=1}^{s} u_i\right)+e. \end{gathered} \tag{19}$$

Although their properties seem to be many but we do not know how to exploit and analyze them, we leave them as open problems. Here we mainly study relative RIP and *coherence* on (15) and (18). In the next section, we give general results.

3.2. Problem Formulation. For (15), our goal is to analyze the effect of the premeasurement noise u and E on its RIP and *coherence.*

Throughout this paper, assume that e is a random noise vector with covariance $\sigma^2 I$, and u is a random noise vector with covariance $\sigma_0^2 I$ independent of e. Under these assumptions, (15) will be proved to be equivalent to $y = \widehat{B}x + w$, where $\widehat{B}$ is a matrix whose *coherence* and RIP constants are very close to that of A, w is whitened noise with variance $(\sigma^2 + (n/m)\sigma_0^2)I$, and I is identity matrix.

3.3. Equivalent Formulation. To set up our conclusion, (15) can be expressed as

$$\begin{gathered} y = (A+E)x+w, \\ w = (A+E)u+e. \end{gathered} \tag{20}$$

By hypothesis of whitened noise, the covariance of effective vector w is $Q: Q = \sigma^2 I + \sigma_0^2 (A+E)(A+E)^T$. Obviously, noise w is not whitened where the recovery process analysis becomes complicated. If w still preserves whitening, one case $\widehat{A} = A + E$ must be proportional to identity matrix. For example, suppose that $A + E$ consists of $r = n/m$ orthogonal basis such as

$$A = [A_1 + E_1, A_2 + E_2, \dots, A_r + E_r] \tag{21}$$

in which $A_i + E_i, i = 1, 2, \dots, r$, are $m \times m$ orthogonal matrices. Therefore, we have

$$\begin{aligned}(A+E)(A+E)^T &= (A_1+E_1)(A_1+E_1)^T + \cdots \\ &\quad + (A_r+E_r)(A_r+E_r)^T = rI \\ &= \frac{n}{m} I,\end{aligned} \tag{22}$$

where noise covariance of w is $Q = \gamma I$, $\gamma = \sigma^2 + (n/m)\sigma_0^2$. Under the special case, $y = (A+E)(x+u) + e$ (or $y = (A+E)x + w$) is equivalent to $y = Ax + e$. Compared with noise covariance of e, noise covariance of w has increased by γ/σ^2. If $\sigma_0^2 \approx \sigma^2$, the noise covariance of w is increased by n/m, which was called *noise folding* [24].

3.4. RIP and Coherence of Our CS Model. We will show that the conclusion holds generally. In other words, if $(A+E)(A+E)^T$ is not proportional to the identity matrix I, (15) and (20) are roughly equivalent really. Now we describe it in detail.

Note that if E is one random matrix, $\widehat{A} = A + E$ is a random matrix. To study RIP and *coherence* of $\widehat{A}$, we must whiten noise w by multiplying $Q_1^{-1/2}$ with $Q_1 = Q/\gamma$ and get the equivalent system:

$$\begin{aligned} y &= \widehat{B}x + v, \\ \text{s.t.} \quad \widehat{B} &= Q_1^{-1/2}(A+E), \\ v &= Q_1^{-1/2} w. \end{aligned} \tag{23}$$

Note that noise vector v is whitened with covariance matrix γI exactly if $(A+E)(A+E)^T$ is proportional to identity matrix. But the biggest difference lies in measurement matrix changing from original matrix $\widehat{A}$ to $\widehat{B}$ by whitening. The changing range is measured with three important indexes: RIP constant, *coherence*, and *stable*. Our theory mainly depends on approximating $(A+E)(A+E)^T$ with $(n/m)I$ and even $\widehat{A}$ is an arbitrary matrix. Let

$$\eta = \left\| I - \frac{m}{n}(A+E)(A+E)^T \right\|_2 \tag{24}$$

measure accuracy of the approximating, in which $\|\cdot\|$ denotes the standard operator norm in R^n. For derivation convenient, assume that η is very small and show that the *coherence* and RIP constant of $\widehat{B}$ are very close to that of A. By convention, the entries of A are i.i.d. mean zero and variance $1/m$ random variables with Gaussian distribution; thus, it is easy to justify that η is always small.

Another useful formula can be formulated:

$$\eta_0 = \left\| I - \frac{m}{n} A A^T \right\|_2 \tag{25}$$

which was introduced in [24]. The fact that η_0 was very small had been proved in [24] with restrictions on only matrix A. It is natural to think whether the difference between η and η_0 is very small. Theorem 5 confirms our conjecture and further inspires us to think whether the difference between $\widehat{B}$'s *coherence* and RIP and $\widehat{A}$'s and A's is very small, respectively. The later related theorems will give us the positive answers.

Theorem 5 shows the relation between η_0 and η under the context $\|E\|_2/\|A\|_2 \le \varepsilon_A$.

Theorem 5. *Assume that sensing matrix $A \in R^{m\times n}$, an unknown random matrix $E \in R^{m\times n}$, $m \ll n$, $\|E\|_2/\|A\|_2 \le \varepsilon_A$, $0 < \varepsilon_A \ll 1$,*

$$\begin{aligned} \eta_0 &= \left\| I - \frac{m}{n} A A^T \right\|_2, \\ \eta &= \left\| I - \frac{m}{n}(A+E)(A+E)^T \right\|_2 \end{aligned} \tag{26}$$

$0 < \eta_0 < 1/2$, $\|A\| = \sigma_1$, where σ_1 is the largest nonzero positive singular value of A; then

$$\eta_0 - \frac{m}{n}\left(2\varepsilon_A + \varepsilon_A^2\right)\sigma_1^2 \le \eta \le \eta_0 + \frac{m}{n}\left(2\varepsilon_A + \varepsilon_A^2\right)\sigma_1^2. \tag{27}$$

Proof. The detailed proof is postponed to the Appendix. □

Remark 6. For (27), by assumption $m \ll n, n \to \infty$, obtain

$$\begin{aligned} &\frac{m}{n} \longrightarrow 0, \\ &\frac{m}{n}\left(2\varepsilon_A + \varepsilon_A^2\right)\sigma_1^2 \longrightarrow 0 \end{aligned} \tag{28}$$

due to $0 < \varepsilon_A \ll 1$, a positive number σ_1. Thus, $\eta_0 \le \eta_1 \le \eta_0$; that is, $\eta_0 = \eta$, such that $m \ll n, n \to \infty, 0 < \varepsilon_A \ll 1$ and there is a positive number σ_1. Theorem 5 shows the relation between η and η_0 which implies that $\eta = \eta_0$ under some special conditions. Therefore, we can let $\eta < 1/2$ like $\eta_0 < 1/2$ in [24].

Theorem 7 shows the RIP of $\widehat{B}$ in the case of $\eta < 1/2$ though $0 < \eta < 1$ is sufficient for the proof of the RIC for $\widehat{B}$.

Theorem 7. *Assume that sensing matrix $A \in R^{m\times n}$, an unknown random matrix $E \in R^{m\times n}$, $m \ll n$. Let $\|E\|_2/\|A\|_2 \le \varepsilon_A$, $0 < \varepsilon_A \ll 1$,*

$$\begin{aligned} \eta &= \left\| I - \frac{m}{n}(A+E)(A+E) \right\|_2, \\ \eta_0 &= \left\| I - \frac{m}{n} A A^T \right\|_2 \end{aligned} \tag{29}$$

$0 < \eta < 1/2$, $0 < \eta_0 < 1/2$, *suppose that* $\widehat{A}$ *satisfies the RIP of order k with* $0 < \widehat{\alpha}_\Lambda \leq \widehat{\beta}_\Lambda$, *and* $\sigma_1 > 0$ *is the largest singular value of matrix A; then* $\widehat{B}$ *satisfies the RIP of order k with different constants below:*

$$\begin{aligned} &\widehat{\alpha}_k \left(1 - \mu_1'\right) \|h\|_2^2 \leq \left\|\widehat{B}_\Lambda h\right\|_2^2 \leq \widehat{\beta}_k \left(1 + \mu_1'\right) \|h\|_2^2, \\ &\mu_1' = \frac{\mu_1}{1 - \mu_1} \\ &\widehat{\alpha}_k \left(1 - \eta_1'\right) \|h\|_2^2 \leq \left\|\widehat{B}_\Lambda h\right\|_2^2 \leq \widehat{\beta}_k \left(1 + \eta_1'\right) \|h\|_2^2, \\ &\eta_1' = \frac{\eta_1}{1 - \eta_1} \\ &\widehat{\alpha}_k \left(1 - \mu_2'\right) \|h\|_2^2 \leq \left\|\widehat{B}_\Lambda h\right\|_2^2 \leq \widehat{\beta}_k \left(1 + \mu_2'\right) \|h\|_2^2, \\ &\mu_2' = \frac{\mu_2}{1 - \mu_2}. \end{aligned} \tag{30}$$

Proof. The detailed proof is postponed to the Appendix. □

Remark 8. In Theorems 5 and 7, the condition $\|E\|/\|A\| \leq \varepsilon_A$ with $0 < \varepsilon_A \ll 1$ can be taken in place of $E = \varepsilon A$ [25], in which E is a simple version of A, so that we can get another result. Due to paper volume, they are omitted here. But their proofs are very simple that researchers can prove them and yield perfect results.

Multiperturbation CS system (18) can be viewed as a generalization of the new proposed CS system (15) so that the general conclusion of the (18) can come from that of (15). Theorems 9 and 11 give us the results.

Theorem 9. *Assume that* $A \in R^{m\times n}$ *is sensing matrix, and* $E_i \in R^{m\times n}$ *is an unknown random matrix with* $m \ll n$. *Let*

$$\begin{aligned} \eta_0 &= \left\|I - \frac{m}{n} AA^T\right\|_2, \\ \widetilde{\eta} &= \left\|I - \frac{m}{n}\left(A + \sum_{i=1}^{s} E_i\right)\left(A + \sum_{i=1}^{s} E_i\right)^T\right\|_2. \end{aligned} \tag{31}$$

$\|E_i\|_2/\|A\|_2 \leq \varepsilon_A$, $0 < \varepsilon_A \ll 1$, *and let* σ_1 *be the largest singular value of matrix A,* $\|A\| = \sigma_1$; *and suppose that matrix A satisfies RIP, a number s is an integer,* $s < n$; *then the relation between* $\widetilde{\eta}$ *and* η_0 *can be formulated as*

$$\begin{aligned} \eta_0 - \frac{m}{n}\left(2s\varepsilon_A\sigma_1 + s^2\varepsilon_A^2\sigma_1\right) &\leq \widetilde{\eta} \\ &\leq \eta_0 + \frac{m}{n}\left(2s\varepsilon_A\sigma_1 + s^2\varepsilon_A^2\sigma_1\right). \end{aligned} \tag{32}$$

Proof. The detailed proof is postponed to the Appendix. □

Remark 10. For (32), since $1 \leq s \ll n$, $m \ll n$, and s, m, n are positive integers and σ_1 is a constant, $(m/n)(2s\varepsilon_A\sigma_1^2 + s^2\varepsilon_A^2\sigma_1^2) \to 0$ when $n \to \infty$ which implies $\widetilde{\eta} \to \eta_0$.

Theorem 11. *Assume that* $A \in R^{m\times n}$ *is sensing matrix, and* $E_i \in R^{m\times n}$ *is an unknown random matrix with* $m \ll n$. *Let*

$$\begin{aligned} \eta_0 &= \left\|\frac{m}{n} AA^T - I\right\|_2, \\ \widetilde{\eta} &= \left\|I - \frac{m}{n}\left(A + \sum_{i=1}^{s} E_i\right)\left(A + \sum_{i=1}^{s} E_i\right)^T\right\|_2, \end{aligned} \tag{33}$$

$\|E_i\|_2/\|A\|_2 \leq \varepsilon_A$, *and* $0 < \varepsilon_A \ll 1$, *and suppose that* $\widetilde{\eta} < 1/2$, σ_1 *is the largest singular value of matrix A, s and n are integers,* $s < n$, *and*

$$\begin{aligned} Q &= \sigma^2 I + \sigma_0^2\left(A + \sum_{i=1}^{s} E_i\right)\left(A + \sum_{i=1}^{s} E_i\right)^T, \\ \gamma &= \sigma^2 + \frac{n}{m}\sigma_0^2, \\ Q_1 &= \frac{Q}{\gamma}; \end{aligned} \tag{34}$$

then,

$$\begin{aligned} &\widehat{\alpha}_k \left(1 - \mu_3'\right) \|h\|_2^2 \leq \left\|\widehat{B}_\Lambda h\right\|_2^2 \leq \widehat{\beta}_k \left(1 + \mu_3'\right) \|h\|_2^2, \\ &\mu_3' = \frac{\mu_3}{1 - \mu_3} \\ &\widehat{\alpha}_k \left(1 - \eta_3'\right) \|h\|_2^2 \leq \left\|\widehat{B}_\Lambda h\right\|_2^2 \leq \widehat{\beta}_k \left(1 + \eta_3'\right) \|h\|_2^2, \\ &\eta_3' = \frac{\eta_3}{1 - \eta_3} \\ &\widehat{\alpha}_k \left(1 - \mu_4'\right) \|h\|_2^2 \leq \left\|\widehat{B}_\Lambda h\right\|_2^2 \leq \widehat{\beta}_k \left(1 + \mu_4'\right) \|h\|_2^2, \\ &\mu_4' = \frac{\mu_4}{1 - \mu_4}. \end{aligned} \tag{35}$$

Proof. The detailed proof is postponed to the Appendix. □

Remark 12. Though $0 < \widetilde{\eta} < 1$ is sufficient for the proof, the RIC for $\widehat{B}$ is positive in the restriction of $\widetilde{\eta} < 1/2$.

Next, we compare the *coherence* of $\widehat{B}$ after whitening $\widehat{A}$ to that of $\widehat{A}$. A_i, $i = 1, \ldots, n$, is used to denoted the ith column vector of a matrix A. Similar to the coherence of A, the coherence of $\widehat{A}$ is first given in Definition 13.

Definition 13. Assume that $A \in R^{m\times n}$ is a random matrix, $E \in R^{m\times n}$ is an unknown random matrix in CS, and $\widehat{A} = A + E$; then *coherence* of $\widehat{A}$, denoted by $\mu(\widehat{A})$, can be formulated as

$$\mu\left(\widehat{A}\right) = \max_{1\leq i<j\leq n} \frac{\left|\widehat{A}_i^T \widehat{A}_j\right|}{\left\|\widehat{A}_i\right\|_2 \left\|\widehat{A}_j\right\|_2}. \tag{36}$$

In fact, $\mu(\widehat{A})$ is the largest absolute inner product between any columns $\widehat{A}_i$, $\widehat{A}_j$, $i \neq j$.

As mentioned above, $\|Q_1 - I\| < 1$, respectively, in some special contexts with $\|E\|_2/\|A\|_2 \le \varepsilon_A$ in Theorem 7. We can take advantage of $\|Q_1 - I\| < 1$ to prove Theorem 14. For the lack of space, we only take $\|Q_1 - I\| < \mu_2$, $\|Q_1^{-1} - I\| < \mu_2'$ as an example with $\eta < 1/2$. The proofs of the rest of the cases, including $\|Q_1 - I\| < \mu_1, \eta_1$, and $\|Q_1^{-1} - I\| < \mu_1', \eta_1'$, are similar, and we leave them to readers. As for the general results $\mu(A + \sum_{i=1}^{n} E_i)$ of the general CS model $y = (A + \sum_{i=1}^{n} E_i)(x+u)+e$, we omit them too due to space constrains. The proof of general coherence of $Q_1^{-1/2}(A + \sum_{i=1}^{n} E_i)$ is similar too. Theorem 14 demonstrates the relation between coherence of $\widehat{B}$ and that of $\widehat{A}$.

Theorem 14. *Assume that $\mu_2 < 3/4$ in $\|Q_1 - I\| \le \mu_2$, $\widehat{B} = Q_1^{-1/2}\widehat{A}$ with $\widehat{A} = A + E$; then,*

$$\mu\left(\widehat{B}\right) \le \frac{(1+\widehat{\mu}_2)}{(1-\widehat{\mu}_2)^2}\mu\left(\widehat{A}\right), \tag{37}$$

where $\widehat{\mu}_2 = (1-\mu_2)^{-1/2} - 1$, $\mu(\widehat{B}) = \max(|\widehat{B}_i^T \widehat{B}_j| / \|\widehat{B}_i\|_2 \|\widehat{B}_j\|_2)$. $\widehat{B}_j$ denotes the jth column vector of whitening matrix $\widehat{B}$; $\widehat{A}_j$ denotes the jth column vector of $\widehat{A}$; that is, $\widehat{A}_j = A_j + E_j$.

Proof. The detailed proof is postponed to the Appendix. □

In [25], E is simply version of random matrix A such as $E = \varepsilon A$ with $0 < \varepsilon \ll 1$. The relation between $\mu(\widehat{A})$ and $\mu(A)$ will be seen from Theorem 15 below in the case of $E = \varepsilon A$, $0 < \varepsilon \ll 1$.

Theorem 15. *Let $\widehat{A} = A + E$, $E = \varepsilon A$, $0 < \varepsilon \ll 1$. The correlation of coherence between A and $\widehat{A}$ proceeds as*

$$\mu\left(\widehat{A}\right) \le \frac{(1+\varepsilon)^2}{(1-\varepsilon)^2}\mu(A) \tag{38}$$

with $0 < \varepsilon \ll 1$.

Proof. The proof of Theorem 15 is similar to that of Theorem 14; here we omit it. □

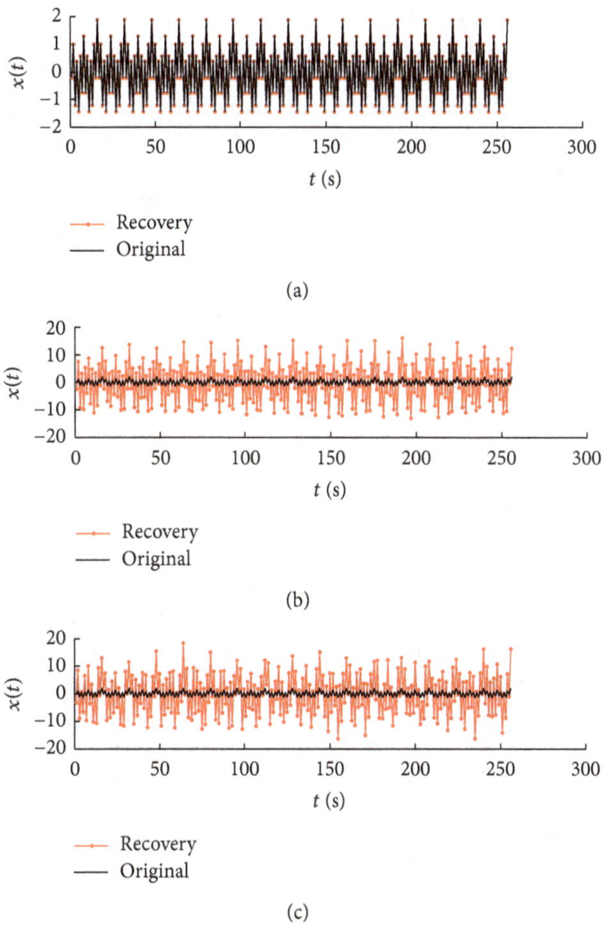

FIGURE 1: (a) Recover x from $y = Ax + e$. The error of the recovered signal value and original signal value is 6.4835e − 004 by the Euclidean norm which indicates that the recovered signal and original signal are almost the same. (b) Recover x from $y = A(x+u)+e$. The error of the recovered signal value and the original signal value is 6.2258 by the Euclidean norm which means that the recovered signal and original signal are quite different. (c) Recover x from $y = (A+E)(x+u)+e$. The error of the recovered signal value and the original signal value is 5.8855 by the Euclidean norm which indicates that the recovered signal is much different from the original signal. Compared with the change (error) between the recovery signal and the original signal in (a), the changes (errors) in (b) and (c) differ little. What is more, comparing the change between the recovery signal and the original signal of (c) with that of (b), the change of (c) is quite a little bigger because E is involved in (c) but E is not involved in (b).

4. Numerical Experiment

Vertical coordinate (sinusoidal) denotes the degree of recovery signal. Horizontal coordinate denotes time whose unit is seconds. Black line denotes original signal and red line denotes recovery signal.

Here we use OMP to give three numerical simulation results which demonstrate that our new proposed generally perturbed CS is feasible. To compare signal recovering with OMP from three figures, signal recovery from measurement noise model $y = Ax + e$ is almost exact because of only noise e in basic CS model $y = Ax + e$. There are a lot of differences between recovery signal and original signal in both $y = (A + E)(x + u) + e$ and $y = A(x + u) + e$ CS models because there are noises u, E, e in the two CS models. Comparing the change between recovered signal and original signal of Figure 1(b) with that of Figure 1(c), the change of Figure 1(c) is a bit bigger than that of Figure 1(b) because perturbation E is involved in Figure 1(c) ($y = (A+E)(x+u)+e$) and E is not involved in Figure 1(b) ($y = A(x+u)+e$), which shows that the different noises E, u, e, have a different impact on signal recovering.

Compared with the change (error) between the recovery signal and the original signal in Figure 1(a), the changes (errors) in Figures 1(b) and 1(c) differ little. Namely, the differences between recovered signal and original signal from $y = (A + E)(x + u) + e$ are almost the same as the differences between recovery signal and original signal from $y = A(x+u)+e$, which indicates that our proposed CS model is feasible.

Comparing the change between recovered signal and original signal in Figure 1(a), the changes in Figures 1(b) and 1(c) are quite different. The fact shows that OMP is not the best algorithm to recover x from $y = (A+E)(x+u)+e$ and $y = A(x+u)+e$ although OMP is used to recover exact original signal from $y = Ax+e$. Thus, it is important to search for a powerful algorithm or more algorithms to recover exactly for original sparse signal from $y = A(x+u)+e$ and $y = (A+E)(x+u)+e$ as open problems. And here leave these problems to the interested researchers to exploit them because the paper cannot focus on searching for optimal algorithms to recover exactly original signal from CS models $y = A(x+u)+e$ and $y = (A+E)(x+u)+e$.

5. Conclusion

We first propose a new CS system (15) by introducing a multiplicative noise E, a signal noise u, and an additive noise e into unperturbed CS model (1). We derive RIP and *coherence* for $\widehat{A} = A+E$ after whitening (15). As a matter of fact, this paper proves that our proposed completely perturbed CS model (15) equals to the classic CS model (2). The only difference is the changed measurement matrix by incorporating a nontrivial perturbation matrix E to measurement matrix and a nontrivial noise u to signal x. And thus this induces noise variance increased by a factor of n/m so that a tighter upper bound and lower bound of RIC is produced. As for *coherence* of deformed measurement matrix $\widehat{A} = A+E$ in CS model (15), the constant is nearly invariant essentially with $n/m \to 0$, $m, n \to \infty$. Finally, we use OMP to give three figures to recover signal from CS model $y = Ax+e$, $y = A(x+u)+e$, $y = (A+E)(x+u)+e$, respectively. Figures 1(b) and 1(c) in our experiment demonstrate that the change between recovered signal and original signal is much bigger than that in Figure 1(a) which indicates our proposed CS model is feasible and OMP is not fit for recovering signal from $y = A(x+u)+e$ and $y = (A+E)(x+u)+e$. Thus, we can try to search one optimal algorithm or more algorithms to recover signal exactly from the two CS model although OMP is the best algorithm to recover exactly original signal from $y = Ax+e$ now.

6. Future Work

Thanks to the features of our proposed CS model (15), there are many works to do. The change between recovered signal and original signal in Figures 1(a), 1(b), and 1(c) indicates that our proposed CS in this paper is feasible although the differences between the recovery signal and original signal in Figures 1(b) and 1(c) are much bigger with OMP than the differences in Figure 1(a). Thus, an obvious problem is to search one algorithm or more optimal algorithms suitable for $y = A(x+u)+e$ and $y = (A+E)(x+u)+e$ to recover signal exactly.

The related RIP of E in [24] further motivates us to think that E as a perturbation sensing matrix could form one perturbed CS model as $y_1 = E(x+u)+e$. Thus, (15) may consist of two similar systems $y_2 = A(x+u)+e$ and $y_1 = E(x+u)+e$. Similarly, our model may be divided into another two models $y_3 = (A+E)x+e$, $y_4 = (A+E)u+e$, or three basic parts $y_1' = Ax+e$, $y_2' = Au+e$, and $y_3' = E(x+u)+e$. If possible, what can we do to reduce or eliminate the influence of an error CS system $y_3' = E(x+u)+e$? Can we recover signal x from the error system $y_3' = E(x+u)+e$? And if can, how to do it? Maybe there exists CS models $y = A(x+u)$ and $y = (A+E)(x+u)$. In addition, maybe we can also consider the impulse noise and use $e+I$ instead of e where I is the impulse noise. If so, maybe it can generalize our model and we can get very good results in impulse noise model. Here we cannot study such impulse noise model and leave it as an open problem, too.

These open problems are worth considering and are to be waited for studying in future work. This paper only does some elementary researches on our proposed CS and we hope that the idea and simple study in this paper will be helpful to study its wide application in the future. We hope that higher level compressed sensing model will be put forward and more and more people explore this areas in the future.

Appendix

Proof of Theorem 5. On the one hand,

$$\begin{aligned}
\eta &= \left\| I - \frac{m}{n}(A+E)(A+E)^T \right\|_2 \\
&= \left\| I - \frac{m}{n}AA^T - \frac{m}{n}\left(AE^T + EA^T + EE^T\right) \right\|_2 \\
&\geq \left\| I - \frac{m}{n}AA^T \right\|_2 - \frac{m}{n}\left\| AE^T \right\|_2 - \frac{m}{n}\left\| EA^T \right\|_2 \\
&\quad - \frac{m}{n}\left\| EE^T \right\|_2 \\
&\geq \eta_0 - \frac{m}{n}\|A\|_2 \left\| E^T \right\|_2 - \frac{m}{n}\|E\|_2 \left\| A^T \right\|_2 \\
&\quad - \frac{m}{n}\|E\|_2 \left\| E^T \right\|_2 \\
&\geq \eta_0 - \frac{m}{n}\|A\|_2 \varepsilon_A \left\| A^T \right\|_2 - \frac{m}{n}\varepsilon_A \|A\|_2 \left\| A^T \right\|_2 \\
&\quad - \frac{m}{n}\varepsilon_A \|A\|_2 \varepsilon_A \left\| A^T \right\|_2 \\
&= \eta_0 - \frac{m}{n}\left(2\varepsilon_A + \varepsilon_A^2\right)\|A\|_2 \left\| A^T \right\|_2 \\
&= \eta_0 - \frac{m}{n}\left(2\varepsilon_A + \varepsilon_A^2\right)\sigma_1^2.
\end{aligned} \tag{A.1}$$

Equation (A.1) holds because of $\|A\|_2 = \sigma_1$ and $\|A\|_2 = \|A^T\|_2$.

On the other hand,

$$\begin{aligned}
\eta &= \left\| I - \frac{m}{n}(A+E)(A+E)^T \right\|_2 = \left\| I - \frac{m}{n}AA^T \right. \\
&\quad \left. - \frac{m}{n}\left(AE^T + EA^T + EE^T\right) \right\|_2 \leq \left\| I - \frac{m}{n}AA^T \right\|_2 \\
&\quad + \frac{m}{n}\left(\left\| AE^T \right\|_2 + \left\| EA^T \right\|_2 + \left\| EE^T \right\|_2 \right) \leq \eta_0 \\
&\quad + \frac{m}{n}\left(\|A\|_2 \left\| E^T \right\|_2 + \|E\|_2 \left\| A^T \right\|_2 + \|E\|_2 \left\| E^T \right\|_2 \right)
\end{aligned}$$

$$\leq \eta_0 + \frac{m}{n}\left(\|A\|_2 \varepsilon_A \left\|A^T\right\|_2 + \varepsilon_A \|A\|_2 \left\|A^T\right\|_2 + \varepsilon_A \|A\|_2 \varepsilon_A \left\|A^T\right\|_2\right) \leq \eta_0 + \frac{m}{n}\left(2\varepsilon_A + \varepsilon_A^2\right)\|A\|_2 \cdot \left\|A^T\right\|_2 = \eta_0 + \frac{m}{n}\left(2\varepsilon_A + \varepsilon_A^2\right)\sigma_1^2. \tag{A.2}$$

The last equation holds because of $\|A\|_2 = \sigma_1$, $\|A\|_2 = \|A^T\|_2$. Combine (A.1) with (A.2) to obtain (32). □

Proof of Theorem 7. The three different inequalities come from different proving processes but in essence they are the same. Here we only prove the first inequality in more detail; the proofs of the second and third inequality are similar. For convenience, we denote them by Cases 1, 2, and 3 related to Cases $1'$, $2'$, and $3'$, respectively. The proofs depend on one fact that $Q_1 = Q/\gamma$ is close to I due to the definition of η. Suppose that

$$\gamma = \sigma^2 + \frac{n}{m}\sigma_0, \quad Q = \sigma^2 I + \sigma_0^2 (A+E)(A+E)^T. \tag{A.3}$$

Assume that (15) can be written as $y = \widehat{B}x + v$, where

$$\widehat{B} = Q_1^{-1/2}(A+E) = Q_1^{-1/2}\widehat{A}, \quad v = Q_1^{-1/2}((A+E)u + e). \tag{A.4}$$

Case 1. Consider

$$\|Q_1 - I\|_2 = \left\|\frac{Q}{\gamma} - I\right\|_2 = \frac{1}{\gamma}\|Q - \gamma I\|_2 = \frac{1}{\gamma}\left\|\sigma^2 I + \sigma_0^2 (A+E)(A+E)^T - \gamma I\right\|_2 = \frac{1}{\gamma}\left\|\sigma^2 I + \sigma_0^2 AA^2 + \sigma_0^2 AE^T + \sigma_0^2 EA^T + \sigma_0^2 EE^T - \sigma^2 I - \frac{n}{m}\sigma_0^2 I\right\|_2 = \frac{(n/m)\sigma_0^2}{\gamma}\left\|\frac{m}{n}AA^T + \frac{m}{n}AE^T + \frac{m}{n}EA^T + \frac{m}{n}\cdot EE^T - I\right\|_2 \leq \frac{(n/m)\sigma_0^2}{\gamma}\left(\frac{m}{n}\left(\|A\|_2\left\|A^T\right\|_2 + 2\varepsilon_A\|A\|_2\cdot\left\|A^T\right\|_2 + \varepsilon_A\|A\|_2\varepsilon_A\left\|A^T\right\|_2\right) + \|I\|_2\right) = \frac{(n/m)\sigma_0^2}{\sigma^2 + (n/m)\sigma_0}\left(\frac{m}{n}\left(\sigma_1^2 + 2\varepsilon_A\sigma_1^2 + \varepsilon_A^2\sigma_1^2\right) + 1\right) \triangleq \mu_1. \tag{A.5}$$

The last equation holds because of $\|A\|_2 = \|A^T\|_2 = \sigma_1$, $\|I\|_2 = 1$.

For (A.5), since $0 < \varepsilon_A \ll 1$, $m \ll n$, σ_1 is positive; thus,

$$\frac{m}{n}\left(\sigma_1^2 + 2\varepsilon_A\sigma_1^2 + \varepsilon_A^2\sigma_1^2\right) + 1 \longrightarrow 1 \tag{A.6}$$

holds when $n \to \infty$; therefore, (A.5) $\to (n/m)\sigma_0^2/(\sigma^2 + (n/m)\sigma_0) < 1$.

That is, $\|Q_1 - I\|_2 < \mu_1 < 1$ when $n \to \infty$.

Case 2. Consider

$$\|Q_1 - I\|_2 = \left\|\frac{Q}{\gamma} - I\right\|_2 = \frac{1}{\gamma}\|Q - \gamma I\|_2 = \frac{1}{\gamma}\left\|\sigma^2 I + \sigma_0^2 (A+E)(A+E)^T - \sigma^2 I - \frac{n}{m}\sigma_0 I\right\|_2 = \frac{(n/m)\sigma_0^2}{\sigma^2 + (n/m)\sigma_0^2}\left\|\frac{m}{n}(A+E)(A+E)^T - I\right\|_2 \leq \frac{(n/m)\sigma_0^2}{\sigma^2 + (n/m)\sigma_0^2}\eta \triangleq \eta_1 < \eta < \frac{1}{2}. \tag{A.7}$$

That is, $\|Q_1 - I\|_2 \leq \eta_1 < \eta < 1/2$.

Case 3. Consider

$$\|Q_1 - I\|_2 = \left\|\frac{Q}{\gamma} - I\right\|_2 = \frac{1}{\gamma}\|Q - \gamma I\|_2 = \frac{1}{\gamma}\left\|\sigma^2 I + \sigma_0^2 (A+E)(A+E)^T - \gamma I\right\|_2 = \frac{1}{\gamma}\left\|\sigma^2 I + \sigma_0^2 AA^2 + \sigma_0^2 AE^T + \sigma_0^2 EA^T + \sigma_0^2 EE^2 - \sigma^2 I - \frac{n}{m}\sigma_0^2 I\right\|_2 \leq \frac{(n/m)\sigma_0^2}{\gamma}\left(\left\|\frac{m}{n}AA^T - I\right\|_2 + \frac{m}{n}\left(\|A\|_2\left\|E^T\right\|_2 + \|E\|_2\left\|A^T\right\|_2 + \|E\|_2\left\|E^T\right\|_2\right)\right) = \frac{(n/m)\sigma_0^2}{\sigma^2 + (n/m)\sigma_0}\left(\eta_0 + \frac{m}{n}\left(2\varepsilon_A\sigma_1^2 + \varepsilon_A^2\sigma_1^2\right)\right) \triangleq \mu_2. \tag{A.8}$$

The last equation holds due to $\|A\|_2 = \|A^T\|_2 = \sigma_1$.

From (A.8), since $0 < \varepsilon_A \ll 1$, $m \ll n$, and $\|A\|_2 = \sigma_1$ is positive,

$$\frac{m}{n}\left(2\varepsilon_A\sigma_1^2 + \varepsilon_A^2\sigma_1^2\right) \longrightarrow 0 \tag{A.9}$$

with $n \to \infty$; thus,

$$\text{(A.8)} \longrightarrow \frac{(n/m)\sigma_0^2}{\sigma^2 + (n/m)\sigma_0}\eta_0. \tag{A.10}$$

That is, $\|Q_1 - I\|_2 < \mu_2 < \eta_0 < 1/2$ with $n \to \infty$.

As mentioned above, $\|Q_1 - I\|_2 < 1$ holds under some conditions. Using Cases 1, 2, and 3, we can obtain three different results, denoted by Cases $1'$, $2'$, and $3'$, respectively.

Here $Q_1^{-1} - I$ can be expressed as follows:

$$Q_1^{-1} - I = \frac{I - Q_1}{Q_1} = \left(\frac{Q_1}{I - Q_1}\right)^{-1} = \frac{I - Q_1}{I - (1 - Q_1)} = \sum_{k\geq 1}(I - Q_1)^k. \tag{A.11}$$

Case 1′. Note that $Q_1^{-1} - I$ converges due to (A.5) $\|I - Q_1\|_2 < \mu_1 < 1$, where $\|\cdot\|_2$ is an operator norm. Take such norm on both sides of the above equality and utilize the triangle inequality to get

$$\begin{aligned}\left\|Q_1^{-1} - I\right\|_2 &= \left\|\sum_{k\geq 1}(I - Q_1)^k\right\|_2 \leq \sum_{k\geq 1}\|I - Q_1\|_2^k \\ &< \sum_{k\geq 1}\mu_1^k = \frac{\mu_1}{1-\mu_1} \triangleq \mu_1'.\end{aligned} \quad (A.12)$$

Let Λ be an index set of size k, $\forall h \in R^k$;

$$\left\|\widehat{B}_\Lambda h\right\|_2^2 - \left\|\widehat{A}_\Lambda h\right\|_2^2 = h^T\widehat{A}_\Lambda^T\left(Q_1^{-1} - I\right)\widehat{A}_\Lambda h \quad (A.13)$$

holds. Since

$$\begin{aligned}\left|h^T\widehat{A}_\Lambda^T\left(Q_1^{-1} - I\right)\widehat{A}_\Lambda h\right| &\leq \left\|Q_1^{-1} - I\right\|_2\left\|\widehat{A}_\Lambda h\right\|_2^2 \\ &\leq \mu_1'\left\|\widehat{A}_\Lambda h\right\|_2^2,\end{aligned} \quad (A.14)$$

we obtain

$$\left|\left\|\widehat{B}_\Lambda h\right\|_2^2 - \left\|\widehat{A}_\Lambda h\right\|_2^2\right| \leq \mu_1'\left\|\widehat{A}_\Lambda h\right\|_2^2. \quad (A.15)$$

Remove the absolute value to get

$$\left(1 - \mu_1'\right)\left\|\widehat{A}_\Lambda h\right\|_2^2 \leq \left\|\widehat{B}_\Lambda h\right\|_2^2 \leq \left(1 + \mu_1'\right)\left\|\widehat{A}_\Lambda h\right\|_2^2. \quad (A.16)$$

Due to

$$\widehat{\alpha}_k\|h\|_2^2 \leq \left\|\widehat{A}h\right\|_2^2 \leq \widehat{\beta}_k\|h\|_2^2, \quad \forall h \in R^k, \quad (A.17)$$

if set $\mu_1' = \mu_1/(1-\mu_1)$, we have

$$\widehat{\alpha}_k\left(1 - \mu_1'\right)\|h\|_2^2 \leq \left\|\widehat{B}_\Lambda h\right\|_2^2 \leq \widehat{\beta}_k\left(1 + \mu_1'\right)\|h\|_2^2. \quad (A.18)$$

Case 2′. Note that $Q_1^{-1} - I$ converges due to (A.7), $\|I - Q_1\|_2 < \eta_1 < 1/2$. Take spectral norm on both sides of $Q_1^{-1} - I$ and utilize the triangle inequality to get

$$\begin{aligned}\left\|Q_1^{-1} - I\right\|_2 &= \left\|\sum_{k\geq 1}(I - Q_1)^k\right\|_2 \leq \sum_{k\geq 1}\|I - Q_1\|_2^k \\ &< \sum_{k\geq 1}\eta_1^k = \frac{\eta_1}{1-\eta_1} \triangleq \eta_1'.\end{aligned} \quad (A.19)$$

The remaining proof of Case $2'$ is similar to that of Case $1'$ except for η_1' instead of μ_1'. At last we have

$$\begin{aligned}&\widehat{\alpha}_k\left(1 - \eta_1'\right)\|h\|_2^2 \leq \left\|\widehat{B}_\Lambda h\right\|_2^2 \leq \widehat{\beta}_k\left(1 + \eta_1'\right)\|h\|_2^2, \\ &\eta_1' = \frac{\eta_1}{1-\eta_1}.\end{aligned} \quad (A.20)$$

Case 3′. Note that $Q_1^{-1} - I$ converges from (A.8) due to $\|I - Q_1\|_2 < \mu_2 < 1/2$. Take spectral norm on both sides of $Q_1^{-1} - I$ and utilize the triangle inequality to get

$$\begin{aligned}\left\|Q_1^{-1} - I\right\|_2 &= \left\|\sum_{k\geq 1}(I - Q_1)^k\right\|_2 \leq \sum_{k\geq 1}\|I - Q_1\|_2^k \\ &< \sum_{k\geq 1}\mu_2^k = \frac{\mu_2}{1-\mu_2} \triangleq \mu_2'.\end{aligned} \quad (A.21)$$

The remaining proof of Case $3'$ is similar to that of Case $1'$ except for μ_2' instead of μ_1'. At last, we obtain

$$\begin{aligned}&\widehat{\alpha}_k\left(1 - \mu_2'\right)\|h\|_2 \leq \left\|\widehat{B}_\Lambda h\right\|_2^2 \leq \widehat{\beta}_k\left(1 + \mu_2'\right)\|h\|_2^2, \\ &\mu_2' = \frac{\mu_2}{1-\mu_2}.\end{aligned} \quad (A.22)$$

□

Proof of Theorem 9. On the one hand,

$$\begin{aligned}\widetilde{\eta} &= \left\|I - \frac{m}{n}\left(A + \sum_{i=1}^s E_i\right)\left(A + \sum_{i=1}^s E_i\right)^T\right\|_2 = \left\|\frac{m}{n}\right. \\ &\cdot AA^T - I + \frac{m}{n}\left(A\sum_{i=1}^s E_i^T + A^T\sum_{i=1}^s E_i\right. \\ &\left.\left.+\left(\sum_{i=1}^s E_i\right)\left(\sum_{i=1}^s E_i\right)^T\right)\right\|_2 \geq \left\|\frac{m}{n}AA^T - I\right\|_2 \\ &- \frac{m}{n}\left(\left\|A\sum_{i=1}^s E_i^T + A^T\sum_{i=1}^s E_i\right.\right. \\ &\left.\left.+\left(\sum_{i=1}^s E_i\right)\left(\sum_{i=1}^s E_i\right)^T\right\|_2\right) \geq \eta_0 - \frac{m}{n}\left(\|A\|_2\right. \\ &\cdot\sum_{i=1}^s\left\|E_i^T\right\|_2 + \left\|A^T\right\|_2\sum_{i=1}^s\|E_i\|_2 + \left(\sum_{i=1}^s\|E_i\|_2\right) \\ &\left.\cdot\left(\sum_{i=1}^s\left\|E_i^T\right\|_2\right)\right) \geq \eta_0 - \frac{m}{n}\left(\|A\|_2\,\varepsilon_A\sum_{i=1}^s\left\|A^T\right\|_2\right. \\ &+ \left\|A^T\right\|_2\sum_{i=1}^s\varepsilon_A\|A\|_2 + \left(\sum_{i=1}^s\varepsilon_A\|A\|_2\right) \\ &\left.\cdot\left(\sum_{i=1}^s\varepsilon_A\left\|A^T\right\|_2\right)\right) = \eta_0 - \frac{m}{n}\left(2s\varepsilon_A\sigma_1^2\right. \\ &\left.+ s^2\varepsilon_A^2\sigma_1^2\right).\end{aligned} \quad (A.23)$$

The last equation holds due to $\|A^T\|_2 = \|A\|_2 = \sigma_1$.

On the other hand,

$$\tilde{\eta} = \left\| I - \frac{m}{n}\left(A + \sum_{i=1}^{s} E_i\right)\left(A + \sum_{i=1}^{s} E_i\right)^T \right\|_2 = \left\| \frac{m}{n} \cdot AA^T - I + \frac{m}{n}\left(A\sum_{i=1}^{s} E_i^T + A^T\sum_{i=1}^{s} E_i + \left(\sum_{i=1}^{s} E_i\right)\left(\sum_{i=1}^{s} E_i\right)^T\right) \right\|_2 \le \left\| \frac{m}{n}AA^T - I \right\|_2 + \frac{m}{n}\left(\left\| A\sum_{i=1}^{s} E_i^T + A^T\sum_{i=1}^{s} E_i + \left(\sum_{i=1}^{s} E_i\right)\left(\sum_{i=1}^{s} E_i\right)^T \right\|_2\right) \le \eta_0 + \frac{m}{n}\left(\|A\|_2 \cdot \sum_{i=1}^{s} \left\|E_i^T\right\|_2 + \left\|A^T\right\|_2 \sum_{i=1}^{s} \|E_i\|_2 + \left(\sum_{i=1}^{s} \|E_i\|_2\right) \cdot \left(\sum_{i=1}^{s} \left\|E_i^T\right\|_2\right)\right) \le \eta_0 + \frac{m}{n}\left(\|A\|_2 \varepsilon_A \sum_{i=1}^{s} \left\|A^T\right\|_2 + \left\|A^T\right\|_2 \sum_{i=1}^{s} \varepsilon_A \|A\|_2 + \left(\sum_{i=1}^{s} \varepsilon_A \|A\|_2\right) \cdot \left(\sum_{i=1}^{s} \varepsilon_A \left\|A^T\right\|_2\right)\right) = \eta_0 + \frac{m}{n}\left(2s\varepsilon_A \sigma_1^2 + s^2\varepsilon_A^2\sigma_1^2\right). \tag{A.24}$$

The last equation holds because of $\|A^T\|_2 = \|A\|_2 = \sigma_1$. As mentioned above, combine (A.23) with (A.24) to get (32). □

Proof of Theorem 11. There are three different results of whitening $A + \sum_{i=1}^{s} E_i$ due to the different proving process. The proof depends on one fact that Q_1 is close to I due to the definition of $\tilde{\eta}$. Assume that (A.1) can be written as $y = \hat{B}x + w$, where $\hat{B} = Q_1^{-1/2}(A + \sum_{i=1}^{s} E_i)$, $w = Q_1^{-1/2}((A + \sum_{i=1}^{s} E_i)u + e)$.

Case 1. Consider

$$\|Q_1 - I\|_2 = \left\| \frac{Q}{\gamma} - I \right\|_2 = \frac{1}{\gamma}\left\| \sigma^2 I + \sigma_0^2\left(A + \sum_{i=1}^{s} E_i\right) \cdot \left(A + \sum_{i=1}^{s} E_i\right)^T - \gamma I \right\|_2 = \frac{1}{\gamma}\left\| \sigma^2 I + \sigma_0^2\left(A + \sum_{i=1}^{s} E_i\right)\left(A + \sum_{i=1}^{s} E_i\right)^T - \left(\sigma^2 + \frac{n}{m}\sigma_0^2\right) I \right\|_2 = \frac{(n/m)\,\sigma_0^2}{\gamma}\left\| \frac{m}{n}\left(A + \sum_{i=1}^{s} E_i\right)\left(A + \sum_{i=1}^{s} E_i\right)^T - I \right\|_2 = \frac{(n/m)\,\sigma_0^2}{\gamma}\left\| \frac{m}{n}AA^T - I + \frac{m}{n}\left(A\sum_{i=1}^{s} E_i^T + A^T\sum_{i=1}^{s} E_i + \left(\sum_{i=1}^{s} E_i\right)\left(\sum_{i=1}^{s} E_i\right)^T\right) \right\|_2 \le \frac{(n/m)\,\sigma_0^2}{\gamma}\left(\left\| \frac{m}{n}AA^T - I \right\|_2 + \frac{m}{n}\left(\|A\|_2 \sum_{i=1}^{s} \left\|E_i^T\right\|_2 + \left\|A^T\right\|_2 \sum_{i=1}^{s} \|E_i\|_2 + \left(\sum_{i=1}^{s} \|E_i\|_2\right)\left(\sum_{i=1}^{s} \left\|E_i^T\right\|_2\right)\right)\right) \le \frac{(n/m)\,\sigma_0^2}{\gamma}\left(\eta_0 + \frac{m}{n}\left(\|A\|_2 \varepsilon_A \sum_{i=1}^{s} \left\|A^T\right\|_2 + \left\|A^T\right\|_2 \sum_{i=1}^{s} \varepsilon_A \|A\|_2 + \varepsilon_A^2\left(\sum_{i=1}^{s} \|A\|_2\right)\left(\sum_{i=1}^{s} \left\|A^T\right\|_2\right)\right)\right) = \frac{(n/m)\,\sigma_0^2}{\gamma}\left(\eta_0 + \frac{m}{n}\left(\varepsilon_A \sum_{i=1}^{s} \sigma_1^2 + \varepsilon_A \sum_{i=1}^{s} \sigma_1^2 + s^2\varepsilon_A^2\sigma_1^2\right)\right) = \frac{(n/m)\,\sigma_0^2}{\sigma^2 + (n/m)\,\sigma_0^2}\left(\eta_0 + \frac{m}{n}\left(2s\varepsilon_A\sigma_1^2 + s^2\varepsilon_A^2\sigma_1^2\right)\right) \triangleq \mu_3. \tag{A.25}$$

The last equation holds because of $\|A\|_2 = \sigma_1$ and $\|A\|_2 = \|A^T\|_2$.

From (A.25), since $0 < \varepsilon_A \ll 1$, $m \ll n$, and $\|A\|_2 = \sigma_1$ is a constant, we have

$$\frac{m}{n}\left(2s\varepsilon_A\sigma_1^2 + s^2\varepsilon_A^2\sigma_1^2\right) \longrightarrow 0 \tag{A.26}$$

when $n \to \infty$; therefore, (A.25) $\to$ $((n/m)\sigma_0^2/(\sigma^2 + (n/m)\sigma_0^2))\eta_0 < \eta_0 < 1/2$. That is, $\|Q_1 - I\|_2 < \mu_3 < 1/2$ when $n \to \infty$.

Case 2. Consider

$$\|Q_1 - I\|_2 = \left\| \frac{Q}{\gamma} - I \right\|_2 = \frac{1}{\gamma}\left\| \sigma^2 I + \sigma_0^2\left(A + \sum_{i=1}^{s} E_i\right)\left(A + \sum_{i=1}^{s} E_i\right)^T - \gamma I \right\|_2 = \frac{1}{\gamma}\left\| \sigma^2 I + \sigma_0^2\left(A + \sum_{i=1}^{s} E_i\right)\left(A + \sum_{i=1}^{s} E_i\right)^T - \left(\sigma^2 + \frac{n}{m}\sigma_0^2\right) I \right\|_2$$

$$= \frac{(n/m)\,\sigma_0^2}{\gamma} \left\| \frac{m}{n} \left(A + \sum_{i=1}^{s} E_i \right) \left(A + \sum_{i=1}^{s} E_i \right)^T - I \right\|_2 = \frac{(n/m)\,\sigma_0^2}{\sigma^2 + (n/m)\,\sigma_0^2} \widetilde{\eta} \triangleq \eta_2 < \widetilde{\eta} < \frac{1}{2}. \tag{A.27}$$

From (A.27), we get $\|Q_1 - I\|_2 < 1$.

Case 3. Consider

$$\begin{aligned}\|Q_1 - I\|_2 &= \left\| \frac{Q}{\gamma} - I \right\|_2 = \frac{1}{\gamma} \left\| \sigma^2 I + \sigma_0^2 \left(A + \sum_{i=1}^{s} E_i \right) \cdot \left(A + \sum_{i=1}^{s} E_i \right)^T - \gamma I \right\|_2 = \frac{1}{\gamma} \left\| \sigma^2 I + \sigma_0^2 \left(A + \sum_{i=1}^{s} E_i \right) \left(A + \sum_{i=1}^{s} E_i \right)^T - \left(\sigma^2 + \frac{n}{m} \sigma_0^2 \right) I \right\|_2 \\
&= \frac{(n/m)\,\sigma_0^2}{\gamma} \left\| \frac{m}{n} \left(A + \sum_{i=1}^{s} E_i \right) \left(A + \sum_{i=1}^{s} E_i \right)^T - I \right\|_2 \le \frac{(n/m)\,\sigma_0^2}{\gamma} \left(\|-I\|_2 + \frac{m}{n} \left(\left\| AA^T \right\|_2 + \|A\|_2 \sum_{i=1}^{s} \left\| E_i^T \right\|_2 + \left\| A^T \right\|_2 \sum_{i=1}^{s} \|E_i\|_2 + \left(\sum_{i=1}^{s} \|E_i\|_2 \right) \left(\sum_{i=1}^{s} \left\| E_i^T \right\|_2 \right) \right) \right) \\
&\le \frac{(n/m)\,\sigma_0^2}{\gamma} \left(1 + \frac{m}{n} \left(\|A\|_2 \left\| A^T \right\|_2 + \|A\|_2 \, \varepsilon_A \sum_{i=1}^{s} \left\| A^T \right\|_2 + \left\| A^T \right\|_2 \sum_{i=1}^{s} \varepsilon_A \|A\|_2 + \left(\sum_{i=1}^{s} \varepsilon_A \|A\|_2 \right) \left(\sum_{i=1}^{s} \varepsilon_A \left\| A^T \right\|_2 \right) \right) \right) \\
&= \frac{(n/m)\,\sigma_0^2}{\gamma} \left(1 + \frac{m}{n} \left(\sigma_1^2 + \varepsilon_A \sum_{i=1}^{s} \sigma_1^2 + \varepsilon_A \sum_{i=1}^{s} \sigma_1^2 + s^2 \varepsilon_A^2 \sigma_1^2 \right) \right) = \frac{\sigma_0^2}{\sigma^2 + (n/m)\,\sigma_0^2} \left(1 + \frac{m}{n} \left(\sigma_1^2 + 2s\varepsilon_A \sigma_1^2 + s^2 \varepsilon_A^2 \sigma_1^2 \right) \right) \triangleq \mu_4.\end{aligned} \tag{A.28}$$

The last equation holds due to $\|A\|_2 = \sigma_1$ and $\|A\|_2 = \|A^T\|_2$.

From (A.28), since $0 < \varepsilon_A \ll 1$, $m \ll n$, and $\|A\|_2 = \sigma_1$ is a constant, then

$$\frac{m}{n} \left(\sigma_1^2 + 2s\varepsilon_A \sigma_1^2 + s^2 \varepsilon_A^2 \sigma_1^2 \right) \longrightarrow 0 \tag{A.29}$$

when $n \to \infty$; thus, (A.28) $\to (n/m)\sigma_0^2/(\sigma^2 + (n/m)\sigma_0^2) < 1$. That is, $\|Q_1 - I\|_2 < \mu_4 < 1$. As mentioned above, $\|Q_1 - I\|_2 < 1$ under some conditions. Using the above three cases (Cases 1, 2, and 3), we can obtain three different results, denoted by Cases $1'$, $2'$, and $3'$, respectively.

Here $Q_1^{-1} - I$ can be expressed as follows:

$$\begin{aligned}Q_1^{-1} - I &= \frac{I - Q_1}{Q_1} = \left(\frac{Q_1}{I - Q_1} \right)^{-1} = \frac{I - Q_1}{I - (1 - Q_1)} \\
&= \sum_{k \ge 1} (I - Q_1)^k.\end{aligned} \tag{A.30}$$

Case $1'$. Note that (A.30) converges due to Case 1: $\|I - Q_1\|_2 < \mu_3 < 1$. Take spectral norm on both sides of the equality and utilize the triangle inequality to get

$$\begin{aligned}\left\| Q_1^{-1} - I \right\|_2 &= \left\| \sum_{k \ge 1} (I - Q_1)^k \right\|_2 \le \sum_{k \ge 1} \|I - Q_1\|_2^k \\
&< \sum_{k \ge 1} \mu_3^k = \frac{\mu_3}{1 - \mu_3} \triangleq \mu_3'.\end{aligned} \tag{A.31}$$

Let Λ be an index set of size k; we have

$$\left\| \widehat{B}_\Lambda h \right\|_2^2 - \left\| \widehat{A}_\Lambda h \right\|_2^2 = h^T \widehat{A}_\Lambda^T \left(Q_1^{-1} - I \right) \widehat{A}_\Lambda h, \quad \forall h \in R^k. \tag{A.32}$$

Since

$$\begin{aligned}\left| h^T \widehat{A}_\Lambda^T \left(Q_1^{-1} - I \right) \widehat{A}_\Lambda h \right| &\le \left\| Q_1^{-1} - I \right\|_2 \left\| \widehat{A}_\Lambda h \right\|_2^2 \\
&\le \mu_3' \left\| \widehat{A}_\Lambda h \right\|_2^2,\end{aligned} \tag{A.33}$$

we obtain that

$$\left| \left\| \widehat{B}_\Lambda h \right\|_2^2 - \left\| \widehat{A}_\Lambda h \right\|_2^2 \right| \le \mu_3' \left\| \widehat{A}_\Lambda h \right\|_2^2. \tag{A.34}$$

Remove the absolute value to get

$$\left(1 - \mu_3' \right) \left\| \widehat{A}_\Lambda h \right\|_2^2 \le \left\| \widehat{B}_\Lambda h \right\|_2^2 \le \left(1 + \mu_3' \right) \left\| \widehat{A}_\Lambda h \right\|_2^2. \tag{A.35}$$

Due to

$$\widehat{\alpha}_k \|h\|_2^2 \le \left\| \widehat{A} h \right\|_2^2 \le \widehat{\beta}_k \|h\|_2^2, \quad \forall h \in R^n, \tag{A.36}$$

we have

$$\begin{aligned}\widehat{\alpha}_k \left(1 - \mu_3' \right) \|h\|_2^2 \le \left\| \widehat{B}_\Lambda h \right\|_2^2 \le \widehat{\beta}_k \left(1 + \mu_3' \right) \|h\|_2^2, \\
\mu_3' = \frac{\mu_3}{1 - \mu_3}.\end{aligned} \tag{A.37}$$

Case 2′. Note that (A.30) converges due to Case 2: $\|I - Q_1\|_2 < \eta_2 < 1$. Take spectral norm on both sides of the equality (A.45) and utilize the triangle inequality to get

$$\begin{aligned}\left\|Q_1^{-1} - I\right\|_2 &= \left\|\sum_{k\geq 1}(I - Q_1)^k\right\|_2 \leq \sum_{k\geq 1}\|I - Q_1\|_2^k \\ &< \sum_{k\geq 1}\eta_2^k = \frac{\eta_2}{1-\eta_2} \triangleq \eta_2'.\end{aligned} \tag{A.38}$$

The remaining proving is similar to that of Case 1′ except for η_2' instead of μ_3'. At last we have

$$\widehat{\alpha}_k\left(1 - \eta_2'\right)\|h\|_2^2 \leq \left\|\widehat{B}_\Lambda h\right\|_2^2 \leq \widehat{\beta}_k\left(1 + \eta_2'\right)\|h\|_2^2. \tag{A.39}$$

Case 3′. Note that (A.45) converges due to Case 3: $\|I - Q_1\|_2 < \mu_4 < 1$. Take spectral norm on both sides of the equality (A.45) and utilize the triangle inequality to get

$$\begin{aligned}\left\|Q_1^{-1} - I\right\|_2 &= \left\|\sum_{k\geq 1}(I - Q_1)^k\right\|_2 \leq \sum_{k\geq 1}\|I - Q_1\|_2^k \\ &< \sum_{k\geq 1}\mu_4^k = \frac{\mu_4}{1-\mu_4} \triangleq \mu_4'.\end{aligned} \tag{A.40}$$

The remaining proving is similar to that of Case 1′ except for μ_4' instead of μ_3'. At last we have

$$\begin{gathered}\widehat{\alpha}_k\left(1 - \mu_4'\right)\|h\|_2^2 \leq \left\|\widehat{B}_\Lambda h\right\|_2^2 \leq \widehat{\beta}_k\left(1 + \mu_4'\right)\|h\|_2^2, \\ \mu_4' = \frac{\mu_4}{1-\mu_4}.\end{gathered} \tag{A.41}$$

□

Proof of Theorem 14. To prove the theorem, we should find out an upper bound of the numerator $|\widehat{B}_i^T\widehat{B}_j|$ of $\mu(\widehat{B})$ and a lower bound of the denominator $\|\widehat{B}_i\|_2$. For $i \neq j$, by assumption, we obtain

$$\begin{aligned}\left|\widehat{B}_i^T\widehat{B}_j\right| &= \left|\widehat{A}_i^T Q_1^{-1}\widehat{A}_j\right| \leq \left|\widehat{A}_i^T\widehat{A}_j\right| + \left|\widehat{A}_i^T\left(Q_1^{-1} - I\right)\widehat{A}_j\right| \\ &\leq \left(1 + \mu_2'\right)\left|\widehat{A}_i^T\widehat{A}_j\right|.\end{aligned} \tag{A.42}$$

Next, we estimate lower bound $\|\widehat{B}_i\|_2$ with restrictions on $\|A_i\|$ and μ_2. Similar to the proof of Theorem 7, $Q_1^{-1/2} - I$ can be expressed as a power series:

$$Q_1^{-1/2} - I = \sum_{k\geq 1} c_k\left(I - Q_1\right), \tag{A.43}$$

where c_k is the coefficients in the Taylor expansion of $(1 - x)^{1/2}$. Both sides of the equality are taken norm obtaining

$$\begin{aligned}\left\|Q_1^{1/2} - I\right\|_2 &\leq \sum_{k\geq 1} c_k\left\|(Q_1 - I)^k\right\|_2^k \leq \sum_{k\geq 1} c_k\mu_2 \\ &= \left(1 - \mu_2\right)^{-1/2} - 1 \triangleq \widehat{\mu}_2.\end{aligned} \tag{A.44}$$

Thus,

$$\begin{aligned}\left\|\widehat{B}_i\right\|_2 &= \left\|Q_1^{-1/2}\widehat{A}_i\right\|_2 \geq \left\|\widehat{A}_i\right\|_2 - \left\|\left(Q_1^{-1/2} - I\right)\widehat{A}_i\right\|_2 \\ &\geq \left(1 - \widehat{\mu}_2\right)\left\|\widehat{A}_i\right\|_2,\end{aligned} \tag{A.45}$$

where $\widehat{\mu}_2 = (1 - \mu_2)^{-1/2} - 1$. Combine (A.42) with (A.45) to get the result in Theorem 14. □

Competing Interests

The authors declare that there is no competing interests regarding the publication of this paper.

Acknowledgments

This work is partially supported by the National Natural Science Foundation of China (NSFC) A3 Foresight Program (no. 61411146001).

References

[1] E. J. Candes, J. Romberg, and T. Tao, "Robust uncertainty principles: exact signal reconstruction from highly incomplete frequency information," *IEEE Transactions on Information Theory*, vol. 52, no. 2, pp. 489–509, 2006.

[2] D. L. Donoho, "Compressed sensing," *IEEE Transactions on Information Theory*, vol. 52, no. 4, pp. 1289–1306, 2006.

[3] R. Baraniuk and P. Steeghs, "Compressive radar imaging," in *Proceedings of the IEEE Radar Conference*, pp. 128–133, Boston, Mass, USA, April 2007.

[4] M. F. Duarte, M. A. Davenport, M. B. Wakin et al., "Sparse signal detection from incoherent projections," in *Proceedings of the IEEE International Conference on Acoustics, Speech and Signal Processing (ICASSP '06)*, vol. 3, p. III, IEEE, Toulouse, France, May 2006.

[5] J. Romberg, "Imaging via compressive sampling: introduction to compressive sampling and recovery via convex programming," *IEEE Signal Processing Magazine*, vol. 25, no. 2, pp. 14–20, 2008.

[6] M. A. Davenport, M. F. Duarte, Y. C. Eldar et al., "Introduction to compressed sensing," in *Compressed Sensing: Theory and Applications*, vol. 93, pp. 1–64, Cambridge University Press, 2011.

[7] S. Foucart and H. Rauhut, *A Mathematical Introduction to Compressive Sensing*, Springer, Berlin, Germany, 2013.

[8] D. P. Wipf and B. D. Rao, "Sparse Bayesian learning for basis selection," *IEEE Transactions on Signal Processing*, vol. 52, no. 8, pp. 2153–2164, 2004.

[9] H. Zayyani, M. Babaie-Zadeh, and C. Jutten, "An iterative Bayesian algorithm for sparse component analysis in presence of noise," *IEEE Transactions on Signal Processing*, vol. 57, no. 11, pp. 4378–4390, 2009.

[10] T. T. Cai, L. Wang, and G. Xu, "New bounds for restricted isometry constants," *IEEE Transactions on Information Theory*, vol. 56, no. 9, pp. 4388–4394, 2010.

[11] J. A. Tropp and A. C. Gilbert, "Signal recovery from random measurements via orthogonal matching pursuit," *IEEE Transactions on Information Theory*, vol. 53, no. 12, pp. 4655–4666, 2007.

[12] M. A. Davenport and M. B. Wakin, "Analysis of orthogonal matching pursuit using the restricted isometry property," *IEEE Transactions on Information Theory*, vol. 56, no. 9, pp. 4395–4401, 2010.

[13] S. Huang and J. Zhu, "Recovery of sparse signals using OMP and its variants:conver-gence analysis based on RIP," *Inverse Problems*, vol. 27, no. 3, 2011.

[14] A. K. Fletcher and S. Rangan, "Orthogonal matching pursuit: a Brownian motion analysis," *IEEE Transactions on Signal Processing*, vol. 60, no. 3, pp. 1010–1021, 2012.

[15] Q. Mo and Y. Shen, "A remark on the restricted isometry property in orthogonal matching pursuit," *IEEE Transactions on Information Theory*, vol. 58, no. 6, pp. 3654–3656, 2012.

[16] D. Needell and J. A. Tropp, "CoSaMP: iterative signal recovery from incomplete and inaccurate samples," *Applied and Computational Harmonic Analysis*, vol. 26, no. 3, pp. 301–321, 2009.

[17] S. Li, F. Gao, G. Ge, and S. Zhang, "Deterministic construction of compressed sensing matrices via algebraic curves," *IEEE Transactions on Information Theory*, vol. 58, no. 8, pp. 5035–5041, 2012.

[18] X. Wang, Z. Zhao, N. Zhao et al., "On the application of compressed sensing in communication networks," in *Proceedings of the 5th International ICST Conference on Communications and Networking in China (CHINACOM '10)*, pp. 1–7, IEEE, Beijing, China, August 2010.

[19] D. Malioutov and M. Malyutov, "Boolean compressed sensing: LP relaxation for group testing," in *Proceedings of the IEEE International Conference on Acoustics, Speech, and Signal Processing (ICASSP '12)*, pp. 3305–3308, Kyoto, Japan, March 2012.

[20] D. Needell and R. Vershynin, "Signal recovery from incomplete and inaccurate measurements via regularized orthogonal matching pursuit," *IEEE Journal on Selected Topics in Signal Processing*, vol. 4, no. 2, pp. 310–316, 2010.

[21] D. L. Donoho, M. Elad, and V. N. Temlyakov, "Stable recovery of sparse overcomplete representations in the presence of noise," *IEEE Transactions on Information Theory*, vol. 52, no. 1, pp. 6–18, 2006.

[22] E. J. Candès, J. K. Romberg, and T. Tao, "Stable signal recovery from incomplete and inaccurate measurements," *Communications on Pure and Applied Mathematics*, vol. 59, no. 8, pp. 1207–1223, 2006.

[23] E. J. Candes and T. Tao, "Decoding by linear programming," *IEEE Transactions on Information Theory*, vol. 51, no. 12, pp. 4203–4215, 2005.

[24] E. Arias-Castro and Y. C. Eldar, "Noise folding in compressed sensing," *IEEE Signal Processing Letters*, vol. 18, no. 8, pp. 478–481, 2011.

[25] M. A. Herman and T. Strohmer, "General deviants: an analysis of perturbations in compressed sensing," *IEEE Journal on Selected Topics in Signal Processing*, vol. 4, no. 2, pp. 342–349, 2010.

[26] J. Ding, L. Chen, and Y. Gu, "Performance analysis of orthogonal matching pursuit under general perturbations," in *Proceedings of the International Conference on Computing, Networking and Communications (ICNC '12)*, pp. 892–896, Maui, Hawaii, USA, February 2012.

[27] J. Ding, L. Chen, and Y. Gu, "Perturbation analysis of orthogonal matching pursuit," *IEEE Transactions on Signal Processing*, vol. 61, no. 2, pp. 398–410, 2013.

4

On the Solution of the Eigenvalue Assignment Problem for Discrete-Time Systems

El-Sayed M. E. Mostafa, Abdallah W. Aboutahoun, and Fatma F. S. Omar

Department of Mathematics and Computer Science, Faculty of Science, Alexandria University, Moharam Bek, Alexandria 21511, Egypt

Correspondence should be addressed to El-Sayed M. E. Mostafa; emostafa@alex-sci.edu.eg

Academic Editor: M. Montaz Ali

The output feedback eigenvalue assignment problem for discrete-time systems is considered. The problem is formulated first as an unconstrained minimization problem, where a three-term nonlinear conjugate gradient method is proposed to find a local solution. In addition, a cut to the objective function is included, yielding an inequality constrained minimization problem, where a logarithmic barrier method is proposed for finding the local solution. The conjugate gradient method is further extended to tackle the eigenvalue assignment problem for the two cases of decentralized control systems and control systems with time delay. The performance of the methods is illustrated through various test examples.

1. Introduction

In this work, we consider the following unconstrained minimization problem:

$$\min_{K \in \mathbb{R}^{p\times r}} \quad f(K) = \rho(A(K)), \tag{1}$$

where $f : \mathbb{R}^{p\times r} \to \mathbb{R}_+$ is generally semismooth and nonconvex and $\rho(\cdot)$ is the spectral radius of the matrix function $A(K)$ that will be defined later on.

It is desirable to have a local solution K_* of (1) such that $\rho(A(K_*)) < 1$. Therefore, we include such a constraint as a cut to the objective function implying the following inequality constrained minimization problem:

$$\begin{aligned} \min \quad & f(K) = \rho(A(K)) \\ \text{s.t.} \quad & g(K) = \rho(A(K)) - 1 + \tau \le 0, \end{aligned} \tag{2}$$

where $\tau \in (0,1)$ is a given constant.

It is well known that, for any matrix norm, it holds that $\rho(M) \le \|M\|$, where M is any square matrix. Then, we can replace the eigenvalue constraint of problem (2) by a regular inequality constraint which yields the following relaxed minimization problem:

$$\begin{aligned} \min_{K} \quad & f(K) = \rho(A(K)) \\ \text{s.t.} \quad & \hat{g}(K) = \|A(K)\| - 1 + \tau \le 0, \end{aligned} \tag{3}$$

where τ is as defined above. This problem can be tackled easily by any constrained optimization solver. However, there is no guarantee in general that a feasible solution exists for this problem.

The three problems (1)–(3) concern the well-known eigenvalue assignment problem (EAP) for discrete-time systems. These problems are generally nonconvex and semismooth optimization problems. Over the past decades, a considerable amount of attention has been given to the EAP where a lot of research can be found in systems and control literature in particular for continuous-time systems (see, e.g., [1–3] and the references therein). In the framework of linear discrete-time systems, a set of eigenvalue assignment algorithms have been developed (see, e.g., [2–11] and the references therein).

A related problem in output feedback control design is the linear-quadratic control problem in which the goal is to design an output feedback gain matrix K that minimizes a certain performance cost function while all eigenvalues of the closed-loop system matrix $A(K)$ must be within the unit circle (see, e.g., [12]). Such a controller K can be calculated to this problem in case of continuous-time systems by available public software packages (e.g., HIFOO) [13]. However, for discrete-time systems, to the best of our knowledge, public software has not yet been developed.

In this work, we focus on the two problems (1) and (2) where the attempt is to minimize the spectral radius of the nonsymmetric real matrix $A(K)$. In this regard, we apply a three-term nonlinear conjugate gradient (CG) method [14] to find a local solution to the minimization problem (1) which is attempting to stabilize the associated control system; see the next section. In addition, a logarithmic barrier interior-point method is proposed to tackle the constrained minimization problem (2) for the same purpose.

Nonlinear conjugate gradient methods are widely studied and comprise a class of unconstrained optimization algorithms which are characterized by low memory requirements and strong global convergence properties (see the survey [15] and later references [3, 14]). We focus on a three-term nonlinear CG method which has a nice performance (see [14]). The CG methods are descent direction methods, which means that, starting from a given point $K_0 \in \mathbb{R}^{p\times r}$, these methods generate a sequence $\{K_k\} \subset \mathbb{R}^{p\times r}$ according to the following relation:

$$K_{k+1} = K_k + \alpha_k \Delta K_k, \tag{4}$$

where the step size $\alpha_k > 0$ satisfies the line search rule and ΔK_k is a descent direction for f at K_k. The update of the new search direction varies from one CG method to another.

The logarithmic barrier interior-point method is one of the standard methods for solving constrained optimization problems (see, e.g., [16]). This method is employed to tackle the constrained problem (2) which converts it into a sequence of unconstrained minimization problems.

This article is organized as follows. In the next section, we state the formulation of the eigenvalue assignment problem and introduce some basic concepts which are needed in the subsequent analysis. In Section 3, we evaluate the required derivatives of the objective function. In Section 4, we introduce the proposed three-term CG method that finds a local solution of the unconstrained minimization problem. In Section 5, we extend the CG method to tackle the output feedback EAP problem for decentralized control systems. In Section 6, we reformulate the discrete-time control system with time delay as an augmented system without any delay so that the EAP can be tackled by the proposed CG method. In Section 7, we introduce a logarithmic barrier method for finding a local solution to problem (2). In Section 8, we test the proposed methods on different test examples from the literature. Then, we end with a conclusion.

Notations. For vectors, the symbol $\|\cdot\|$ is the 2-norm, while for matrices $\|\cdot\|$ denotes the Frobenius norm defined by $\|M\| = \sqrt{\langle M, M\rangle}$, where $\langle\cdot,\cdot\rangle$ is the inner product given by $\langle M_1, M_2\rangle = \mathrm{Tr}(M_2^{*^T} M_1)$ and $\mathrm{Tr}(\cdot)$ is the trace operator. The eigenvalues of a matrix $M \in \mathbb{R}^{n\times n}$ are denoted by $\lambda_i(M)$, $i = 1, \ldots, n$. The Greek letter $\rho(M)$ denotes the spectral radius of a square matrix M and $\Lambda(M) := \mathrm{diag}(\lambda_1, \ldots, \lambda_n)$ is a diagonal matrix with eigenvalues on its main diagonal. Moreover, $\mathrm{Re}(\cdot)$ and $\mathrm{Im}(\cdot)$ denote the real and imaginary parts of a complex number, respectively. Sometimes and for the sake of simplicity, we skip the arguments of the considered functions; for example, we use f_k to denote $f(K_k)$.

2. Problem Formulation and Preliminary

The static output feedback eigenvalue assignment problem for discrete-time systems can be stated as follows (see, e.g., the abovementioned citations and the references therein). Consider the linear time-invariant discrete-time system with the following state space realization:

$$\begin{aligned} x(k+1) &= Ax(k) + Bu(k), \\ y(k) &= Cx(k), \\ & k = 0, 1, 2, \ldots, \end{aligned} \tag{5}$$

where $x(k) \in \mathbb{R}^n$, $u(k) \in \mathbb{R}^p$, and $y(k) \in \mathbb{R}^r$ are the state, the control input, and the measured output vectors, respectively. Moreover, $A \in \mathbb{R}^{n\times n}$, $B \in \mathbb{R}^{n\times p}$, and $C \in \mathbb{R}^{r\times n}$ are given constant matrices. Such a system is often closed by the control law $u(k) = Ky(k)$ which yields

$$\begin{aligned} x(k+1) = (A + BKC)\, x(k) = A(K)\, x(k), \\ k = 0, 1, 2, \ldots, \end{aligned} \tag{6}$$

where $A(K) := A + BKC$ and $K \in \mathbb{R}^{p\times r}$ is the output feedback gain matrix which represents the unknown.

The following definitions are needed for later use.

Definition 1. The spectral radius of a matrix $A \in \mathbb{C}^{n\times n}$ with eigenvalues $\lambda_1, \ldots, \lambda_n$ is defined as

$$\rho(A) = \max\{|\lambda_i| : i \in \{1, 2, \ldots, n\}\}. \tag{7}$$

Definition 2. The discrete-time control system (6) is asymptotically stable (i.e., $x(k) \to 0$ as $k \to \infty$ for any initial $x(0)$) if and only if $\rho(A(K)) < 1$.

The eigenvalue assignment problem is to design an output feedback gain matrix K providing a closed-loop system in a satisfactory stage by shifting controllable eigenvalues to desirable locations in the complex plane. In particular, the EAP requires the spectral radius of the closed-loop matrix $A(K)$ to be strictly within the unit circle in the complex plane.

A necessary condition for the EAP by constant output feedback is given in [9]. Fu [7] has also shown that the EAP via static output feedback is NP-hard. Systems with a symmetric state space realization, namely, $A = A^T, C = B^T$, occur in different applications such as RC networks [10]. The symmetric EAP might be stated as follows: find a matrix K such that

$$\rho\left(A + BKB^T\right) < 1. \tag{8}$$

It is well known that the eigenvalues of a real symmetric matrix are not everywhere differentiable. A classical theorem (see, e.g., [17]) states that each eigenvalue of a symmetric matrix is the difference of two convex functions, which implies that the eigenvalues are semismooth functions. This fact allows us to use the theory of semismoothness to establish convergence results for the proposed methods.

Definition 3. A functional $\psi : \mathbb{R}^{p\times r} \to \mathbb{R}$ is said to be locally Lipschitz continuous at $K \in \mathbb{R}^{p\times r}$ with a constant $L > 0$ if and only if there exists a number $\sigma > 0$ such that

$$\|\psi(K_1) - \psi(K_2)\| \le L\|K_1 - K_2\|, \quad \forall K_1, K_2 \in \mathscr{B}_\sigma(K), \tag{9}$$

where $\mathscr{B}_\sigma(K)$ is an open ball with center at K and radius σ.

Let $\psi : \mathbb{R}^{p\times r} \to \mathbb{R}$ be a locally Lipschitz continuous function. Radmacher's theorem (see, e.g., [18]) implies that such a mapping ψ is differentiable almost everywhere. Let $\mathscr{D}_\psi$ be the set of all K at which ψ is differentiable and let $\nabla\psi^T$ be its Jacobian whenever it exists. Let $\mathscr{D} \subseteq \mathbb{R}^{p\times r}$ be the set of all K at which f is differentiable. Such a set is open. Therefore, it is convenient to replace $\mathscr{D}$ by the following level set:

$$\Omega := \{K \in \mathscr{D} : f(K) \le c\}, \tag{10}$$

where $c > 0$ is a given constant. This set is assumed to be bounded.

3. Derivative of the Objective Function

The eigenvalues of the matrix $A(K)$ of (1) are not in general differentiable at a point K where $A(K)$ has repeated eigenvalues. Therefore, let us consider the following assumption on eigenvalues of $A(K)$.

Assumption 4. Assume that $A(K)$ has no multiple eigenvalues for all $K \in \mathbb{R}^{p\times r}$.

Let $A(K)$ be diagonalizable and let $U(K) \in \mathbb{C}^{n\times n}$ and $V(K) \in \mathbb{C}^{n\times n}$ be a couple of matrices whose columns are the left and right eigenvectors of $A(K)$; that is, $U(K)$, $V(K)$, and $\Lambda(A(K))$ satisfy

$$\begin{aligned} U(K)^T A(K) &= \Lambda U(K)^T, \\ A(K)V(K) &= V(K)\Lambda. \end{aligned} \tag{11}$$

The columns of $U(K)$ and $V(K)$ satisfy

$$U_i^* V_i \ne 0, \quad i = 1, \dots, n. \tag{12}$$

The following lemma provides the first-order derivatives of the objective function $f(K)$ of the minimization problem (1) required by the CG methods.

Lemma 5. *Suppose that $A(K)$ satisfies Assumption 4 and $f(K)$ is differentiable at $K \in \mathbb{R}^{p\times r}$. Let $\hat\lambda = \mathrm{Re}(\hat\lambda) + i\,\mathrm{Im}(\hat\lambda)$ be the largest in magnitude eigenvalue of $A(K)$. Then, the entries of the gradient of the objective function f are given by*

$$\begin{aligned} \frac{\partial f(K)}{\partial K_{kl}} &= \frac{\mathrm{Re}(\hat\lambda)}{|\hat\lambda|}\left(u_1^T B E_{kl} C v_1 + u_2^T B E_{kl} C v_2\right) \\ &\quad + \frac{\mathrm{Im}(\hat\lambda)}{|\hat\lambda|}\left(u_1^T B E_{kl} C v_2 - u_2^T B E_{kl} C v_1\right), \\ &\qquad k = 1, \dots, p, \; l = 1, \dots, r, \end{aligned} \tag{13}$$

where $(u_1 + iu_2)$ and $(v_1 + iv_2)$ are the left and right eigenvectors associated with $\hat\lambda$ and $E_{kl} \in \mathbb{R}^{p\times r}$ is a matrix with zero entries except at the (k, l) position where its value is one.

Proof (see [19, Lemma 6.3.10]). The two eigenvectors are normalized such that

$$(u_1 - iu_2)^T (v_1 + iv_2) = 1, \tag{14}$$

which by differentiation yields

$$(u_1' - iu_2')^T (v_1 + iv_2) + (u_1 - iu_2)^T (v_1' + iv_2') = 0, \tag{15}$$

where the dash denotes the first-order derivative with respect to the entries of K. The eigenvalue $\hat\lambda$ also satisfies that

$$\hat\lambda = (u_1 - iu_2)^T A(K)(v_1 + iv_2), \tag{16}$$

which by differentiation gives

$$\begin{aligned} \hat\lambda' &= (u_1 - iu_2)^T A(K)' (v_1 + iv_2) \\ &= \left(u_1^T A(K)' v_1 + u_2^T A(K)' v_2\right) \\ &\quad + i\left(u_1^T A(K)' v_2 - u_2^T A(K)' v_1\right). \end{aligned} \tag{17}$$

Suppose that $\hat\lambda = \mathrm{Re}(\hat\lambda) + i\,\mathrm{Im}(\hat\lambda)$ is the largest in magnitude eigenvalue of $A(K)$. Then,

$$\rho(A(K)) = |\hat\lambda| = \sqrt{\mathrm{Re}(\hat\lambda)^2 + \mathrm{Im}(\hat\lambda)^2}, \tag{18}$$

and by the chain rule we have

$$\begin{aligned} \frac{\partial\rho(A(K))}{\partial K_{kl}} &= \frac{\partial\rho(A(K))}{\partial\mathrm{Re}(\hat\lambda)}\frac{\partial\mathrm{Re}(\hat\lambda)}{\partial K_{kl}} + \frac{\partial\rho(A(K))}{\partial\mathrm{Im}(\hat\lambda)}\frac{\partial\mathrm{Im}(\hat\lambda)}{\partial K_{kl}} \\ &= \frac{\mathrm{Re}(\hat\lambda)}{|\hat\lambda|}\left(u_1^T\frac{\partial(A(K))}{\partial K_{kl}}v_1 + u_2^T\frac{\partial(A(K))}{\partial K_{kl}}v_2\right) \\ &\quad + \frac{\mathrm{Im}(\hat\lambda)}{|\hat\lambda|}\left(u_1^T\frac{\partial(A(K))}{\partial K_{kl}}v_2 - u_2^T\frac{\partial(A(K))}{\partial K_{kl}}v_1\right), \\ &\qquad k = 1, \dots, p, \; l = 1, \dots, r. \end{aligned} \tag{19}$$

Since $\partial(A(K))/\partial K_{kl} = BE_{kl}C$, then the result follows. □

4. Three-Term CG Method for the EAP

In this section, we analyze and study a three-term nonlinear CG method for computing the local solution of the minimization problem (1) (see [14]). For a given starting $K_0 \in \mathbb{R}^{p\times r}$, the CG method generates a sequence of iterates according to the relation

$$K_{k+1} = K_k + \alpha_k \Delta K_k, \quad (20)$$

where $\Delta K_k \in \mathbb{R}^{p\times r}$ is supposed to be a descent direction for f at K_k and $\alpha_k > 0$ is the step size. The new search direction for the CG method is given by the following relation:

$$\begin{aligned}\Delta K_{k+1} &= -\nabla f_{k+1} - \delta_k S_k - \eta_k Y_k, \\ \Delta K_0 &= -\nabla f_0,\end{aligned} \quad (21)$$

where ∇f_{k+1} is the gradient of f at K_{k+1}, $S_k = K_{k+1} - K_k$, $Y_k = \nabla f_{k+1} - \nabla f_k$, and

$$\begin{aligned}\delta_k &= \left(1 + 2\frac{\|Y_k\|^2}{\operatorname{Tr}(Y_k^T S_k)}\right)\frac{\operatorname{Tr}(S_k^T \nabla f_{k+1})}{\operatorname{Tr}(Y_k^T S_k)} \\ &\quad - \frac{\operatorname{Tr}(Y_k^T \nabla f_{k+1})}{\operatorname{Tr}(Y_k^T S_k)},\end{aligned} \quad (22)$$

$$\eta_k = \frac{\operatorname{Tr}(S_k^T \nabla f_{k+1})}{\operatorname{Tr}(Y_k^T S_k)}.$$

In order to globalize this CG method, we recall Wolfe conditions (see, e.g., [16]) to update a suitable step size α_k for the new iterate (20):

$$f(K_k + \alpha_k \Delta K_k) \le f(K_k) + \gamma \alpha_k \operatorname{Tr}(\nabla f_k^T \Delta K_k), \quad (23)$$

$$\operatorname{Tr}(\nabla f_{k+1}^T \Delta K_k) \ge \hat{\gamma} \operatorname{Tr}(\nabla f_k^T \Delta K_k), \quad (24)$$

where $0 < \gamma < \hat{\gamma} < 1$. The strong Wolfe conditions replace (24) by the following condition:

$$\left|\operatorname{Tr}(\nabla f_{k+1}^T \Delta K_k)\right| \le \hat{\gamma}\left|\operatorname{Tr}(\nabla f_k^T \Delta K_k)\right|. \quad (25)$$

The following theorem shows that the search direction ΔK_k is a descent direction to the objective function.

Theorem 6. *Let $K_k \in \mathcal{D}$ be generated by (20) and let the step size α_k satisfy Wolfe conditions (23)-(24). Then, ΔK_{k+1} evaluated by (21)-(22) is a descent direction to f for all $k = 0, 1, 2, \ldots$.*

Proof (see also [14, Proposition 3.1]). From Wolfe's second condition (24), we obtain the curvature condition $\operatorname{Tr}(Y_k^T S_k) > 0$. Moreover, from (21)-(22), it follows that

$$\begin{aligned}&\operatorname{Tr}(\nabla f_{k+1}^T \Delta K_{k+1}) \\ &\quad = -\|\nabla f_{k+1}\|^2 \\ &\quad - \left(1 + 2\frac{\|Y_k\|^2}{\operatorname{Tr}(Y_k^T S_k)}\right)\frac{\left(\operatorname{Tr}(S_k^T \nabla f_{k+1})\right)^2}{\operatorname{Tr}(Y_k^T S_k)} \le 0.\end{aligned} \quad (26)$$

□

The three-term nonlinear CG algorithm is stated in the following lines.

Algorithm 7 (three-term CG method for the output feedback EAP).

(0) Let $K_0 \in \mathbb{R}^{p\times r}$, $0 < \gamma < \hat{\gamma} < 1$, $0 < \tau < 1$ be given constants and $\epsilon \in (0, 1)$ be the tolerance. Moreover, let A, B, C be given constant matrices. Choose $\alpha_0 > 0$, compute ∇f_0, and set $\Delta K_0 = -\nabla f_0$. If $\|\nabla f_0\| \le \epsilon$ or $f_0 < 1 - \tau$, stop; otherwise, set $k \leftarrow 0$ and go to the next step.

While $\|\nabla f_k\| > \epsilon$ or $f_k \ge 1 - \tau$, do

(1) Compute $\alpha_k > 0$ that satisfies Wolfe conditions (23)-(24); set $K_{k+1} = K_k + \alpha_k \Delta K_k$ and then calculate the gradient ∇f_{k+1}.

(2) If $\|\nabla f_{k+1}\| \le \epsilon$ or $f_{k+1} < 1 - \tau$, stop; otherwise, go to the next step.

(3) Calculate a new search direction ΔK_{k+1} by (21)-(22).

(4) Set $k \leftarrow k + 1$ and repeat.

End (do)

Remark 8. For the considered linear control system, one of the major tasks is to achieve a stabilizing output feedback controller, that is. to calculate K such that $\rho(A(K))$ is strictly less than one. However, there is no relationship between achieving such a controller and finding K, a local minimum of problem (1). Therefore, it is reasonable to stop the CG method as soon as a local solution is achieved or a stabilizing K is reached with a sufficient stability margin.

To prove global convergence of the CG algorithm, we assume that $\nabla f_k \ne 0$ for all k; otherwise, a stationary point is found.

Assumption 9. The following assumptions are assumed to hold:

(a) The level set Ω as defined in (10) is bounded; that is, there exists a constant $\beta > 0$ such that, for all $K \in \Omega$, $\|K\| < \beta$.

(b) In some neighborhood $\mathcal{N}$ of the level set (10), the gradient $\nabla f(K)$ of the objective function f is Lipschitz continuous.

From Assumption 9(a), one has $\|S_k\| \le \|K_{k+1}\| + \|K_k\| \le 2\beta$ and from Assumption 9(b) we deduce that there exists a constant $\Gamma > 0$ such that

$$\|\nabla f(K)\| \le \Gamma \quad \forall K \in \Omega. \quad (27)$$

The following lemma provides a lower bound to the step size α_k; see [14, Proposition 5.1].

Lemma 10. *Let $\{K_k\}$ be generated by Algorithm 7 and suppose that ΔK_k is a descent direction for f_k. In addition, let Assumption 9 hold. Then,*

$$\alpha_k \geq \frac{(1-\hat{\gamma})\left|\operatorname{Tr}\left(\nabla f_k^T \Delta K_k\right)\right|}{L\left\|\Delta K_k\right\|^2}. \tag{28}$$

Proof. By subtracting $\operatorname{Tr}(\nabla f_k^T \Delta K_k)$ from both sides of (24) and using Lipschitz condition,

$$\begin{aligned}(\hat{\gamma}-1)\operatorname{Tr}\left(\nabla f_k^T \Delta K_k\right) &\leq \operatorname{Tr}\left(\left(\nabla f_{k+1}-\nabla f_k\right)\Delta K_k\right)\\ &= \operatorname{Tr}\left(Y_k^T \Delta K_k\right) \leq \left|\operatorname{Tr}\left(Y_k^T \Delta K_k\right)\right| \\ &\leq \left\|Y_k\right\|\left\|\Delta K_k\right\| \leq \alpha_k L\left\|\Delta K_k\right\|.\end{aligned} \tag{29}$$

Since ΔK_k is a descent direction for f and $\hat{\gamma}$ is less than one, then (28) follows. □

The result of the following lemma is used in proving the main global convergence theorem.

Lemma 11. *Let $\{K_k\}$ be generated by Algorithm 7 and assume that ΔK_k is a descent direction. Furthermore, let Assumption 9 hold. Then,*

$$\sum_{k=1}^{\infty}\frac{\operatorname{Tr}\left(\nabla f_k^T \Delta K_k\right)^2}{\left\|\Delta K_k\right\|^2} < +\infty. \tag{30}$$

Proof. From Wolfe condition (23) and Lemma 10, one has

$$\begin{aligned}f(K_k) - f(K_k+\alpha_k\Delta K_k) &\geq -\gamma\alpha_k\left|\operatorname{Tr}\left(\nabla f_k^T \Delta K_k\right)\right|\\ &\geq \frac{-\gamma(1-\hat{\gamma})\operatorname{Tr}\left(\nabla f_k^T \Delta K_k\right)^2}{L\left\|\Delta K_k\right\|^2}.\end{aligned} \tag{31}$$

Then, Assumption 9 implies condition (30). □

Next, we have the following result for the three-term CG method (see, e.g., [14, Proposition 5.3]).

Lemma 12. *Let Assumption 9 hold. Consider $\{K_k\}$ to be generated by Algorithm 7 where ΔK_k is a descent direction for f and α_k satisfies Wolfe conditions (23) and (24). If*

$$\sum_{k>1}\frac{1}{\left\|\Delta K_k\right\|^2} = \infty, \tag{32}$$

then

$$\liminf_{k\to\infty}\left\|\nabla f_k\right\| = 0. \tag{33}$$

Proof (see also [20, Lemma 2.2]). Suppose (33) is not true. Then, there exists a constant $\varepsilon > 0$ such that

$$\left\|\nabla f_k\right\| \geq \varepsilon \quad \forall k. \tag{34}$$

Therefore, from (32) and Lemma 11, we have

$$\sum_{k=0}^{\infty}\frac{\varepsilon^4}{\left\|\Delta K_k\right\|^2} \leq \sum_{k=0}^{\infty}\frac{\left\|\nabla f_k\right\|^4}{\left\|\Delta K_k\right\|^2} \leq \sum_{k=0}^{\infty}\frac{\operatorname{Tr}\left(\nabla f_k^T \Delta K_k\right)^2}{\left\|\Delta K_k\right\|^2} = \infty. \tag{35}$$

This contradicts Lemma 12 which completes the proof. □

Under condition (30), the following global convergence result is obtained.

Theorem 13. *Let ΔK_k be generated by Algorithm 7. Assume that Assumption 9 holds. Assume further that there exists a constant $\Gamma_1 > 0$ such that $\operatorname{Tr}(Y_k^T S_k) \geq \Gamma_1$ for any $k \geq 1$. Then, it holds that*

$$\liminf_{k\to\infty}\left\|\nabla f_k\right\| = 0. \tag{36}$$

Proof (see also [14, Theorem 5.2]). Since $\operatorname{Tr}(\nabla f_k^T S_k) < 0$ for all k, then

$$\begin{aligned}\operatorname{Tr}\left(S_k^T \nabla f_{k+1}\right) &= \operatorname{Tr}\left(Y_k^T S_k\right) + \operatorname{Tr}\left(\nabla f_k^T S_k\right)\\ &< \operatorname{Tr}\left(Y_k^T S_k\right).\end{aligned} \tag{37}$$

By direct computation, we have

$$\begin{aligned}\Delta K_{k+1} = -\nabla f_{k+1} &+ \Bigg(\frac{\operatorname{Tr}\left(Y_k^T \nabla f_{k+1}\right)}{\operatorname{Tr}\left(Y_k^T S_k\right)}\\ &- \left(1+2\frac{\left\|Y_k\right\|^2}{\operatorname{Tr}\left(Y_k^T S_k\right)}\right)\frac{\operatorname{Tr}\left(S_k^T \nabla f_{k+1}\right)}{\operatorname{Tr}\left(Y_k^T S_k\right)}\Bigg)S_k\\ &- \frac{\operatorname{Tr}\left(S_k^T \nabla f_{k+1}\right)}{\operatorname{Tr}\left(Y_k^T S_k\right)}Y_k.\end{aligned} \tag{38}$$

Consequently, one has

$$\begin{aligned}\left\|\Delta K_{k+1}\right\| &\leq \left\|\nabla f_{k+1}\right\| + \frac{\left|\operatorname{Tr}\left(Y_k^T \nabla f_{k+1}\right)\right|}{\left|\operatorname{Tr}\left(Y_k^T S_k\right)\right|}\left\|S_k\right\|\\ &\quad + \left(1+2\frac{\left\|Y_k\right\|^2}{\left|\operatorname{Tr}\left(Y_k^T S_k\right)\right|}\right)\left\|S_k\right\| + \left\|Y_k\right\|\\ &< \Gamma + \left(1+\frac{2L\Gamma\beta+8L^2\beta^2}{\Gamma_1}\right)2\beta + 2L\beta.\end{aligned} \tag{39}$$

Then, from Lemma 12, (36) follows. □

5. The EAP for Decentralized Control Systems

Consider the linear time-invariant decentralized control system with ν control stations:

$$\begin{aligned}x(k+1) &= Ax(k) + \sum_{i=1}^{\nu}B_i u_i(k),\\ y_i(k) &= C_i x(k),\\ &\qquad i = 0,1,2,\ldots,\nu,\end{aligned} \tag{40}$$

where $x(k) \in \mathbb{R}^n$, $u_i(k) \in \mathbb{R}^{p_i}$, and $y_i(k) \in \mathbb{R}^{r_i}$ are the state, the control input, and the measured output vectors, respectively. $A \in \mathbb{R}^{n\times n}$, $B_i \in \mathbb{R}^{n\times p_i}$, and $C_i \in \mathbb{R}^{r_i\times n}$ are given constant matrices, $i = 1,\ldots,\nu$.

The output feedback EAP for the decentralized system (40) is to find output feedback gain matrices $K_i \in \mathbb{R}^{p_i \times r_i}$ that place, by using the control law

$$u_i(k) = K_i y_i(k), \quad i = 1, \ldots, \nu, \tag{41}$$

the eigenvalues of the closed-loop system matrix $A(K_1, \ldots, K_\nu) = A + \sum_{i=1}^{\nu} B_i K_i C_i$ to strictly lie within the unit disk.

By introducing the following augmented matrices, it is straightforward to rewrite the decentralized control system (40) in the original structure (5):

$$B = \begin{bmatrix} B_1 & \cdots & B_\nu \end{bmatrix}, \quad C = \begin{bmatrix} C_1^T & \cdots & C_\nu^T \end{bmatrix}^T. \tag{42}$$

The corresponding closed-loop system matrix is $A(K) = A + BKC$, where the output feedback gain matrix K is given by the block-diagonal matrix

$$K = \operatorname{diag}(K_1, \ldots, K_\nu), \tag{43}$$

and the corresponding unconstrained minimization problem takes the form

$$\min_{K_1, \ldots, K_\nu} f(K_1, \ldots, K_\nu) = \rho(A(K)). \tag{44}$$

The three-term CG method of the last section can be applied to tackle the minimization problem (44). Let $\widehat{\mathcal{D}} \subseteq \mathbb{R}^{p_1 \times r_1} \times \cdots \times \mathbb{R}^{p_\nu \times r_\nu}$ be the set of all

$$K = \operatorname{diag}(K_1, \ldots, K_\nu) \in \mathbb{R}^{p_1 \times r_1} \times \cdots \times \mathbb{R}^{p_\nu \times r_\nu} \tag{45}$$

at which f is differentiable. For a given starting K_0, the three-term CG method generates a sequence of the form

$$K_{k+1} = K_k + \alpha_k \Delta K_k, \quad k = 0, 1, 2, \ldots, \tag{46}$$

where $\alpha_k > 0$ is the step size that must satisfy Wolfe conditions (23)-(24), while the search direction $\Delta K_k = \operatorname{diag}(\Delta K_k^1, \ldots, \Delta K_k^\nu)$ is a descent direction for f_k at K_k. The new search direction ΔK_{k+1} is updated by using (21)-(22).

Algorithm 7 is restated in the following lines to tackle the output feedback EAP problem for decentralized control systems.

Algorithm 14 (three-term CG method for the decentralized output feedback EAP).

(0) Let $K_0 \in \mathbb{R}^{p_1 \times r_1} \times \cdots \times \mathbb{R}^{p_\nu \times r_\nu}$, $\epsilon \in (0, 1)$, $0 < \tau < 1$, and $0 < \gamma < \widehat{\gamma} < 1$ be given constants. Moreover, let $A, B_1, \ldots, B_\nu, C_1, \ldots, C_\nu$ be given constant matrices. Choose $\alpha_0 > 0$. Compute ∇f_0 and set $\Delta K_0 = -\nabla f_0$. If $\|\nabla f_0\| \le \epsilon$ or $f_0 < 1 - \tau$, stop; otherwise, set $k := 0$ and go to the next step.

While $\|\nabla f_k\| > \epsilon$ or $f_k < 1 - \tau$, do

(1) Calculate a step size $\alpha_k > 0$ that satisfies Wolfe conditions (23)-(24); set $K_{k+1} = K_k + \alpha_k \Delta K_k$ and then calculate the gradient ∇f_{k+1}.

(2) If $\|\nabla f_{k+1}\| \le \epsilon$ or $f_{k+1} < 1 - \tau$, stop; otherwise, go to the next step.

(3) Calculate the new search direction ΔK_{k+1} according to (21)-(22).

(4) Set $k \leftarrow k + 1$ and repeat.

End (do)

Remark 15. The way in which the CG method is designed allows us to maintain the block-diagonal structure of the unknown matrix K efficiently without skipping any information of the data matrices of the control system.

6. The EAP for Time-Delay Systems

Consider the following linear discrete-time and time-delay system (see, e.g., [21, 22]):

$$\begin{aligned} \widehat{x}(k+1) &= \sum_{i=0}^{d} A_i \widehat{x}(k-i) + \widehat{B} u(k), \\ y(k) &= \sum_{i=0}^{d} C_i \widehat{x}(k-i), \\ & \qquad k = 0, 1, 2, \ldots, \end{aligned} \tag{47}$$

where $A_i \in \mathbb{R}^{n \times n}$, $\widehat{B} \in \mathbb{R}^{n \times p}$, and $C_i \in \mathbb{R}^{r \times n}$ are given constant matrices and $d \in \mathbb{N}$ is the time delay in the state vector.

The above time-delay system has a specific feature that it can be converted into an augmented linear system without any delay (see, e.g., [22]). This is achieved by introducing the augmented state vector

$$x^T(k) = \begin{bmatrix} \widehat{x}^T(k) & \cdots & \widehat{x}^T(k-d) \end{bmatrix}. \tag{48}$$

The time-delay system (47) is equivalently rewritten as

$$\begin{aligned} x(k+1) &= Ax(k) + Bu(k), \\ y(k) &= Cx(k), \end{aligned} \tag{49}$$

where

$$A = \begin{bmatrix} A_0 & A_1 & \cdots & A_{d-1} & A_d \\ I_n & 0 & \cdots & 0 & 0 \\ 0 & I_n & \ddots & 0 & 0 \\ \vdots & \ddots & \ddots & \vdots & \vdots \\ 0 & 0 & \cdots & I_n & 0 \end{bmatrix}, \quad B = \begin{bmatrix} \widehat{B} \\ 0 \\ 0 \\ \vdots \\ 0 \end{bmatrix}, \quad C = (C_0, C_1, \ldots, C_{d-1}, C_d). \tag{50}$$

By using the control law $u(k) = Ky(k)$, we close system (49) as

$$x(k+1) = (A + BKC)\,x(k) =: A(K)\,x(k), \quad (51)$$

where $A(K) = A + BKC$. For such a system, one can apply Algorithm 7 to compute an output feedback gain matrix K_*, a local solution to the minimization problem (1). It is known that the eigenvalues of the time-delay system coincide with the eigenvalues of the augmented linear system (49).

We emphasize that the same approach can be considered in case of control systems with input delay.

7. Logarithmic Barrier Method

The attempt of this section is to improve formulation (1), where we aim to achieve a stabilizing K together with minimizing the spectral radius of the closed-loop system matrix. Therefore, let us consider the inequality constrained minimization problem (2). By following the idea of the logarithmic barrier method, we consider the following unconstrained minimization problem:

$$\min_{K\in\mathbb{R}^{p\times r}} \phi^{\mu}(K) = f(K) - \mu \log(-g(K)), \quad (52)$$

where $\mu > 0$ is the barrier parameter and the functions f and g are as defined in (2). According to the theory of barrier methods, the minimizer of ϕ^{μ} approaches a local solution of the original problem (2) as $\mu \to 0$ under certain conditions.

Let us define the following strict feasible region:

$$\mathcal{F} := \{K \in \mathbb{R}^{p\times r} : g(K) \le 0\}. \quad (53)$$

We assume that $\mathcal{F} \neq \emptyset$.

The Lagrangian functions for problem (2) are defined as

$$\ell(K, \upsilon) = f(K) + \upsilon g(K), \quad (54)$$

where f and g are as defined in (2) and $\upsilon \ge 0$ is the associated Lagrange multiplier.

First-order derivative of ϕ^{μ} with respect to the entries of K is obtained in the following lemma.

Lemma 16. *Suppose that $A(K)$ satisfies Assumption 4 and $f(K)$ is differentiable at $K \in \mathbb{R}^{p\times r}$. Let $\widehat{\lambda} = \mathrm{Re}(\widehat{\lambda}) + i\,\mathrm{Im}(\widehat{\lambda})$ be the largest in magnitude eigenvalue of $A(K)$. Then, the first derivative of ϕ^{μ} is given by*

$$\begin{aligned}\frac{\partial \phi^{\mu}(K)}{\partial K_{kl}} &= \left(1 + \frac{\mu}{g(K)}\right) \cdot \frac{\mathrm{Re}(\widehat{\lambda})}{|\widehat{\lambda}|}\left(u_1^T B E_{kl} C v_1 + u_2^T B E_{kl} C v_2\right) \\ &+ \left(1 + \frac{\mu}{g(K)}\right) \cdot \frac{\mathrm{Im}(\widehat{\lambda})}{|\widehat{\lambda}|}\left(u_1^T B E_{kl} C v_2 - u_2^T B E_{kl} C v_1\right), \\ &\qquad k = 1, \dots, p,\ l = 1, \dots, r,\end{aligned} \quad (55)$$

where $(u_1 + iu_2)$ and $(v_1 + iv_2)$ are the left and right eigenvectors associated with $\widehat{\lambda}$ and $E_{kl} \in \mathbb{R}^{p\times r}$ is a matrix with zero entries except at the (k, l) position where it has a value of one.

Proof. By differentiating ϕ^{μ} with respect to the entries of K and utilizing the derivative of the spectral radius $\rho(A(K))$ obtained in Lemma 5, the result follows. □

As can be seen in (55), all terms of the first derivative of ϕ^{μ} depend on μ, which leads to an ill-conditioned Hessian matrix of the barrier function ϕ^{μ} at the solution. Therefore, second-order methods are not recommended to compute a local solution of (52), but rather first-order methods such as nonlinear CG methods of Section 4 are recommended.

The gradient of ℓ takes the form

$$\begin{aligned}\nabla_K \ell(K, \upsilon) &= \nabla f(K) + \upsilon \nabla g(K) \\ &= (1 + \upsilon)\,\nabla \rho(A(K)).\end{aligned} \quad (56)$$

From (54) and (55), we see that if υ satisfies

$$\upsilon_* = \frac{\mu^*}{g(K_*)}, \quad (57)$$

then the solution obtained by the proposed log-barrier method satisfies the stationary requirement of the Karush–Kuhn–Tucker conditions.

The logarithmic barrier interior-point method is stated in the following lines.

Algorithm 17 (logarithmic barrier method for the output feedback EAP).

(0) Choose a starting barrier parameter $\mu_0 > 0$, outer and inner-loop tolerances ϵ, $\epsilon^{\mathrm{in}} > 0$, a starting feasible point $K_0^s \in \mathbb{R}^{p\times r}$. Let $a, b, \tau \in (0, 1)$ be given constants.

For $j = 0, 1, 2, \dots,$ do

(1) Find an approximate local minimizer K_j of $\phi^{\mu_j}(K)$ starting from K_j^s; terminate if $\|\nabla \phi^{\mu_j}\| \le \epsilon_j^{\mathrm{in}}$.

(2) If the final stopping test $\|\nabla \phi^{\mu_j}\| \le \epsilon$ is satisfied or $\rho(A(K_j)) < 1 - \tau$, stop; otherwise, go to the next step.

(3) Choose a new barrier parameter $\mu_{j+1} \in (0, a\mu_j)$ and a new inner-loop tolerance $\epsilon_{j+1}^{\mathrm{in}} \in (0, b\epsilon_j^{\mathrm{in}})$.

(4) Choose a new starting point K_{j+1}^s and set $j \leftarrow j + 1$.

End (do)

Remark 18. The major task of Algorithm 17 lies in Step (1) where an unconstrained minimization method has to be employed to obtain an approximate local solution up to the prescribed accuracy represented by the inner-loop tolerance ϵ^{in}. In the implementation, we consider the three-term CG method of Section 4 to calculate such a local solution.

8. Illustrative Numerical Examples

In this section, various test examples are provided to illustrate the performance of the proposed methods. Among the considered test problems are two examples of decentralized control systems and two examples for time-delay systems. A starting feasible point is required for the log-barrier method which might be obtained by the CG method applied on problem (1). The log-barrier method aims to obtain a local solution for the constrained problem (2) or at least achieves stabilizing output feedback that strictly lies within the unit circle.

The methods are implemented using Matlab and all computations were carried out on a Laptop with 3.07 GHz and 1 GB RAM. Some of the considered test problems are for continuous-time systems. The Matlab function `c2d` from the control toolbox is employed to provide the corresponding discrete-time data matrices.

As mentioned in Remark 8, we stop the method as soon as the objective function f_k is strictly less than one or a local solution is achieved. The CG method uses the sufficient decrease condition (23) with $\gamma = 10^{-4}$ for the backtracking line search.

Example 1. This test problem is borrowed from [23]. The constant data matrices for the corresponding discrete-time model are

$$A = \begin{bmatrix} 1.0000 & 0.1000 & 0.0050 \\ 0 & 1.0000 & 0.1000 \\ 0 & 0 & 1.0000 \end{bmatrix},$$

$$B = \begin{bmatrix} 0.1052 & 0.0002 \\ 0.1050 & 0.0050 \\ 0.1000 & 0.1000 \end{bmatrix}, \quad (58)$$

$$C^T = \begin{bmatrix} 1 & 0 \\ 0 & 1 \\ 0 & 0 \end{bmatrix},$$

where the spectral radius of the system matrix A is 1.0. Starting from K_0, a matrix of ones, the proposed CG method successfully converges to a local minimizer for the unconstrained problem (1). The achieved output feedback gain matrix is

$$K_* = \begin{bmatrix} -7.7969 & -7.9723 \\ -3.3475 & -2.7390 \end{bmatrix}, \quad (59)$$

and the corresponding objective function value is $f(K_*) = 0.9387$.

Table 1 shows the convergence behavior of the considered CG method to a local solution of problem (1), which is also a stabilizing output feedback gain to the corresponding control system (5).

Example 2. This test problem is the aircraft model in cruise flight conditions [24, AC1] for a continuous-time control system. The function `c2d` is used to have the discrete-time counterpart with the following data matrices:

Table 1: Convergence behavior of the CG method to a local solution for Example 1.

k	$f(K_k)$	$\|\nabla f(K_k)\|$
1	$1.2783e+000$	$1.8255e-001$
2	$9.4060e-001$	$5.9548e-002$
3	$9.4044e-001$	$5.1829e-002$
⋮	⋮	⋮
26	$9.3874e-001$	$1.5049e-002$
27	$9.3874e-001$	$1.0949e-005$
28	$9.3874e-001$	$6.4275e-006$

$$\begin{aligned} A &= \begin{bmatrix} 1.0000 & 0.0014 & 0.1132 & 0.0005 & -0.0967 \\ 0 & 0.9945 & -0.0171 & -0.0005 & 0.0068 \\ 0 & 0.0003 & 1.0000 & 0.0957 & -0.0048 \\ 0 & 0.0060 & -0.0000 & 0.9131 & -0.0936 \\ 0 & -0.0277 & 0.0002 & 0.0973 & 0.9287 \end{bmatrix}, \\ B &= \begin{bmatrix} -0.0076 & 0.0000 & 0.0003 \\ -0.0115 & 0.0997 & 0.0000 \\ 0.0212 & 0.0000 & -0.0081 \\ 0.4152 & 0.0003 & -0.1589 \\ 0.1742 & -0.0014 & -0.0154 \end{bmatrix}, \\ C &= \begin{bmatrix} 1 & 0 & 0 & 0 & 0 \\ 0 & 1 & 0 & 0 & 0 \\ 0 & 0 & 1 & 0 & 0 \end{bmatrix}. \end{aligned} \quad (60)$$

The spectral radius of the system matrix A is 1. Starting from K_0, a matrix of ones, the CG method achieves the following stabilizing output feedback gain matrix after 12 iterations:

$$K_* = \begin{bmatrix} 0.7473 & 0.7490 & 0.0293 \\ -0.2442 & -1.2308 & -2.2948 \\ 1.0950 & 1.1358 & 1.2428 \end{bmatrix}. \quad (61)$$

The corresponding objective function value at both points is $f(K_0) = 1.0841$ and $f(K_*) = 0.9836$.

Example 3. This test problem represents a chemical reactor model [24, REA1]. The data matrices of the corresponding discrete-time system are as follows:

$$A = \begin{bmatrix} 1.1782 & 0.0015 & 0.5116 & -0.4033 \\ -0.0515 & 0.6619 & -0.0110 & 0.0613 \\ 0.0762 & 0.3351 & 0.5606 & 0.3824 \\ -0.0006 & 0.3353 & 0.0893 & 0.8494 \end{bmatrix},$$

$$B = \begin{bmatrix} 0.0045 & -0.0876 \\ 0.4672 & 0.0012 \\ 0.2132 & -0.2353 \\ 0.2131 & -0.0161 \end{bmatrix},$$

$$C = \begin{bmatrix} 1 & 0 & 1 & -1 \\ 0 & 1 & 0 & 0 \\ 0 & 0 & 1 & -1 \end{bmatrix}. \tag{62}$$

The spectral radius of the system matrix A is 1.2203. The system is clearly Schur unstable. By starting with K_0, the zero matrix, the CG method converges to a local solution K_* after 25 iterations, where

$$K_* = \begin{bmatrix} 0.3847 & -0.2828 & 0.3315 \\ 2.2667 & 0.4446 & 1.0987 \end{bmatrix}, \tag{63}$$

and the corresponding objective function value is $f(K_*) = 0.9382$.

8.1. The CG Method for Decentralized Control Systems

Example 4. This test problem is borrowed from [25] of a decentralized control system with two control stations; each station has one input and one measured output. The given constant data matrices are

$$A = \begin{bmatrix} 1.0 & 0 & -1.6 \\ -1.0 & 1.0 & -0.3 \\ 0 & 0.4 & 1.0 \end{bmatrix},$$

$$B_1 = \begin{bmatrix} 1 \\ 0 \\ 0 \end{bmatrix},$$

$$B_2 = \begin{bmatrix} 0 \\ 1 \\ 1 \end{bmatrix}, \tag{64}$$

$$C_1 = \begin{bmatrix} 1 & 0 & 0 \end{bmatrix},$$

$$C_2 = \begin{bmatrix} 0 & 1 & 1 \end{bmatrix}.$$

The uncontrolled system is discrete-time Schur unstable, where $\rho(A) = 1.8154 > 1$. Starting with the following K_0, the CG method requires only 6 iterations to converge to a stabilizing output feedback gain matrix K_*, where

$$K_0 = \begin{bmatrix} 0.1 & 0 \\ 0 & 0.1 \end{bmatrix},$$

$$K_* = \begin{bmatrix} -1.0060 & 0 \\ 0 & -0.4585 \end{bmatrix}. \tag{65}$$

The objective function value at the two points is $f(K_0) = 1.9924$ and $f(K_*) = 0.6996 < 1$, respectively. Although $f(K_*)$ seems to be relatively small, a stationary point fails to be achieved.

Example 5. This is a fifth-order decentralized control system with two control stations. The data matrices are randomly generated which are as follows:

$$A = \begin{bmatrix} 0.8755 & -0.1324 & -0.0105 & 0.1762 & 0.0484 \\ 0.0289 & 0.9512 & -0.0056 & -0.0257 & 0.0048 \\ -0.0516 & -0.0153 & 1.0470 & -0.0900 & 0.0207 \\ -0.0915 & 0.2109 & -0.0292 & 1.0087 & -0.0915 \\ 0.1305 & 0.0608 & 0.0939 & -0.1345 & 0.8591 \end{bmatrix},$$

$$B_1 = \begin{bmatrix} -0.1251 & 0.0472 \\ -0.0569 & 0.0821 \\ -0.0322 & 0.1360 \\ -0.1195 & 0.0421 \\ -0.0267 & 0.1732 \end{bmatrix}, \tag{66}$$

$$B_2^T = \begin{bmatrix} 0.0583 & 0.0432 & -0.1143 & -0.0340 & 0.0179 \\ 0.1186 & 0.0153 & 0.1403 & -0.1515 & -0.0926 \end{bmatrix},$$

$$C_1 = \begin{bmatrix} 0 & 1 & 0 & 0 & 0 \\ 0 & 0 & 1 & 0 & 0 \end{bmatrix},$$

$$C_2 = \begin{bmatrix} 0 & 0 & 0 & 0 & 1 \\ 0 & 0 & 0 & 0 & 0 \end{bmatrix}.$$

The uncontrolled system is discrete-time Schur unstable, where $\rho(A) = 1.1114 > 1$. Starting from the following randomly generated K_0, the CG method reaches a stabilizing K_* after 8 iterations, where

$$K_0 = \begin{bmatrix} -0.9513 & -1.8038 & 0 & 0 \\ 0.6090 & 1.2671 & 0 & 0 \\ 0 & 0 & 0.3870 & 0.3936 \\ 0 & 0 & 0.2648 & 0.4900 \end{bmatrix},$$

$$K_* = \begin{bmatrix} 9.4329 & -12.4352 & 0 & 0 \\ -23.6142 & 1.2606 & 0 & 0 \\ 0 & 0 & 1.1031 & 1.1097 \\ 0 & 0 & 1.2748 & 1.5000 \end{bmatrix}. \tag{67}$$

The corresponding objective function value at both points is $f(K_0) = 1.2056$ and $f(K_*) = 0.9669$, respectively.

8.2. The CG Method for Time-Delay Systems

Example 6. This test problem represents a time-delay system (see [22]) which has the following constant data matrices:

$$A_0 = \begin{bmatrix} 0.95 & 0.78 \\ -1.76 & -0.87 \end{bmatrix},$$

$$A_1 = \begin{bmatrix} 0.12 & 0.89 \\ -0.95 & -2.97 \end{bmatrix},$$

$$\widehat{B} = \begin{bmatrix} 0.5 \\ -0.95 \end{bmatrix},$$

$$C_0^T = \begin{bmatrix} 0.95 \\ -2.98 \end{bmatrix},$$

$$C_1 = \begin{bmatrix} 0 & 0 \end{bmatrix}, \tag{68}$$

where $d = 1$. The data matrices for this system after converting it from continuous to discrete are as follows:

$$A = \begin{bmatrix} 1.0928 & 0.0823 & 0.0087 & 0.0811 \\ -0.1805 & 0.8957 & -0.0914 & -0.2903 \\ 0.1047 & 0.0040 & 1.0005 & 0.0042 \\ -0.0090 & 0.0951 & -0.0046 & 0.9854 \end{bmatrix},$$

$$B = \begin{bmatrix} 0.0485 \\ -0.0948 \\ 0.0025 \\ -0.0047 \end{bmatrix}, \tag{69}$$

$$C^T = \begin{bmatrix} 0.9500 \\ -2.9800 \\ 0 \\ 0 \end{bmatrix}.$$

The spectral radius of the system matrix A is 1.0098. Starting from the following K_0, the three-term CG method successfully converges to a local solution of the minimization problem (1). The starting and achieved local solutions are

$$K_0 = [-0.4838],$$
$$K_* = [-1.3741]. \tag{70}$$

The objective function value at both points is $f(K_0) = 1.0008$ and $f(K_*) = 0.9887$, respectively.

Example 7. This test problem represents a time-delay system (see [22]) which has the following constant data matrices:

$$A_0 = \begin{bmatrix} 1 & -0.6 \\ 0.4 & 0.5 \end{bmatrix},$$

$$A_1 = \begin{bmatrix} 0.5 & 0.2 \\ 0.6 & 0.4 \end{bmatrix},$$

$$\widehat{B} = \begin{bmatrix} 0.1 & 1 \\ 0 & 0.1 \end{bmatrix}, \tag{71}$$

$$C_0 = \begin{bmatrix} 1 & 0 \\ 0 & 1 \end{bmatrix},$$

where $d = 1$. By converting the system from continuous to discrete-time, the corresponding data matrices are

$$A = \begin{bmatrix} 1.1065 & -0.0637 & 0.0507 & 0.0198 \\ 0.0463 & 1.0520 & 0.0626 & 0.0415 \\ 0.1052 & -0.0031 & 1.0025 & 0.0010 \\ 0.0022 & 0.1026 & 0.0031 & 1.0020 \end{bmatrix},$$

$$B = \begin{bmatrix} 0.0105 & 0.0207 \\ 0.0002 & 0.0107 \\ 0.0005 & 0.0010 \\ 0.0000 & 0.0005 \end{bmatrix}, \tag{72}$$

$$C^T = \begin{bmatrix} 1 & 0 & 0 & 0 \\ 0 & 1 & 0 & 0 \end{bmatrix}.$$

The spectral radius of the system matrix A is 1.1183. Starting from K_0, the three-term CG method successfully converges to a local solution for the minimization problem (1). The starting and achieved local solutions are as follows:

$$K_0 = \begin{bmatrix} -147.2142 & 16.2595 \\ 72.9190 & -39.6464 \end{bmatrix},$$
$$K_* = \begin{bmatrix} -162.5446 & 4.7207 \\ 52.8133 & -71.3982 \end{bmatrix}. \tag{73}$$

The objective function value at both points is $f(K_0) = 1.1136$ and $f(K_*) = 0.9989$, respectively.

8.3. The Logarithmic Barrier Method. The following two examples quite show the performance of the logarithmic barrier interior-point method for finding a local solution to problem (2). A starting feasible point for the method might be obtained by executing the CG method with some iterations on problem (1) until a stabilizing K is obtained.

Example 8. This test problem is borrowed from [23]. The constant data matrices for the corresponding discrete-time model are as follows:

$$A = \begin{bmatrix} 1.0000 & 0.1000 & 0.0050 \\ 0 & 1.0000 & 0.1000 \\ 0 & 0 & 1.0000 \end{bmatrix},$$

$$B = \begin{bmatrix} 0.1052 & 0.0002 \\ 0.1050 & 0.0050 \\ 0.1000 & 0.1000 \end{bmatrix}, \tag{74}$$

$$C^T = \begin{bmatrix} 1 & 0 \\ 0 & 1 \\ 0 & 0 \end{bmatrix},$$

where the spectral radius of the system matrix A is 1.0. Starting from the following feasible K_0, the interior-point

TABLE 2: Convergence behavior of the log-barrier interior-point method to a local solution for Example 9.

j	ϵ_j^{in}	μ_j	k	$f(K_k)$	$\|\nabla\phi^{\mu_j}(K_k)\|$
0	9.0e − 001	1.0e − 001	0	9.4856e − 001	2.9173e − 001
			1	9.4063e − 001	3.6122e − 001
1	9.0e − 002	4.0e − 003	0	9.4063e − 001	3.6122e − 001
			1	9.3240e − 001	1.0758e − 001
			2	9.3240e − 001	5.3765e − 001
			⋮	⋮	⋮
			61	8.2291e − 001	3.5946e − 001
			62	8.2283e − 001	1.0708e − 002
2	9.0e − 003	6.4e − 006	0	8.2283e − 001	1.0708e − 002
			⋮	⋮	⋮
			5	8.2282e − 001	6.5932e − 003
3	9.0e − 004	4.1e − 010	0	8.2282e − 001	6.5932e − 003
			⋮	⋮	⋮
			6	8.2281e − 001	7.3226e − 004
4	9.0e − 005	1.0e − 015	0	8.2281e − 001	7.3226e − 004
			⋮	⋮	⋮
			3	8.2281e − 001	4.5061e − 005

method successfully converges to a local solution K_*. The two matrices are as follows:

$$K_0 = \begin{bmatrix} -7.79 & -7.97 \\ -3.34 & -2.73 \end{bmatrix},$$
$$K_* = \begin{bmatrix} -7.7592 & -7.9516 \\ -3.3145 & -2.7147 \end{bmatrix}, \quad (75)$$

where the corresponding objective function values are $f(K_0) = 0.9430$ and $f(K_*) = 0.9388$, respectively.

Example 9. This test problem represents a chemical reactor model [24, REA2]. The data matrices for the corresponding discrete-time system are as follows:

$$A = \begin{bmatrix} 1.1805 & 0.0014 & 0.5122 & -0.4038 \\ -0.0515 & 0.6619 & -0.0110 & 0.0613 \\ 0.0763 & 0.3351 & 0.5606 & 0.3823 \\ -0.0006 & 0.3353 & 0.0893 & 0.8494 \end{bmatrix},$$
$$B = \begin{bmatrix} 0.0045 & -0.0876 \\ 0.4672 & 0.0012 \\ 0.2132 & -0.2353 \\ 0.2131 & -0.0161 \end{bmatrix}, \quad (76)$$
$$C = \begin{bmatrix} 1 & 0 & 1 & -1 \\ 0 & 1 & 0 & 0 \end{bmatrix}.$$

The spectral radius of the system matrix A is 1.2227. Starting from the following feasible K_0, the log-barrier method converges after 4 outer-loop iterations and 77 inner-loop iterations to a local solution K_*. The starting and achieved local solutions are as follows:

$$K_0 = \begin{bmatrix} 0.07 & -0.49 \\ 1.53 & -0.09 \end{bmatrix},$$
$$K_* = \begin{bmatrix} -0.5680 & -0.0131 \\ 5.8580 & -0.8531 \end{bmatrix}. \quad (77)$$

Table 2 shows the convergence behavior of the log-barrier interior-point method while converging to the local solution.

The above results quite show the convergence behavior of the proposed three-term CG method and the logarithmic barrier interior-point method. For all the considered test problems, the CPU time is less than 3 sec. If a local solution fails to be achieved, the CG method at least finds a stabilizing output feedback controller for the closed-loop system.

9. Conclusion

The eigenvalue assignment problem for discrete-time systems formulated as an unconstrained optimization is considered. A three-term CG method is proposed to find a local solution of the minimization problem or at least to achieve a stabilizing output feedback gain matrix.

The conjugate gradient method is extended to tackle the eigenvalue assignment problem for the case of decentralized control systems and the case of control systems with time delay. Global convergence is established for the conjugate gradient method.

In order to maintain a selected stability margin for the control system, the optimization problem is reformulated as an inequality constraint problem by adding a cut to the objective function. A logarithmic barrier method is proposed for tackling this problem. The performance of the CG method as well as the logarithmic barrier method is shown on various test problems. It is quite interesting to see the performance of some variants of quasi-Newton and inexact Newton methods on the eigenvalue assignment problem for discrete-time systems.

Conflicts of Interest

The authors declare that there are no conflicts of interest regarding the publication of this paper.

References

[1] E. K. Chu, "Optimization and pole assignment in control system design," *International Journal of Applied Mathematics and Computer Science*, vol. 11, no. 5, pp. 1035–1053, 2001.

[2] E. M. Mostafa, M. A. Tawhid, and E. Elwan, "Inexact Newton and quasi-Newton methods for the output feedback pole assignment problem," *Computational & Applied Mathematics*, vol. 33, no. 3, pp. 517–542, 2014.

[3] E. M. Mostafa, M. A. Tawhid, and E. . Elwan, "Nonlinear conjugate gradient methods for the output feedback pole assignment problem," *Pacific Journal of Optimization. An International Journal*, vol. 12, no. 1, pp. 56–85, 2016.

[4] A. T. Alexandridis and P. N. Paraskevopoulos, "A new approach to eigenstructure assignment by output feedback," *Institute of Electrical and Electronics Engineers. Transactions on Automatic Control*, vol. 41, no. 7, pp. 1046–1050, 1996.

[5] O. Bachelier, J. Bosche, and D. Mehdi, "On pole placement via eigenstructure assignment approach," *Institute of Electrical and Electronics Engineers. Transactions on Automatic Control*, vol. 51, no. 9, pp. 1554–1558, 2006.

[6] C. I. Byrnes, "Pole assignment by output feedback," in *Three decades of mathematical system theory*, vol. 135 of *Lect. Notes Control Inf. Sci.*, pp. 31–78, Springer, Berlin, 1989.

[7] M. Fu, "Pole placement via static output feedback is NP-hard," *Institute of Electrical and Electronics Engineers. Transactions on Automatic Control*, vol. 49, no. 5, pp. 855–857, 2004.

[8] S. M. Karbassi, "An algorithm for minimizing the norm of state feedback controllers in eigenvalue assignment," *Computers & Mathematics with Applications. An International Journal*, vol. 41, no. 10-11, pp. 1317–1326, 2001.

[9] K. H. Kiritsis, "A necessary condition for pole assignment by constant output feedback," *Systems and Control Letters*, vol. 45, no. 4, pp. 317–320, 2002.

[10] R. E. Mahony and U. Helmke, "System assignment and pole placement for symmetric realisations," *Journal of Mathematical Systems, Estimation, and Control*, vol. 5, no. 2, pp. 1–32, 1995.

[11] M. Tarokh, "A unified approach to exact, approximate, optimized and decentralized output feedback pole assignment," *International Journal of Control, Automation and Systems*, vol. 6, pp. 939–947, 2008.

[12] Y. Chen and Y. Zhu, "Indefinite LQ optimal control with equality constraint for discrete-time uncertain systems," *Japan Journal of Industrial and Applied Mathematics*, vol. 33, no. 2, pp. 361–378, 2016.

[13] J. Burke, D. Henrion, A. Lewis, and M. Overton, "HIFOO: A MATLAB package for fixed–order controller design and H-infinity optimization," *IFAC Proceedings Volumes*, vol. 39, no. 9, pp. 339–344, 2006.

[14] N. Andrei, "On three-term conjugate gradient algorithms for unconstrained optimization," *Applied Mathematics and Computation*, vol. 219, no. 11, pp. 6316–6327, 2013.

[15] W. W. Hager and H. Zhang, "A survey of nonlinear conjugate gradient methods," *Pacific Journal of Optimization. An International Journal*, vol. 2, no. 1, pp. 35–58, 2006.

[16] J. Nocedal and S. J. Wright, *Numerical Optimization*, Springer, New York, NY, USA, 1999.

[17] D. Sun and J. Sun, "Strong semismoothness of eigenvalues of symmetric matrices and its application to inverse eigenvalue problems," *SIAM Journal on Numerical Analysis*, vol. 40, pp. 2352–2367, 2003.

[18] F. H. Clarke, *Optimization and Nonsmooth Analysis*, Wiley, New York, NY, USA, 1983.

[19] R. A. Horn and C. R. Johnson, *Matrix Analysis*, Cambridge University Press, Cambridge, UK, 1985.

[20] Y. Narushima, H. Yabe, and J. A. Ford, "A three-term conjugate gradient method with sufficient descent property for unconstrained optimization," *SIAM Journal on Optimization*, vol. 21, no. 1, pp. 212–230, 2011.

[21] H. A. Tehrani and N. Ramroodi, "Eigenvalue assignment of discretetime linear systems with state and input timedelays," in *Proceedings of the N. Ramroodi. Eigenvalue assignment of discrete–time linear systems with state and input time–delays*, vol. 45, pp. 23–30, 2013.

[22] B. Zhou, J. Lam, and G.-R. Duan, "Full delayed state feedback pole assignment of discrete-time time-delay systems," *Optimal Control Applications & Methods*, vol. 31, no. 2, pp. 155–169, 2010.

[23] E. J. Davison and S. H. Wang, "On Pole Assignment in Linear Multivariable Systems Using Output Feedback," *IEEE Transactions on Automatic Control*, vol. 20, no. 4, pp. 516–518, 1975.

[24] F. Leibfritz, "COMPlib: COnstraint Matrix-optimization Problem librarya collection of test examples for nonlinear semidefinite programs," Tech. Rep., 2004, http://www.complib.de/.

[25] K. H. Lee, J. H. Lee, and W. H. Kwon, "Sufficient LMI conditions for H8 output feedback stabilization of linear discretetime systems," *IEEE Transactions on Automatic Control*, vol. 51, no. 4, pp. 675–680, 2006.

5

Numerical Procedures for Random Differential Equations

Mohamed Ben Said,[1] Lahcen Azrar,[2,3] and Driss Sarsri[4]

[1]*Mathematical Modeling and Control, Department of Mathematics, Faculty of Sciences and Techniques of Tangier, Abdelmalek Essaadi University, Tangier, Morocco*
[2]*Research Center STIS, Department of Applied Mathematics and Informatics, ENSET, Mohammed V University, Rabat, Morocco*
[3]*Department of Mechanical Engineering, Faculty of Engineering, King Abdulaziz University, Jeddah, Saudi Arabia*
[4]*LTI, Ecole Nationale des Sciences Appliquées de Tanger, Abdelmalek Essaadi University, Tangier, Morocco*

Correspondence should be addressed to Lahcen Azrar; l.azrar@um5s.net.ma

Academic Editor: M. Shariyat

Some methodological approaches based on generalized polynomial chaos for linear differential equations with random parameters following various types of distribution laws are proposed. Mainly, an internal random coefficients method 'IRCM' is elaborated for a large number of random parameters. A procedure to build a new polynomial chaos basis and a connection between the one-dimensional and multidimensional polynomials are developed. This allows handling easily random parameters with various laws. A compact matrix formulation is given and the required matrices and scalar products are explicitly presented. For random excitations with an arbitrary number of uncertain variables, the IRCM is couplet to the superposition method leading to successive random differential equations with the same main random operator and right-hand sides depending only on one random parameter. This methodological approach leads to equations with a reduced number of random variables and thus to a large reduction of CPU time and memory required for the numerical solution. The conditional expectation method is also elaborated for reference solutions as well as the Monte-Carlo procedure. The applicability and effectiveness of the developed methods are demonstrated by some numerical examples.

1. Introduction

Stochastic and random differential equations constitute a growing field of great scientific interest. There are mainly three categories of random differential equations. The first and the simplest class is one where only the initial conditions are random. The second class is characterized by the presence of random nonhomogeneous or input terms and the third one is the differential equations with random coefficients. To deal with errors and uncertainties, random coefficients have been increasingly used in the last few decades.

This paper focuses on the combined second and third classes because this type of equations offers a natural and rational approach for mathematical modeling of many physical phenomena. The last decades have witnessed an enormous effort in the fields of parameters uncertainty and random or stochastic differential processes. This is due to the fact that any physical system contains uncertainties and its real phenomena may be modeled by stochastic differential equations with random or stochastic process coefficients. These equations take into account the approximate knowledge of the numerical values of the physical parameters on which the system depends and have been a matter of intensive investigation.

A number of techniques are available for uncertainty sensitivity and propagation such as Monte-Carlo procedure [1, 2], sensitivity analysis methods [3], and polynomial chaos [4, 5] among others. Monte-Carlo (MC) method has been the mainstream uncertainty quantification technique for decades. It is the most used method and is valid for a wide range of problems. However, it is very computationally expensive since it requires a large number of simulations using the full model.

An alternative approach is based on the expansion of the response in terms of a series of polynomials that are orthogonal with respect to mean value operations. Polynomial chaos was first introduced by Wiener [6] where Hermite polynomials were used to model stochastic processes with Gaussian random variables. A number of other expansions have been proposed in the literature for representing non-Gaussian process [7, 8]. Recent review papers by Stefanou [9] and by Schuëller and Pradlwarter [10] summarized the assessment of the past and current status of the procedure for stochastic structural analysis.

This polynomial representation provides a framework suitable for computational simulation and then widespread in mathematical and numerical analysis of many engineering problems. Various problems have been solved based on this approximation such as solution of stochastic differential equations [11], linear structural dynamics [4, 5], and nonlinear random vibration [12, 13]; soil-structure interaction [14], structural reliability [15], and identification [16, 17]. More recently, Trcala used polynomial chaos for nonlinear diffusion problems of moisture transfer in wood [18]. The accuracy of the PC approximation has been evaluated by Field and Grigoriu [19]. A convergence of the decomposition of the solution into the polynomials chaos is studied by Dvurecenskij et al. [20] and the conditions associated with the distribution function of the random vector appearing in the solution for a convergence toward the solution are given by Ernst et al. [21].

The polynomial chaos has been used in many finite elements problems [4]. Accurate discrete modeling of complex industrial structures leads to a large finite element model. To reduce the CPU time, reducing the order of the models is very useful. Component mode synthesis (CMS) is a well-established method for efficiently constructing models to analyze the dynamics of large and complex structures that are often described by separate substructure (or components) models. Sarsri et al. [22] have coupled the CMS methods with the projection chaos polynomials methods in the first and second orders to compute the frequency transfer functions of stochastic structures. This coupling methodological approach has been used by Sarsri and Azrar in time domain [23] as well as a coupling with the perturbation method [24].

The polynomial chaos methods are well suited for the random differential equations, RDE, with a very few number of random variables defining their main coefficients. It is well known that if the number of the considered random variables increases, the needed number of unknowns to be determined for solving the random systems increases very rapidly with the degree of the polynomials. Thus, for accurate solution, the CPU time and memory required may be prohibitive. This greatly limits these methods to random differential equations with very few numbers of random parameters.

An alternative approach called internal random coefficients method (IRCM) is developed in this paper. A careful presentation is given in the frame of higher-order random differential equations. This method is based on generalized polynomial chaos and the superposition principle. It can be used to solve random differential equations with a large number of random variables and an input right-hand side decomposed in an arbitrary number of random coefficients. The considered random parameters may follow various distribution laws.

A procedure to build a new polynomial chaos basis and a connection between the one-dimensional and multidimensional polynomials is established. Different distribution laws can be easily considered. Based on the superposition principle, the random differential equation with an input depending on several random variables is decomposed on a sequence of RDE with the same main random operator and reduced right-hand sides. A series of RDEs with reduced number of random variables have thus to be solved based on the generalized polynomial chaos decomposition. The global system is then solved by a drastic reduction of the CPU time and memory space. For the sake of comparison, the conditional expectation method is developed for the considered random differential equations as well as the Monte-Carlo method. The applicability and effectiveness of the presented methodological approach have been demonstrated by numerically solving various examples.

2. Mathematical Formulation

In this work, various methodological approaches are elaborated to solve higher-order initial value problems with linear and nonlinear random variables subjected to a random input right-hand side. For this aim, the following stochastic differential equation is considered:

$$\mathscr{L} \cdot U(t,\omega) = F(t) \tag{1}$$

with deterministic initial conditions, where U is the stochastic process response and $\mathscr{L}$ is a linear random operator of order m defined by

$$\mathscr{L} = \sum_{i=0}^{m} C_i(\omega_1)\frac{d^i}{dt^i}. \tag{2}$$

It is assumed that the random coefficients C_i depend on the random vector ω_1 which is defined in a probability space $(\Omega^1, \mathfrak{F}^1, P^1)$. The input right-hand side, F, is assumed to be dependent on the random vector ω_2 that is defined in the probability space $(\Omega^2, \mathfrak{F}^2, P^2)$. Explicit expression of $F(t)$ is given later.

For numerical solution of (1), a new procedure based on the general polynomial chaos, GPC, expansion procedure is elaborated. Herein, the classical GPC procedure is reviewed in a clear manner.

In the present work, the random variables, component of the vector (ω_1, ω_2), are assumed to be independent but may have general distinct distributions f_j. If f_j are classical distributions, such as normal, uniform, gamma, and beta, the associated known polynomial chaos can be used. Otherwise, the procedure, developed in this paper, will be used to build the needed polynomial basis. Explicit expressions of this basis for general cases are given. In addition, the number of random variables $(n + 1)$ and the differential order m are arbitrary.

The concept of internal random coefficients is introduced and combined with the superposition principle and generalized polynomial chaos expansion. For the sake of comparison, conditional expectation and Monte-Carlo methods are also elaborated.

2.1. General Polynomial Chaos Formulation

2.1.1. General Formulation. For general purpose, let us consider random vectors ω_1 and ω_2, presented in the following forms:

$$\omega_1 = (\xi_0, \xi_1, \dots, \xi_p)$$
$$\omega_2 = (\xi_{p+1}, \xi_{p+2}, \dots, \xi_n), \tag{3}$$

where ξ_i for $i = 0$ to n are random variables defined from the probabilistic field $(\Omega_i, \mathcal{F}_i, P_i)$ to $\mathbb{R}$. The random vectors ω_1 and ω_2 are assumed to be independent and gathered in the vector ω:

$$\omega = (\omega_1, \omega_2). \tag{4}$$

We assume that the random vector ω has a distribution function with respect to the Lebesgue measure denoted by f. $L_f^2(\mathbb{R}^{n+1})$ denotes the set of square-integrable functions with respect to the weight measure f:

$$L_f^2(\mathbb{R}^{n+1}) = \left\{H : \int_{\mathbb{R}^{n+1}} H^2(x) f(x)\,dx < +\infty\right\} \tag{5}$$

with the following associated inner product:

$$\langle H, G\rangle = \int_{\mathbb{R}^{n+1}} H(x) G(x) f(x)\,dx. \tag{6}$$

Let us note that the distribution f may be Gaussian or non-Gaussian. In the present analysis, various types of distribution functions may be considered.

The general polynomial chaos associated with the random vector ω is denoted by $\{\phi_k;\ k \in \mathbb{N}\}$. These polynomials coincide with the orthogonal polynomials associated with the inner product defined in (6) and verify

$$\langle \phi_k, \phi_j\rangle = \boldsymbol{\alpha}_k \delta_{kj}, \tag{7}$$

where $\boldsymbol{\alpha}_k$ are given by

$$\boldsymbol{\alpha}_k = \langle \phi_k, \phi_k\rangle. \tag{8}$$

The solution U of the main equation (1) is a time stochastic process depending on the random vectorω and decomposed in the polynomial chaos basis $\{\phi_k;\ k \in \mathbb{N}\}$:

$$\mathbf{U}(t, \omega) = \sum_{k\ge 0} \phi_k(\omega) U_k(t). \tag{9}$$

For general purpose, the random coefficients C_i are assumed to depend linearly and nonlinearly on the random variables ξ_l; $0 \le l \le p$ and written in the following general form:

$$C_i(\omega_1) = \sum_{k=0}^{q_i} \sum_{|\alpha|=k} c_\alpha^i \prod_{l=0}^{p} (\xi_l)^{\alpha_l}, \tag{10}$$

in which $\alpha = (\alpha_0, \dots, \alpha_p)$, $\alpha_j \in \mathbb{N}$ and $|\alpha| = \sum_{j=0}^{p} \alpha_j$.

The right-hand side of (1), $F(t)$, is assumed to be a time-dependent random function that depends linearly on ω_2 and expressed by

$$F(t, \omega_2) = \sum_{j=p+1}^{n} (a_{0j} + a_{1j}\xi_j)\, g_j(t), \tag{11}$$

where $g_j(t)$ and a_{0j}, a_{1j} are considered deterministic function and constants.

Note that a more general right-hand side excitation can be decomposed in the form (11) using the Karhunen-Loéve expansion [25].

Based on a reduced decomposition using the $(N+1)$ first terms, the stochastic process $U(t,\omega)$ can be approximated by

$$\mathbf{U}_N(t, \omega) \cong \sum_{l=0}^{N} \phi_l(\omega) U_l(t), \tag{12}$$

where $\phi_l(\omega)$ is a multidimensional general polynomial chaos depending on the random vector $\omega = (\xi_0, \xi_1, \dots, \xi_n)$.

The insertion of these expressions in (1) leads to the following m^{th} order random differential equation:

$$\sum_{i=0}^{m} \sum_{l=0}^{N} \left(\sum_{s=0}^{q_i} \sum_{|\alpha|=s} c_\alpha^i \prod_{d=0}^{p} (\xi_d)^{\alpha_d} \right) \phi_l(\omega) \frac{d^i U_l(t)}{dt^i} = \sum_{j=p+1}^{n} (a_{0j} + a_{1j}\xi_j)\, g_j(t). \tag{13}$$

Projecting this equation with respect to ϕ_k for $k = 0$ to N, the following deterministic differential system is then obtained:

$$\sum_{i=0}^{m} \sum_{l=0}^{N} \left(\sum_{s=0}^{q_i} \sum_{|\alpha|=s} c_\alpha^i \left\langle \prod_{d=0}^{p} (\xi_d)^{\alpha_d} \phi_l, \phi_k \right\rangle \right) \frac{d^i U_l(t)}{dt^i} = \sum_{j=p+1}^{n} \langle (a_{0j} + a_{1j}\xi_j), \phi_k\rangle\, g_j(t). \tag{14}$$

Note that the first- and mainly the second-order differential equations of the above kind, $m = 1$ or 2, have been investigated by many authors when the number n of random variables is too small. When the random variables ξ_j are Gaussian, Hermite-chaos polynomials in $]-\infty, +\infty[$ are used in [4]. This standard approach is very often used in structural dynamics. Various other works are elaborated when ξ_j are uniform, gamma, or beta and thus Legendre-chaos in $[a, b]$, Laguerre-chaos in $[0, +\infty[$, and Jacobi-chaos in $[a, b]$ are, respectively, used [8].

The expansion on polynomial basis of the vector ω is related to the polynomial basis associated wih each random variable ξ_j in the case of independent variables. This relation is clarified and the correspondence is clearly established herein. The procedure allowing clarifying the inner product used in the differential system (14) is established and an explicit simple procedure is given, in the next subsection.

Firstly, this procedure is established in the next paragraphs for independent random variables, based on the relationships between the random variables ξ_j and random vector ω. Secondly, new variables are introduced that are not necessarily independent and the procedure is established for general cases.

2.1.2. Condensed Formulation. In order to formulate the considered problem in a condensed form, the following mathematical developments will be used. As the variables $(\xi_i,\ 0 \le i \le n)$ are assumed to be pairwise independent, the joint distribution function f is then given by

$$f(\omega) = \prod_{k=0}^{n} f_k(\xi_k), \tag{15}$$

where f_k, $k = 0$ to n are functions of marginal distributions associated with each variable ξ_k.

The general polynomial chaos associated with each variable ξ_k is denoted by $\{\phi_{kj},\ j \ge 0\}$. These polynomials coincide with the orthogonal polynomials associated with the inner product defined in $L^2_{f_k}(\mathbb{R})$ with respect to the weight function f_k:

$$L^2_{f_k}(\mathbb{R}) = \left\{h : \int_{\mathbb{R}} h^2(x) f_k(x)\, dx < +\infty\right\} \tag{16}$$

with the associated inner product given by

$$\langle h, g\rangle_k = \int_{\mathbb{R}} h(x) g(x) f_k(x)\, dx. \tag{17}$$

The set of orthogonal polynomials satisfies the orthogonal conditions:

$$\left\langle \phi_{ki}, \phi_{kj}\right\rangle_k = \alpha_i^k \delta_{ij}, \tag{18}$$

where

$$\alpha_i^k = \langle \phi_{ki}, \phi_{ki}\rangle_k. \tag{19}$$

In order to make a correspondence between the set of polynomial chaos associated with each variable ξ_i and that associated with the random vector ω, the following total order is introduced on the set $\mathbb{N}^{n+1}$ by

$$\begin{gathered}
\forall\left((i_0,\ldots,i_n),(j_0,\ldots,j_n)\right) \in \mathbb{N}^{n+1}\times\mathbb{N}^{n+1}, \\
(i_0,\ldots,i_n) = (j_0,\ldots,j_n) \Longleftrightarrow \\
i_k = j_k, \quad \forall k \in [\![0,n]\!] \\
(i_0,\ldots,i_n) < (j_0,\ldots,j_n) \\
\text{if } \sum_{k=0}^{n} i_k < \sum_{k=0}^{n} j_k \\
\text{or if } \sum_{k=0}^{n} i_k = \sum_{k=0}^{n} j_k \\
\exists l \in [\![0,n]\!], \\
(i_0,\ldots,i_{l-1}) = (j_0,\ldots,j_{l-1}), \quad i_l > j_l.
\end{gathered} \tag{20}$$

This order allows considering a bijection ψ_n from $\mathbb{N}^{n+1}$ to $\mathbb{N}$, defined by

$$\begin{aligned}
\psi_n : \mathbb{N}^{n+1} &\longrightarrow \mathbb{N} \\
(i_0,\ldots,i_n) &\longmapsto \psi_n(i_0,\ldots,i_n) = i.
\end{aligned} \tag{21}$$

This bijection relates each element $(i_0,\ldots,i_n)$ in $\mathbb{N}^{n+1}$ by its order i defined in (20). Let L be a nonzero integer; the integer N used in the decomposition (12) is taken as

$$N = \psi_n(0,0,\ldots,0,L). \tag{22}$$

The choice of N allows decomposing the solution in a set of polynomial chaos associated with the vector ω of degree less than or equal to L. So, for all integers k between 0 and N, there is a single $(k_0,\ldots,k_n) \in \mathbb{N}^{n+1}$ such that

$$k = \psi_n(k_0,\ldots,k_n), \tag{23}$$

$$\sum_{j=0}^{n} k_j \le L. \tag{24}$$

This one-to-one correspondence allows writing the multidimensional polynomial chaos associated with the random vector ω as a function of the one-dimensional polynomial chaos corresponding to each variable ξ_i by

$$\phi_k(\omega) = \prod_{d=0}^{n} \phi_{dk_d}, \quad \forall k \in [\![0,N]\!], \tag{25}$$

where $(k_0,\ldots,k_n)$ is the multi-index associated with the integer k, introduced by (23).

For all integers i and h between 0 and n the square matrix β^{ih} of order $L+1$ is defined by

$$\beta^{ih}(l+1,k+1) = \left\langle \xi_i^h \phi_{il}, \phi_{ik}\right\rangle_i \quad \forall (l,k) \in [\![0,L]\!]^2. \tag{26}$$

Let $\wp_{Li} = \mathbb{R}_L[\xi_i]$, the set of polynomials of degree less than or equal to L. Then, the set of the classical polynomial chaos $B_{CP}^{Li} = \{\phi_{ik},\ k \in [\![0,L]\!]\}$ is an orthogonal basis of $\wp_{Li}$ for the inner product defined by (17). Let $B_c^{Li} = \{1,\xi_i,\ldots,\xi_i^L\}$ be the canonical basis of $\wp_{Li}$ and P_L^i be the passage matrix from the canonical basis B_C^{Li} to the chaos basis B_{CP}^{Li}. This matrix can be obtained in a standard way, using the Gram-Schmidt procedure or a recursive method.

Let the vectors

$$\begin{aligned}
V_C^{iL} &= \left\{1,\ldots,\xi_i^L\right\}^T, \\
V_{CP}^{iL} &= \left\{\phi_{i0},\ldots,\phi_{iL}\right\}^T.
\end{aligned} \tag{27}$$

Then, one has

$$V_{CP}^{iL} = \left(P_L^i\right)^T V_C^{iL}. \tag{28}$$

For $h \in \mathbb{N}$ and i between 0 and n, the moment m_{ih} of order h of the random variable ξ_i is defined by

$$m_{ih} = E\left(\xi_i^h\right) = \int_{-\infty}^{+\infty} x^h f_i(x)\, dx. \tag{29}$$

The square matrices of order $(L+1)$, B_{iL}^h, are defined by

$$B_{iL}^h(l+1,k+1) = m_{i(h+k+l)} \quad \forall (l,k) \in [\![0,L]\!]^2. \tag{30}$$

This gives explicitly matrix β^{ih} by

$$\beta^{ih} = {P_L^i}^T B_{iL}^h P_L^i. \tag{31}$$

Let k and l be two integers between 0 and N. From (23) we have, respectively, unique elements $(k_0,\ldots,k_n)$ and $(l_0,\ldots,l_n)$ in $\mathbb{N}^{n+1}$ such that $k = \psi_n(k_0,\ldots,k_n)$ and $l = \psi_n(l_0,\ldots,l_n)$. Rewriting ϕ_k using expression (26), the following expression is explicitly obtained:

$$\left\langle \prod_{d=0}^{p} (\xi_d)^{\alpha_d} \phi_l, \phi_k \right\rangle = \prod_{d=0}^{p} \left(\beta^{d\alpha_d} (l_d+1, k_d+1)\right) \cdot \prod_{d=p+1}^{n} \left(\beta^{d0} (l_d+1, k_d+1)\right). \tag{32}$$

The expression in the right-hand side of (14) is thus given by

$$\left\langle \left(a_{0j} + a_{1j}\xi_j\right), \phi_k \right\rangle = \left(a_{0j}\beta^{j0}\left(1, k_j+1\right) + a_{1j}\beta^{j1}\left(1, k_j+1\right)\right) \cdot \prod_{\substack{d=0 \\ d\neq j}}^{n} \left(\beta^{d0}\left(1, k_d+1\right)\right). \tag{33}$$

For independent random variables, a relationship between the multidimensional and associated one-dimensional generalized polynomials is established. This leads to explicit and closed forms of the used scalar product and needed terms to be numerically computed. Using these relationships resulted in the following deterministic differential system:

$$\sum_{i=0}^{m} \sum_{l=0}^{N} \sum_{s=0}^{q_i} \sum_{|\alpha|=s} c_\alpha^i \left(\prod_{d=0}^{p} \left(\beta^{d\alpha_d}(l_d+1, k_d+1)\right) \prod_{d=p+1}^{n} \left(\beta^{d0}(l_d+1, k_d+1)\right) \right) \frac{d^i U_l(t)}{dt^i} = \sum_{j=p+1}^{n} \left(a_{0j}\beta^{j0}\left(1,k_j+1\right) + a_{1j}\beta^{j1}\left(1,k_j+1\right)\right) \prod_{\substack{d=0 \\ d\neq j}}^{n} \left(\beta^{d0}\left(1,k_d+1\right)\right) g_j(t). \tag{34}$$

For a compact formulation and using notation (26), the square matrices T^i of order $(N+1)$ and the time-dependent vector D of dimension $(N+1)$ are introduced for all integers $k, l \in [0, N]$ by

$$T^i(l+1, k+1) = \sum_{s=0}^{q_i} \sum_{|\alpha|=s} c_\alpha^i \left(\prod_{d=0}^{p} \left(\beta^{d\alpha_d}(l_d+1, k_d+1)\right) \prod_{d=p+1}^{n} \left(\beta^{d0}(l_d+1, k_d+1)\right) \right), \tag{35}$$

$$D_{k+1}(t) = \sum_{j=p+1}^{n} \left(a_{0j}\beta^{j0}\left(1,k_j+1\right) + a_{1j}\beta^{j1}\left(1,k_j+1\right)\right) \prod_{\substack{d=0 \\ d\neq j}}^{n} \left(\beta^{d0}\left(1,k_d+1\right)\right) g_j(t). \tag{36}$$

Using these notations, differential system (34) is rewritten in the following closed form system:

$$\sum_{i=0}^{m} \sum_{l=0}^{N} T^i(l+1, k+1) \frac{d^i U_l(t)}{dt^i} = D_{k+1}(t); \quad k = 0, \ldots, N. \tag{37}$$

Using the presented methodological approach, the truncated solution (12) can be numerically obtained. Its mean and variance are given by

$$m_N(t) = U_0(t), \qquad V_N(t) = \sum_{k=1}^{N} \boldsymbol{\alpha}_k U_k^2(t). \tag{38}$$

Equation (14) is usually given for Hermite polynomials when random variables are Gaussian. This kind of projection is classically done and many authors follow this procedure.

In this paper, the random variables ξ_i may follow various types of laws. The presented generalized formalism allows one to handle multilaws by using the canonical basis and the orthogonalization principle with respect to a scalar product associated with the distribution function of the random vector. Explicit and closed forms of the used scalar products and needed terms to be numerically computed are given.

It should to be noted that the major problem of this classical decomposition into polynomials chaos expansion is that the number of unknowns to estimate increases very rapidly when the degree of the polynomial chaos and the number of random parameters increase. More clearly, for n random variables, the number of unknown coefficients in the polynomial chaos of orders less than or equal to L is $\binom{n+L}{L}$.

Table 1 presents the numbers of the needed unknown terms for various L and n. This very fast growth of dimensionality is the main limitation of this classical approach.

To overcome this drawback, a concept of internal random coefficients is introduced herein. This allows reducing drastically the number of random variables, especially if the initially considered number of random variables is large.

2.2. Internal Random Coefficients Method (IRCM). Let us recall that m is the derivative order of the random operator $\mathscr{L}$ and p is the number of the random variables defining $\mathscr{L}$.

If $p \le m$, then better use the methodological approach presented in the previous sections.

If $m < p$, then the coefficients C_i will be considered as the new random variables and called here internal random coefficients. In fact, the following procedure will be of big interest when the number p of random variables ξ_j is too large with respect to the differential order m. A large reduction will result based on the procedures developed herein.

Let us consider the worst case: $m < p$. The coefficients $(C_i)_{0\le i\le m}$ and the random vector ω_2 can be gathered in a new random vector:

$$Y^1 = (C_0, C_1, \ldots, C_m, \omega_2), \tag{39}$$

which depends on the initial random vector $\omega = (\omega_1, \omega_2)$. It has to be noted that the coefficients C_i are not independent of each other but independent of ω_2.

In this section, it is assumed that the distribution function f of the random vector ω is a continuous function in $\mathbb{R}^{n+1}$ and has a compact support denoted by K_1.

This hypothesis ensures the quadratic convergence of the series formed by polynomials chaos to the solution [20, 21]. Further, we assume that there exists a diffeomorphism h from K_1 to $K_2 = h(K_1)$, defined by

$$\begin{aligned} h : K_1 &\longrightarrow K_2 \\ \omega &\longmapsto h(\omega). \end{aligned} \tag{40}$$

Such that, for all i between 0 and m, one has

$$h_i(\omega) = C_i \tag{41}$$

and, for all i between $p+1$ and n, one has

$$h_i(\omega) = \xi_i. \tag{42}$$

The other components of h, from $m+1$ to p, are chosen from the components of ω_1 to complete the construction of h.

Let us put $Y = h(\omega)$. The random vector Y, so defined, has a distribution function with respect to the Lebesgue measure, denoted by g. This function is given over the distribution function associated with the random vector ω by

$$g(y) = f\left(h^{-1}(y)\right)\left|J_{h^{-1}(y)}\right|, \tag{43}$$

where $|J_{h^{-1}(y)}|$ is the determinant of the Jacobean matrix of $h^{-1}(y)$.

Let g^1 be the distribution function related to the considered Lebesgue measure of the random vector Y^1 given by

$$g^1\left(y^1\right) = \int_{\mathbb{R}^{p-m}} g(y)\, dy_{m+1} \cdots dy_p, \tag{44}$$

where $y^1 = (y_0, \ldots, y_m, y_{p+1}, \ldots, y_n)$ is an observation of the random vector Y^1.

For normalized parameters, the following reduced variables are introduced:

$$X_i^1 = \frac{C_i - E(C_i)}{\sigma_i} \quad \text{For } 0 \le i \le m, \tag{45}$$

where $\sigma_i = \sigma(C_i)$ and

$$X^1 = \left(X^{11}, \omega_2\right); \quad X^{11} = \left(X_0^1, \ldots, X_m^1\right). \tag{46}$$

Then, for all integers i between 0 and m, one has

$$C_i = E(C_i) + \sigma_i X_i^1. \tag{47}$$

Using these new expressions, the main random equation (1) is reduced to the following simplified random differential equation:

$$\begin{aligned} &\sum_{i=0}^{m} \left(E(Ci) + \sigma_i X_i^1\right) \frac{d^i U\left(t, X^1\right)}{dt^i} \\ &\quad = \sum_{j=p+1}^{n} \left(a_{0j} + a_{1j}\xi_j\right) g_j(t). \end{aligned} \tag{48}$$

It has to be noted here that the number of random variables is reduced and the resulting random differential equation is thus easier to handle.

Recall that the coefficients and the random variables, used in the right-hand side of (48), are independent of each other as well as of X_i^1. This is due to the fact that the random vector X^{11} depends only on the random vector ω_1. X^{11} and ω_2 are then independent.

Let g^{11} and g^{12} be, respectively, the distribution functions of the random vectors X^{11} and ω_2 with respect to the associated Lebesgue measures. Taking into account the independence of random variables $(\xi_i)_{p+1\le i\le n}$ one has

$$g^{12}\left(x_{p+1}, \ldots, x_n\right) = \prod_{k=p+1}^{n} f_k(x_k). \tag{49}$$

The decomposition of the solution $U(t, X^1)$ of the random equation (48), according to the multidimensional polynomial chaos basis, is given by

$$U\left(t, X^1\right) = \sum_{k\ge 0} \Psi_k^1\left(X^1\right) u_k(t). \tag{50}$$

Using M^{th} first terms of the series (50), the process $U(t, X^1)$ can be approximated by

$$U_M\left(t, X^1\right) \cong \sum_{k=0}^{M} \Psi_k^1\left(X^1\right) u_k(t). \tag{51}$$

TABLE 1: Number of needed unknown terms in the polynomial chaos.

n	L			
	5	10	20	30
5	252	3,003	53,130	32,4632
10	3,003	184,756	30,045,015	847,660,528
20	53,130	30,045,015	137,846,528,820	47,129,212,243,960
30	324,632	847,660,528	47,129,212,243,960	11,826,458,156,4861,424

Inserting these expressions in (48), the following random differential equation resulted:

$$\sum_{k=0}^{M}\sum_{i=0}^{m}\left(E(Ci)+\sigma_i^1 X_i^1\right)\mathbf{\Psi}_k^1\left(X^1\right)\frac{d^i u_k(t)}{dt^i} = \sum_{j=p+1}^{n}\left(a_{0j}+a_{1j}\xi_j\right)g_j(t). \tag{52}$$

Projecting this equation with respect to $\mathbf{\Psi}_l^1$ for $l = 0, M$ leads to

$$\sum_{k=0}^{M}\sum_{i=0}^{m}\left\langle\left(E(Ci)+\sigma_i^1 X_i^1\right)\mathbf{\Psi}_k^1,\mathbf{\Psi}_l^1\right\rangle\frac{d^i u_k(t)}{dt^i} = \sum_{j=p+1}^{n}\left\langle\left(a_{0j}+a_{1j}\xi_j\right),\mathbf{\Psi}_l^1\right\rangle g_j(t). \tag{53}$$

Let $\{\mathbf{\Psi}_j^{11}, j \in \mathbb{N}\}$ be polynomial chaos associated with the random vector X^{11}.

Using bijection (21), there exist two elements in $\mathbb{N}^{n-p+m+1}(k_0,\ldots,k_{n+m-p})$ and $(l_0,\ldots,l_{n+m-p})$, such that $(k_0,\ldots,k_{n+m-p}) = \psi_{n+m-p}^{-1}(k)$, $(l_0,\ldots,l_{n+m-p}) = \psi_{n+m-p}^{-1}(l)$ for all integers k, l and

$$\begin{aligned} k^1 &= \psi_m\left(k_0,\ldots,k_m\right),\\ l^1 &= \psi_m\left(l_0,\ldots,l_m\right). \end{aligned} \tag{54}$$

The multidimensional polynomial chaos $\mathbf{\Psi}_k^1$ associated with the random vector X^1 can be expressed by the polynomial chaos associated with the random vector X^{11} and the polynomial chaos associated with the random variables $(\xi_j,\ p+1 \le j \le n)$ using the notations (23) and (54) by

$$\mathbf{\Psi}_k^1\left(X^1\right) = \mathbf{\Psi}_{k^1}^{11}\left(X^{11}\right)\prod_{q=p+1}^{n}\phi_{qk_{q+m-p}}\left(\xi_q\right). \tag{55}$$

Inserting this relationship into (53), the following random differential equation resulted:

$$\begin{aligned}\sum_{k=0}^{M}\sum_{i=0}^{m}&\left\langle\left(E\left(C_i\right)+\sigma_i^1 X_i^1\right)\mathbf{\Psi}_{k^1}^{11}\left(X^{11}\right)\prod_{q=p+1}^{n}\phi_{qk_{q+m-p}}\left(\xi_q\right),\right.\\ &\left.\mathbf{\Psi}_{l^1}^{11}\left(X^{11}\right)\prod_{q=p+1}^{n}\phi_{ql_{q+m-p}}\left(\xi_q\right)\right\rangle\frac{d^i u_k(t)}{dt^i}\\ &= \sum_{j=p+1}^{n}\left\langle\left(a_{0j}+a_{1j}\xi_j\right),\mathbf{\Psi}_{l^1}^{11}\left(X^{11}\right)\prod_{q=p+1}^{n}\phi_{ql_{q+m-p}}\right.\\ &\left.\cdot\left(\xi_b\right)\right\rangle g_j(t).\end{aligned} \tag{56}$$

Based on the independence of the random vector X^{11} and the random variables $(\xi_j,\ p+1 \le j \le n)$, the inner products in (56) are rearranged as

$$\begin{aligned}&\left\langle\left(E\left(C_i\right)+\sigma_i^1 X_i^1\right)\right.\\ &\left.\cdot\mathbf{\Psi}_{k^1}^{11}\prod_{q=p+1}^{n}\phi_{qk_{q+m-p}}\left(\xi_q\right),\mathbf{\Psi}_{l^1}^{11}\prod_{q=p+1}^{n}\phi_{ql_{q+m-p}}\left(\xi_q\right)\right\rangle\\ &= \left(m_i^1\left\langle\mathbf{\Psi}_{k^1}^{11},\mathbf{\Psi}_{l^1}^{11}\right\rangle_{11}+\sigma_i^1\left\langle X_i^1\mathbf{\Psi}_{k^1}^{11},\mathbf{\Psi}_{l^1}^{11}\right\rangle_{11}\right)\\ &\cdot\prod_{q=p+1}^{n}\left(\alpha_{k_{q+m-p}}^{q}\delta_{k_{q+m-p}l_{q+m-p}}\right)\\ &\left\langle\left(a_{0j}+a_{1j}\xi_j\right),\mathbf{\Psi}_{l^1}^{11}\prod_{q=p+1}^{n}\phi_{ql_{q+m-p}}\left(\xi_q\right)\right\rangle\\ &= \delta_{0l^1}\left(\prod_{\substack{q=p+1\\ q\neq j}}^{n}\delta_{0l_{q+m-p}}\right)\left(a_{0j}\delta_{0l_{j+m-p}}\right.\\ &\left.+a_{1j}\delta_{j+1+m-pl_{j+m-p}}\right),\end{aligned} \tag{57}$$

where δ_{pq} is the Kronecker symbol.

Let B_c be the canonical basis of $\mathbb{R}[X_0,\ldots,X_m]$. This basis can be ordered in the following form:

$$B_c = \left\{\prod_{k=0}^{m}(X_k)^{(\alpha_k-\alpha_{k+1})} : \text{where } \alpha_0 = q,\ 0 \le \alpha_{k+1} \le \alpha_k,\ \alpha_{m+1} = 0;\ q \in \mathbb{N}\right\}. \tag{58}$$

The orthogonalization of the basis B_c gives the polynomial chaos $\{\boldsymbol{\Psi}_k^{\mathbf{11}}, k \in \mathbb{N}\}$ associated with the random vector X^{11}. This orthogonalization is performed using a classical method such as the Gram-Schmidt procedure in $\mathbb{R}[X_0,\ldots,X_m]$ and the following inner product:

$$\langle H,K\rangle_{11} = \int_{\mathbb{R}^{m+1}} H(x)\,K(x)\,g^{11}(x)\,dx \quad \forall (H,K) \in \left(\mathbb{R}\left[X^{11}\right]\right)^2. \tag{59}$$

Let B^1 be the tensor defined for all $\alpha \in \mathbb{N}^{m+1}$, $\alpha = (\alpha_0,\ldots,\alpha_m)$ by

$$B^1(\alpha) = E\left(\prod_{k=0}^{m}(X_k)^{\alpha_k}\right) = \int_{\mathbb{R}^{m+1}}\prod_{k=0}^{m}(x_k)^{\alpha_k} g^{11}(x)\,dx. \tag{60}$$

This integral can be computed using the distribution function of the vector ω_1 and the diffeomorphism h by

$$B^1(\alpha) = \int_{\mathbb{R}^{p+1}}\prod_{k=0}^{m}\left(\frac{h_k(\omega_1)-m_k}{\sigma_k}\right)^{\alpha_k} f^1(\omega_1)\,d\omega_1. \tag{61}$$

The integer M, considered in decomposition (51), is chosen such that the considered polynomials have the degree less than or equal to L and expressed by

$$M = \psi_{(n+m-p)}(0,0,\ldots,0,L) \quad \forall L \in \mathbb{N}. \tag{62}$$

Let B_{cL} be the canonical basis of $\mathbb{R}_L[X_0,\ldots,X_m]$, then

$$B_{cL} = \left\{\prod_{k=0}^{m}(X_k)^{(\alpha_k-\alpha_{k+1})} : \text{where } \alpha_0 = q,\ 0 \le \alpha_{k+1} \le \alpha_k,\ \alpha_{m+1} = 0;\ q \in \mathbb{N},\ q \le L\right\}. \tag{63}$$

Let us denote by $\mathbf{P^1}$ the passage matrix from the canonical basis B_{cL} to the polynomial chaos basis associated with the vector X^1 of degree less than or equal to L. Using the following notations,

$$\begin{aligned} NL &= \psi_m(0,0,\ldots,0,L), \\ Q_l &= \prod_{i=0}^{m}(X_i)^{l_i} \\ &\quad \text{where } (l_0,\ldots,l_m) = \psi_m^{-1}(l) \text{ for } 0 \le l \le NL, \\ \mathbf{V}_c^{11} &= [Q_0,\ldots,Q_{NL}]^T, \\ \mathbf{V}_{CP}^{11} &= \left[\boldsymbol{\Psi}_0^{\mathbf{11}},\ldots,\boldsymbol{\Psi}_{NL}^{\mathbf{11}}\right]^T, \end{aligned} \tag{64}$$

one gets

$$\mathbf{V}_{CP}^{11} = \left(\mathbf{P^1}\right)^{\mathbf{T}}\mathbf{V}_c^{11}. \tag{65}$$

The matrices $\boldsymbol{\Upsilon}$, $\mathbf{T}^i$ and D, defined for all integers k and l between 0 and NL, are introduced for $0 \le i \le m$, by

$$\boldsymbol{\Upsilon}(k+1,l+1) = \left\langle \boldsymbol{\Psi}_k^{\mathbf{11}},\boldsymbol{\Psi}_l^{\mathbf{11}}\right\rangle_{11} = \boldsymbol{\alpha}_k^{11}\delta_{kl} \tag{66}$$

$$\mathbf{T}^i(k+1,l+1) = \left\langle X_i^1\boldsymbol{\Psi}_k^{\mathbf{11}},\boldsymbol{\Psi}_l^{\mathbf{11}}\right\rangle_{11} \tag{67}$$

$$D(k+1,l+1) = B^1\left(\psi_m^{-1}(k)+\psi_m^{-1}(l)\right) \quad \text{for } 0 \le k \le NL,\ 0 \le l \le NL. \tag{68}$$

Let $(e_i)_{0\le i\le m}$, $(m+1)$ elements in $\mathbb{N}^{m+1}$ such that

$$\begin{aligned} e_0 &= (1,0,\ldots,0); \\ e_1 &= (0,1,\ldots,0); \\ &\vdots \\ e_m &= (0,0,\ldots,1). \end{aligned} \tag{69}$$

The matrices $\mathbf{D}^i$ of order $(NL+1)$ for $0 \le i \le m$ are defined by

$$\mathbf{D}^i(k+1,l+1) = B^1\left(\psi_m^{-1}(k)+\psi_m^{-1}(l)+e_i\right) \quad \text{for } 0 \le k \le NL,\ 0 \le l \le NL. \tag{70}$$

This allows defining the matrices $\boldsymbol{\Upsilon}$ and $\mathbf{T}^i$ by

$$\begin{aligned} \boldsymbol{\Upsilon} &= \mathbf{P^1}\mathbf{D}\left(\mathbf{P^1}\right)^{\mathbf{T}}, \\ \mathbf{T}^i &= \mathbf{P^1}\mathbf{D}^i\left(\mathbf{P^1}\right)^{\mathbf{T}}. \end{aligned} \tag{71}$$

The random differential system (52) is thus reduced to the following deterministic one:

$$\begin{aligned}\sum_{k=0}^{M}\sum_{i=0}^{m}&\left(E\left(C_i\right)\mathbf{\Upsilon}\left(k^1+1,l^1+1\right)\right.\\&\left.+\sigma_i^1\mathbf{T}^{\mathrm{i}}\left(k^1+1,l^1+1\right)\right)\\&\cdot\prod_{q=p+1}^{n}\left(\alpha_{k_{q+m-p}}^{q}\delta_{k_{q+m-p}l_{q+m-p}}\right)\frac{d^iu_k\left(t\right)}{dt^i}\\&=\sum_{j=p+1}^{n}\delta_{0l^1}\left(\prod_{\substack{q=p+1\\q\neq j}}^{n}\delta_{0l_{q+m-p}}\right)\left(a_{0j}\delta_{0l_{j+m-p}}\right.\\&\left.+a_{1j}\delta_{j+1+m-pl_{j+m-p}}\right)g_j\left(t\right),\end{aligned}\tag{72}$$

where k^1 and l^1 are integers related to k and l and introduced in (54).

Finally, the last differential system is rewritten in the following condensed matrix form:

$$\sum_{k=0}^{M}\sum_{i=0}^{m}H^i\left(k^1+1,l^1+1\right)\frac{d^iu_k\left(t\right)}{dt^i}=E_l\left(t\right),\tag{73}$$

where

$$\begin{aligned}H^i\left(k^1+1,l^1+1\right)&=\left(E\left(C_i\right)\mathbf{\Upsilon}\left(k^1+1,l^1+1\right)\right.\\&\left.+\sigma_i^1\mathbf{T}^{\mathrm{i}}\left(k^1+1,l^1+1\right)\right)\prod_{q=p+1}^{n}\left(\alpha_{k_{q+m-p}}^{q}\delta_{k_{q+m-p}l_{q+m-p}}\right).\end{aligned}\tag{74}$$

$$\begin{aligned}E_l\left(t\right)=&\sum_{j=p+1}^{n}\delta_{0l^1}\left(\prod_{\substack{q=p+1\\q\neq j}}^{n}\delta_{0l_{q+m-p}}\right)\\&\cdot\left(a_{0j}\delta_{0l_{j+m-p}}+a_{1j}\delta_{j+1+m-pl_{j+m-p}}\right)g_j\left(t\right)\end{aligned}\tag{75}$$

Based on the presented mathematical procedures, the truncated solution (51) can be numerically obtained.

An equivalent form of (73) has been introduced in (37) based on original random variables. These two equations give the approximate solution of $\mathbf{U}$. The coefficients of these two equations are given, respectively, by the inner products in (57) for (73) and (32)-(33) for (37).

It should be noted, on one hand, that these coefficients are hard to be determined for (37) but easier for (73). On the other hand, the number of the random variables used in (73) is largely reduced. This reduces the number of unknowns used to determine the solution $\mathbf{U}$ and greatly simplifies the required numerical computation.

Note that when uncertainties come from random parameters of the system parameters, they can be efficiently handled by the previous mathematical procedures. When the right-hand side excitation depends on many random variables, the previous internal coefficients procedure can be largely improved by the so-called superposition method.

2.3. Superposition Method. Thanks to the linear behavior of the considered system, the superposition principle can be applied to the standard random differential equation (13) and to the resulting equation from the internal random coefficients procedure (52). In both cases, the considered random right-hand side function $F(t)$ depends linearly on $(n-p)$ random variables ξ_j. The superposition method consists of solving $(n-p)$ differential equations with elementary right-hand sides given by

$$\overline{\mathscr{L}}\cdot U\left(t,\omega\right)=\left(a_{0j}+a_{1j}\xi_j\right)g_j\left(t\right)\qquad\text{for } j=p+1,\ldots,n,\tag{76}$$

where $\overline{\mathscr{L}}$ may be the left-hand side differential operator of (13) or (52). In fact, this superposition is adding only one random variable ξ_j to the system and the resulting random vectors are (ω_1,ξ_j) for (13) and (X^{11},ξ_j) for (52) for each j.

As the internal random coefficient procedure leads to a reduced number of random variables, the superposition principle is applied to (52). Our focus is then on the resulting random differential equation:

$$\begin{aligned}\sum_{i=0}^{m}&\left(E\left(C_i\right)+\sigma_i^1X_i^1\right)\frac{d^iU\left(t,X^1\right)}{dt^i}\\&=\sum_{j=p+1}^{n}\left(a_{0j}+a_{1j}\xi_j\right)g_j\left(t\right),\end{aligned}\tag{77}$$

with the deterministic initials conditions given by

$$\left.\frac{d^iU\left(t\right)}{dt^i}\right|_{t=0}=u_0^i\quad\text{for } 0\le i\le m-1.\tag{78}$$

The polynomial chaos associated with the random vector ω_2 is considered and denoted by $\{\phi_k^2;\ k\in\mathbb{N}\}$. The solution U is in this case a process that depends on t, on the random vector X^{11}, and on the random vector ω_2. The solution is then written in the following form:

$$U\left(t,X^1\right)=U\left(t,X^{11},\omega_2\right).\tag{79}$$

Since random vectors X^{11} and ω_2 are independent, the factorization by the polynomial chaos associated with the random vector ω_2 is used in the decomposition (50). The solution can be rewritten as follows:

$$U\left(t,X^{11},\omega_2\right)=\sum_{k\ge0}U_k\left(t,X^{11}\right)\phi_k^2\left(\omega_2\right).\tag{80}$$

The insertion of this decomposition in the differential equation (77) and the projection of the equation and of the initial conditions over each polynomial chaos ϕ_l^2 for $l\ge0$ lead to the following simplified system of random differential equations:

For $k = 0$,

$$\sum_{i=0}^{m} \left(E(C_i) + \sigma_i^1 X_i^1\right) \frac{d^i U_0 (t, X^{11})}{dt^i} = \sum_{j=p+1}^{n} a_{0j} g_j (t)$$
$$\left. \frac{d^i U_0 (t, X^{11})}{dt^i} \right|_{t=0} = u_0^i \quad \text{for } 0 \le i \le m-1. \tag{81a}$$

For $1 \le k \le n - p$,

$$\sum_{i=0}^{m} \left(E(C_i) + \sigma_i^1 X_i^1\right) \frac{d^i U_0 (t, X^{11})}{dt^i} = a_{1p+k} g_{p+k} (t)$$
$$\left. \frac{d^i U_k (t, X^{11})}{dt^i} \right|_{t=0} = 0 \quad \text{for } 0 \le i \le m-1. \tag{81b}$$

For $k \ge n - p + 1$,

$$\sum_{i=0}^{m} \left(E(C_i) + \sigma_i^1 X_i^1\right) \frac{d^i U_k (t, X^{11})}{dt^i} = 0$$
$$\left. \frac{d^i U_k (t, X^{11})}{dt^i} \right|_{t=0} = 0 \quad \text{for } 0 \le i \le m-1. \tag{81c}$$

It has to be noted that the case $k \ge n - p + 1$ leads to null initial conditions and null right-hand side. In this case, the solution is then null.

The solution given in (80) is then reduced to

$$U\left(t, X^{11}, \omega_2\right) = \sum_{k=0}^{n-p} \phi_k^2 (\omega_2) U_k \left(t, X^{11}\right). \tag{82}$$

For $0 \le k \le n - p$ and considering the decomposition of the solution $U_k(t, X^{11})$ according to the polynomials chaos associated with the random vector X^{11}, one gets

$$U_k \left(t, X^{11}\right) = \sum_{d=0}^{NL} u_{kd} (t) \mathbf{\Psi}_d^{\mathbf{11}} \left(X^{11}\right). \tag{83}$$

Based on this decomposition, the differential equations (81a)–(81c) are then reduced to the following:

For $k = 0$,

$$\sum_{i=0}^{m} \sum_{d=0}^{NL} \mathbf{\Psi}_d^{\mathbf{11}} \left(E(C_i) + \sigma_i^1 X_i^1\right) \frac{d^i u_{0d} (t)}{dt^i} = \sum_{j=p+1}^{n} a_{0j} g_j (t)$$
$$\left. \frac{d^i u_{0d} (t, X^{11})}{dt^i} \right|_{t=0} = u_{0d}^i \quad \text{for } 0 \le i \le m-1. \tag{84a}$$

For $1 \le k \le n - p$,

$$\sum_{i=0}^{m} \sum_{d=0}^{NL} \mathbf{\Psi}_d^{\mathbf{11}} \left(E(C_i) + \sigma_i^1 X_i^1\right) \frac{d^i u_{kd} (t)}{dt^i} = a_{1p+k} g_{p+k} (t)$$
$$\left. \frac{d^i u_{kd} (t, X^{11})}{dt^i} \right|_{t=0} = 0 \quad \text{for } 0 \le i \le m-1. \tag{84b}$$

Projecting this expression over each polynomial chaos $\mathbf{\Psi}_q^{\mathbf{11}}$ for $q = 0$ to NL, one gets

$$\sum_{i=0}^{m} \sum_{d=0}^{NL} \left\langle \mathbf{\Psi}_d^{\mathbf{11}} \left(E(C_i) + \sigma_i^1 X_i^1\right), \mathbf{\Psi}_q^{\mathbf{11}} \right\rangle \frac{d^i u_{0d} (t)}{dt^i} = \delta_{0q} \left(\sum_{j=p+1}^{n} a_{0j} g_j (t) \right) \tag{85}$$

and, for $1 \le k \le n - p$,

$$\sum_{i=0}^{m} \sum_{d=0}^{NL} \left\langle \mathbf{\Psi}_d^{\mathbf{11}} \left(E(C_i) + \sigma_i^1 X_i^1\right), \mathbf{\Psi}_q^{\mathbf{11}} \right\rangle \frac{d^i u_{kd} (t)}{dt^i} = \delta_{kq} a_{1p+k} g_{p+k} (t) \tag{86}$$

Using the matrices $\mathbf{Y}$ and $\mathbf{T}^i$ introduced in (71), the last equations become the following.

For $q = 0$ to NL,

$$\sum_{i=0}^{m} \sum_{d=0}^{NL} \left(E(C_i) \mathbf{Y}\left(d^1 + 1, q^1 + 1\right) + \sigma_i^1 \mathbf{T}^i \left(d^1 + 1, q^1 + 1\right)\right) \frac{d^i u_{0d} (t)}{dt^i} = \delta_{0q} \left(\sum_{j=p+1}^{n} a_{0j} g_j (t) \right) \tag{87}$$

and, for $1 \le k \le n - p$,

$$\sum_{i=0}^{m} \sum_{d=0}^{NL} \left(E(C_i) \mathbf{Y}\left(d^1 + 1, q^1 + 1\right) + \sigma_i^1 \mathbf{T}^i \left(d^1 + 1, q^1 + 1\right)\right) \frac{d^i u_{kd} (t)}{dt^i} = \delta_{kq} a_{1p+k} g_{p+k} (t), \tag{88}$$

where the integers d^1 and q^1 are related to d and q by notations (54).

These deterministic systems are then rewritten in the following compact matrix form.

For $q = 0$ to NL,

$$\sum_{i=0}^{m}\sum_{d=0}^{NL}\mathbf{F}^{i}\left(d^{1}+1,q^{1}+1\right)\frac{d^{i}u_{0d}(t)}{dt^{i}} = \delta_{0q}\left(\sum_{j=p+1}^{n}a_{0j}g_{j}(t)\right) \quad (89)$$

$$\left.\frac{d^{i}u_{0d}\left(t,X^{11}\right)}{dt^{i}}\right|_{t=0} = u_{0d}^{i} \quad \text{for } 0 \le i \le m-1$$

and, for $1 \le k \le n-p$,

$$\sum_{i=0}^{m}\sum_{d=0}^{NL}\mathbf{F}^{i}\left(d^{1}+1,q^{1}+1\right)\frac{d^{i}u_{kd}(t)}{dt^{i}} = a_{1p+k}g_{p+k}(t)\,\delta_{kq}$$

$$\left.\frac{d^{i}u_{kd}\left(t,X^{11}\right)}{dt^{i}}\right|_{t=0} = 0 \quad \text{for } 0 \le i \le m-1, \quad (90)$$

where

$$\mathbf{F}^{i} = E\left(C_{i}\right)\mathbf{\Upsilon} + \sigma_{i}^{1}\mathbf{T}^{i}. \quad (91)$$

The final truncated solution is then given by

$$U(t) = \sum_{k=0}^{n-p}\phi_{k}^{2}\left(\omega^{2}\right)\sum_{d=0}^{NL}u_{kd}(t)\,\mathbf{\Psi}_{d}^{11}\left(X^{11}\right). \quad (92a)$$

Its mean and variance are given:

$$m_{N}(t) = u_{00}(t),$$

$$V_{N}(t) = \sum_{k=1}^{n-p}\sum_{d=0}^{N}\boldsymbol{\alpha}_{\mathbf{d}}^{11}u_{kd}^{2}(t) + \sum_{d=1}^{N}\boldsymbol{\alpha}_{\mathbf{d}}^{11}u_{0d}^{2}(t). \quad (92b)$$

This solution is the main approximate solution of the random initial value problem (1) elaborated herein. Following (89) and (90), one has to solve NL$(n-p+1)$ simple initial values differential systems. These systems can be solved by standard methods such as the Runge-Kutta method.

The main advantages of this methodological approach are as follows:

(i) The deterministic differential systems have the same left-hand side that is built only once and in a compact matrix form.

(ii) Simple right-hand sides are obtained for an arbitrary number of random variables defining the excitation.

(iii) A large reduction of the number of unknowns based on polynomial chaos can result.

(iv) The CPU time and memory needed can be largely reduced.

(v) The numerical solution can be done in parallel manner and particularly for a large number of uncertain parameters.

2.4. Conditional Expectation Method. For the sake of comparison, a methodological approach based on the conditional expectation method is also elaborated here to solve (1). Remember that the solution U of (1) is a stochastic process that depends on the random vectors ω_1 and ω_2 and time t. Let $\omega_0 = (\omega_{10}, \omega_{20})$ be an observation of the random vector $\omega = (\omega_1, \omega_2)$. The conditional expectation of U such that $\omega = \omega_0$, noted by $E(U \mid \omega = \omega_0)$, is the solution of the deterministic equation:

$$\mathscr{L}_{0}u(t,\omega_{0}) = F(t,\omega_{20}), \quad (93)$$

where $\mathscr{L}_0$ is a deterministic operator defined by

$$\mathscr{L}_{0} = \sum_{i=0}^{m}C_{i}(\omega_{10})\frac{d^{i}}{dt^{i}}. \quad (94)$$

Like in the first section, the random vector ω is assumed to have a density function relative to the Lebesgue measure denoted by f. The moments of order p $(p \in \mathbb{N}^*)$ of the solution of (1) are then given by the Bayes formulas:

$$m_{p}(t) = \int_{\mathbb{R}^{n+1}}\left(E\left(U \mid \omega = \omega_{0}\right)\right)^{p}f(\omega_{0})\,d\omega_{0}. \quad (95)$$

The numerical computation of the integral (95) needs the solution of (93) for any observation vector ω_0. The analytical solution can be obtained in general cases. Assume that $\omega_0 \in \Omega$, where Ω is a bounded domain in $\mathbb{R}^{n+1}$.

The integral can be approximated by the Gauss-Legendre quadrature method. Let the sets $\{\overline{\omega}_{0i},\ 1 \le i \le q\}$ and $\{\overline{p}_i,\ 1 \le i \le q\}$ be Gauss points and associated Gauss weights. The moments $m_p(t)$ can be approximated by

$$m_{p}(t) \cong \sum_{i=1}^{q}\overline{p}_{i}\left(E\left(U \mid \omega = \overline{\omega}_{0i}\right)\right)^{p}f(\omega_{0i}). \quad (96)$$

The main advantage of this method is that it allows obtaining a general solution that is considered here as a reference solution as well as the obtained one based on the Monte-Carlo method. But, the inconvenience of these two methods is that a very large CPU time is needed for accurate numerical solutions.

3. Applications

For the sake of clarity, the developed methodological approaches are explicitly presented for the most standard initial value problem. Let us consider the following second-order differential equation modeling the dynamical forced behavior of Euler-Bernoulli beams with the following uncertain parameters: thickness h, mass density ρ, and Young modulus E:

$$C_{2}\ddot{U}(t) + C_{1}\dot{U}(t) + C_{0}U(t) = F(t),$$
$$U(0) = U_{O}, \quad (97)$$
$$\dot{U}(0) = \dot{U}_{0}$$

where

$$\begin{aligned} C_2 &= \rho h \\ C_1 &= \alpha C_2 + \beta C_0 \\ C_0 &= c_0 E h + c_1 E h^3 \end{aligned} \tag{98a}$$

$$F(t) = \sum_{i=3}^{n} (a_{0i} + a_{1i}\xi_i) \sin(\theta_i t), \tag{98b}$$

in which c_0, c_1, α, β, a_{0i}, a_{1i}, θ_i, U_0, and $\dot{U}_0$ are deterministic constants. The considered physical random variables ρ, h, and E are written in the following forms:

$$\begin{aligned} \rho &= \rho_0 + \xi_0 \rho_1 \\ h &= h_0 + \xi_1 h_1 \\ E &= E_0 + \xi_2 E_1, \end{aligned} \tag{98c}$$

where $(\xi_i,\ 0 \le i \le n)$ are random variables with zero means and one variance, two by two independent, and possessing distribution functions relative to the Lebesgue measure, respectively, denoted by f_i, $0 \le i \le n$.

It has to be noted that the internal random coefficients C_0, C_1, and C_2 depend nonlinearly on the random variables ξ_0, ξ_1, and ξ_2 and n is an arbitrary integer. For a clear presentation, the three methodological approaches, elaborated here, are developed for the solution of (97).

3.1. Generalized Polynomial Chaos Method. The random vector ω_1, $\omega_1 = (\xi_0, \xi_1, \xi_2)$, the order of linear random operator $\mathscr{L}$ is $m = 2$, and the internal coefficients in this case are $C_0(\omega_1)$, $C_1(\omega_1)$, $C_2(\omega_1)$. Note that these coefficients depend nonlinearly on ξ_0, ξ_1, and ξ_2 and the order of nonlinearity is 2 for C_2 and 4 for C_0 and C_1. Based on the notations of the first section, these coefficients can be expanded as follows:

$$\begin{aligned} C_0(\omega_1) &= c^0_{000} + c^0_{010}\xi_1 + c^0_{001}\xi_2 + c^0_{020}\xi_1^2 + c^0_{011}\xi_1\xi_2 \\ &\quad + c^0_{030}\xi_1^3 + c^0_{021}\xi_1^2\xi_2 + c^0_{031}\xi_1^3\xi_2 \\ C_1(\omega_1) &= c^1_{000} + c^1_{100}\xi_0 + c^1_{010}\xi_1 + c^1_{001}\xi_2 + c^1_{110}\xi_0\xi_1 \\ &\quad + c^1_{020}\xi_1^2 + c^1_{011}\xi_1\xi_2 + c^1_{030}\xi_1^3 + c^1_{021}\xi_1^2\xi_2 \\ &\quad + c^1_{031}\xi_1^3\xi_2, \\ C_2(\omega_1) &= c^2_{000} + c^2_{100}\xi_0 + c^2_{010}\xi_1 + c^2_{110}\xi_0\xi_1, \end{aligned} \tag{99}$$

where these coefficients are given by

$$\begin{aligned} c^0_{000} &= c_0 E_0 h_0 + c_1 E_0 h_0^3, \\ c^0_{010} &= c_0 E_0 h_1 + 3c_1 E_0 h_0^2 h_1, \\ c^0_{001} &= c_0 E_1 h_0 + c_1 E_1 h_0^3, \\ c^0_{020} &= 3c_1 E_0 h_0 h_1^2, \\ c^0_{011} &= c_0 E_1 h_1 + 3c_1 E_1 h_0^2 h_1, \\ c^0_{003} &= c_1 E_0 h_1^3, \\ c^0_{021} &= 3c_1 E_1 h_0 h_1^2, \\ c^0_{031} &= c_1 E_1 h_1^3; \\ c^2_{000} &= \rho_0 h_0, \\ c^2_{100} &= \rho_1 h_0, \\ c^2_{010} &= \rho_0 h_1, \\ c^2_{110} &= \rho_1 h_1; \\ c^1_{000} &= \left(\alpha c^2_{000} + \beta c^0_{000}\right), \\ c^1_{100} &= \alpha c^2_{100}, \\ c^1_{010} &= \left(\alpha c^2_{010} + \beta c^0_{010}\right), \\ c^1_{001} &= \beta c^0_{001}, \\ c^1_{110} &= \alpha c^2_{110}, \\ c^1_{020} &= \beta c^0_{020}, \\ c^1_{011} &= \beta c^0_{011}, \\ c^1_{003} &= \beta c^0_{003}, \\ c^1_{021} &= \beta c^0_{021}, \\ c^1_{031} &= \beta c^0_{031}. \end{aligned} \tag{100}$$

Let k and $l = 0$ to N, and, using the notation (23), one gets $(k_0, \ldots, k_n) = \psi_n^{-1}(k)$ and $(l_0, \ldots, l_n) = \psi_n^{-1}(l)$, where $(k_0, \ldots, k_n)$ and $(l_0, \ldots, l_n)$ are in $\mathbb{N}^{n+1}$.

The matrices T^i for $i = 0$ to 2, introduced in (35), are then given by the following.

For $k, l = 0 : N$,

$$\begin{aligned} T^0(l+1, k+1) &= c^0_{000} \prod_{d=0}^{n} \left(\beta^{d0}(l_d+1, k_d+1)\right) \\ &\quad + c^0_{010}\beta^{11}(l_1+1, k_1+1) \prod_{\substack{d=0 \\ d\neq 1}}^{n} \left(\beta^{d0}(l_d+1, k_d+1)\right) \\ &\quad + c^0_{001}\beta^{21}(l_2+1, k_2+1) \prod_{\substack{d=0 \\ d\neq 2}}^{n} \left(\beta^{d0}(l_d+1, k_d+1)\right) \\ &\quad + c^0_{020}\beta^{12}(l_1+1, k_1+1) \prod_{\substack{d=0 \\ d\neq 1}}^{n} \left(\beta^{d0}(l_d+1, k_d+1)\right) \end{aligned}$$

$$+ c^0_{011}\beta^{11}(l_1+1,k_1+1)\beta^{21}(l_2+1,k_2+1)$$
$$\cdot\prod_{\substack{d=0\\d\neq1,2}}^{n}\left(\beta^{d0}(l_d+1,k_d+1)\right)$$
$$+ c^0_{030}\beta^{13}(l_1+1,k_1+1)\prod_{\substack{d=0\\d\neq1}}^{n}\left(\beta^{d0}(l_d+1,k_d+1)\right)$$
$$+ c^0_{021}\beta^{12}(l_1+1,k_1+1)\beta^{21}(l_2+1,k_2+1)$$
$$\cdot\prod_{\substack{d=0\\d\neq1,2}}^{n}\left(\beta^{d0}(l_d+1,k_d+1)\right)$$
$$+ c^0_{031}\beta^{13}(l_1+1,k_1+1)\beta^{21}(l_2+1,k_2+1)$$
$$\cdot\prod_{\substack{d=0\\d\neq1,2}}^{n}\left(\beta^{d0}(l_d+1,k_d+1)\right)$$
$$T^2(l+1,k+1) = c^2_{000}\prod_{d=0}^{n}\left(\beta^{d0}(l_d+1,k_d+1)\right)$$
$$+ c^2_{100}\beta^{01}(l_1+1,k_1+1)\prod_{d=1}^{n}\left(\beta^{d0}(l_d+1,k_d+1)\right)$$
$$+ c^2_{010}\beta^{11}(l_1+1,k_1+1)\prod_{\substack{d=0\\d\neq1}}^{n}\left(\beta^{d0}(l_d+1,k_d+1)\right)$$
$$+ c^2_{110}\beta^{01}(l_1+1,k_1+1)\beta^{11}(l_1+1,k_1+1)$$
$$\cdot\prod_{d=2}^{n}\left(\beta^{d0}(l_d+1,k_d+1)\right),$$
$$T^1 = \alpha T^0 + \beta T^2. \tag{101}$$

The variable vector D of dimension $(N+1)$, introduced in (36), is given by the following.

For $k = 0$ to N,

$$D_{k+1}(t) = \sum_{j=3}^{n}\left(a_{0j}\beta^{j0}(1,k_j+1) + a_{1j}\beta^{j1}(1,k_j+1)\right) \cdot\prod_{\substack{d=0\\d\neq j}}^{n}\left(\beta^{d0}(1,k_d+1)\right)\sin(\theta_j t). \tag{102}$$

The passage from the index for $0 \le l \le N$ to the multi-indices $(l_0, l_1, \ldots, l_n)$, used for the determination the matrices T^0, T^1, and T^2 needs the bijection ψ_n. This bijection is given with Algorithm 1.

The determination of T^0, T^1, and T^2, in this case, needs the computation of the inner products $\langle \xi_i^p \phi_k, \phi_l\rangle$ for $p \le 3$, by (32).

```
Read L and n
L non null integer
t = 0.
For l_0 = 0 : L; l_1 = 0 : l_0; l_2 = 0 : l_1; ... ; l_n = 0 : l_{n-1}.
ψ_n(l_0 − l_1, l_1 − l_2, ..., l_{n-1} − l_n, l_n) = t + 1;
t = t + 1;
end.
```

Algorithm 1

If the random coefficients ξ_i are not identically distributed then $\langle \xi_i^p \phi_k, \phi_l\rangle$ have to be computed for each $i = 0$ to n. Otherwise, these scalar products have to be computed only once. This difficulty as well as the number of unknowns can be largely reduced by using the internal random coefficients method.

3.2. Internal Random Coefficients Method. Based on this method, the internal coefficients of (97) are rewritten in the following form:

$$\begin{aligned} C_0(\omega_1) &= E(C_0) + \sigma_0^1 X_0^1 \\ C_2(\omega_1) &= E(C_0) + \sigma_2^1 X_2^1 \\ C_1(\omega_1) &= \alpha C_0(\omega_1) + \beta C_2(\omega_1). \end{aligned} \tag{103}$$

The random vector X^{11}, introduced by this method, is then given by

$$X^{11} = \left(X_0^1, X_2^1\right). \tag{104}$$

The number of random variables used here is $(n-1)$ variables, two variables for the coefficients C_j, and $(n-3)$ variables for the right-hand side.

For $k, l = 0, \ldots, NL$; and let k^1, l^1 be the integers associated with integers k, l, introduced by the notations (23)–(54); one has

$$\begin{aligned} (k_0,\ldots,k_{n-1}) &= \psi_{n-1}^{-1}(k); \\ (l_0,\ldots,l_{n-1}) &= \psi_{n-1}^{-1}(l), \\ k^1 &= \psi_2(k_0,k_1), \\ l^1 &= \psi_2(l_0,l_1). \end{aligned} \tag{105}$$

The matrices H^i, introduced in (74), are given in this case by

$$H^0(k^1,l^1) = \left(E(C_0)\Upsilon(k^1+1,l^1+1) + \sigma_0^1\mathbf{T}^0(k^1+1,l^1+1)\right)\prod_{d=2}^{n-1}\left(\beta^{d0}(l_d+1,k_d+1)\right)$$

$$H^2(k^1, l^1) = \left(E(C_1)\Upsilon(k^1+1, l^1+1) + \sigma_1^1\mathbf{T}^2(k^1+1, l^1+1)\right)\prod_{d=2}^{n-1}\left(\beta^{d0}(l_d+1, k_d+1)\right),$$
$$H^1 = \alpha H^0 + \beta H^2 \quad (106)$$

and the variable vector E of dimension $(NL + 1)$, introduced in (75), is given by

$$E_l(t) = \sum_{j=3}^{n}\delta_{0l^1}\left(\prod_{\substack{q=3\\ q\neq j}}^{n}\delta_{0l_{q-1}}\right)\left(a_{0j}\delta_{0l_{j-1}} + a_{1j}\delta_{jl_{j-1}}\right) \cdot \sin(\theta_j t) \quad l = 1, NL+1. \quad (107)$$

The number of random variables, n, is arbitrary considered. For more clarity about the needed operations, let us consider $n = 5$. Based on the classical polynomial chaos and if the decomposition of the solution into the polynomials chaos of degree less than or equal to seven is considered then the needed unknowns to be determined are of number $\binom{13}{7} = 1716$, while the internal random coefficient method requires $\binom{12}{7} = 792$ unknowns to be determined. The number unknowns of is thus diminished by half. In addition the determination of the tensors T^0, T^1, T^2 needs the computation of the scalar products $\langle\prod_{d=0}^{p}(\xi_d^{\alpha_d})\psi_k, \psi_l\rangle$ where $\sum_{d=0}^{p}\alpha_d \leq q_i$ and q_i is the degree of a random polynomial coefficient $C_i(\omega_1)$ given by relationship (10), while the tensors H^0, H^1, H^2 need only the calculation of the inner products $\langle(X_i^1)^p\mathbf{\Psi}_k^1, \mathbf{\Psi}_l^1\rangle$ for $p = 0$ or 1. This demonstrates an extra reduction in the computation effort. For more reduction in the CPU time, the internal random coefficient method is coupled with the superposition method.

3.3. Superposition Method. Here, (97) is replaced with an equivalent system with $(n-2)$ equations; each equation is a random equation depending on the random vector X^{11} introduced in (46). These equations are given by

$$\sum_{i=0}^{2}C_i(\omega_1)\frac{d^iU_0(t, X^{11})}{dt^i} = \sum_{j=3}^{n}a_{0j}\sin(\theta_j t) \quad (108a)$$

with the initial conditions given by $(dU_0(t, X^{11})/dt)|_{t=0} = \dot{u}_0$. $U_0(0, X^{11}) = u_0$,
And, for $1 \leq k \leq n-2$,

$$\sum_{i=0}^{2}C_i(\omega_1)\frac{d^iU_k(t, X^{11})}{dt^i} = a_{1(2+k)}\sin(\theta_{2+k}t) \quad (108b)$$

with the initial conditions given by

$$\left.\frac{dU_k(t, X^{11})}{dt}\right|_{t=0} = 0, \quad U_k(0, X^{11}) = 0. \quad (109)$$

The associated matrices, introduced in (91), are given by

$$\mathbf{F}^0 = E(C_0)\Upsilon + \sigma_0^1\mathbf{T}^0$$
$$\mathbf{F}^2 = E(C_2)\Upsilon + \sigma_2^1\mathbf{T}^2, \quad (110)$$
$$\mathbf{F}^1 = \alpha\mathbf{F}^0 + \beta\mathbf{F}^2.$$

These matrices are replaced with their expressions in systems (108a) and (108b). In this case, the determination of the tensors $\mathbf{F}^0, \mathbf{F}^1, \mathbf{F}^2$ needs the determination of inner products $\langle(X_i^{11})^p\mathbf{\Psi}_{k^1}^{11}, \mathbf{\Psi}_{l^1}^{11}\rangle$ for $p = 0$ or 1 where k^1 and l^1 are the integers associated with integers k and l given by notation (54). In this case, the polynomial chaos $\mathbf{\Psi}_{k^1}^{11}$ associated with the random vector X^{11} is simple. The solution of the deterministic system defined by (86)-(87) determines an approximate solution U of the initial random equation (97).

This system largely reduces the number of unknowns to be determined for the approximate solution U and makes the use of polynomial chaos possible even when the classical projection method is unable to determine the solution. The CPU time and memory required to solve this problem by classical projection methods are significantly reduced by this method.

It is well known that, for a stochastic excitation and with Karhunen-Loéve expansion, the classical projection methods are limited to a small number of random variables used in the decomposition. This will affect the accuracy of the obtained solution. This kind of problems is completely solved the by the presented method.

4. Numerical Results

The dynamical system defined in (97) is considered. To analyze the effect of the uncertain parameters, the free system $(F(t) = 0)$ is first considered. For free systems, two variants of the internal random coefficients method are used. The first one consists of using two random coefficients C0 and C2 called here IRCM2; the second one consists of dividing by C2 and then using only one random coefficient (C2/C0). This variant is referred to here as IRCM1.

4.1. Free Dynamics Response. For numerical analysis, the following values are used:

$$\begin{aligned}
\rho_0 &= 7800,\\
\rho_1 &= \alpha_0\rho_0,\\
h_0 &= 5 * 10^{-5},\\
h_1 &= \alpha_1 h_0,\\
E_0 &= 21 * 10^{10},\\
E_1 &= \alpha_2 E_0,\\
\alpha &= 0.01,\\
\beta &= 0.001.
\end{aligned} \quad (111)$$

The random variables ξ_0, ξ_1, ξ_2 are assumed to follow various laws and α_0, α_1, and α_2 are the ratios of the standard deviation and the mean of the random variables ρ, h, and E.

Figures 1(a) and 1(b) present the standard deviation and mean of the random solution U, where ξ_i are random variables with normalized truncated exponential laws in $[0, 10]$, with 5% of standard deviation for parameters ($\alpha_0 = \alpha_1 = \alpha_2 = 5\%$). The solution $U(t)$ is computed using the chaos methods associated with the random variables ξ_i, the IRCM1, IRCM2, the conditional expectation method, and the Monte-Carlo method with 1000 simples.

In the case of the classical chaos method, solution U is expanded in the polynomial chaos of degree less than or equal to 7 and these polynomials are dependent on the three random variables ξ_i for $i = 0, 1$, and 2. For the first internal random coefficient method, IRCM1, the coefficients of the random equation (97) are divided by the random variables $C_0(\omega_1)$ and the random equation depends only on one new random variable $\eta = C_2(\omega_1)/C_0(\omega_1)$. The solution U is then expanded into the polynomials chaos associated with the random variable η of degree less than or equal to 12. For the second internal random coefficient method (IRCM2), the random equation depends on the random variables $C_0(\omega_1), C_2(\omega_2)$. Solution $U$is here expanded into the polynomials chaos associated with these two random variables of degree less than or equal to 2 and also less than 7. The number of unknowns used by the classical chaos method is 120 variables for degrees ≤ 7. The ICRM2 needs 6 unknowns for degrees ≤ 2 and 36 unknowns for degrees ≤ 7 while the IRCM1 needs only 13 unknowns for degrees ≤ 12 as presented in Table 2. This simple case demonstrates clearly how the IRCM reduces the number of unknowns.

The mean and standard deviation of the solution U obtained by IRCM1, IRCM2, chaos, conditional expectation, and Monte-Carlo methods, presented in Figures 1(a) and 1(b), demonstrate a good accuracy between the predictions obtained by these methods.

Figures 2(a) and 2(b) present the standard deviation and mean of the solution U, when the random variables ξ_i are uniform in $[-4, 4]$ with 5% standard deviation for each parameter ($\alpha_0 = \alpha_1 = \alpha_2 = 5\%$). Solution U is computed using the previous methods. For the case of chaos method, the solution U is expanded into polynomials chaos of degree less than or equal to seven and these polynomials are dependent on the three random variables ξ_i for $i = 0, 1$, and 2. For the ICRM1, solution U is expanded into the polynomial chaos associated with the random variable η of degree less than or equal to 12. For ICRM2, solution U is expanded into the polynomial chaos associated with two random variables of degree less than or equal to 7.

For large standard deviations, these methods are also tested. Figures 3(a) and 3(b) present the standard deviation and mean of the solution U, when the random variables ξ_i are uniform in $[-4, 4]$ with 20% standard deviation ($\alpha_0 = \alpha_1 = \alpha_2 = 20\%$). For the case of chaos method, solution U is expanded in polynomial chaos of degree less than or equal to 7 and these polynomials are dependent on the three random variables ξ_i for $i = 0, 1$, and 2. ICRM1 and ICRM2 are also used with polynomial of degree ≤ 7. The number of coefficients to compute is presented in Table 3. Results obtained by the conditional expectation and Monte-Carlo methods are also presented. It is observed that, even for large standard deviation, 20%, a good agreement is observed between the predictions obtained by these methods.

Figures 4(a) and 4(b) present the standard deviation and mean of solution U, where the random variables ξ_i follow a truncated exponential law in $[0, 10]$ with 20% standard deviation ($\alpha_0 = \alpha_1 = \alpha_2 = 20\%$). The degree of the used polynomials is ≤ 7 for chaos, ≤ 12 for IRCM1, and ≤ 2 and ≤ 7 for IRCM2 and the number of coefficients to compute is presented in Table 4. It is demonstrated that, for the mean value, all the methods agree with each other but there are some discrepancies for the standard deviation. The IRCM2 is better than IRCM1 and particularly for large standard deviations. The division procedure, used in IRCM1, reduces the number of random parameters by one but introduces additional nonlinear random effect. This effect affects the accuracy of the IRCM1, as clearly presented in Figure 3.

4.2. Dynamical System with a Random Excitation. To test the flexibility of the presented methodological approach to handle dynamical systems with random variables following different laws, (97) is considered with a random excitation given by $F(t) = (a_0 + a_1\xi_3)\sin(\omega_0 t)$. a_0, a_1, and ω_0 are considered deterministic. The random coefficients ξ_0, ξ_1, ξ_2, and ξ_3 are assumed here to follow different laws.

Figures 5(a) and 5(b) present the mean and standard deviation of $\mathbf{U}(t)$ obtained by the presented methods. ξ_0 and ξ_2 are truncated normal variables in $[-4, 4]$,ξ_1 is a uniform variable in $[-1, 1]$, and ξ_3 is a truncated exponential variable in $[0, 10]$ with 5% of standard deviations.

The presented results are obtained by the chaos method where solution $\mathbf{U}$ is expanded in the polynomials chaos of degree less than or equal to 2 and 5 and these polynomials depend on the four random variables ξ_i for $i = 0, 1, 2$, and 3. For internal random coefficients method, solution $\mathbf{U}$ is expanded in polynomial chaos associated with the two random variables $C_0(\omega_1)$,$C_2(\omega_1)$ of degree less than or equal to 2 and 5. The results predicted by the IRCM are better than those obtained by the chaos method. In addition, the number of unknowns used by the chaos method is 15 for polynomial chaos of degrees ≤ 2 and 126 for polynomial chaos of degree ≤ 5 while by the IRCM combined to the superposition method we have to solve a system with two differential equations and each one with 6 unknowns for expansion into polynomials of degrees ≤ 2 and 21 unknowns for expansion into the polynomial of degree ≤ 7 as presented in Table 5.

To analyze the standard deviation effect on the obtained results, the parameters of the system and of the excitation are random variables with 20% standard deviations. The obtained results are given in Figures 6(a) and 6(b) for mean and standard deviation. It is shown that chaos of degree ≤ 2 does not predict good results in comparison with reference results, obtained by conditional expectation and Monte-Carlo 2000 methods. The IRCM with degree ≤ 2 predicts better results than chaos with degree ≤ 5 and IRCM with degree ≤ 5 coincides perfectly with the reference results. The random coefficients ξ_0, ξ_1, ξ_2, and ξ_3 are assumed here to follow

TABLE 2: Number of coefficients to compute for IRCM1-IRCM2 and PC at the considered degrees.

Method	IRCM1	IRCM2	IRCM2	Chaos
Degree	12	2	7	7
Unknowns	13	6	36	120

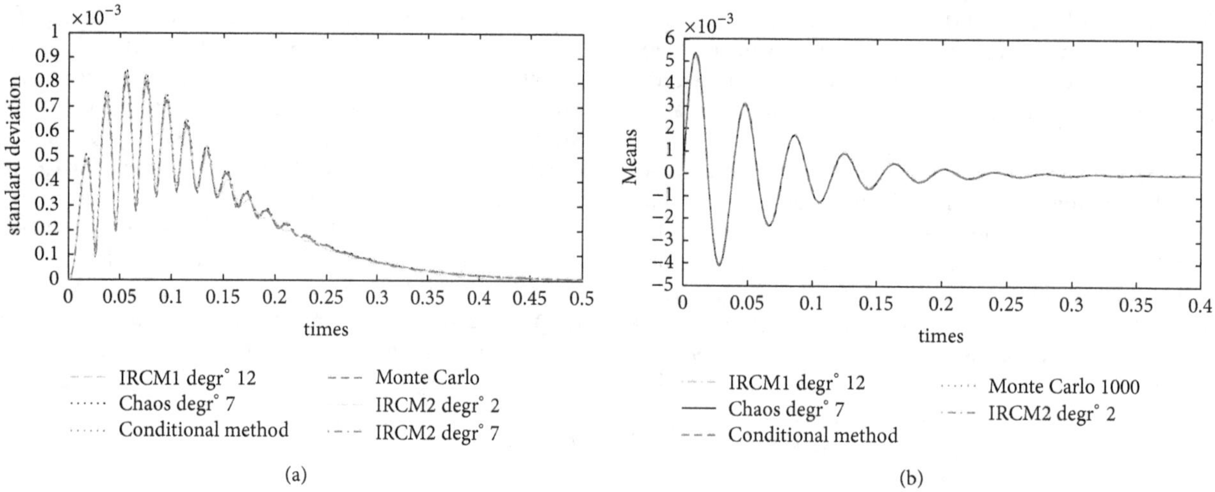

FIGURE 1: (a) Standard deviation of U when ξ_i follow the truncated exponential law in $[0, 10]$ and $\sigma_i = 5\%$. (b) Means of U when ξ_i follow the truncated exponential law in $[0, 10]$ and $\sigma_i = 5\%$.

TABLE 3: Number of coefficients to compute for IRCM1-IRCM2 and PC at degree 7 for the considered number of variables.

Method	IRCM1	IRCM2	Chaos
Degree	7	7	7
Unknowns	8	36	120

different laws. The required CPU time for the used methods is presented in Table 6. These results demonstrate clearly the efficiency and accuracy of the proposed IRCM.

The error between the used methods is clearly demonstrated in Figures 7, 8, and 9. It is clearly seen that the error is lower for IRCM than for PC related to the conditional solution assumed to be the reference solution.

5. Conclusion

Methodological approaches based on generalized polynomial chaos for random differential equation with an arbitrary number of random parameters following different types of distribution laws are developed. These random parameters can be linear or nonlinear. The conditional expectation method for the considered random differential equations is well clarified. The obtained results based on this method are considered to be exact results and used as reference results as well as those obtained by the Monte-Carlo method.

A procedure to build a new polynomial chaos basis and a connection between the one-dimensional and multidimensional polynomials is well-established. Firstly, this procedure is given for independent variables and then generalized for variables that may be dependent. This procedure permits considering a basis given by the general polynomial chaos associated with the considered parameters and the solution is expanded into this new basis. A new methodological approach called internal random coefficient method is developed. Compact matrix representations are elaborated and the required matrices, vectors, and scalar products are explicitly given. These projection methods, combined with the superposition method, lead to large CPU time and memory space reductions required for the solution. It is demonstrated that the developed methods are efficient and accurate and these methods may lead to spectacular memory space and CPU time reductions and particularly when a large number of random parameters are considered.

Appendix

For the sake of clarity about the used indices, let us consider the case of two random variables, $n = 1$. The canonical basis of the polynomial space generated by the random variables (ξ_1, ξ_2) can be ordered in the given following form:

$$B = \{1, \xi_1, \xi_2, \xi_1^2, \xi_1\xi_2, \xi_2^2, \xi_1^3, \xi_1^2\xi_2, \xi_1\xi_2^2, \xi_2^3, \xi_1^4, \xi_1^3\xi_2, \xi_1^2\xi_2^2, \xi_1\xi_2^3, \dots \text{etc}\}. \tag{A.1}$$

For this order, the function ψ_1 given in (21) is defined by

$$\begin{gathered}\psi_1 : \mathbb{N}^2 \longrightarrow \mathbb{N} \\ (i_0, i_1) \longmapsto \psi_1(i_0, i_1) = i.\end{gathered} \tag{A.2}$$

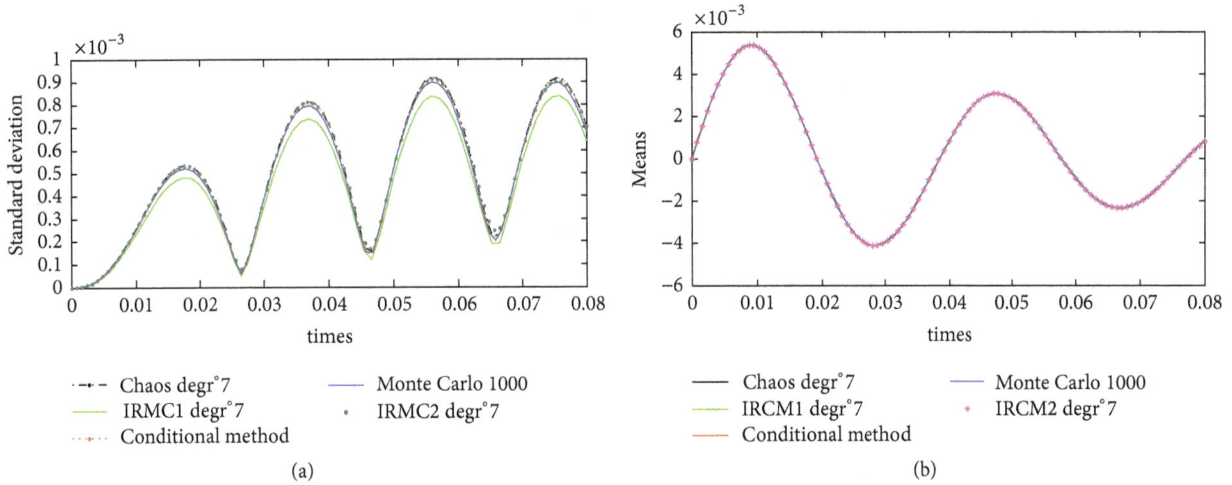

FIGURE 2: (a) Standard deviation of U when ξ_i follow the uniform law in $[-4, 4]$ and $\sigma_i = 5\%$. (b) Means of U when ξ_i follow the uniform law in $[-4, 4]$ and $\sigma_i = 5\%$.

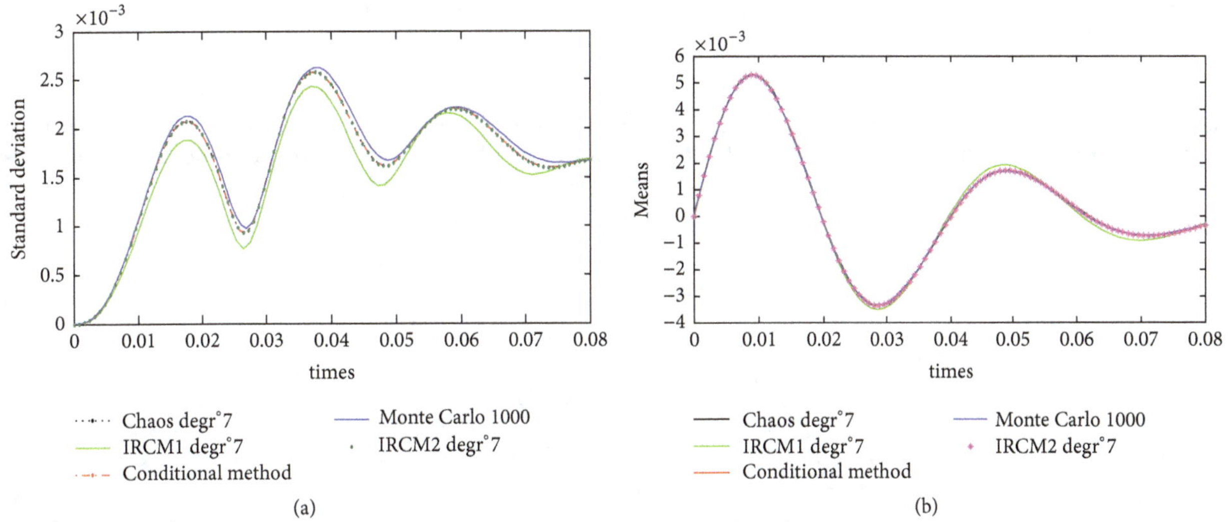

FIGURE 3: (a) Standard deviation of U when ξ_i follow the uniform law in $[-4, 4]$ and $\sigma_i = 20\%$. (b) Means of U when ξ_i follow the uniform law in $[-4, 4]$ and $\sigma_i = 20\%$.

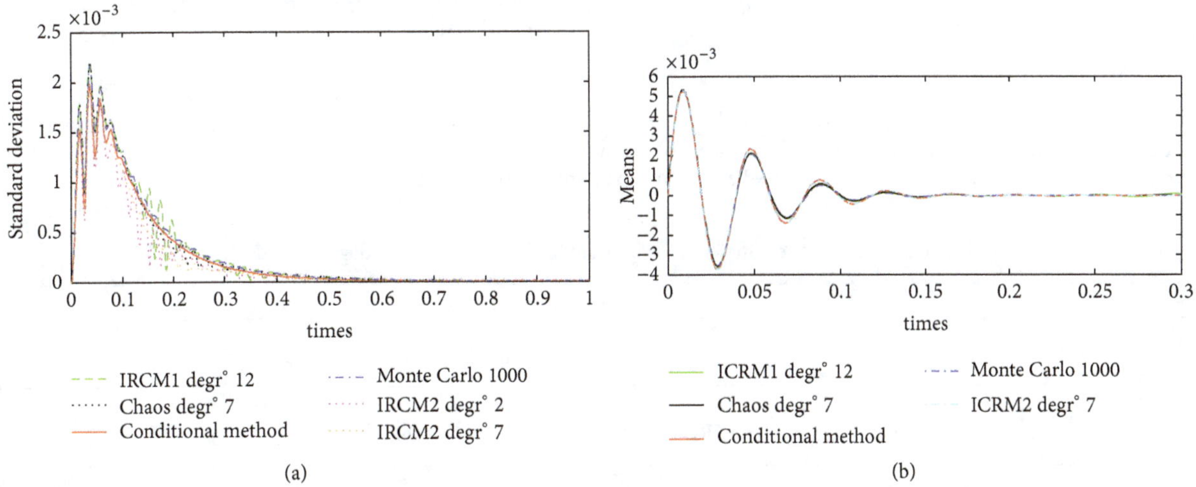

FIGURE 4: (a) Standard deviation of U when ξ_i follow the truncated exponential law in $[0, 10]$ and $\sigma_i = 20\%$. (b) Means of U when ξ_i follow the truncated exponential law in $[0, 10]$ and $\sigma_i = 20\%$.

Table 4: Number of coefficients to compute for IRCM1-IRCM2 and PC at different degrees.

Method	IRCM1	IRCM2	IRCM2	Chaos
Degree	12	2	7	7
Unknowns	13	6	36	120

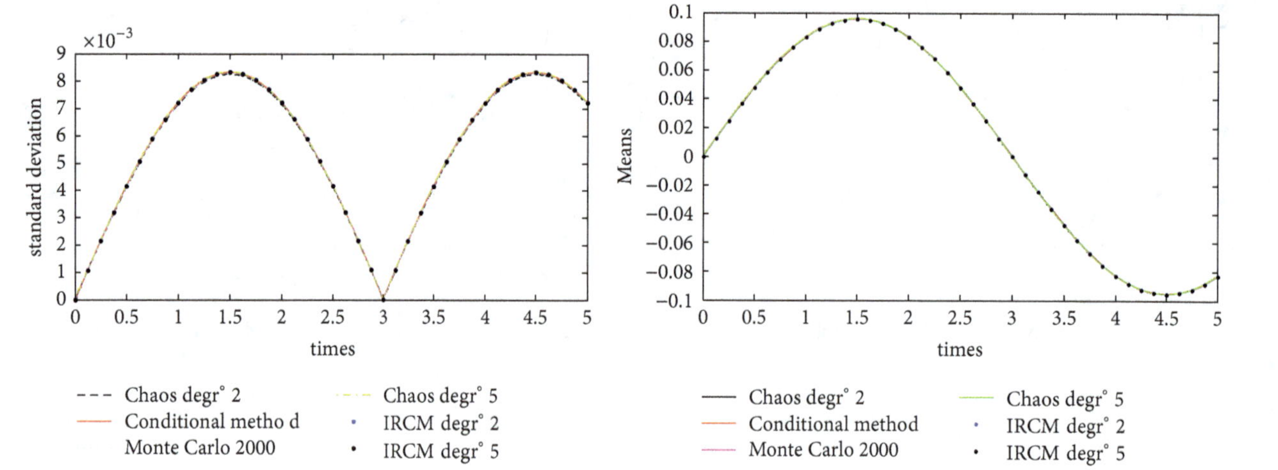

(a) (b)

Figure 5: (a) Standard deviation of solution where ξ_0 and ξ_2 are truncated normal variables in $[-4, 4]$, ξ_1 is a uniform variable in $[-1, 1]$, and ξ_3 is a truncated exponential variable in $[0, 10]$ and $\sigma_i = 5\%$. (b) Means of solution where ξ_0 and ξ_2 are truncated normal variables in $[-4, 4]$, ξ_1 is a uniform variable in $[-1, 1]$, and ξ_3 is a truncated exponential variable in $[0, 10]$ and $\sigma_i = 5\%$.

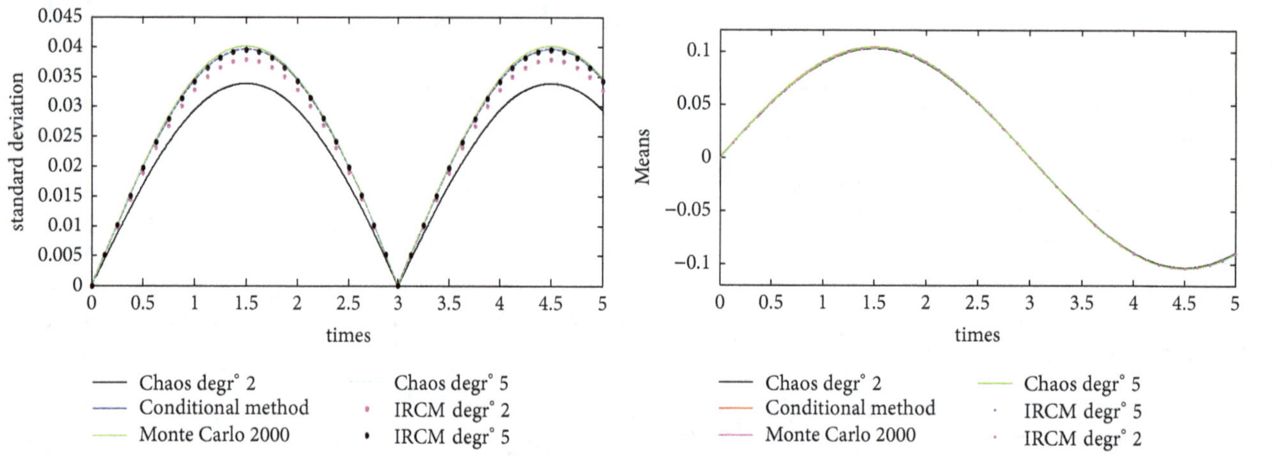

(a) (b)

Figure 6: (a) Standard deviation of solution where ξ_0 and ξ_2 are truncated normal variables in $[-4, 4]$, ξ_1 is a uniform variable in $[-1, 1]$, and ξ_3 is a truncated exponential variable in $[0, 10]$ and $\sigma_i = 20\%$. (b) Means of solution where ξ_0 and ξ_2 are truncated normal variables in $[-4, 4]$, ξ_1 is a uniform variable in $[-1, 1]$, and ξ_3 is a truncated exponential variable in $[0, 10]$ and $\sigma_i = 20\%$.

Table 5: Number of coefficients to compute for IRCM and PC at degrees 2 and 5.

Method	IRCM	IRCM	Chaos	Chaos
Degree	2	5	2	5
Unknowns	12	42	15	126

Table 6: CPU times to compute for IRCM, PC at degrees 12, MC (2000), and conditional method.

Method	IRCM d° 12	PC d° 12	Conditional	MC (2000)
CPUT	6.52	$1.47e + 002$	$2.98e + 003$	36.43

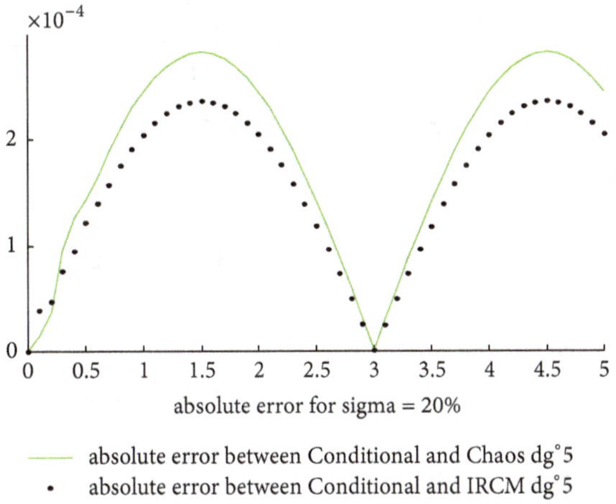

FIGURE 7: Absolute error between PC of degree 5 and IRCM of degree 5 and reference methods where ξ_0 and ξ_2 are truncated normal variables in $[-4, 4]$, ξ_1 is a uniform variable in $[-1, 1]$, and ξ_3 is a truncated exponential variable in $[0, 10]$ and $\sigma_i = 20\%$.

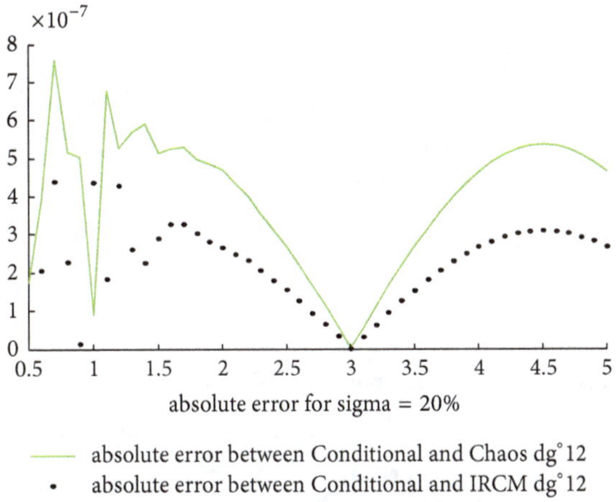

FIGURE 8: Absolute error between PC of degree 12 and IRCM of degree 12 and reference methods where ξ_0 and ξ_2 are truncated normal variables in $[-4, 4]$, ξ_1 is a uniform variable in $[-1, 1]$ and ξ_3 is a truncated exponential variable in $[0, 10]$ and $\sigma_i = 20\%$.

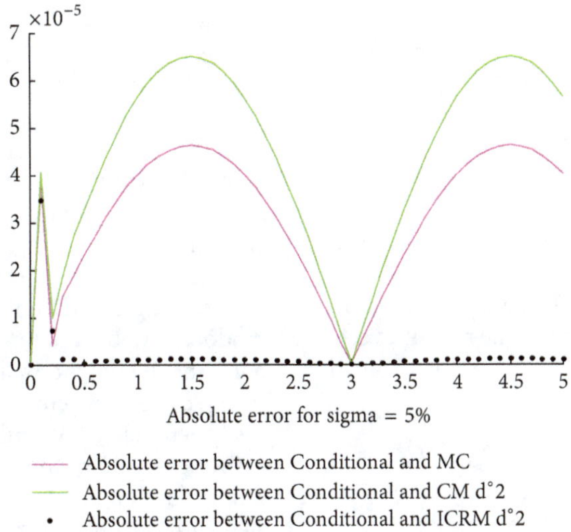

FIGURE 9: Absolute error between PC of degree 2 and IRCM of degree 2 MC (2000) and reference methods where ξ_0 and ξ_2 are truncated normal variables in $[-4, 4]$, ξ_1 is a uniform variable in $[-1, 1]$, and ξ_3 is a truncated exponential variable in $[0, 10]$ and $\sigma_i = 5\%$.

The explicit expression of ψ_1 is thus given by

$$\begin{aligned}
\psi_1(0,0) &= 0,\\
\psi_1(1,0) &= 1,\\
\psi_1(0,1) &= 2;\\
\psi_1(2,0) &= 3,\\
\psi_1(1,1) &= 4,\\
\psi_1(0,2) &= 5\\
\psi_1(3,0) &= 6\\
&\vdots\\
&\text{etc.}
\end{aligned} \tag{A.3}$$

This bijection can be easily condensed in the following form:

$$\psi_1(i_0, i_1) = \binom{2}{i_0 + i_1 + 1} + i_1 - 1. \tag{A.4}$$

The orthogonalization of the canonical basis, using the scalar product defined by (6), coincides with the classical chaos polynomial defined in the literature.

For the two random variables case, the relationship between the indices and the construction of the used basis is straightforward. Three independent random variables are more instructive. For this aim, let us consider the simple random differential equation with three random variables defined by

$$\ddot{U} + C_0\dot{U} + C_1 U = 0, \tag{A.5}$$

where C_1 and C_2 are random variables given by

$$\begin{aligned}
C_0 &= C_{00} + C_{01}\xi_0,\\
C_1 &= (C_{10} + C_{11}\xi_1)(C_{20} + C_{21}\xi_2),
\end{aligned} \tag{A.6}$$

in which ξ_0, ξ_1, and ξ_2 are independent random variables having the same distribution function f and

$$\begin{aligned}
E(\xi_0) &= E(\xi_1) = E(\xi_2) = 0,\\
\sigma(\xi_0) &= \sigma(\xi_1) = \sigma(\xi_2) = 1.
\end{aligned} \tag{A.7}$$

The procedure giving the polynomial chaos associated with the random vector (ξ_0, ξ_1, ξ_2) is first given. The order defined by (20) is considered and the canonical basis associated with polynomial spaced generated by the random vector (ξ_0, ξ_1, ξ_2) is considered. In this case, the basis is explicitly given by

$$\begin{aligned}
B = \{&1, \xi_0, \xi_1, \xi_2, \xi_0^2, \xi_0\xi_1, \xi_0\xi_2, \xi_1^2, \xi_1\xi_2, \xi_2^2, \xi_0^3, \xi_0^2\xi_1, \xi_0^2\xi_2,\\
&\xi_0\xi_1^2, \xi_0\xi_1\xi_2, \xi_0\xi_2^2, \xi_1^3, \xi_1^2\xi_2, \xi_1\xi_2^2, \xi_2^3, \xi_0^4, \xi_0^3\xi_1, \xi_0^3\xi_2, \xi_0^2\xi_1^2,\\
&\xi_0^2\xi_1\xi_2, \xi_0^2\xi_2^2, \xi_0\xi_1^3, \xi_0\xi_1^2\xi_2, \xi_0\xi_1\xi_2^2, \xi_0\xi_2^3, \xi_1^4, \xi_1^3\xi_2, \xi_1^2\xi_2^2,\\
&\xi_1\xi_2^3, \xi_2^4, \text{ect}\ldots\}.
\end{aligned} \tag{A.8}$$

Assuming that the random variables ξ_i follow the normal law, the orthogonalization procedure leads to the following classical polynomial chaos of degree less than 4:

$$\begin{aligned}
&[1, \xi_0, \xi_1, \xi_2, \xi_0^2 - 1, \xi_0\xi_1, \xi_0\xi_2, \xi_1^2 - 1, \xi_1\xi_2, \xi_2^2 - 1, \xi_0^3\\
&- 3\xi_0, \xi_0^2\xi_1 - \xi_1, \xi_0^2\xi_2 - \xi_2, \xi_0\xi_1^2 - \xi_0, \xi_0\xi_1\xi_2, \xi_0\xi_2^2\\
&- \xi_0, \xi_1^3 - 3\xi_1, \xi_1^2\xi_2 - \xi_2, \xi_1\xi_2^2 - \xi_1, \xi_2^3 - 3\xi_2, \xi_0^4\\
&- 6\xi_0^2 + 3, \xi_0^3\xi_1 - 3\xi_0\xi_1, \xi_0^3\xi_2 - 3\xi_0\xi_2, \xi_0^2\xi_1^2 - \xi_0^2\\
&- \xi_1^2 + 1, \xi_0^2\xi_1\xi_2 - \xi_1\xi_2, \xi_0^2\xi_2^2 - \xi_0^2 - \xi_2^2 + 1, \xi_0\xi_1^3\\
&- 3\xi_0\xi_1, \xi_0\xi_1^2\xi_2 - \xi_0\xi_2, \xi_0\xi_1\xi_2^2 - \xi_0\xi_1, \xi_0\xi_2^3\\
&- 3\xi_0\xi_2, \xi_1^4 - 6\xi_1^2 + 3, \xi_1^3\xi_2 - 3\xi_1\xi_2, \xi_1^2\xi_2^2 - \xi_1^2 - \xi_2^2\\
&+ 1, \xi_1\xi_2^3 - 3\xi_1\xi_2, \xi_2^4 - 6\xi_2^2 + 3].
\end{aligned} \tag{A.9}$$

It is then demonstrated that, in this case, these polynomials chaos coincide with the classical Hermit polynomial chaos.

The connection between the 1D and multi-D polynomial chaos is given by the bijection ψ_2. To make this connection explicit, let us consider the index (0, 2, 2). Using Algorithm 1, one gets

$$\begin{aligned}
\psi_2(0,2,2) &= 32,\\
\Phi_{32}(\xi_0, \xi_1, \xi_2) &= \phi_0^0(\xi_0)\,\phi_2^1(\xi_1)\,\phi_2^2(\xi_2),
\end{aligned} \tag{A.10}$$

where

$$\begin{aligned}
\Phi_{32}(\xi_0, \xi_1, \xi_2) &= \xi_1^2\xi_2^2 - \xi_1^2 - \xi_2^2 + 1,\\
\phi_0^0(\xi_0) &= 1,\\
\phi_2^1(\xi_1) &= \xi_1^2 - 1,\\
\phi_2^2(\xi_1) &= \xi_2^2 - 1.
\end{aligned} \tag{A.11}$$

Based on this algorithm, the 1D and multi-D indices connection can be easily established.

It should be noted that there are two procedures to build the multi-D polynomial chaos. The first one is to use a numerical method to orthogonalize base B defined above. It is well known that this procedure is numerically expensive and particularly when a large number of random variables are considered.

The second one, elaborated in this paper, consists of determining the 1D polynomial chaos associated with each random variable ξ_k, $0 \le k \le 2$ and using the bijection ψ_2 to build the 3D polynomial chaos related to the random vector (ξ_0, ξ_1, ξ_2). As presented in Section 2.1, this procedure can be easily generalized to multivariables following different laws.

Response U of the considered random differential equation is developed into the multi-D polynomial chaos associated with the random vector (ξ_0, ξ_1, ξ_2) and given by

$$U = \sum_{k=0}^{N} u_k(t)\,\Phi_k(\xi_0, \xi_1, \xi_2), \tag{A.12}$$

where integer N is defined such that this expansion is to be considered for the polynomial chaos of degree less than a given integer L. Thus, N can be written as follows:

$$N = \psi_2(0, 0, L) \quad \text{(A.13)}$$

This expansion can be given in the following form:

$$U = \sum_{0 \le k_0 + k_1 + k_2 \le L} u_{\psi_2(k_0,k_1,k_2)} \Phi_{\psi_2(k_0,k_1,k_2)}(\xi_0, \xi_1, \xi_2) \quad \text{(A.14)}$$

Based on the relationship between the 1D and the multi-D polynomial chaos, this expression is given by

$$U = \sum_{0 \le k_0 + k_1 + k_2 \le L} u_{\psi_2(k_0,k_1,k_2)}(t)\, \Phi_{k_0}(\xi_0)\, \Phi_{k_1}(\xi_1) \cdot \Phi_{k_2}(\xi_2) \quad \text{(A.15)}$$

Introducing this expansion in the first random deferential equation, one gets

$$\begin{aligned}&\sum_{0 \le k_0 + k_1 + k_2 \le L} \ddot{u}_{\psi_2(k_0,k_1,k_2)}(t)\, \Phi_{k_0}(\xi_0)\, \Phi_{k_1}(\xi_1)\, \Phi_{k_2}(\xi_2) \\ &+ \sum_{0 \le k_0 + k_1 + k_2 \le L} \dot{u}_{\psi_2(k_0,k_1,k_2)}(t)\, C_1 \Phi_{k_0}(\xi_0)\, \Phi_{k_1}(\xi_1) \\ &\cdot \Phi_{k_2}(\xi_2) + \sum_{0 \le k_0 + k_1 + k_2 \le L} u_{\psi_2(k_0,k_1,k_2)}(t)\, C_2 \Phi_{k_0}(\xi_0) \\ &\cdot \Phi_{k_1}(\xi_1)\, \Phi_{k_2}(\xi_2) = 0.\end{aligned} \quad \text{(A.16)}$$

Projecting this random differential equation with respect to the polynomial chaos of degree less than L, $\Phi_{\psi_2(l_0,l_1,l_2)}(\xi_0, \xi_1, \xi_2)$ where $0 \le l_0 + l_1 + l_2 \le L$, and using the relation between 1D and the multi-D polynomial chaos and the square matrix β^{ih} defed by (26), the following deterministic differential system resulted:

$$\begin{aligned}&\ddot{X}_{\psi_2(l_0,l_1,l_2)}(t)\, \alpha_{l_0}^0 \alpha_{l_1}^1 \alpha_{l_2}^2 + \dot{X}_{\psi_2(l_0,l_1,l_2)}(t)\, C_{00} \alpha_{l_0}^0 \alpha_{l_1}^1 \alpha_{l_2}^2 \\ &+ \dot{X}_{\psi_2(l_0+1,l_1,l_2)}(t)\, C_{01} \beta^{01}(l_0+1, l_0+2)\, \alpha_{l_1}^1 \alpha_{l_2}^2 \\ &+ \dot{X}_{\psi_2(l_0,l_1,l_2)}(t)\, C_{01} \beta^{01}(l_0+1, l_0+1)\, \alpha_{l_1}^1 \alpha_{l_2}^2 \\ &+ \dot{X}_{\psi_2(l_0-1,l_1,l_2)}(t)\, C_{01} \beta^{01}(l_0, l_0+1)\, \alpha_{l_1}^1 \alpha_{l_2}^2 \\ &+ X_{\psi_2(l_0,l_1,l_2)}(t)\, C_{10} C_{20} \alpha_{l_0}^0 \alpha_{l_1}^1 \alpha_{l_2}^2 + X_{\psi_2(l_0,l_1,l_2)}(t) \\ &\cdot C_{11} C_{20} \alpha_{l_0}^0 \beta^{11}(l_1+1, l_1+1)\, \alpha_{l_2}^2 \\ &+ X_{\psi_2(l_0,l_1+1,l_2)}(t)\, C_{11} C_{20} \alpha_{l_0}^0 \beta^{11}(l_1+1, l_1+2)\, \alpha_{l_2}^2 \\ &+ X_{\psi_2(l_0,l_1-1,l_2)}(t)\, C_{11} C_{20} \alpha_{l_0}^0 \beta^{11}(l_1, l_1+1)\, \alpha_{l_2}^2 \\ &+ X_{\psi_2(l_0,l_1,l_2)}(t)\, C_{10} C_{21} \alpha_{l_0}^0 \alpha_{l_1}^1 \beta^{21}(l_2+1, l_2+1) \\ &+ X_{\psi_2(l_0,l_1,l_2+1)}(t)\, C_{10} C_{21} \alpha_{l_0}^0 \alpha_{l_1}^1 \beta^{21}(l_2+1, l_2+2) \\ &+ X_{\psi_2(l_0,l_1,l_2-1)}(t)\, C_{10} C_{21} \alpha_{l_0}^0 \alpha_{l_1}^1 \beta^{21}(l_2, l_2+1) \\ &+ X_{\psi_2(l_0,l_1,l_2)}(t)\, C_{11} C_{21} \alpha_{l_0}^0 \beta^{11}(l_1+1, l_1+1) \\ &\cdot \beta^{21}(l_2+1, l_2+1) + X_{\psi_2(l_0,l_1,l_2+1)}(t) \\ &\cdot C_{11} C_{21} \alpha_{l_0}^0 \beta^{11}(l_1+1, l_1+1)\, \beta^{21}(l_2+1, l_2+2) \\ &+ X_{\psi_2(l_0,l_1,l_2-1)}(t)\, C_{11} C_{21} \alpha_{l_0}^0 \beta^{11}(l_1+1, l_1+1) \\ &\cdot \beta^{21}(l_2, l_2+1) + X_{\psi_2(l_0,l_1-1,l_2)}(t) \\ &\cdot C_{11} C_{21} \alpha_{l_0}^0 \beta^{11}(l_1, l_1+1)\, \beta^{21}(l_2+1, l_2+1) \\ &+ X_{\psi_2(l_0,l_1-1,l_2+1)}(t) \\ &\cdot C_{11} C_{21} \alpha_{l_0}^0 \beta^{11}(l_1, l_1+1)\, \beta^{21}(l_2+1, l_2+2) \\ &+ X_{\psi_2(l_0,l_1-1,l_2-1)}(t)\, C_{11} C_{21} \alpha_{l_0}^0 \beta^{11}(l_1, l_1+1) \\ &\cdot \beta^{21}(l_2, l_2+1) + X_{\psi_2(l_0,l_1+1,l_2)}(t) \\ &\cdot C_{11} C_{21} \alpha_{l_0}^0 \beta^{11}(l_1+1, l_1+2)\, \beta^{21}(l_2+1, l_2+1) \\ &+ X_{\psi_2(l_0,l_1+1,l_2+1)}(t) \\ &\cdot C_{11} C_{21} \alpha_{l_0}^0 \beta^{11}(l_1+1, l_1+2)\, \beta^{21}(l_2+1, l_2+2) \\ &+ X_{\psi_2(l_0,l_1+1,l_2-1)}(t)\, C_{11} C_{21} \alpha_{l_0}^0 \beta^{11}(l_1+1, l_1+2) \\ &\cdot \beta^{21}(l_2, l_2+1) = 0.\end{aligned} \quad \text{(A.17)}$$

In this last expression if some indices are nonpositive then the term vanishes.

In this case that the expansion of the response of response U of the differential equation defined above is considered, this random vector is expanded using the polynomial chaos associated with the random vector defined by the IRCM method; for this, the given random variables are introduced:

$$X_i^1 = \frac{C_i - E(C_i)}{\sigma_i} \quad \text{For } i = 0 \text{ or } 1. \quad \text{(A.18)}$$

Let the canonical basis associated with the polynomial space spanned by these new random variables X_i^1 this canonical be defined by:

$$\begin{aligned}B = \{&1, X_0^1, X_1^1, (X_0^1)^2, X_0^1 X_1^1, (X_1^1)^2, (X_0^1)^3, (X_0^1)^2 \\ &\cdot X_1^1, X_0^1 (X_1^1)^2, (X_1^1)^3, (X_0^1)^4, (X_0^1)^3 X_1^1, (X_0^1)^2 \\ &\cdot (X_1^1)^2, X_0^1 (X_1^1)^3, (X_1^1)^4, \dots \text{etc}\}.\end{aligned} \quad \text{(A.19)}$$

The orthogonalization of this basis gives the polynomial chaos $\{\mathbf{\Psi}_k^1, k \in \mathbb{N}\}$ associated with these new random variables X_i^1; the solution is expanded into this new polynomial chaos and for this the following expression is obtained:

$$U = \sum_{0 \le k \le N} u_k(t) \mathbf{\Psi}_k^1 \left(X_0^1, X_1^1\right), \quad (A.20)$$

where N is choosing that the polynomial chaos must be of degree less than a given integer L. Then N is choosing that

$$N = \psi_1(0, L). \quad (A.21)$$

Introducing this expression in differential equation defined by (A.5), then the following differential equation is given:

$$\sum_{0 \le k \le N} \ddot{u}_k(t) \mathbf{\Psi}_k^1 + \sum_{0 \le k \le N} \dot{u}_k(t) C_1 \mathbf{\Psi}_k^1 + \sum_{0 \le k \le N} u_k(t) C_2 \mathbf{\Psi}_k^1 = 0. \quad (A.22)$$

Using the bijection ψ_1 defined by (21), then deterministic differential system is obtained:

$$\sum_{0 \le k_0 + k_1 \le L} \ddot{u}_{\psi_1(k_0,k_1)}(t) \mathbf{\Psi}^1_{\psi_1(k_0,k_1)} + \sum_{0 \le k_0 + k_1 \le L} \dot{u}_{\psi_1(k_0,k_1)}(t) C_1 \mathbf{\Psi}^1_{\psi_1(k_0,k_1)} + \sum_{0 \le k_0 + k_1 \le L} u_{\psi_1(k_0,k_1)}(t) C_2 \mathbf{\Psi}^1_{\psi_1(k_0,k_1)} = 0. \quad (A.23)$$

Projecting this equation with respect to the polynomial chaos $\mathbf{\Psi}^1_{\psi_1(l_0,l_1)}$ for $0 \le l_0 + l_1 \le L$ and using the orthogonalization properties of the polynomial chaos $\mathbf{\Psi}^1_{\psi_1(l_0,l_1)}$ and the matrix introduced in (67), then the deterministic system resulted:

$$\begin{aligned}
&\ddot{u}_{\psi_1(k_0,k_1)}(t)\, \boldsymbol{\alpha}^1_{\psi_1(k_0,k_1)} + \dot{u}_{\psi_1(k_0,k_1)} \boldsymbol{\alpha}^1_{\psi_1(k_0,k_1)} C_{00} \\
&+ \dot{u}_{\psi_1(k_0,k_1)} \mathbf{T}^0 \left(\psi_1(k_0,k_1) + 1, \psi_1(k_0,k_1) + 1\right) \\
&\cdot C_{01} + \dot{u}_{\psi_1(k_0,k_1)} \mathbf{T}^0 \left(\psi_1(k_0+1,k_1)\right. \\
&\left.+ 1, \psi_1(k_0,k_1) + 1\right) C_{01} \\
&+ \dot{u}_{\psi_1(k_0,k_1)} \mathbf{T}^0 \left(\psi_1(k_0-1,k_1) + 1, \psi_1(k_0,k_1)\right. \\
&\left.+ 1\right) C_{00} + u_{\psi_1(k_0,k_1)}(t)\, E(C_1)\, \boldsymbol{\alpha}^1_{\psi_1(k_0,k_1)} \\
&+ u_{\psi_1(k_0,k_1)} \mathbf{T}^0 \left(\psi_1(k_0,k_1) + 1, \psi_1(k_0,k_1) + 1\right) \sigma_1 \\
&+ u_{\psi_1(k_0,k_1)} \mathbf{T}^0 \left(\psi_1(k_0+1,k_1) + 1, \psi_1(k_0,k_1)\right. \\
&\left.+ 1\right) \sigma_1 + u_{\psi_1(k_0,k_1)} \mathbf{T}^0 \left(\psi_1(k_0-1,k_1)\right. \\
&\left.+ 1, \psi_1(k_0,k_1) + 1\right) \sigma_1 = 0.
\end{aligned} \quad (A.24)$$

Conflicts of Interest

The authors declare that they have no conflicts of interest.

Acknowledgments

This project was funded by the CNRST and the Moroccan Ministry of High Education and Scientific Research with Project PPR2 coordinated by Pr. Azrar and the Deanship of Scientific Research (DSR) at King Abdulaziz University, Jeddah, Saudi Arabia. The authors, therefore, acknowledge thankfully the DSR-KAU and PPR2 financial supports.

References

[1] M. Shinozuka, "Monte Carlo solution of structural dynamics," *Computers & Structures*, vol. 2, no. 5-6, pp. 855–874, 1972.

[2] S. Weinzierl, "Introduction to Monte Carlo methods," Tech. Rep. NIKHEF-00-012, NIKHEF, TheoryGroup, The Netherlands, 2000.

[3] A. Saltelli, K. Chan, and E. M. Scott, *Sensitivity Analysis*, Wiley, 2000.

[4] R. Ghanem and P. Spanos, *Stochastic Finite Elements: A Spectral Approach*, Springer, New York, NY, USA, 1991.

[5] D. Xiu, *Numerical Methods for Stochastic Computations: A Spectral Method Approach*, Princeton University Press, 2010.

[6] N. Wiener, "The homogeneous chaos," *American Journal of Mathematics*, vol. 60, no. 4, pp. 897–936, 1938.

[7] H. G. Matthies, C. E. Brenner, C. G. Bucher, and C. G. Soares, "Uncertainties in probabilistic numerical analysis of structures and solids - Stochastic finite elements," *Structural Safety*, vol. 19, no. 3, pp. 283–336, 1997.

[8] D. Xiu and G. E. Karniadakis, "The wiener-askey polynomial chaos for stochastic differential equations," *SIAM Journal on Scientific Computing*, vol. 24, no. 2, pp. 619–644, 2002.

[9] G. Stefanou, "The stochastic finite element method: past, present and future," *Computer Methods Applied Mechanics and Engineering*, vol. 198, no. 9–12, pp. 1031–1051, 2009.

[10] G. I. Schuëller and H. J. Pradlwarter, "Uncertain linear systems in dynamics: Retrospective and recent developments by stochastic approaches," *Engineering Structures*, vol. 31, no. 11, pp. 2507–2517, 2009.

[11] T. Banek, "Chaos expansion for the solutions of stochastic differential equations," *Systems & Control Letters*, vol. 36, no. 5, pp. 351–358, 1999.

[12] R. Ghanem and P. D. Spanos, "A stochastic Galerkin expansion for nonlinear random vibration analysis," *Probabilistic Engineering Mechanics*, vol. 8, no. 3-4, pp. 255–264, 1993.

[13] R. Li and R. Ghanem, "Adaptive polynomial chaos expansions applied to statistics of extremes in nonlinear random vibration," *Probabilistic Engineering Mechanics*, vol. 13, no. 2, pp. 125–136, 1998.

[14] H. G. Matthies, C. Brenner, C. Bucher, and C. Soares, "Uncertainties in probabilistic numerical analysis of structures and solids - Stochastic finite elements," *Structural Safety*, vol. 19, no. 3, pp. 283–336, 1997.

[15] J. Zhang and B. Ellingwood, "Orthogonal series expansions of random fields in reliability analysis," *Journal of Engineering Mechanics*, vol. 120, no. 12, pp. 2660–2677, 1994.

[16] R. Ghanem, S. Masri, M. Pellissetti, and R. Wolfe, "Identification and prediction of stochastic dynamical systems in a polynomial chaos basis," *Computer Methods Applied Mechanics and Engineering*, vol. 194, no. 12–16, pp. 1641–1654, 2005.

[17] C. Soize, "Identification of high-dimension polynomial chaos expansions with random coefficients for non-Gaussian tensor-valued random fields using partial and limited experimental data," *Computer Methods Applied Mechanics and Engineering*, vol. 199, no. 33-36, pp. 2150–2164, 2010.

[18] M. Trcala, "Spectral stochastic modeling of uncertainties in nonlinear diffusion problems of moisture transfer in wood," *Applied Mathematical Modelling*, vol. 39, no. 5-6, pp. 1740–1748, 2015.

[19] R. V. Field Jr. and M. Grigoriu, "On the accuracy of the polynomial chaos approximation," *Probabilistic Engineering Mechanics*, vol. 19, no. 1, pp. 65–80, 2004.

[20] A. Dvurecenskij, P. Lahti, and K. Ylinen, *The Uniqueness Question in The Multididimensional Moment Problem with Applications to Phase Space Observables*, Bratislava preprint series, Mathematical Institute Slovak Academy of Sciences, 2001.

[21] O. G. Ernst, A. Mugler, H.-J. Starkloff, and E. Ullmann, "On the convergence of generalized polynomial chaos expansions," *ESAIM: Mathematical Modelling and Numerical Analysis*, vol. 46, no. 2, pp. 317–339, 2012.

[22] D. Sarsri, L. Azrar, A. Jebbouri, and A. El Hami, "Component mode synthesis and polynomial chaos expansions for stochastic frequency functions of large linear FE models," *Computers & Structures*, vol. 89, no. 3-4, pp. 346–356, 2011.

[23] D. Sarsri and L. Azrar, "Time response of structures with uncertain parameters under stochastic inputs based on coupled polynomial chaos expansion and component mode synthesis methods," *Mechanics of Advanced Materials and Structures*, vol. 23, no. 5, pp. 596–606, 2016.

[24] D. Sarsri and L. Azrar, "Dynamic analysis of large structures with uncertain parameters based on coupling component mode synthesis and perturbation method," *Ain Shams Engineering Journal*, vol. 7, no. 1, pp. 371–381, 2016.

[25] R. G. Jaimez and M. J. V. Bonnet, "On the Karhunen-Loeve expansion for transformed processes," *Trabajos de Estadistica*, vol. 2, no. 2, pp. 81–90, 1987.

6

Numerical Solution to Coupled Burgers' Equations by Gaussian-Based Hermite Collocation Scheme

Nissaya Chuathong[1] and Sayan Kaennakham[2,3]

[1] *Faculty of Science Environment and Energy, King Mongkut's University of Technology North Bangkok (Rayong Campus), Rayong 21120, Thailand*
[2] *School of Mathematics, Institute of Science, Suranaree University of Technology, Nakhon Ratchasima 30000, Thailand*
[3] *Centre of Excellence in Mathematics, Bangkok 10400, Thailand*

Correspondence should be addressed to Sayan Kaennakham; sayan_kk@g.sut.ac.th

Academic Editor: Mustafa Inc

One of the most challenging PDE forms in fluid dynamics namely Burgers equations is solved numerically in this work. Its transient, nonlinear, and coupling structure are carefully treated. The Hermite type of collocation mesh-free method is applied to the spatial terms and the 4th-order Runge Kutta is adopted to discretize the governing equations in time. The method is applied in conjunction with the Gaussian radial basis function. The effect of viscous force at high Reynolds number up to 1,300 is investigated using the method. For the purpose of validation, a conventional global collocation scheme (also known as "Kansa" method) is applied parallelly. Solutions obtained are validated against the exact solution and also with some other numerical works available in literature when possible.

1. Introduction

Fluid is known to dominate most part of the planet and can be modelled mathematically by the well-known Navier-Stoke equation. To understand its behavior under a certain set of conditions, one seeks solutions (if any or if possible) of the equation and this is not an easy task. The combination of challenging factors, nonlinearity, coupling, and transient nature, of the equation makes it one of the most complex systems both analytically and numerically.

Acknowledged as a simple case of the Navier-Stoke equation, the famous "Burgers" equations, a system of equations that describes the interaction between two crucial physical manners of nature, convection, and diffusion, are used to model variety of applications. This includes flows through a shock wave traveling in viscous fluid, the phenomena of turbulence, sedimentation of two kinds of particles in fluid suspensions under the effect of gravity, and also dispersion of pollutants in rivers (see [1–3]).

The standard form of this type of equation is expressed as

$$\frac{\partial u}{\partial t} + u\frac{\partial u}{\partial x} + v\frac{\partial u}{\partial y} = \frac{1}{\mathrm{Re}}\left(\frac{\partial^2 u}{\partial x^2} + \frac{\partial^2 u}{\partial y^2}\right) \tag{1}$$

$$\text{and } \frac{\partial v}{\partial t} + u\frac{\partial v}{\partial x} + v\frac{\partial v}{\partial y} = \frac{1}{\mathrm{Re}}\left(\frac{\partial^2 v}{\partial x^2} + \frac{\partial^2 v}{\partial y^2}\right) \tag{2}$$

The ratio of inertial forces to viscous forces is represented by the Reynolds number, Re, and the x- and y-velocity components are noted by $u(x, y, t)$ and $v(x, y, t)$, respectively.

For a domain $\Omega = \{(x, y) : a \le x, y \le b\}$ with its boundary Γ, the above system is subject to the initial conditions

$$u(x, y, 0) = \beta_1(x, y), \quad (x, y) \in \Omega \tag{3}$$

$$v(x, y, 0) = \beta_2(x, y), \quad (x, y) \in \Omega \tag{4}$$

and the boundary conditions

$$u(x,y,t) = \gamma_1(x,y,t), \quad x, y \in \Gamma, \ t > 0, \tag{5}$$

$$v(x,y,t) = \gamma_2(x,y,t), \quad x, y \in \Gamma, \ t > 0, \tag{6}$$

Here $\beta_1, \beta_2, \gamma_1$, and γ_2 are known functions.

In 1983, Fletcher [4] applied the Hopf-Cole transformation and successfully solved the equation analytically. Ever since, the coupled two-dimensional Burgers' Equations have been attracting a number of investigations both numerically and analytically. Some of the recent computational researches include the application of a fully implicit finite-difference (see [5]), by Adomain-Pade technique (see [6]), by adopting variational iteration method (see [7]), and by applying a mesh-free technique (see [8, 9]). Even more recently, more numerical works have been successfully carried out and nicely documented in literature: the application of the dual reciprocity boundary element method (see [10, 11]), POD/discrete empirical interpolation method- (DEIM-) reduced-order model (ROM) (see [12]), and a combination between the homotopy analysis method and finite differences (see [13]).

Amongst the well-known numerical scheme, finite volume, finite difference, and finite element method that have been invented, developed, and applied in a wide range of science and engineering problems, a rather young idea was discovered and has been categorized as "meshless/mesh-free" methods. The methods under this category have recently become promising alternative tools for numerically solving variety of science and engineering problems. Generally, these meshless schemes can be grouped into two main classes:

(i) Strong forms: the finite point method (see [14]), The hp-meshless cloud method (see [15]), the collocation method (see [16]), and references therein.

(ii) Weak forms: the diffuse element method (see [17]), the element-free Galerkin method (EFGM; see [18]), the point interpolation method (see [19]), and references therein.

This depends on the requirement for integration. In this work, particularly, the main attention is paid to the development of the so-called "Collocation Method" characterized in the strong form family. The method is truly mesh-free meaning that no mesh is at all required at any point of the whole computing process. Amongst several versions of its further improvements, this work focuses on an application of the Hermite form of the conventional collocation where the involved differential operator is applied twice, explained in more detail in Section 2.

The paper is organized as follows. Section 2 provides the necessary mathematical background concerning the Gaussian radial basis function and also the fundamental idea of collocation mesh-free schemes in general. The implementation of the schemes to the couple-Burgers equations is then detailed in Section 3. Section 4 demonstrates the numerical experiments and discusses general aspect of the application before some interesting findings are listed in Section 5.

2. The Gaussian-Based Collocation Meshless Method

Letting $\mathbf{x}_c$ be a point in a d-dimensional Euclidean space $\mathbb{R}^d$, for any point $\mathbf{x} \in \mathbb{R}^d$ that depends only on the distance, r, from the fixed $\mathbf{x}_c$, then the function $\varphi : \mathbb{R}^d \longrightarrow \mathbb{R}$ that is written as

$$\varphi(r,\varepsilon) = \varphi(\|\mathbf{x} - \mathbf{x}_c\|) \tag{7}$$

is called a Radial Basis Function (RBF), with $\mathbf{x}_c$ being called the center and some nonnegative parameter ε.

Among different types of RBFs invented, improved, and applied nowadays, in this work, one of the most widely used forms of radial functions, known as "Gaussian," is focused on. The standard form of this type of RBF is expressed as

$$\varphi(r) = \exp\left(-(\varepsilon r)^2\right) \tag{8}$$

where $r \in \mathbb{R}$ and ε is called "shape parameter," that is related to the variance of the normal distribution function by $\varepsilon^2 = 1/(2\delta^2)$. Under the context of collocation schemes, this relation is linked to the Euclidean distance function, $\|.\|_2$, via the following multivariate function:

$$\Xi_k(\mathbf{x}) = e^{-\chi} \quad \text{with } \chi = \varepsilon^2 \left(\|\mathbf{x} - \mathbf{x}_k\|_2\right)^2 \tag{9}$$

for a fixed center $\mathbf{x} \in \mathbb{R}^d$ and $\mathbf{x}_k \in \mathbb{R}^d$. It can then be seen that the connection between and is as follows:

$$\Xi_k(\mathbf{x}) = \varphi\left(\|\mathbf{x} - \mathbf{x}_k\|_2\right) \tag{10}$$

The collocation scheme starts with considering the following elliptical partial differential equation defined on a bounded and connected domain Ω:

$$\Phi[u(\mathbf{x})] = f(x) \quad \text{for } \mathbf{x} \in \Omega \subset \mathbb{R}^n \tag{11}$$

$$B_1 u(\mathbf{x}) = g(x) \quad \text{for } \mathbf{x} \in \Gamma_1 \tag{12}$$

$$B_1 u(\mathbf{x}) = h(x) \quad \text{for } \mathbf{x} \in \Gamma_1 \tag{13}$$

where $\partial\Omega$ is the domain boundary containing two nonoverlap sections: Γ_1 and Γ_2, with $\Gamma_1 \cap \Gamma_2 = \phi$. These differential operators, Φ and B_1, B_2 are applied on the domain and the two boundary sections, respectively. Three known functions $f(\mathbf{x}), g(\mathbf{x})$, and $h(\mathbf{x})$ can well be dependent of space and/or time. Let $\mathbf{X}^c = \{\mathbf{x}_j\}_{j=1}^N$ be a set of randomly selected points, known as "collocation" or "centers", on the domain, where $\{\mathbf{x}_j\}_{j=1}^{N_i}$ are those contained within, where $\{\mathbf{x}_j\}_{j=N_i+1}^{N_i+N_1}$ and $\{\mathbf{x}_j\}_{j=N_i+N_1+1}^{N}$ are those on the boundaries Γ_1 and Γ_2, respectively.

2.1. The Conventional Kansa Collocation Method. Initially, a collocation scheme writes the approximate solution, $\tilde{u}(\mathbf{x})$, as a linear combination of the basis function $\{\varphi(: -)\}_j^N$, shown in the following form:

$$u(\mathbf{x}) \simeq \tilde{u}(\mathbf{x}) = \sum_{j=1}^{N} \alpha_j \varphi\left(\|\mathbf{x} - \mathbf{x}_j\|_2\right) \tag{14}$$

where α_j are coefficients and $\|.\|_2$ is the Euclidean norm. The basis function used now is the radial type as defined previously. Applying Φ and B_1, B_2 to both domain and boundary sections, satisfying the governing system of equations, allows the system to arrive at

$$\mathbf{A}\boldsymbol{\alpha} = \mathbf{b} \tag{15}$$

where $\boldsymbol{\alpha} = [\alpha_1 \ \alpha_2 \ \ldots \ \alpha_N]$, and the known $\mathbf{b}$ vector expressed is as follows:

$$\mathbf{b} = \left[f(\mathbf{x}_1) \ f(\mathbf{x}_2) \ \cdots \ f(\mathbf{x}_{N_i}) \ g(\mathbf{x}_{N_i+1}) \ \cdots \ g(\mathbf{x}_{N_i+N_1}) \ h(\mathbf{x}_{N_i+N_1+1}) \ \cdots \ h(\mathbf{x}_N) \right]^T \tag{16}$$

and by setting $\boldsymbol{\varphi}$ to be a matrix with entries $\varphi_{ij} = \varphi(\|\mathbf{x}_i - \mathbf{x}_j\|_2)$ for $i, j = 1, 2, \ldots, N$ we have

$$\mathbf{A} = \begin{bmatrix} \Phi[\boldsymbol{\varphi}] \\ B_1[\boldsymbol{\varphi}] \\ B_2[\boldsymbol{\varphi}] \end{bmatrix} \tag{17}$$

Once α_j are obtained by (15), the approximate solution is straightforward yielded. This method is known as "Kansa," in honor of a great mathematician E. J. Kansa [22] who discovered the idea in 1990. The method has been applied to a wide range of problems ever since [23, 24]. It, nevertheless, has shortcomings as it is known to suffer the problem of unsymmetric interpolation matrix, $\mathbf{A}$, and the rigorous mathematical proof of its solvability is still not available [25], and very often produces low quality results particularly in boundary-adjacent region [26, 27].

2.2. The Hermite Collocation Method. In order to alleviate the difficulty mentioned earlier, in 1997, Fasshauer [28] proposed a new way of interpolating by applying the self-adjoint operators Φ^* and B_1^*, B_2^* to the governing system of equations and rewriting the approximate solution (14) as

$$\begin{aligned} u(\mathbf{x}) &\simeq \tilde{u}(\mathbf{x}) \\ &= \sum_{j=1}^{N_i} \alpha_j \Phi^* \varphi\left(\|\mathbf{x} - \mathbf{x}_j\|_2\right) \\ &\quad + \sum_{j=N_i+1}^{N_i+N_1} \alpha_j B_1^* \varphi\left(\|\mathbf{x} - \mathbf{x}_j\|_2\right) \\ &\quad + \sum_{j=N_i+N_1+1}^{N} \alpha_j B_2^* \varphi\left(\|\mathbf{x} - \mathbf{x}_j\|_2\right) \end{aligned} \tag{18}$$

This leads to a new interpolation matrix $\mathbf{A}$, shown as follows:

$$\mathbf{A} = \begin{bmatrix} \Phi\Phi^*[\boldsymbol{\varphi}] & \Phi B_1^*[\boldsymbol{\varphi}] & \Phi B_2^*[\boldsymbol{\varphi}] \\ B_1\Phi^*[\boldsymbol{\varphi}] & B_1 B_1^*[\boldsymbol{\varphi}] & B_1 B_2^*[\boldsymbol{\varphi}] \\ B_2\Phi^*[\boldsymbol{\varphi}] & B_2 B_1^*[\boldsymbol{\varphi}] & B_2 B_2^*[\boldsymbol{\varphi}] \end{bmatrix} \tag{19}$$

In this study, nevertheless, the problem domain has only one continuous boundary with no differential operator. An application of this Hermite type of collocation method to elastostatic problem was successfully investigated by Leitao [29]. Additionally, some interesting implementations of the scheme to the transient and nonlinear plate problems can be found in [30–32].

3. Implementation to Burgers' Equations

As previously mentioned, the coupled Burgers' equations can be seen as a simple model that mimics the coupling of fluid flow to thermal dynamics, as well as some other scientific fields. High and low Reynolds numbers play important roles in both theoretically modeling and numerical simulation. In this section, the Hermite collocation is now applied based on the Gaussian type of radial basis function.

Since the principle implementation of the conventional collocation method to nonlinear and time dependent problem has already been documented in some of our previous studies [33, 34], this section is therefore dedicated to the application of the Hermite version of collocation as explained in details in the following subsections.

3.1. Space Discretization with the Hermite Scheme. With Hermite interpolation technique, it begins with writing the approximation of solution for $\tilde{u}(\mathbf{x})$ and $\tilde{v}(\mathbf{x})$, respectively, with the same radial basis function $\varphi(\|\mathbf{x} - \mathbf{x}_j\|_2)$, as follows:

$$\begin{aligned} u(\mathbf{x}, t) &\simeq \tilde{u}(\mathbf{x}, t) \\ &= \sum_{j=1}^{N_i} \alpha_j^t \Phi^* \varphi\left(\|\mathbf{x} - \mathbf{x}_j\|_2\right) \\ &\quad + \sum_{j=N_i+1}^{N} \alpha_j^t B^* \varphi\left(\|\mathbf{x} - \mathbf{x}_j\|_2\right) \end{aligned} \tag{20}$$

and

$$\begin{aligned} v(\mathbf{x}, t) &\simeq \tilde{v}(\mathbf{x}, t) \\ &= \sum_{j=1}^{N_i} \beta_j^t \Phi^* \varphi\left(\|\mathbf{x} - \mathbf{x}_j\|_2\right) \\ &\quad + \sum_{j=N_i+1}^{N} \beta_j^t B^* \varphi\left(\|\mathbf{x} - \mathbf{x}_j\|_2\right) \end{aligned} \tag{21}$$

Note that it is now assumed that there is only one section of boundary $\Gamma_1 = \Gamma_2 = \Gamma$. Therefore, at n^{th}– derivative, this leads to

$$\frac{\partial^{(n)} u(\mathbf{x},t)}{\partial x^{(n)}} \simeq \sum_{j=1}^{N_i} \alpha_j^t \frac{\partial^{(n)}\left[\Phi^* \varphi\left(\left\|\mathbf{x}-\mathbf{x}_j\right\|_2\right)\right]}{\partial x^{(n)}} + \sum_{j=N_i+1}^{N} \alpha_j^t \frac{\partial^{(n)}\left[B^* \varphi\left(\left\|\mathbf{x}-\mathbf{x}_j\right\|_2\right)\right]}{\partial x^{(n)}} \tag{22}$$

and

$$\frac{\partial^{(n)} v(\mathbf{x},t)}{\partial y^{(n)}} \simeq \sum_{j=1}^{N_i} \beta_j^t \frac{\partial^{(n)}\left[\Phi^* \varphi\left(\left\|\mathbf{x}-\mathbf{x}_j\right\|_2\right)\right]}{\partial y^{(n)}} + \sum_{j=N_i+1}^{N} \beta_j^t \frac{\partial^{(n)}\left[B^* \varphi\left(\left\|\mathbf{x}-\mathbf{x}_j\right\|_2\right)\right]}{\partial y^{(n)}} \tag{23}$$

Therefore, the governing equations are now rewritten, at a ith–center node, as

$$\begin{aligned}\left[\frac{\partial u}{\partial t}\right]_i &= \frac{1}{\mathrm{Re}}\left[\left(\sum_{j=1}^{N_i}\alpha_j\Phi^*\left(H_{ij}^{xx}\right)+\sum_{j=N_i+1}^{N}\alpha_jB^*\left(H_{ij}^{xx}\right)\right)\right. \\ &\quad \left.+\left(\sum_{j=1}^{N_i}\alpha_j\Phi^*\left(H_{ij}^{yy}\right)+\sum_{j=N_i+1}^{N}\alpha_jB^*\left(H_{ij}^{yy}\right)\right)\right] \\ &\quad - u_i\left(\sum_{j=1}^{N_i}\alpha_j\Phi^*\left(H_{ij}^{x}\right)+\sum_{j=N_i+1}^{N}\alpha_jB^*\left(H_{ij}^{x}\right)\right) \\ &\quad - v_i\left(\sum_{j=1}^{N_i}\alpha_j\Phi^*\left(H_{ij}^{y}\right)+\sum_{j=N_i+1}^{N}\alpha_jB^*\left(H_{ij}^{y}\right)\right)\end{aligned} \tag{24}$$

$$\begin{aligned}\left[\frac{\partial v}{\partial t}\right]_i &= \frac{1}{\mathrm{Re}}\left[\left(\sum_{j=1}^{N_i}\beta_j\Phi^*\left(H_{ij}^{xx}\right)+\sum_{j=N_i+1}^{N}\beta_jB^*\left(H_{ij}^{xx}\right)\right)\right. \\ &\quad \left.+\left(\sum_{j=1}^{N_i}\beta_j\Phi^*\left(H_{ij}^{yy}\right)+\sum_{j=N_i+1}^{N}\beta_jB^*\left(H_{ij}^{yy}\right)\right)\right] \\ &\quad - u_i\left(\sum_{j=1}^{N_i}\beta_j\Phi^*\left(H_{ij}^{x}\right)+\sum_{j=N_i+1}^{N}\beta_jB^*\left(H_{ij}^{x}\right)\right) \\ &\quad - v_i\left(\sum_{j=1}^{N_i}\beta_j\Phi^*\left(H_{ij}^{y}\right)+\sum_{j=N_i+1}^{N}\beta_jB^*\left(H_{ij}^{y}\right)\right)\end{aligned} \tag{25}$$

where

$$\begin{aligned}H_{ij}^{xx} &= \frac{\partial^2\varphi\left(\left\|\mathbf{x}_i-\mathbf{x}_j\right\|\right)}{\partial x^2}, \\ H_{ij}^{yy} &= \frac{\partial^2\varphi\left(\left\|\mathbf{x}_i-\mathbf{x}_j\right\|\right)}{\partial y^2}, \\ H_{ij}^{x} &= \frac{\partial\varphi\left(\left\|\mathbf{x}_i-\mathbf{x}_j\right\|\right)}{\partial x}, \\ \text{and } H_{ij}^{y} &= \frac{\partial\varphi\left(\left\|\mathbf{x}_i-\mathbf{x}_j\right\|\right)}{\partial y}\end{aligned} \tag{26}$$

leading their matrix forms expresses as below:

$$\begin{bmatrix} u_1^t \\ u_2^t \\ \vdots \\ u_N^t \end{bmatrix} = \left[\begin{bmatrix} \psi_{11} & \psi_{12} & \cdots & \psi_{1N} \\ \psi_{21} & \psi_{22} & \cdots & \psi_{1N} \\ \vdots & \vdots & \ddots & \vdots \\ \psi_{N1} & \psi_{N2} & \cdots & \psi_{NN} \end{bmatrix} - \begin{bmatrix} u_1 & 0 & \cdots & 0 \\ 0 & u_2 & \cdots & 0 \\ \vdots & \vdots & \ddots & \vdots \\ 0 & 0 & \cdots & u_N \end{bmatrix}\begin{bmatrix} \zeta_{11} & \zeta_{12} & \cdots & \zeta_{1N} \\ \zeta_{21} & {}_{22} & \cdots & \zeta_{2N} \\ \vdots & \vdots & \ddots & \vdots \\ \zeta_{N1} & \zeta_{N2} & \cdots & \zeta_{NN} \end{bmatrix} - \begin{bmatrix} v_1 & 0 & \cdots & 0 \\ 0 & v_2 & \cdots & 0 \\ \vdots & \vdots & \ddots & \vdots \\ 0 & 0 & \cdots & v_N \end{bmatrix}\begin{bmatrix} \vartheta_{11} & \vartheta_{12} & \cdots & \vartheta_{1N} \\ \vartheta_{21} & \vartheta_{22} & \cdots & \vartheta_{2N} \\ \vdots & \vdots & \ddots & \vdots \\ \vartheta_{N1} & \vartheta_{N2} & \cdots & \vartheta_{NN}^{y} \end{bmatrix}\right] \cdot \begin{bmatrix} \alpha_1 \\ \alpha_2 \\ \vdots \\ \alpha_N \end{bmatrix} \tag{27}$$

Similarly,

$$\begin{bmatrix} v_1^t \\ v_2^t \\ \vdots \\ v_N^t \end{bmatrix} = \left[\begin{bmatrix} \psi_{11} & \psi_{12} & \cdots & \psi_{1N} \\ \psi_{21} & \psi_{22} & \cdots & \psi_{1N} \\ \vdots & \vdots & \ddots & \vdots \\ \psi_{N1} & \psi_{N2} & \cdots & \psi_{NN} \end{bmatrix} - \begin{bmatrix} u_1 & 0 & \cdots & 0 \\ 0 & u_2 & \cdots & 0 \\ \vdots & \vdots & \ddots & \vdots \\ 0 & 0 & \cdots & u_N \end{bmatrix}\begin{bmatrix} \zeta_{11} & \zeta_{12} & \cdots & \zeta_{1N} \\ \zeta_{21} & {}_{22} & \cdots & \zeta_{2N} \\ \vdots & \vdots & \ddots & \vdots \\ \zeta_{N1} & \zeta_{N2} & \cdots & \zeta_{NN} \end{bmatrix}\right.$$

$$-\begin{bmatrix} v_1 & 0 & \cdots & 0 \\ 0 & v_2 & \cdots & 0 \\ \vdots & \vdots & \ddots & \vdots \\ 0 & 0 & \cdots & v_N \end{bmatrix}\begin{bmatrix} \vartheta_{11} & \vartheta_{12} & \cdots & \vartheta_{1N} \\ \vartheta_{21} & \vartheta_{22} & \cdots & \vartheta_{2N} \\ \vdots & \vdots & \ddots & \vdots \\ \vartheta_{N1} & \vartheta_{N2} & \cdots & \vartheta^{y}_{NN} \end{bmatrix}\Bigg] \cdot \begin{bmatrix} \beta_1 \\ \beta_2 \\ \vdots \\ \beta_N \end{bmatrix} \tag{28}$$

where

$$\psi_{ij} = \begin{cases} \frac{1}{\mathrm{Re}}\left(\Phi^* H_{ij}^{xx} + \Phi^* H_{ij}^{yy}\right); & 1 \le j \le N_i \\ \frac{1}{\mathrm{Re}}\left(B^* H_{ij}^{xx} + B^* H_{ij}^{yy}\right); & N_i + 1 \le j \le N, \end{cases}$$

$$\text{and } \zeta_{ij} = \begin{cases} \left(\Phi^* H_{ij}^{x} + B^* H_{ij}^{x}\right); & 1 \le j \le N_i \\ \left(\Phi^* H_{ij}^{x} + B^* H_{ij}^{x}\right); & N_i + 1 \le j \le N, \end{cases} \tag{29}$$

$$\text{and } \vartheta_{ij} = \begin{cases} \left(\Phi^* H_{ij}^{y} + B^* H_{ij}^{y}\right); & 1 \le j \le N_i \\ \left(\Phi^* H_{ij}^{y} + B^* H_{ij}^{y}\right); & N_i + 1 \le j \le N \end{cases}$$

Consequently, the systematics of matrices involved in this coupling process is

$$\mathbf{U^t} = \left(\boldsymbol{\psi} - \operatorname{diag}(\mathbf{U})\boldsymbol{\zeta} - \operatorname{diag}(\mathbf{V})\vartheta\right)\boldsymbol{\alpha} = \widetilde{\mathbf{A}}\boldsymbol{\alpha} \tag{30}$$

$$\mathbf{V^t} = \left(\boldsymbol{\psi} - \operatorname{diag}(\mathbf{U})\boldsymbol{\zeta} - \operatorname{diag}(\mathbf{V})\vartheta\right)\boldsymbol{\beta} = \widetilde{\mathbf{B}}\boldsymbol{\beta} \tag{31}$$

3.2. Time Discretization. In order to carry out the time-dependent terms, the 4th-order Runge Kutta is applied to both terms appearing in the above equations, that is, by setting

$$\tilde{\mathbf{u}} = \left[\tilde{u}(\mathbf{x}_1), \tilde{u}(\mathbf{x}_2), \ldots, \tilde{u}(\mathbf{x}_N)\right]^T$$
$$\text{and } \tilde{\mathbf{v}} = \left[\tilde{v}(\mathbf{x}_1), \tilde{v}(\mathbf{x}_2), \ldots, \tilde{v}(\mathbf{x}_N)\right]^T \tag{32}$$

That is,

$$\frac{d\tilde{\mathbf{u}}}{dt} = \mathbf{U^t} = \widetilde{\mathbf{A}}\boldsymbol{\alpha} = F(\tilde{\mathbf{u}}) \tag{33}$$

$$\frac{d\tilde{\mathbf{v}}}{dt} = \mathbf{V^t} = \widetilde{\mathbf{B}}\boldsymbol{\beta} = G(\tilde{\mathbf{v}}) \tag{34}$$

The time progressing of variables involved can be estimated as follows:

$$u^{t+1} = u^t + \frac{1}{6}\left(k_1 + 2k_2 + 2k_3 + k_4\right) \tag{35}$$

where

$$\begin{aligned} k_1 &= \Delta t F\left(u^i, t^i\right), \\ k_2 &= \Delta t F\left(u^i + \frac{k_1}{2}, t^i + \frac{\Delta t}{2}\right), \\ k_3 &= \Delta t F\left(u^i + \frac{k_2}{2}, t^i + \frac{\Delta t}{2}\right), \\ k_4 &= \Delta t F\left(u^i + k_3, t^i + \Delta t\right) \end{aligned} \tag{36}$$

where the similar manner is applied for the other case, $d\tilde{\mathbf{v}}/dt$, and the systems are treated and advanced in time with an ODE method. Then the approximate solution obtained by this final process is expressed in terms of the radial basis function at N collocation nodes as follows:

$$\tilde{u}(\mathbf{x}) = \sum_{j=1}^{N_i} \alpha_j \Phi^* \varphi\left(\left\|\mathbf{x} - \mathbf{x}_j\right\|_2\right) + \sum_{j=N_i+1}^{N_i+N} \alpha_j B^* \varphi\left(\left\|\mathbf{x} - \mathbf{x}_j\right\|_2\right) \tag{37}$$

$$\text{and } \tilde{v}(\mathbf{x}) = \sum_{j=1}^{N_i} \beta_j \Phi^* \varphi\left(\left\|\mathbf{x} - \mathbf{x}_j\right\|_2\right) + \sum_{j=N_i+1}^{N_i+N} \beta_j B^* \varphi\left(\left\|\mathbf{x} - \mathbf{x}_j\right\|_2\right) \tag{38}$$

3.3. Gaussian-Based Hermite Collocation. Letting $\varphi(r)$ be a radial basis function depending on a distance function, $r_{ij} = \|\mathbf{x}_i - \mathbf{x}_j\|_2$, where $\mathbf{x}_i, \mathbf{x}_j \in \mathbb{R}^2$, it can be seen that

$$\begin{aligned} \frac{\partial \varphi}{\partial x} &= \frac{\left(x_i - x_j\right)}{r}\frac{d\varphi}{dr}, \\ \frac{\partial \varphi}{\partial y} &= \frac{\left(y_i - y_j\right)}{r}\frac{d\varphi}{dr}, \\ \frac{\partial^2 \varphi}{\partial x^2} &= \frac{\left(x_i - x_j\right)^2}{r^2}\frac{d^2\varphi}{dr^2} + \frac{\left(y_i - y_j\right)^3}{r^3}\frac{d\varphi}{dr}, \\ \text{and } \frac{\partial^2 \varphi}{\partial y^2} &= \frac{\left(y_i - y_j\right)^2}{r^2}\frac{d^2\varphi}{dr^2} + \frac{\left(x_i - x_j\right)^3}{r^3}\frac{d\varphi}{dr}, \end{aligned} \tag{39}$$

This leads to the Laplacian expressed as follows:

$$\left(\frac{\partial^2}{\partial x^2} + \frac{\partial^2}{\partial y^2}\right)\varphi(r) = \frac{d^2}{dr^2}\varphi(r) + \frac{1}{r}\frac{d}{dr}\varphi(r) \tag{40}$$

Furthermore, when the Hermite concept of collocation is employed, it is necessary that the Lapacian be applied twice resulting in the following so-called fourth-order biharmonic form;

$$\left(\frac{\partial^4}{\partial x^4}+2\frac{\partial^4}{\partial x^2\partial y^2}+\frac{\partial^4}{\partial y^4}\right)\varphi(r) = \frac{d^4}{dr^4}\varphi(r)+\frac{2}{r}\frac{d^3}{dr^3}\varphi(r)-\frac{1}{r^2}\frac{d^2}{dr^2}\varphi(r) + \frac{1}{r^3}\frac{d}{dr}\varphi(r) \tag{41}$$

For this work, Gaussian type of radial basis function, is used and its first four orders of derivatives can, be respectively, expressed as

$$\varphi\left(\left\|\mathbf{x}-\mathbf{x}_j\right\|_2\right)=\exp\left(-(\varepsilon r)^2\right) = \exp\left(-\left(\varepsilon\left\|\mathbf{x}-\mathbf{x}_j\right\|_2\right)^2\right) \tag{42}$$

$$\text{Thus, } \frac{d}{dr}\varphi(r)=-2r\varepsilon^2\exp\left(-(\varepsilon r)^2\right), \tag{43}$$

$$\frac{d^2}{dr^2}\varphi(r)=2\varepsilon^2\exp\left(-(\varepsilon r)^2\right)\left[2(\varepsilon r)^2-1\right], \tag{44}$$

$$\frac{d^3}{dr^3}\varphi(r)=2\varepsilon^2\exp\left(-\varepsilon^2r^2\right)\left[6\varepsilon^2r-4\varepsilon^4r^3\right], \tag{45}$$

$$\text{and } \frac{d^4}{dr^4}\varphi(r) = 2\varepsilon^2\exp\left(-\varepsilon^2r^2\right)\left[8\varepsilon^6r^4-24\varepsilon^4r^2+6\varepsilon^2\right] \tag{46}$$

for each pair of center nodes: $\mathbf{x}_i,\mathbf{x}_j$.

What follows is the brief demonstration of how the self-adjoint operator works.

Hermite starts with the following solution approximation:

$$u(\mathbf{x},t)\simeq\tilde{u}(\mathbf{x},t) = \sum_{j=1}^{N_i}\alpha_j^t\Phi^*\varphi\left(\left\|\mathbf{x}-\mathbf{x}_j\right\|_2\right) + \sum_{j=N_i+1}^{N}\alpha_j^tB^*\varphi\left(\left\|\mathbf{x}-\mathbf{x}_j\right\|_2\right) \tag{47}$$

From (24),(25), (27), and (28), when considering for $1\le j\le N_i$, we have $\psi_{ij}=(1/\mathrm{Re})(\Phi^*H_{ij}^{xx}+\Phi^*H_{ij}^{yy})$. That is,

$$\begin{aligned}\left(\frac{\partial^2}{\partial x^2}+\frac{\partial^2}{\partial y^2}\right)u(\mathbf{x},t)&\simeq\left(\frac{\partial^2}{\partial x^2}+\frac{\partial^2}{\partial y^2}\right)\tilde{u}(\mathbf{x},t)\\ &=\left(\frac{\partial^2}{\partial x^2}+\frac{\partial^2}{\partial y^2}\right)\left(\sum_{j=1}^{N_i}\alpha_j\Phi^*\varphi\left(\left\|\mathbf{x}-\mathbf{x}_j\right\|_2\right) + \sum_{j=N_i+1}^{N}\alpha_jB^*\varphi\left(\left\|\mathbf{x}-\mathbf{x}_j\right\|_2\right)\right)\\ &=\sum_{j=1}^{N_i}\alpha_j\left(\frac{\partial^2}{\partial x^2}+\frac{\partial^2}{\partial y^2}\right)\Phi^*\varphi\left(\left\|\mathbf{x}-\mathbf{x}_j\right\|_2\right) + \sum_{j=N_i+1}^{N}\alpha_j\left(\frac{\partial^2}{\partial x^2}+\frac{\partial^2}{\partial y^2}\right)B^*\varphi\left(\left\|\mathbf{x}-\mathbf{x}_j\right\|_2\right)\\ &=\sum_{j=1}^{N_i}\alpha_j\Phi^*\left(\frac{\partial^2}{\partial x^2}+\frac{\partial^2}{\partial y^2}\right)\varphi\left(\left\|\mathbf{x}-\mathbf{x}_j\right\|_2\right) + \sum_{j=N_i+1}^{N}\alpha_jB^*\left(\frac{\partial^2}{\partial x^2}+\frac{\partial^2}{\partial y^2}\right)\varphi\left(\left\|\mathbf{x}-\mathbf{x}_j\right\|_2\right)\end{aligned} \tag{48}$$

For this particular example we have $\Phi^*=(\partial^2/\partial x^2+\partial^2/\partial y^2)$ and there is no derivative operation taking place on the boundary. Therefore, the above equation becomes

$$\begin{aligned}\left(\frac{\partial^2}{\partial x^2}+\frac{\partial^2}{\partial y^2}\right)\tilde{u}(\mathbf{x},t)&=\sum_{j=1}^{N_i}\alpha_j\left(\frac{\partial^2}{\partial x^2}+\frac{\partial^2}{\partial y^2}\right)\left(\frac{\partial^2}{\partial x^2}+\frac{\partial^2}{\partial y^2}\right)\varphi\left(\left\|\mathbf{x}-\mathbf{x}_j\right\|_2\right)+\sum_{j=N_i+1}^{N}\alpha_j\left(\frac{\partial^2}{\partial x^2}+\frac{\partial^2}{\partial y^2}\right)\cdot\varphi\left(\left\|\mathbf{x}-\mathbf{x}_j\right\|_2\right)\\ &=\sum_{j=1}^{N_i}\alpha_j\left(\frac{\partial^4}{\partial x^4}+2\frac{\partial^4}{\partial x^2\partial y^2}+\frac{\partial^4}{\partial y^4}\right)\cdot\varphi\left(\left\|\mathbf{x}-\mathbf{x}_j\right\|_2\right)+\sum_{j=N_i+1}^{N}\alpha_j\left(\frac{d^2}{dr^2}\varphi(r)+\frac{1}{r}\frac{d}{dr}\cdot\varphi(r)\right)\\ &=\sum_{j=1}^{N_i}\alpha_j\left[\frac{d^4}{dr^4}\varphi\left(\left\|\mathbf{x}-\mathbf{x}_j\right\|_2\right)+\frac{2}{r}\frac{d^3}{dr^3}\cdot\varphi\left(\left\|\mathbf{x}-\mathbf{x}_j\right\|_2\right)-\frac{1}{r^2}\frac{d^2}{dr^2}\varphi\left(\left\|\mathbf{x}-\mathbf{x}_j\right\|_2\right)+\frac{1}{r^3}\frac{d}{dr}\cdot\varphi\left(\left\|\mathbf{x}-\mathbf{x}_j\right\|_2\right)\right]\\ &\quad+\sum_{j=N_i+1}^{N}\alpha_j\left(\frac{d^2}{dr^2}\varphi\left(\left\|\mathbf{x}-\mathbf{x}_j\right\|_2\right)+\frac{1}{r}\frac{d}{dr}\varphi\left(\left\|\mathbf{x}-\mathbf{x}_j\right\|_2\right)\right)\\ &=\sum_{j=1}^{N_i}\alpha_j\left[2\varepsilon^2\cdot\exp\left(-\varepsilon^2\left\|\mathbf{x}-\mathbf{x}_j\right\|_2^2\right)\left[8\varepsilon^6\left\|\mathbf{x}-\mathbf{x}_j\right\|_2^4-24\varepsilon^4\left\|\mathbf{x}-\mathbf{x}_j\right\|_2^2+6\varepsilon^2+2\left(2\varepsilon^2\cdot\exp\left(-\varepsilon^2\left\|\mathbf{x}-\mathbf{x}_j\right\|_2^2\right)\left[6\varepsilon^2-4\varepsilon^4\left\|\mathbf{x}-\mathbf{x}_j\right\|_2^2\right]\right)\right.\right.\\ &\quad-\frac{1}{\left\|\mathbf{x}-\mathbf{x}_j\right\|_2^2}\left(2\varepsilon^2\exp\left(-\left(\varepsilon\left\|\mathbf{x}-\mathbf{x}_j\right\|_2\right)^2\right)\cdot\left[2\left(\varepsilon\left\|\mathbf{x}-\mathbf{x}_j\right\|_2\right)^2-1\right]\right)+\frac{1}{\left\|\mathbf{x}-\mathbf{x}_j\right\|_2^2}\left(-2\varepsilon^2\right.\end{aligned}$$

$$\cdot\exp\left(-\left(\varepsilon\left\|\mathbf{x}-\mathbf{x}_j\right\|_2\right)^2\right)\Big)\Big]+\sum_{j=N_i+1}^{N}\alpha_j\Big[\Big(2\varepsilon^2$$

$$\cdot\exp\left(-\left(\varepsilon\left\|\mathbf{x}-\mathbf{x}_j\right\|_2\right)^2\right)\left[2\left(\varepsilon\left\|\mathbf{x}-\mathbf{x}_j\right\|_2\right)^2-1\right]\Big)$$

$$+\left(-2\varepsilon^2\exp\left(-\left(\varepsilon\left\|\mathbf{x}-\mathbf{x}_j\right\|_2\right)^2\right)\right)\Big] \tag{49}$$

For $1 \le j \le N_i$, we have $\zeta_{ij}=(\Phi^* H_{ij}^x + B^* H_{ij}^x)$

$$u(\mathbf{x},t) \simeq \tilde{u}(\mathbf{x},t) = \sum_{j=1}^{N_i}\alpha_j^t\Phi^*\varphi\left(\left\|\mathbf{x}-\mathbf{x}_j\right\|_2\right) + \sum_{j=N_i+1}^{N}\alpha_j^t B^*\varphi\left(\left\|\mathbf{x}-\mathbf{x}_j\right\|_2\right)$$

$$\frac{\partial}{\partial x}u(\mathbf{x},t) \simeq \frac{\partial}{\partial x}\tilde{u}(\mathbf{x},t) = \frac{\partial}{\partial x}\left(\sum_{j=1}^{N_i}\alpha_j\Phi^*\varphi\left(\left\|\mathbf{x}-\mathbf{x}_j\right\|_2\right) + \sum_{j=N_i+1}^{N}\alpha_j B^*\varphi\left(\left\|\mathbf{x}-\mathbf{x}_j\right\|_2\right)\right) \tag{50}$$

$$= \sum_{j=1}^{N_i}\alpha_j\frac{\partial}{\partial x}\left[\Phi^*\varphi\left(\left\|\mathbf{x}-\mathbf{x}_j\right\|_2\right)\right] + \sum_{j=N_i+1}^{N}\alpha_j\frac{\partial}{\partial x}\left[B^*\varphi\left(\left\|\mathbf{x}-\mathbf{x}_j\right\|_2\right)\right]$$

$$= \sum_{j=1}^{N_i}\alpha_j\Phi^*\frac{\partial}{\partial x}\varphi\left(\left\|\mathbf{x}-\mathbf{x}_j\right\|_2\right) + \sum_{j=N_i+1}^{N}\alpha_j B^*\frac{\partial}{\partial x}\varphi\left(\left\|\mathbf{x}-\mathbf{x}_j\right\|_2\right)$$

Now, we have $\Phi^* = (\partial/\partial x)(\,)$ and with the absence of the derivative operation on the boundary, B^*, the remaining term of the above equation can be written as

$$\frac{\partial}{\partial x}u(\mathbf{x},t) \simeq \frac{\partial}{\partial x}\tilde{u}(\mathbf{x},t) = \sum_{j=1}^{N_i}\alpha_j\frac{\partial}{\partial x}\left(\frac{\partial\varphi\left(\left\|\mathbf{x}-\mathbf{x}_j\right\|_2\right)}{\partial x}\right) + \sum_{j=N_i+1}^{N}\alpha_j\frac{\partial}{\partial x}\varphi\Big(\Big\|\mathbf{x}$$

$$-\mathbf{x}_j\Big\|_2\Big) = \sum_{j=1}^{N_i}\alpha_j\left(\frac{\left(x_i-x_j\right)^2}{\left\|\mathbf{x}-\mathbf{x}_j\right\|_2^2}\frac{d^2\varphi}{dr^2} + \frac{\left(y_i-y_j\right)^3}{\left\|\mathbf{x}-\mathbf{x}_j\right\|_2^3}\cdot\frac{d\varphi}{dr}\right) + \sum_{j=N_i+1}^{N}\alpha_j\left(\frac{\left(x_i-x_j\right)}{r}\frac{d\varphi}{dr}\right)$$

$$= \sum_{j=1}^{N_i}\alpha_j\left[\frac{\left(x_i-x_j\right)^2}{\left\|\mathbf{x}-\mathbf{x}_j\right\|_2^2}\Big(2\varepsilon^2\cdot\exp\left(-\left(\varepsilon\left\|\mathbf{x}-\mathbf{x}_j\right\|_2\right)^2\right)\left[2\left(\varepsilon\left\|\mathbf{x}-\mathbf{x}_j\right\|_2\right)^2-1\right]\Big) + \frac{\left(y_i-y_j\right)^3}{\left\|\mathbf{x}-\mathbf{x}_j\right\|_2^2}\left(-2\varepsilon^2\exp\left(-\left(\varepsilon\left\|\mathbf{x}-\mathbf{x}_j\right\|_2\right)^2\right)\right)\right] + \sum_{j=N_i+1}^{N}\alpha_j\Big(\left(x_i-x_j\right)\Big(-2\varepsilon^2\cdot\exp\left(-\left(\varepsilon\left\|\mathbf{x}-\mathbf{x}_j\right\|_2\right)^2\right)\Big)\Big) \tag{51}$$

Similarly,

$$\frac{\partial}{\partial y}u(\mathbf{x},t) \simeq \frac{\partial}{\partial y}\tilde{u}(\mathbf{x},t) = \sum_{j=1}^{N_i}\alpha_j\left[\frac{\left(y_i-y_j\right)^2}{\left\|\mathbf{x}-\mathbf{x}_j\right\|_2^2}\Big(2\varepsilon^2\cdot\exp\left(-\left(\varepsilon\left\|\mathbf{x}-\mathbf{x}_j\right\|_2\right)^2\right)\left[2\left(\varepsilon\left\|\mathbf{x}-\mathbf{x}_j\right\|_2\right)^2-1\right]\Big) + \frac{\left(x_i-x_j\right)^3}{\left\|\mathbf{x}-\mathbf{x}_j\right\|_2^2}\left(-2\varepsilon^2\exp\left(-\left(\varepsilon\left\|\mathbf{x}-\mathbf{x}_j\right\|_2\right)^2\right)\right)\right] + \sum_{j=N_i+1}^{N}\alpha_j\Big(\left(y_i-y_j\right)\Big(-2\varepsilon^2\cdot\exp\left(-\left(\varepsilon\left\|\mathbf{x}-\mathbf{x}_j\right\|_2\right)^2\right)\Big)\Big) \tag{52}$$

By substituting (49), (51), and (52) back into the governing equations

$$\frac{\partial u}{\partial t} = \frac{1}{\mathrm{Re}}\left(\frac{\partial^2 u}{\partial x^2}+\frac{\partial^2 u}{\partial y^2}\right) - u\frac{\partial u}{\partial x} - v\frac{\partial u}{\partial y} \tag{53}$$

Therefore, for each center $\mathbf{x}_k$, the following is reached:

$$\left[\frac{\partial u}{\partial t}\right]_k = \frac{1}{\mathrm{Re}}\left[\sum_{j=1}^{N_i}\alpha_j\left[2\varepsilon^2\exp\left(-\varepsilon^2\left\|\mathbf{x}_k-\mathbf{x}_j\right\|_2^2\right)\left[8\varepsilon^6\left\|\mathbf{x}_k-\mathbf{x}_j\right\|_2^4 - 24\varepsilon^4\left\|\mathbf{x}_k-\mathbf{x}_j\right\|_2^2 + 6\varepsilon^2\right.\right.\right.$$

$$\left. + 2\left(2\varepsilon^2\exp\left(-\varepsilon^2\left\|\mathbf{x}_k-\mathbf{x}_j\right\|_2^2\right)\left[6\varepsilon^2 - 4\varepsilon^4\left\|\mathbf{x}_k-\mathbf{x}_j\right\|_2^2\right]\right)\right.$$

$$
\begin{aligned}
&- \frac{1}{\left\|\mathbf{x}_k - \mathbf{x}_j\right\|_2^2} \left(2\varepsilon^2 \exp\left(-\left(\varepsilon \left\|\mathbf{x}_k - \mathbf{x}_j\right\|_2\right)^2\right)\left[2\left(\varepsilon\left\|\mathbf{x}_k - \mathbf{x}_j\right\|_2\right)^2 - 1\right]\right) + \frac{1}{\left\|\mathbf{x}_k - \mathbf{x}_j\right\|_2^2}\left(-2\varepsilon^2 \exp\left(-\left(\varepsilon\left\|\mathbf{x}_k - \mathbf{x}_j\right\|_2\right)^2\right)\right)\Bigg] \\
&+ \sum_{j=N_i+1}^{N} \alpha_j \left(\left(2\varepsilon^2 \exp\left(-\left(\varepsilon\left\|\mathbf{x} - \mathbf{x}_j\right\|_2\right)^2\right)\left[2\left(\varepsilon\left\|\mathbf{x} - \mathbf{x}_j\right\|_2\right)^2 - 1\right]\right) + \left(-2\varepsilon^2 \exp\left(-\left(\varepsilon\left\|\mathbf{x} - \mathbf{x}_j\right\|_2\right)^2\right)\right)\right)\Bigg] \\
&- u_k \left[\sum_{j=1}^{N_i} \alpha_j \left[\frac{\left(x_i - x_j\right)^2}{\left\|\mathbf{x}_k - \mathbf{x}_j\right\|_2^2}\left(2\varepsilon^2 \exp\left(-\left(\varepsilon\left\|\mathbf{x}_k - \mathbf{x}_j\right\|_2\right)^2\right)\left[2\left(\varepsilon\left\|\mathbf{x}_k - \mathbf{x}_j\right\|_2\right)^2 - 1\right]\right)\right.\right. \\
&\left.\left. + \frac{\left(y_i - y_j\right)^3}{\left\|\mathbf{x}_k - \mathbf{x}_j\right\|_2^2}\left(-2\varepsilon^2 \exp\left(-\left(\varepsilon\left\|\mathbf{x}_k - \mathbf{x}_j\right\|_2\right)^2\right)\right)\right] + \sum_{j=N_i+1}^{N} \alpha_j \left(\left(x_i - x_j\right)\left(-2\varepsilon^2 \exp\left(-\left(\varepsilon\left\|\mathbf{x} - \mathbf{x}_j\right\|_2\right)^2\right)\right)\right)\right] \\
&- v_k \left[\sum_{j=1}^{N_i} \alpha_j \left[\frac{\left(y_i - y_j\right)^2}{\left\|\mathbf{x}_k - \mathbf{x}_j\right\|_2^2}\left(2\varepsilon^2 \exp\left(-\left(\varepsilon\left\|\mathbf{x}_k - \mathbf{x}_j\right\|_2\right)^2\right)\left[2\left(\varepsilon\left\|\mathbf{x}_k - \mathbf{x}_j\right\|_2\right)^2 - 1\right]\right)\right.\right. \\
&\left.\left. + \frac{\left(x_i - x_j\right)^3}{\left\|\mathbf{x}_k - \mathbf{x}_j\right\|_2^2}\left(-2\varepsilon^2 \exp\left(-\left(\varepsilon\left\|\mathbf{x}_k - \mathbf{x}_j\right\|_2\right)^2\right)\right)\right] + \sum_{j=N_i+1}^{N} \alpha_j \left(\left(y_i - y_j\right)\left(-2\varepsilon^2 \exp\left(-\left(\varepsilon\left\|\mathbf{x} - \mathbf{x}_j\right\|_2\right)^2\right)\right)\right)\right]
\end{aligned} \tag{54}
$$

The whole process explained above is applied to the other equation of the coupled governing equations. Note that all the terms with $1/\|\mathbf{x}_k - \mathbf{x}_j\|_2$ shall be cancelling each other out so there is no zero-division problem in computing process.

3.4. Computing Setup and Algorithm. Each computation under this work was completed following the computing steps detailed below.

Step 1. Choose N collocation or center nodes $\mathbf{X}^c = \{\mathbf{x}_j\}_{j=1}^{N}$, on the domain Ω.

Step 2. Specify the desired values of

(2.1) The Reynolds number, Re.

(2.2) The final time t.

(2.3) The time step Δt.

(2.4) The Gaussian shape parameter ε.

Step 3. Compute the initial collocation matrices $\widetilde{\mathbf{A}}$ and $\widetilde{\mathbf{B}}$ using the initial Hermite approximation equations (20)-(21).

Step 4. Apply the initial conditions, $\mathbf{U}^{(0)}, \mathbf{V}^{(0)}$, to get $\boldsymbol{\alpha}$ and $\boldsymbol{\beta}$ from (30), (31), (33), (34), (35), (36), (37), (38), (40), (41), (42), and (43).

Step 5. Compute the solutions for the next time step via the time-discretization explained above.

Step 6. Construct the collocation matrices $\widetilde{\mathbf{A}}$ and $\widetilde{\mathbf{B}}$ in the forms expressed in (30), (31), (33), (34), (35), (36), (37), (38), (40), (41), (42), and (43) using the solutions previously obtained in Step 5.

Step 7. Carry on Steps 5 and 6 until reaching the final time step t.

All computing experiments carried out in this study were executed on a computer notebook: Intel(R) Core(TM) i7-5500U CPU @ 2.40GHz with 8.00 GB of RAM and 64-bit operating system.

4. Numerical Experiments and General Discussion

The test case chosen for numerical investigation carried out in this study is the most popular form of Burgers' equations where the corresponding exact solutions for validation are provided by using a Hopf-Cole transformation nicely carried out by Fletcher [4] in 1983 and are expressed as follows:

$$u(x,y,t) = \frac{3}{4} - \frac{1}{4}\left[1 + \exp\left(\left(-4x + 4y - t\right)\frac{\text{Re}}{32}\right)\right]^{-1} \tag{55}$$

$$v(x,y,t) = \frac{3}{4} + \frac{1}{4}\left[1 + \exp\left(\left(-4x + 4y - t\right)\frac{\text{Re}}{32}\right)\right]^{-1} \tag{56}$$

where both the initial and the boundary conditions to be imposed to the equation system are generated directly from the above exact forms over the domain $\Omega = \{(x,y) : 0 \le x \le 1, 0 \le y \le 1\}$. With their rich of challenging features, the equations have received interest from several researchers and been treated with variety of numerical techniques as mentioned in Section 1. Nevertheless, it can be clearly seen that, with the reasons previously mentioned, most studies are concerned only with only cases of relatively low Reynolds numbers, i.e., $1 \leq \mathrm{Re} \leq 500$. In this work, the investigation covers a wide range of Reynolds number from low to moderate $\mathrm{Re} \in [1, 800]$ and, moreover, goes above $\mathrm{Re} = 1,000$, where no numerical works have ever reached before.

Solutions obtained from applying the Hermite collocation method are validated by comparing to both those documented in literatures and those produced by the exact formula. Error measurements are carried out using the following error norms:

(1) Maximum error (L_∞);

$$L_\infty = \max_{1\leq i\leq N} \left|\tilde{u}(\mathbf{x}_i) - u(\mathbf{x}_i)\right| \tag{57}$$

(2) Root-mean-square error (L_{rms});

$$L_{rms} = \left(\frac{1}{N}\sum_{j=1}^{N}\left(\tilde{u}(\mathbf{x}_j) - u(\mathbf{x}_j)\right)^2\right)^{1/2} \tag{58}$$

In addition to these norms, the results presented below are sometimes referred to as "*Average-Error*" and it is to be understood as

$$Average - L_\infty = \frac{L_\infty(U) + L_\infty(V)}{2} \tag{59}$$

where $L_\infty(U)$ and $L_\infty(V)$ are of U- and V-velocity component, respectively.

The whole study is split into two main parts, one concerned with low- to moderate-Reynolds number cases, $1 \leq \mathrm{Re} \leq 500$, and that with relatively high value of Reynolds number, $500 < \mathrm{Re} \leq 1,300$. The so-called Reynolds number plays crucial rules in determining the ratio of forces exerted in the system and this is remained as one of the most challenging tasks to simulate and mimic both numerically and experimentally.

In order to demonstrate the overall effectiveness of the Hermite type of collocation scheme, investigations using the conventional global collocation known as "Kansa" method, (14), are also studied parallelly. The results obtained from using the Hermite type and Kansa are referred to hereafter as "**Hermite**" and "**Conventional Kansa/Kansa,**" respectively.

4.1. Low- to Moderate-Reynolds Number Cases. As it is known to be one of the most challenging tasks for any collocation method, finding a suitable shape parameter, ε, is to be firstly presented. In the past, some attempts to pinpoint the optimal value of ε involve the classic work of Hardy [35] where it was shown that, by fixing the shape at $\varepsilon = 1/(0.815d)$, where $d = (1/N)\sum_{i=1}^{N} d_i$ and d_i is the distance from the node to its nearest neighbor, good results should be anticipated. Also, in the work of Franke and Schaback [36] where the choice of a fixed shape of the form $\varepsilon = 0.8\sqrt{N}/D$ with D is the diameter of the smallest circle containing all data nodes, can also be a good alternative. In 2000, Zhang et al. [37] demonstrated and concluded that the optimal shape parameter is highly problem dependent. In 2002, Wang and Lui [38] pointed out that, by analyzing the condition number of the collocation matrix, a suitable range of derivable values of ε can be found. Later in 2003, Lee et al. [39] suggested that the final numerical solutions obtained are found to be less affected by the method when the approximation is applied locally rather than globally. A rather recent work is the selection of an interval for variable shape parameter by Biazar and Hosami in 2016 [40], where a novel algorithm for determining and interval was proposed.

It is to be noted, however, that even though this topic has long been widely studied (see also in [41–44]), the majority of studies are done on "Multiquadric RBF" or "Inverse-Multiquadric RBF," whereas works on Gaussian type are rather rare. A confirmation on problem-dependent nature of Gaussian RBF shape value was confirmed in the work of Davydov and Oanh [45]. A more recent work is the application of the traditional Kansa collocation to linear PDE based on a variable shape parameter called symmetric variable shape parameter (SVSP) carried out by Ranjbar [46].

Even though Gaussian RBF was focused on in these two later works, the scheme does not contain any self-adjoint operator. Moreover, the complications introduced by the structure of the governing equation at hand are known to greatly affect the final choice of the shape. Taking these facts into consideration, with the transient, nonlinear, and coupling nature of Burgers' equations, this work chose to directly measure the overall errors generated in the system. To complete this task, a large number of numerical experiments were carried out in this work aiming to reach the optimal value of ε. The impact of shape parameter on the quality of the computed solution is demonstrated in Figures 1–4. Figures 1 and 2 show that, within the range of $0 < \varepsilon \leq 1.5$, both methods reveal slight differences in the trends of L_{rms}, with some strong fluctuations being found for the case of Kansa. For this particular set of conditions, at $\mathrm{Re} = 10$ with $t = 0.5, \Delta t = 0.005, N = 10 \times 10$, it can be noted that a more suitable shape parameter may be found at $\varepsilon < 0.1$ for the Hermite scheme. This observation can also be useful for Kansa case as indicated in the Figure. Figures 3 and 4 indicate clearly that both of the methods become less sensitive to the shape value when it goes beyond $\varepsilon \approx 300$. Above this point, it should be observed that L_{rms} remains under $10E - 01$ when using Hermite scheme whereas L_{rms} stays constantly above $10E - 01$.

Information provided previously on the effect of shape parameter leads to decision made for the case with $\mathrm{Re} = 50$. At randomly chosen points, Table 1 provides the solutions of both U- and V-velocity components compared against the corresponding exact values. It can be seen that the solutions are in good agreement for both choices of shape parameter; $\varepsilon = 0.001$ and $\varepsilon = 0.1$. A bigger picture of results obtained in this work is shown in Table 2 and Table 3, with cases of Reynolds number starting from 1 up to 400.

From these two Tables, it can be clearly seen that, with different values of time step Δt and the number of collocation nodes N, both methods produce approximately the same quality of results at both points of measured times: $t = 0.01$ and $t = 2.0$.

TABLE 1: Numerical solutions for Re = 50, $\Delta t = 0.001$, $t = 0.5$, and $N = 12 \times 12$.

		$\varepsilon = 0.001$				$\varepsilon = 0.1$			
x	y	U-velocity		V-velocity		U-velocity		V-velocity	
		Hermite	Exact	Hermite	Exact	Hermite	Exact	Hermite	Exact
0	0.3	0.6873	0.6872	0.8127	0.8128	0.6873	0.6872	0.8127	0.8128
0.1	0.2	0.6653	0.6628	0.8347	0.8372	0.6553	0.6631	0.8247	0.8367
0.3	0.8	0.7381	0.7395	0.7719	0.7605	0.7281	0.7398	0.7719	0.7601
0.4	0.6	0.6938	0.6943	0.8162	0.8057	0.6838	0.6946	0.8162	0.8052
0.6	0.7	0.6553	0.6628	0.8347	0.8372	0.6653	0.6631	0.8247	0.8367
0.7	0.2	0.5149	0.5105	0.9951	0.9895	0.5049	0.5109	0.9951	0.9890
0.7	0.6	0.5892	0.5872	0.9008	0.9128	0.5892	0.5875	0.9108	0.9124
0.8	0.1	0.5014	0.5031	0.9986	0.9969	0.5014	0.5034	0.9986	0.9964
1	0.4	0.5027	0.5027	0.9973	0.9973	0.5027	0.5027	0.9973	0.9973
1	0.7	0.5164	0.5164	0.9836	0.9836	0.5164	0.5164	0.9836	0.9836

TABLE 2: Averaged L_{rms}-error obtained with $N = 100$ and $\varepsilon = 0.001$ for low to moderate Reynolds number.

Re	N	Δt	$t = 0.01$		$t = 2.0$	
			Kansa	Hermite	Kansa	Hermite
1	100	5E-04	1.11E-06	1.56E-05	3.55E-05	3.12E-03
10	100	0.0050	1.08E-03	1.58E-04	5.65E+26	2.91E-02
200	900	0.0010	8.97E-03	1.19E-03	3.66E-01	1.32E-01
400	900	0.01	1.55E-02	1.77E-03	4.48E-01	1.39E-01

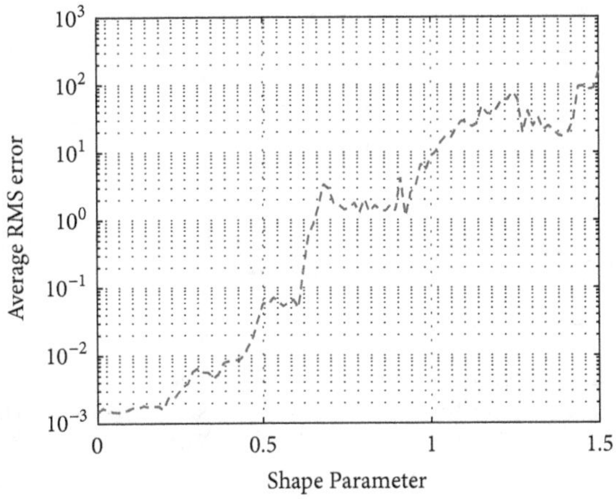

FIGURE 1: Effect of the Gaussian shape parameter ($\varepsilon \leq 1.5$) on the result accuracy for the Hermite collocation scheme computed at Re = 10 with $t = 0.5$, $\Delta t = 0.005$, and $N = 10 \times 10$.

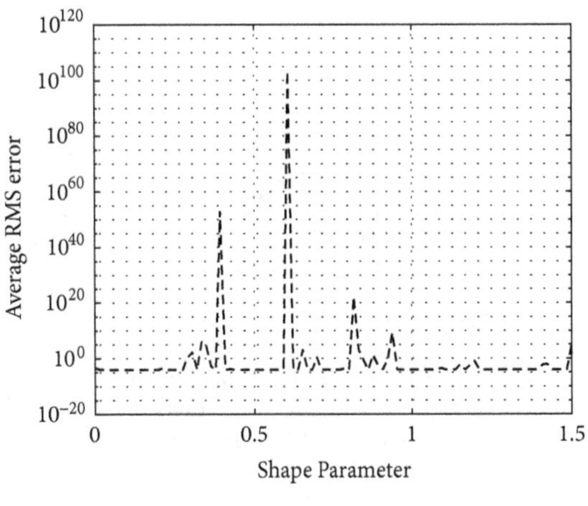

FIGURE 2: Effect of the Gaussian shape parameter ($\varepsilon \leq 1.5$) on the result accuracy for the conventional Kansa scheme computed at Re = 10 with $t = 0.5$, $\Delta t = 0.005$, and $N = 10 \times 10$.

For a validation purpose, a comparison of computed solution obtained from using Hermite to other numerical works available in literature is done and presented in Tables 4 and 5. In this case, the number of nodes is 441 and the measurement is taken at Re = 100, $\Delta t = 0.0001$, and $t = 2.0$. The comparison clearly shows that good agreement in terms of solution quality is reached using the method under investigation.

At Reynolds number of 500, solutions produced with $t = 0.5$, $\Delta t = 0.001$, $N = 30 \times 30$ are depicted in Figures 5 and 6.

TABLE 3: Averaged L_∞-error obtained with N = 100, ε = 0.001 for low to moderate Reynolds number.

Re	N	Δt	t = 0.01		t = 2.0	
			Kansa	Hermite	Kansa	Hermite
1	100	5E-04	2.10E-06	1.95E-05	7.51E-05	3.91E-03
10	100	0.0050	2.20E-03	1.95E-04	1.10E+27	3.87E-02
200	900	0.0010	2.38E-02	3.90E-03	8.36E-01	2.49E-01
400	900	0.01	4.11E-02	7.80E-03	1.03E+00	2.50E-01

TABLE 4: Computed value of U-velocity at for Re = 100, Δt = 0.0001, t = 2.0, and N = 21 × 21.

x	y	Exact	Hermite	[20]	[8]	[21]	[5]
0.1	0.1	0.500482	0.502321	0.500470	0.50035	0.50012	0.49983
0.3	0.3	0.500482	0.504311	0.500441	0.50042	0.50042	0.49977
0.5	0.5	0.500482	0.502553	0.500414	0.50046	0.50041	0.49973
0.3	0.7	0.555675	0.552258	0.554805	0.55609	0.55587	0.55429
0.1	0.9	0.744256	0.741302	0.744197	0.74409	0.74416	0.74340
0.5	0.9	0.555675	0.556811	0.554489	0.55604	0.55637	0.55413

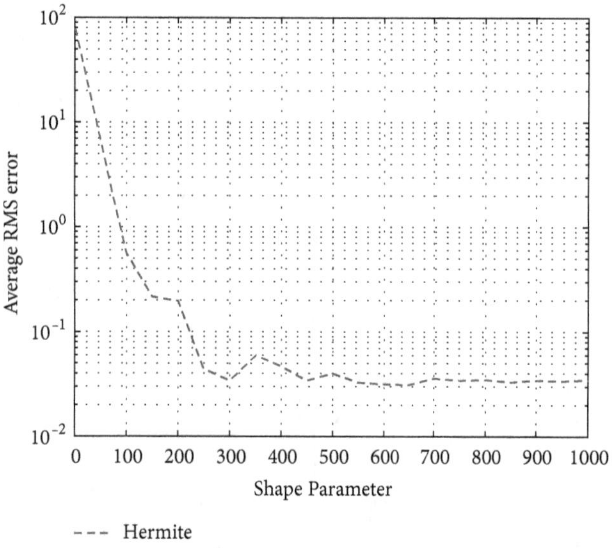

FIGURE 3: Effect of the Gaussian shape parameter ($\varepsilon \gg 1.5$) on the result accuracy for the Hermite collocation scheme computed at Re = 10 with t = 0.5, Δt = 0.005, and N = 10 × 10.

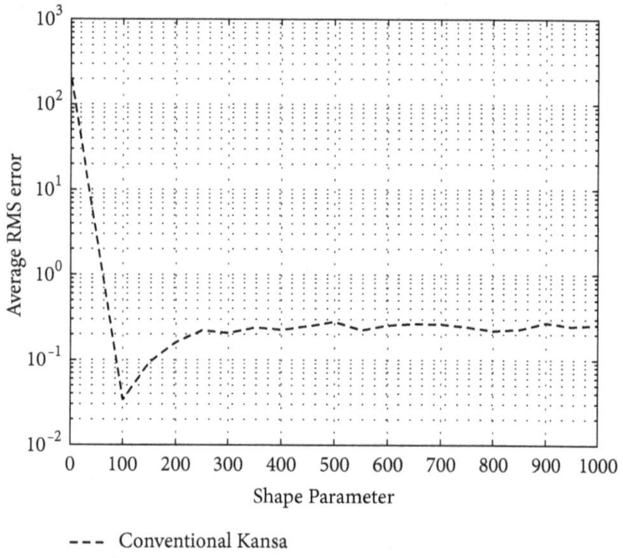

FIGURE 4: Effect of the Gaussian shape parameter ($\varepsilon \gg 1.5$) on the result accuracy for the conventional Kansa scheme computed at Re = 10 with t = 0.5, Δt = 0.005, and N = 10 × 10.

Once again, reasonably good quality of computed solutions is obtained.

For all the information presented so far, it strongly indicates that Hermite is now successfully applied to this type of PDEs. This gives some confidence to move to more challenging cases with higher Reynolds number and the results are to be presented in the following sections.

4.2. Moderate- to High-Reynolds Number Cases. It is well known that the problem becomes more difficult when the Reynolds number increases as the instability in the numerical process can completely ruin the final solution [47]. This problem is generally encountered by any study that concerns Reynolds number above 800. In this work, the numerical investigation starts from Re = 600 and moves to the highest of 1,300. With using only 900 collocation nodes and a fixed value of shape ε = 0.001, Tables 6 and 7, respectively, show L_{rms} and L_∞ of solutions computed by Hermite and Kansa at two different times t = 0.01 and t = 2.0. At the smaller time t = 0.01, both tables reveal that both Kansa and Hermite produce approximately the same results quality with those obtained from Hermite being only slightly better

TABLE 5: Computed value of V-velocity at Re = 100 with $\Delta t = 0.0001$, $t = 2.0$, and $N = 21 \times 21$.

x	y	Exact	Hermite	[20]	[8]	[21]	[5]
0.1	0.1	0.999518	0.99695	0.999530	0.99936	0.99946	0.99826
0.3	0.3	0.999518	0.99771	0.999559	0.99951	0.99938	0.99861
0.5	0.5	0.999518	0.99882	0.999586	0.99958	0.99941	0.99821
0.3	0.7	0.944325	0.94502	0.945195	0.94387	0.94387	0.94409
0.1	0.9	0.755744	0.75499	0.755803	0.75592	0.75558	0.75500
0.5	0.9	0.944325	0.94102	0.945511	0.94392	0.94345	0.94345

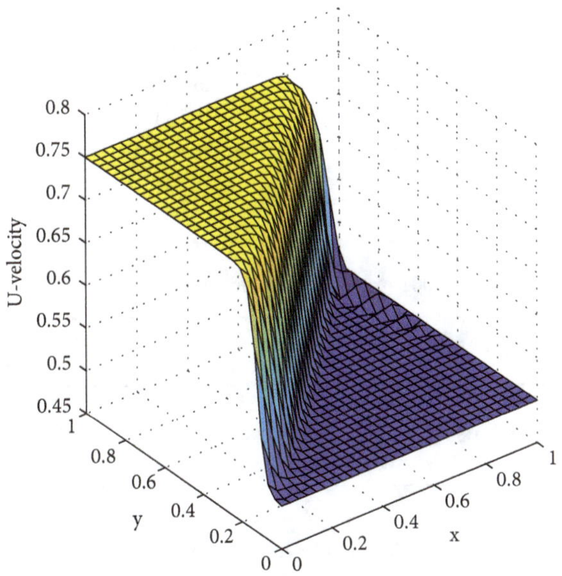

FIGURE 5: U-velocity profile produced by (left) the Hermite collocation scheme and (right) the exact, at Re = 500 with $t = 0.5, \Delta t = 0.001$, and $N = 30 \times 30$.

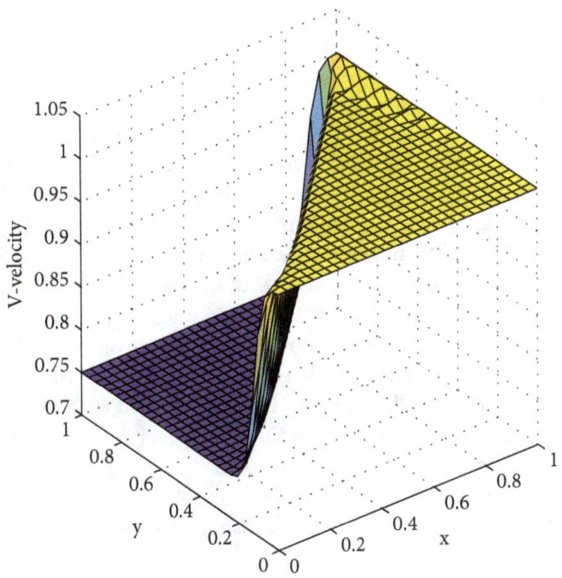

FIGURE 6: V-velocity profile Average produced by (left) the Hermite collocation scheme and (right) The exact, at Re = 500 with $t = 0.5, \Delta t = 0.001$, and $N = 30 \times 30$.

TABLE 6: Averaged L_{rms}- error obtained with $\varepsilon = 0.001$ for moderate to high Reynolds number.

Re	N	Δt	$t = 0.01$ Kansa	$t = 0.01$ Hermite	$t = 2.0$ Kansa	$t = 2.0$ Hermite
600	900	0.001	1.52E-02	2.27E-03	4.23E-01	1.41E-01
800	900	0.001	4.94E-02	2.87E-03	1.04E+04	1.42E-01
1000	900	0.001	4.88E-02	3.54E-03	1.11E+04	1.42E-01
1200	900	0.001	4.87E-02	4.21E-03	1.20E+04	1.42E-01

TABLE 7: Averaged L_∞- error obtained with $\varepsilon = 0.001$ for moderate to high Reynolds number.

Re	N	Δt	$t = 0.01$ Kansa	$t = 0.01$ Hermite	$t = 2.0$ Kansa	$t = 2.0$ Hermite
600	900	0.001	4.37E-02	1.17E-02	9.55E-01	2.50E-01
800	900	0.001	1.50E-01	1.55E-02	2.87E+04	2.50E-01
1000	900	0.001	1.52E-01	1.94E-02	3.09E+04	2.50E-01
1200	900	0.001	1.55E-01	2.32E-02	3.34E+04	2.50E-01

TABLE 8: Averaged L_{rms} - error obtained with $N = 100$, $\Delta t = 0.001$ for high Reynolds number.

ε	Re	$t = 0.01$ Kansa	$t = 0.01$ Hermite	$t = 2.0$ Kansa	$t = 2.0$ Hermite
0.01	1000	6.70E-03	5.49E-03	4.22E+01	1.22E-01
0.1	1100	7.55E-03	6.02E-03	2.90E+02	1.22E-01
1	1200	3.91E-02	6.58E-03	2.27E+08	1.85E-01

TABLE 9: Averaged L_∞- error obtained with $N = 100$, $\Delta t = 0.001$ for high Reynolds number.

ε	Re	$t = 0.01$ Kansa	$t = 0.01$ Hermite	$t = 2.0$ Kansa	$t = 2.0$ Hermite
0.01	1000	2.49E-02	1.94E-02	1.31E+02	2.50E-01
0.1	1100	2.91E-02	2.13E-02	1.03E+03	2.50E-01
1	1200	1.35E-01	2.38E-02	8.82E+08	5.86E-01

by an order of magnitude. This is no longer the case when the computation reaches the time of $t = 2.0$ where it can clearly been seen from both tables that Kansa has a strong growth in errors with $L_{rms} \approx$ 1.0E-02 growing to $L_{rms} \approx$ 1.0E+04 and $L_\infty \approx$ 1.0E-02 to $L_\infty \approx$ 1.0E+04. However, the Hermite loses only slightly its accuracy with $L_{rms} \approx$ 1.0E-03 to $L_{rms} \approx$ 1.0E-01 and $L_\infty \approx$ 1.0E-02 to $L_\infty \approx$ 1.0E-01.

The influence of the shape parameter to the solution accuracy is further studied also for these cases of high Reynolds number and the general observation can be made based on information given in Tables 8 and 9. At a very small number of collocation nodes, $N = 100$, and a fixed time step $\Delta t = 0.001$, both tables reveal the same conclusion as previously shown. Kansa is capable of reproducing good results quality only for the case with small time, i.e., $t = 0.01$. At $t = 2.0$, on the other hand, while Kansa is seen to fail to provide good solutions, Hermite clearly manages to keep the quality of numerical solutions around 1.00E-01 for both error norms and for all values of shapes under investigation, $\varepsilon \in \{0.01, 0.1, 1.0\}$. At a particular Re = 1,000, a wider range of shape parameter, $0 < \varepsilon \leq 3$, is further investigated and the measurement of L_{rms} is shown in Figures 7 and 8 where a clear observation on the effect of shape parameter is illustrated. The figures indicate clearly that the Hermite scheme is much less sensitive to shape parameter when the Reynolds number increases when compared with the conventional Kansa method. In addition, it is also observed that the Hermite scheme studied and applied in this work is comparatively less sensitive to the progression of time even when the Reynolds number is taken up to 1,300. This observation is illustrated in Figures 9 and 10 with surface plots of solutions obtained from Kansa and Hermite being depicted in Figure 11 together with the exact ones. It must be noted here that results of U-velocity component are not shown here as they are in similar trend to those for V-velocity component.

5. Conclusion

In this investigation, a further developed version of Kansa collocation method called "Hermite" scheme was applied to

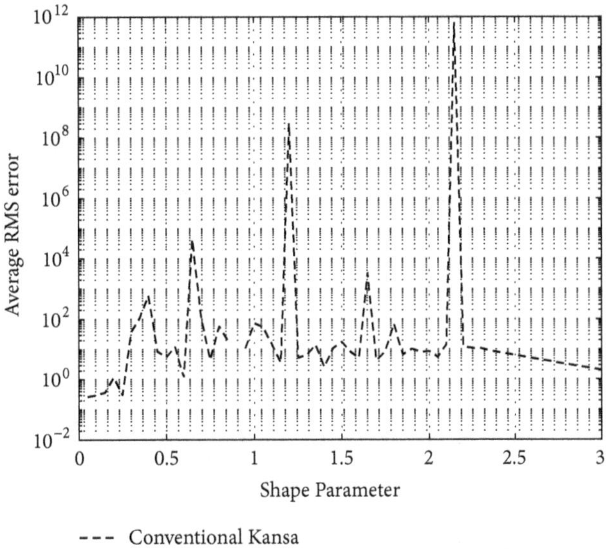

FIGURE 7: Average L_{rms}-error obtained from using Kansa collocation scheme at Re = 1,000 with $t = 0.5, \Delta t = 0.005$, and $N = 10 \times 10$.

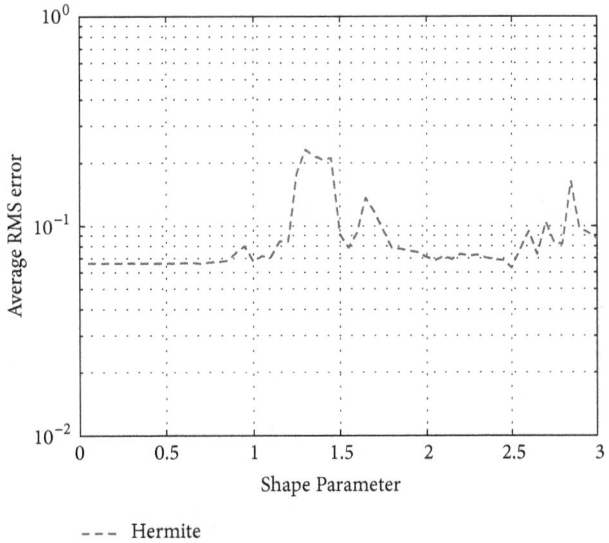

FIGURE 8: Average L_{rms}- error obtained from using Hermite collocation scheme at Re = 1,000 with $t = 0.5, \Delta t = 0.005$, and $N = 10 \times 10$.

one of the most complex types of PDEs. To evaluate the general effectiveness of the mesh-free method, the Burgers' equations are chosen. The Gaussian type of radial basis function was chosen for this task. With applying the differential operator twice to the governing equation before being approximated via a linear summation of the basis function, as in the Hermite scheme, the solutions obtained have revealed some interesting aspects and some findings can be listed as follows:

(i) Judging from the low Reynolds number cases, the Hermite is successfully applied to the transient, nonlinear, and coupled-Burgers' equations with a good agreement in results' accuracy, compared with the exact solution and other numerical works.

(ii) It is found that, with using the Hermite method, the computing process is much less sensitive to the change of the shape parameter ε making it more practical when in use.

(iii) By using the Hermite scheme, the growth of error accumulated in time is found to be significantly smaller than the conventional Kansa method particularly at high Reynolds number.

(iv) With a suitable value of shape parameter ε, at moderate to very high Reynolds number, the numerical method is found to produce reasonable results.

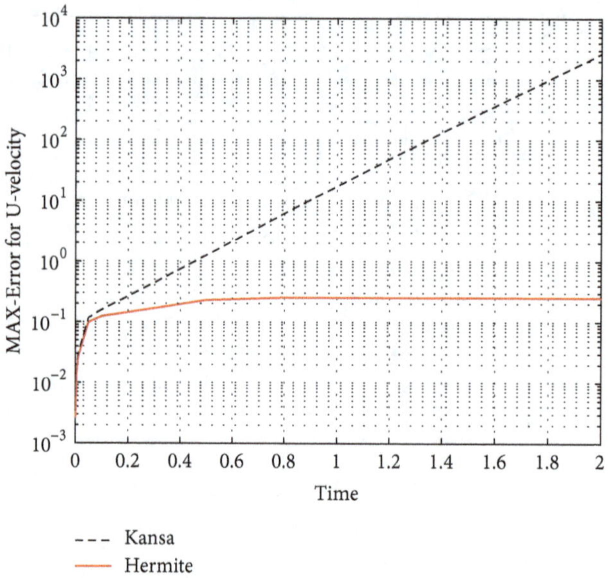

FIGURE 9: L_{∞}-error progression in time $(0.00 \le t \le 2.00)$of the U-maximum error comparison computed at Re = 1,300 with $\varepsilon = 0.005$, $\Delta t = 0.0005$, and $N = 10 \times 10$.

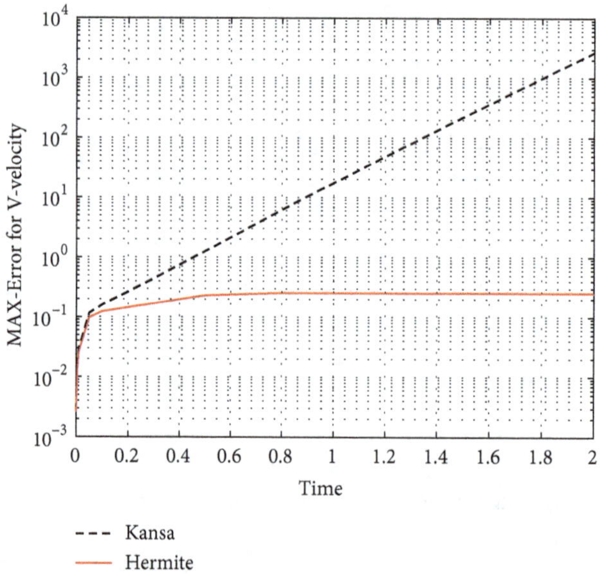

FIGURE 10: L_{∞}-error progression in time $(0.00 \le t \le 2.00)$of the V-maximum error comparison computed at Re = 1,300 with $\varepsilon = 0.005$, $\Delta t = 0.0005$, and $N = 10 \times 10$.

(v) Across the whole range of Reynolds number under investigation, $1 \le \text{Re} \le 1,300$, the Hermite scheme is found to outperform the conventional version of collocation or Kansa method.

Even though some desirable aspects have been discovered in this study regarding the use of collocation mesh-free methods, there are still some interesting questions to be further addressed and investigated, for instance, the process of choosing the optimal shape parameter where several factors shall be taken into consideration: the effect of viscous force, the number of centers, the time-dependent physics, and so on. This is all subject to further studies of the authors.

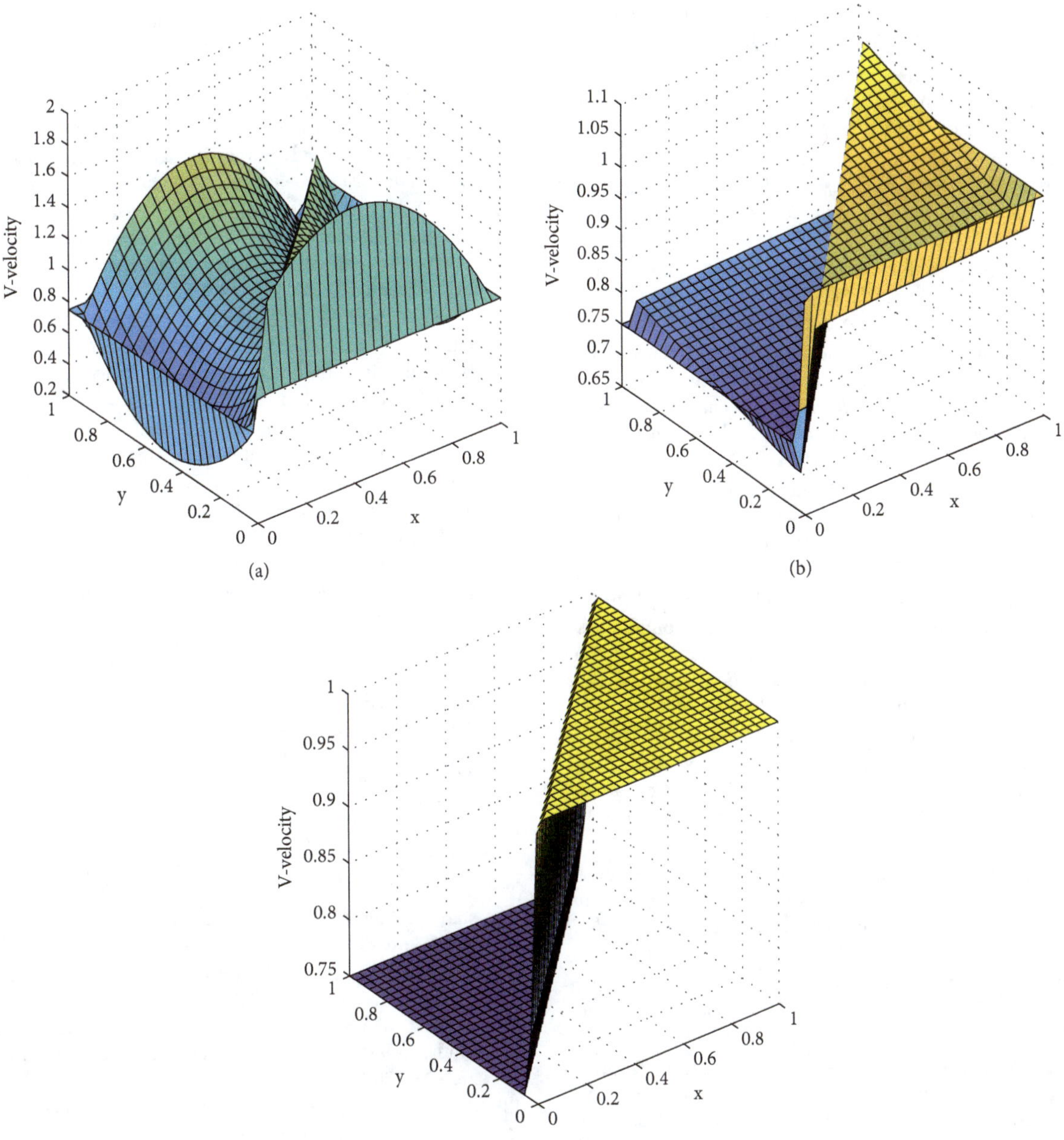

FIGURE 11: V-velocity profile at Re = 1,300 with $\varepsilon = 0.005, t = 0.5, \Delta t = 0.0005, N = 30 \times 30$; (a) Conventional Kansa, (b) Hermite scheme, and (c) exact solution.

Conflicts of Interest

The authors declare that there are no conflicts of interest regarding the publication of this paper.

Acknowledgments

The corresponding author would like to express his sincere appreciation to the Center of Excellence in Mathematics, Thailand, for their kind support for this project.

References

[1] J. M. Burgers, "A mathematical model illustrating the theory of turbulence," *Advances in Applied Mechanics*, vol. 1, pp. 171–199, 1948.

[2] J. Nee and J. Duan, "Limit set of trajectories of the coupled viscous Burgers' equations," *Applied Mathematics Letters*, vol. 11, no. 1, pp. 57–61, 1998.

[3] S. E. Esipov, "Coupled Burgers equations: a model of polydispersive sedimentation," *Physical Review E*, vol. 52, no. 4, pp. 3711–3718, 1995.

[4] C. A. J. Fletcher, "Generating exact solutions of the two-dimensional burgers' equations," *International Journal for Numerical Methods in Fluids*, vol. 3, no. 3, pp. 213–216, 1982.

[5] A. R. Bahadır, "A fully implicit finite-difference scheme for two-dimensional Burgers' equations," *Applied Mathematics and Computation*, vol. 137, no. 1, pp. 131–137, 2003.

[6] M. Dehghan, A. Hamidi, and M. Shakourifar, "The solution of coupled Burgers' equations using Adomian-Pade technique," *Applied Mathematics and Computation*, vol. 189, no. 2, pp. 1034–1047, 2007.

[7] A. A. Soliman, "On the solution of two-dimensional coupled Burgers' equations by variational iteration method," *Chaos, Solitons & Fractals*, vol. 40, no. 3, pp. 1146–1155, 2009.

[8] A. Ali, S. Islam, and S. Haqand, "A computational meshfree technique for the numerical solution of the two-dimensional coupled Burgers' equations," *International Journal for Computational Methods in Engineering Science and Mechanics*, vol. 10, no. 5, pp. 406–422, 2009.

[9] H. Zhu, H. Shu, and M. Ding, "Numerical solutions of two-dimensional Burgers' equations by discrete adomian decomposition method," *Computers & Mathematics with Applications*, vol. 60, no. 3, pp. 840–848, 2010.

[10] S. Kaennakham, K. Chanthawara, and W. Toutip, "Optimal radial basis function (rbf) for dual reciprocity boundary element method (drbem) applied to coupled burgers equations with increasing reynolds number," *Australian Journal of Basic and Applied Sciences*, vol. 8, no. 7, pp. 426–476, 2014.

[11] K. Chanthawara, S. Kaennakham, and W. Toutip, "The dual reciprocity boundary element method (DRBEM) with multiquadric radial basis function for coupled burgers' equations," *International Journal of Multiphysics*, vol. 8, no. 2, pp. 123–143, 2014.

[12] Y. Wang, I. M. Navon, X. Wang, and Y. Cheng, "2D Burgers equation with large Reynolds number using POD/DEIM and calibration," *International Journal for Numerical Methods in Fluids*, vol. 82, no. 12, pp. 909–931, 2016.

[13] I. A. Cristescu, "Numerical resolution of coupled two-dimensional Burgers' equation," *Romanian Journal of Physics*, vol. 62, p. 103, 2017.

[14] E. Oñate, S. Idelsohn, O. C. Zienkiewicz, and R. L. Taylor, "A finite point method in computational mechanics. Applications to convective transport and fluid flow," *International Journal for Numerical Methods in Engineering*, vol. 39, no. 22, pp. 3839–3866, 1996.

[15] T. J. Liszka, C. A. M. Duarte, and W. W. Tworzydlo, "*hp*-Meshless cloud method," *Computer Methods Applied Mechanics and Engineering*, vol. 139, no. 1–4, pp. 263–288, 1996.

[16] X. Zhang, X.-H. Liu, K.-Z. Song, and M.-W. Lu, "Least-squares collocation meshless method," *International Journal for Numerical Methods in Engineering*, vol. 51, no. 9, pp. 1089–1100, 2001.

[17] B. Nayroles, G. Touzot, and P. Villon, "Generalizing the finite element method: diffuse approximation and diffuse elements," *Computational Mechanics*, vol. 10, no. 5, pp. 307–318, 1992.

[18] T. Belytschko, L. Gu, and Y. Y. Lu, "Fracture and crack growth by element free Galerkin methods," *Modelling and Simulation in Materials Science and Engineering*, vol. 2, no. 3A, pp. 519–534, 1994.

[19] G. R. Liu and Y. T. Gu, "A point interpolation method for two-dimensional solids," *International Journal for Numerical Methods in Engineering*, vol. 50, no. 4, pp. 937–951, 2001.

[20] C. Gaspar, "Multi-level meshless methods based on direct multi-elliptic interpolation," *Journal of Computational and Applied Mathematics and Computer*, vol. 226, no. 2, pp. 259–267, 2009.

[21] D. L. Young, C. M. Fan, S. P. Hu, and S. N. Atluri, "The Eulerian-Lagrangian method of fundamental solutions for two-dimensional unsteady Burgers' equations," *Engineering Analysis with Boundary Elements*, vol. 32, no. 5, pp. 395–412, 2009.

[22] E. J. Kansa, "Multiquadrics-A scattered data approximation scheme with applications to computational fluid-dynamics-II," *Computers Mathematics with Applications*, vol. 19, no. 8/9, pp. 147–161, 1990.

[23] Y. Liu, K. M. Liew, Y. C. Hon, and X. Zhang, "Numerical simulation and analysis of an electroactuated beam using a radial basis function," *Smart Materials and Structures*, vol. 14, no. 6, pp. 1163–1171, 2005.

[24] J. Li, Y. Chen, and D. Pepper, "Radial basis function method for 1-D and 2-D groundwater contaminant transport modeling," *Computational Mechanics*, vol. 32, no. 1-2, pp. 10–15, 2003.

[25] Y. C. Hon and R. Schaback, "On unsymmetric collocation by radial basis functions," *Applied Mathematics and Computation*, vol. 119, no. 2-3, pp. 177–186, 2001.

[26] J. Li and C. S. Chen, "Some observations on unsymmetric radial basis function collocation methods for convection-diffusion problems," *International Journal for Numerical Methods in Engineering*, vol. 57, no. 8, pp. 1085–1094, 2003.

[27] S. Islam, B. Šarler, and R. Vertnik, "Radial basis function collocation method for the numerical solution of the two-dimensional transient nonlinear coupled Burgers equations," *Applied Mathematical Modelling*, vol. 36, no. 3, pp. 1148–1160, 2012.

[28] G. E. Fasshauer, "Solving partial differential equations by collocation with radial basis functions," *Proceedings of Chamonix*, vol. 1997, pp. 1–8, 1996.

[29] V. M. A. Leitão, "RBF-based meshless methods for 2D elastostatic problems," *Engineering Analysis with Boundary Elements*, vol. 28, no. 10, pp. 1271–1281, 2004.

[30] A. La Rocca, H. Power, V. La Rocca, and M. Morale, "A meshless approach based upon radial basis function hermite collocation method for predicting the cooling and the freezing times of foods," *Computers, Materials and Continua*, vol. 2, no. 4, pp. 239–250, 2005.

[31] A. H. Rosales, A. La Rocca, and H. Power, "Radial basis function Hermite collocation approach for the numerical simulation of the effect of precipitation inhibitor on the crystallization process of an over-saturated solution," *Numerical Methods for Partial Differential Equations*, vol. 22, no. 2, pp. 361–380, 2006.

[32] M. Naffa and H. J. Al-Gahtani, "RBF-based meshless method for large deflection of thin plates," *Engineering Analysis with Boundary Elements*, vol. 31, no. 4, pp. 311–317, 2007.

[33] N. K. S. Chuathong and W. Toutip, "Numerical solutions of 2D nonlinear PDEs using Kansa's meshless method and the search for optimal radial basis function," in *Proceedings of the 19th International Annual Symposium on Computational Science and Engineering*, vol. 19, pp. 17–19, 2015.

[34] N. K. Chuathong and W. Toutip, "The Kansa meshless method for convection diffusion problems using various radial basis functions," in *Proceeding on the 11th IMT-GT International Conference on Mathematics, Statistics and Its Applications (ICMSA 2015)*, vol. 11, pp. 23–25, 2015.

[35] R. L. Hardy, "Multiquadratic equation of topology and other irregular surface," *Journal of Geophysical Research*, vol. 76, p. 1905, 1971.

[36] C. Franke and R. Schaback, "Convergence order estimates of meshless collocation methods using radial basis functions," *Advances in Computational Mathematics*, vol. 93, no. 4, pp. 73–82, 1998.

[37] X. Zhang, K. Z. Song, M. W. Lu, and X. Liu, "Meshless methods based on collocation with radial basis functions," *Computational Mechanics*, vol. 26, no. 4, pp. 333–343, 2000.

[38] J. G. Wang and G. R. Liu, "On the optimal shape parameters of radial basis functions used for 2-D meshless methods," *Computer Methods Applied Mechanics and Engineering*, vol. 191, no. 23-24, pp. 2611–2630, 2002.

[39] C. K. Lee, X. Liu, and S. C. Fan, "Local multiquadric approximation for solving boundary value problems," *Computational Mechanics*, vol. 30, no. 5-6, pp. 396–409, 2003.

[40] J. Biazar and M. Hosami, "Selection of an Interval for Variable Shape Parameter in Approximation by Radial Basis Functions," *Advances in Numerical Analysis*, vol. 2016, Article ID 1397849, 11 pages, 2016.

[41] K. Chanthawara and S. Kaennakham, "A Numerical Experiment on Optimal Inverse Multiquadric RBF Shape Parameter in the Dual Reciprocity Boundary Element Method for Convective-Dominated Problems," *Thai Journal of Mathematics*, pp. 119–131, 2016, Special Issue on ICMSA2015.

[42] S. Kaennakham and N. Chuathong, "Solution to a convection-diffusion problem using a new variable inverse-multiquadric parameter in a collocation meshfree scheme," *International Journal of Multiphysics*, vol. 11, no. 4, pp. 359–374, 2017.

[43] N. Chuatoing and S. Kaennakham, "A numerical investigation on variable shape parameter schemes in a meshfree method applied to a convection-diffusion problem," *International Journal of Applied Engineering Research*, vol. 12, no. 14, pp. 4162–4170, 2017.

[44] M. Yaghouti and H. Ramezannezhad Azarboni, "Determining optimal value of the shape parameter c in RBF for unequal distances topographical points by Cross-Validation algorithm," *Journal of Mathematical Modeling*, vol. 5, no. 1, pp. 53–60, 2017.

[45] O. Davydov and D. T. Oanh, "On the optimal shape parameter for Gaussian radial basis function finite difference approximation of the Poisson equation," *Computers and Mathematics with Applications*, vol. 62, no. 5, pp. 2143–2161, 2011.

[46] M. Ranjbar, "A new variable shape parameter strategy for Gaussian radial basis function approximation methods," *Annals of the University of Craiova, Mathematics and Computer Science Series*, vol. 42, no. 2, pp. 260–272, 2015.

[47] K. Chanthawara, S. Kaennakham, and W. Toutip, "Inverse Multiquadric RBF in the Dual Reciprocity Boundary Element Method(DRBEM) for Coupled 2D Burgers' Equations at high Reynolds numbers," in *Proceedings of the 19th International Annual Symposium on Computational Science and Engineering (ANSCSE19)*, vol. 19, pp. 2–22, 2015.

7

Generalized Hybrid One-Step Block Method Involving Fifth Derivative for Solving Fourth-Order Ordinary Differential Equation Directly

Mohammad Alkasassbeh and Zurni Omar

Department of Mathematics, School of Quantitative Sciences, College of Arts and Sciences, Universiti Utara Malaysia, Kedah, Malaysia

Correspondence should be addressed to Mohammad Alkasassbeh; mfkasassbeh@yahoo.com

Academic Editor: Mehmet Sezer

A general one-step three-hybrid (off-step) points block method is proposed for solving fourth-order initial value problems of ordinary differential equations directly. A power series approximate function is employed for deriving this method. The approximate function is interpolated at $\{x_n, x_{n+r}, x_{n+s}, x_{n+t}\}$ while its fourth and fifth derivatives are collocated at all points $\{x_i\}$, $i = \{0, r, s, t, 1\}$, in the interval of approximation. Several fourth-order initial value problems of ordinary differential equations are then solved to compare the performance of the proposed method with the derived methods. The analysis of the method reveals that the method is consistent and zero stable concluding that the method is also convergent. The numerical results demonstrate the superiority of the new method over the existing ones in terms of error.

1. Introduction

In this article, we consider the following general fourth-order initial value problems (IVPs) as shown:

$$\begin{aligned} y^{(iv)} &= f\left(x, y, y', y'', y'''\right), \\ y(a) &= \alpha_1, \\ y_0'(a) &= \alpha_2, \\ y_0''(a) &= \alpha_3, \\ y_0'''(a) &= \alpha_4, \\ & \qquad x \in [a,b], \end{aligned} \tag{1}$$

with the assumption that $f \in C^5[a,b]$. Several phenomena in physical fields such as neural networks, electric circuits, and ship dynamics can be expressed in differential equations (DEs) forms (see [1–3]).

Block method is one of the efficient methods proposed in 1953 by [4] to enhance the performance of the numerical methods. In 1967, [5] employed block method to provide the essential starting values needed for the predictor schemes. Subsequently, hybrid methods were initiated by [6], which involve the evaluation of functions off-step (nonstep) points. The introduction of hybrid (off-step) points in block methods has many advantages such as the ability to change step size, utilizing data off-step points, and the most important feature according to [7] is the capability to circumvent zero stability barrier condition ([8]).

To increase the accuracy further and to solve the stiffness problem in ordinary differential equations (ODEs), [9] derived second derivative multistep methods for stiff ODEs. Recently, authors like in [10, 11] proposed high derivative methods for the same reason. The former developed block hybrid-second derivative method for stiff systems, while the latter introduced a Simpson's type second derivative method for the solution of a first-order stiff system of IVP. In addition, [12] proposed a continuous fourth derivative method for third-order boundary value problems. Following these scholars' footsteps a new generalized three-hybrid one-step fifth derivative method for solving fourth-order ODEs directly using the approach of interpolation and collocation will be proposed.

This article consists of five sections: Section 1 is for the introduction, while Section 2 illustrates the method derivation, where we consider three-off-step points and employing the collocation approach. The analysis of the approach is discussed in Section 3 which includes zero stability, order, consistency, and convergence. Section 4 covers the solution of some mathematical problems to show the performance of the developed method. Finally, a brief conclusion is performed in Section 5.

2. Development of the Method

Let the following power series polynomial be the approximate solution of (1)

$$\psi(x) = \sum_{j=0}^{2m+l-1} a_j \left(\frac{x-x_n}{h}\right)^j, \tag{2}$$

where $l = 4$ and $m = 5$, the number of interpolation and collocation points. Now, differentiating (2) four and five times yields

$$\begin{aligned} \psi^{iv}(x) &= \sum_{j=4}^{2m+l-1} \frac{a_j j!}{h^4 (j-4)!} \left(\frac{x-x_n}{h}\right)^{j-4} \\ &= f\left(x, y, y', y'', y'''\right), \\ \psi^{v}(x) &= \sum_{j=5}^{2m+l-1} \frac{a_j j!}{h^5 (j-5)!} \left(\frac{x-x_n}{h}\right)^{j-5} \\ &= g\left(x, y, y', y'', y'''\right). \end{aligned} \tag{3}$$

Interpolate (2) at $x_{n+i} = x_n + ih$, $i = \{0, r, s, t\}$, and collocate (3) at all points $x_{n+i} = x_n + ih$, $i = \{0, r, s, t, 1\}$, where $\{r, s, t\} \in (0, 1)$ produces a system of equations in matrix form as below

$$\begin{bmatrix} 1 & 0 & 0 & 0 & 0 & 0 & 0 & \cdots & 0 \\ 1 & r & r^2 & r^3 & r^4 & r^5 & r^6 & \cdots & r^{\hat{v}} \\ 1 & s & s^2 & s^3 & s^4 & s^5 & s^6 & \cdots & s^{\hat{v}} \\ 1 & t & t^2 & t^3 & t^4 & t^5 & t^6 & \cdots & t^{\hat{v}} \\ 0 & 0 & 0 & 0 & \frac{4!}{0!h^4} & 0 & 0 & \cdots & 0 \\ 0 & 0 & 0 & 0 & \frac{4!}{0!h^4} & \frac{5!r}{1!h^4} & \frac{6!r^2}{2!h^4} & \cdots & \frac{\hat{v}!r^{(\hat{v}-4)}}{(\hat{v}-4)!h^4} \\ 0 & 0 & 0 & 0 & \frac{4!}{0!h^4} & \frac{5!s}{1!h^4} & \frac{6!s^2}{2!h^4} & \cdots & \frac{\hat{v}!s^{(\hat{v}-4)}}{(\hat{v}-4)!h^4} \\ 0 & 0 & 0 & 0 & \frac{4!}{0!h^4} & \frac{5!t}{1!h^4} & \frac{6!t^2}{2!h^4} & \cdots & \frac{\hat{v}!t^{(\hat{v}-4)}}{(\hat{v}-4)!h^4} \\ 0 & 0 & 0 & 0 & \frac{4!}{0!h^4} & \frac{5!}{1!h^4} & \frac{6!}{2!h^4} & \cdots & \frac{\hat{v}!}{(\hat{v}-4)!h^4} \\ 0 & 0 & 0 & 0 & 0 & \frac{5!}{0!h^5} & 0 & \cdots & 0 \\ 0 & 0 & 0 & 0 & 0 & \frac{5!}{0!h^5} & \frac{6!r}{1!h^5} & \cdots & \frac{(\hat{v})!r^{(\hat{v}-5)}}{(\hat{v}-5)!h^5} \\ 0 & 0 & 0 & 0 & 0 & \frac{5!}{0!h^5} & \frac{6!s}{1!h^5} & \cdots & \frac{(\hat{v})!s^{(\hat{v}-5)}}{(\hat{v}-5)!h^5} \\ 0 & 0 & 0 & 0 & 0 & \frac{5!}{0!h^5} & \frac{6!t}{1!h^5} & \cdots & \frac{(\hat{v})!t^{(\hat{v}-5)}}{(\hat{v}-5)!h^5} \\ 0 & 0 & 0 & 0 & 0 & \frac{5!}{0!h^5} & \frac{6!}{1!h^5} & \cdots & \frac{(\hat{v})!}{(\hat{v}-5)!h^5} \end{bmatrix} \begin{bmatrix} a_0 \\ a_1 \\ a_2 \\ a_3 \\ a_4 \\ a_5 \\ a_6 \\ a_7 \\ a_8 \\ a_9 \\ a_{10} \\ a_{11} \\ a_{12} \\ a_{13} \end{bmatrix} = \begin{bmatrix} y_n \\ y_{n+r} \\ y_{n+s} \\ y_{n+t} \\ f_n \\ f_{n+r} \\ f_{n+s} \\ f_{n+t} \\ f_{n+1} \\ g_n \\ g_{n+r} \\ g_{n+s} \\ g_{n+t} \\ g_{n+1} \end{bmatrix}, \tag{4}$$

where $\hat{v} = 2m + l - 1$. Matrix manipulation is then employed to solve the resulting system (4) for the unknown coefficients $a_j^{\prime}s$, $j = \{0, 1, \dots, 13\}$. Substituting the obtained values of the coefficients back into (2) yields

$$\begin{aligned} \psi(x) &= \sum_{i=0,r,s,t} \alpha_i y(x+ih) \\ &+ h^4 \left(\sum_{i=r,s,t} \beta_i f(x+ih) + \sum_{i=0}^{1} \beta_i f(x+ih) \right) \\ &+ h^5 \left(\sum_{i=r,s,t} \gamma_i g(x+ih) + \sum_{i=0}^{1} \gamma_i g(x+ih) \right), \end{aligned} \tag{5}$$

where $n = 0, 1, 2, \dots, N-1$, $h = x_n - x_{n-1}$ is the constant step size for the partition π_N of the interval $[a, b]$ which is given by $\pi_N = [a = x_0 < x_1 < \cdots < x_{N-1} < x_N = b]$, α_i, β_i, and γ_i are undetermined constants listed in Appendix I in the Supplementary Material (see Supplementary Material available online at https://doi.org/10.1155/2017/7637651). For simplicity we shall use $y(x+ih) = y_{n+i}$, $f(x+ih) = f_{n+i}$, $g(x+ih) = g_{n+i}$, and $g_{n+i} = f_{n+i}^{\prime}$.

Calculating the first, second, and third derivatives of (5) produces

$$\begin{aligned} \psi'(x) &= \frac{d}{dx} \left(\sum_{i=0,r,s,t} \alpha_i y_{n+i} \right) \\ &+ h^4 \frac{d}{dx} \left(\sum_{i=r,s,t} \beta_i f_{n+i} + \sum_{i=0}^{1} \beta_i f_{n+i} \right) \\ &+ h^5 \frac{d}{dx} \left(\sum_{i=r,s,t} \gamma_i g_{n+i} + \sum_{i=0}^{1} \gamma_i g_{n+i} \right), \end{aligned} \tag{6}$$

$$\psi''(x) = \frac{d^2}{dx^2}\left(\sum_{i=0,r,s,t} \alpha_i y_{n+i}\right) + h^4 \frac{d^2}{dx^2}\left(\sum_{i=r,s,t} \beta_i f_{n+i} + \sum_{i=0}^{1} \beta_i f_{n+i}\right) + h^5 \frac{d^2}{dx^2}\left(\sum_{i=r,s,t} \gamma_i g_{n+i} + \sum_{i=0}^{1} \gamma_i g_{n+i}\right), \quad (7)$$

$$\psi'''(x) = \frac{d^3}{dx^3}\left(\sum_{i=0,r,s,t} \alpha_i y_{n+i}\right) + h^4 \frac{d^3}{dx^3}\left(\sum_{i=r,s,t} \beta_i f_{n+i} + \sum_{i=0}^{1} \beta_i f_{n+i}\right) + h^5 \frac{d^3}{dx^3}\left(\sum_{i=r,s,t} \gamma_i g_{n+i} + \sum_{i=0}^{1} \gamma_i g_{n+i}\right). \quad (8)$$

Evaluating (5) at the noninterpolating points $\{x_{n+t}, x_{n+1}\}$ with (6) and (7) at all points $\{x_{n+i}\}$, $i = \{0, r, s, t, 1\}$, gives the following general equations in block form:

$$A^{[0]} Y_m^{[1]} = A^{[1]} Y_m^{[0]} + \sum_{i=0}^{1} B^{[i]} F_m^{[i]} + \sum_{i=0}^{1} D^{[i]} G_m^{[i]}, \quad (9)$$

where $A^{[0]}$ is an identity matrix of order 12 and

$$A^{[1]} = \begin{bmatrix}
0&0&0&1&0&0&0&rh&0&0&0&\frac{r^2h^2}{2}&0&0&0&\frac{r^3h^3}{6}\\
0&0&0&1&0&0&0&sh&0&0&0&\frac{s^2h^2}{2}&0&0&0&\frac{s^3h^3}{6}\\
0&0&0&1&0&0&0&th&0&0&0&\frac{t^2h^2}{2}&0&0&0&\frac{t^3h^3}{6}\\
0&0&0&1&0&0&0&h&0&0&0&\frac{h^2}{2}&0&0&0&\frac{h^3}{6}\\
0&0&0&0&0&0&0&1&0&0&0&rh&0&0&0&\frac{r^2h^2}{2}\\
0&0&0&0&0&0&0&1&0&0&0&sh&0&0&0&\frac{s^2h^2}{2}\\
0&0&0&0&0&0&0&1&0&0&0&th&0&0&0&\frac{t^2h^2}{2}\\
0&0&0&0&0&0&0&1&0&0&0&h&0&0&0&\frac{h^2}{2}\\
0&0&0&0&0&0&0&0&0&0&0&1&0&0&0&rh\\
0&0&0&0&0&0&0&0&0&0&0&1&0&0&0&sh\\
0&0&0&0&0&0&0&0&0&0&0&1&0&0&0&th\\
0&0&0&0&0&0&0&0&0&0&0&1&0&0&0&h\\
0&0&0&0&0&0&0&0&0&0&0&0&0&0&0&1\\
0&0&0&0&0&0&0&0&0&0&0&0&0&0&0&1\\
0&0&0&0&0&0&0&0&0&0&0&0&0&0&0&1\\
0&0&0&0&0&0&0&0&0&0&0&0&0&0&0&1
\end{bmatrix},$$

$$B^{[0]} = \begin{bmatrix}
0&0&0&0&0&0&0&0&0&0&0&0&0&0&0&B_{116}^{[0]}\\
0&0&0&0&0&0&0&0&0&0&0&0&0&0&0&B_{216}^{[0]}\\
0&0&0&0&0&0&0&0&0&0&0&0&0&0&0&B_{316}^{[0]}\\
0&0&0&0&0&0&0&0&0&0&0&0&0&0&0&B_{416}^{[0]}\\
0&0&0&0&0&0&0&0&0&0&0&0&0&0&0&B_{516}^{[0]}\\
0&0&0&0&0&0&0&0&0&0&0&0&0&0&0&B_{616}^{[0]}\\
0&0&0&0&0&0&0&0&0&0&0&0&0&0&0&B_{716}^{[0]}\\
0&0&0&0&0&0&0&0&0&0&0&0&0&0&0&B_{816}^{[0]}\\
0&0&0&0&0&0&0&0&0&0&0&0&0&0&0&B_{916}^{[0]}\\
0&0&0&0&0&0&0&0&0&0&0&0&0&0&0&B_{1016}^{[0]}\\
0&0&0&0&0&0&0&0&0&0&0&0&0&0&0&B_{1116}^{[0]}\\
0&0&0&0&0&0&0&0&0&0&0&0&0&0&0&B_{1216}^{[0]}\\
0&0&0&0&0&0&0&0&0&0&0&0&0&0&0&B_{1316}^{[0]}\\
0&0&0&0&0&0&0&0&0&0&0&0&0&0&0&B_{1416}^{[0]}\\
0&0&0&0&0&0&0&0&0&0&0&0&0&0&0&B_{1516}^{[0]}\\
0&0&0&0&0&0&0&0&0&0&0&0&0&0&0&B_{1616}^{[0]}
\end{bmatrix},$$

$$B^{[1]} = \begin{bmatrix}
B_{11}^{[1]}&B_{12}^{[1]}&B_{13}^{[1]}&B_{14}^{[1]}\\
B_{21}^{[1]}&B_{22}^{[1]}&B_{23}^{[1]}&B_{24}^{[1]}\\
B_{31}^{[1]}&B_{32}^{[1]}&B_{33}^{[1]}&B_{34}^{[1]}\\
B_{41}^{[1]}&B_{42}^{[1]}&B_{43}^{[1]}&B_{44}^{[1]}\\
B_{51}^{[1]}&B_{52}^{[1]}&B_{53}^{[1]}&B_{54}^{[1]}\\
B_{61}^{[1]}&B_{62}^{[1]}&B_{63}^{[1]}&B_{64}^{[1]}\\
B_{71}^{[1]}&B_{72}^{[1]}&B_{73}^{[1]}&B_{74}^{[1]}\\
B_{81}^{[1]}&B_{82}^{[1]}&B_{83}^{[1]}&B_{84}^{[1]}\\
B_{91}^{[1]}&B_{92}^{[1]}&B_{93}^{[1]}&B_{94}^{[1]}\\
B_{101}^{[1]}&B_{102}^{[1]}&B_{103}^{[1]}&B_{104}^{[1]}\\
B_{111}^{[1]}&B_{112}^{[1]}&B_{113}^{[1]}&B_{114}^{[1]}\\
B_{121}^{[1]}&B_{122}^{[1]}&B_{123}^{[1]}&B_{124}^{[1]}\\
B_{131}^{[1]}&B_{132}^{[1]}&B_{133}^{[1]}&B_{134}^{[1]}\\
B_{141}^{[1]}&B_{142}^{[1]}&B_{143}^{[1]}&B_{144}^{[1]}\\
B_{151}^{[1]}&B_{152}^{[1]}&B_{153}^{[1]}&B_{154}^{[1]}\\
B_{161}^{[1]}&B_{162}^{[1]}&B_{163}^{[1]}&B_{164}^{[1]}
\end{bmatrix},$$

$$D^{[0]} = \begin{bmatrix} 0 & 0 & 0 & 0 & 0 & 0 & 0 & 0 & 0 & 0 & 0 & 0 & 0 & 0 & 0 & D^{[0]}_{116} \\ 0 & 0 & 0 & 0 & 0 & 0 & 0 & 0 & 0 & 0 & 0 & 0 & 0 & 0 & 0 & D^{[0]}_{216} \\ 0 & 0 & 0 & 0 & 0 & 0 & 0 & 0 & 0 & 0 & 0 & 0 & 0 & 0 & 0 & D^{[0]}_{316} \\ 0 & 0 & 0 & 0 & 0 & 0 & 0 & 0 & 0 & 0 & 0 & 0 & 0 & 0 & 0 & D^{[0]}_{416} \\ 0 & 0 & 0 & 0 & 0 & 0 & 0 & 0 & 0 & 0 & 0 & 0 & 0 & 0 & 0 & D^{[0]}_{516} \\ 0 & 0 & 0 & 0 & 0 & 0 & 0 & 0 & 0 & 0 & 0 & 0 & 0 & 0 & 0 & D^{[0]}_{616} \\ 0 & 0 & 0 & 0 & 0 & 0 & 0 & 0 & 0 & 0 & 0 & 0 & 0 & 0 & 0 & D^{[0]}_{716} \\ 0 & 0 & 0 & 0 & 0 & 0 & 0 & 0 & 0 & 0 & 0 & 0 & 0 & 0 & 0 & D^{[0]}_{816} \\ 0 & 0 & 0 & 0 & 0 & 0 & 0 & 0 & 0 & 0 & 0 & 0 & 0 & 0 & 0 & D^{[0]}_{916} \\ 0 & 0 & 0 & 0 & 0 & 0 & 0 & 0 & 0 & 0 & 0 & 0 & 0 & 0 & 0 & D^{[0]}_{1016} \\ 0 & 0 & 0 & 0 & 0 & 0 & 0 & 0 & 0 & 0 & 0 & 0 & 0 & 0 & 0 & D^{[0]}_{1116} \\ 0 & 0 & 0 & 0 & 0 & 0 & 0 & 0 & 0 & 0 & 0 & 0 & 0 & 0 & 0 & D^{[0]}_{1216} \\ 0 & 0 & 0 & 0 & 0 & 0 & 0 & 0 & 0 & 0 & 0 & 0 & 0 & 0 & 0 & D^{[0]}_{1316} \\ 0 & 0 & 0 & 0 & 0 & 0 & 0 & 0 & 0 & 0 & 0 & 0 & 0 & 0 & 0 & D^{[0]}_{1416} \\ 0 & 0 & 0 & 0 & 0 & 0 & 0 & 0 & 0 & 0 & 0 & 0 & 0 & 0 & 0 & D^{[0]}_{1516} \\ 0 & 0 & 0 & 0 & 0 & 0 & 0 & 0 & 0 & 0 & 0 & 0 & 0 & 0 & 0 & D^{[0]}_{1616} \end{bmatrix},$$

$$D^{[1]} = \begin{bmatrix} D^{[1]}_{11} & D^{[1]}_{12} & D^{[1]}_{13} & D^{[1]}_{14} \\ D^{[1]}_{21} & D^{[1]}_{22} & D^{[1]}_{23} & D^{[1]}_{24} \\ D^{[1]}_{31} & D^{[1]}_{32} & D^{[1]}_{33} & D^{[1]}_{34} \\ D^{[1]}_{41} & D^{[1]}_{42} & D^{[1]}_{43} & D^{[1]}_{44} \\ D^{[1]}_{51} & D^{[1]}_{52} & D^{[1]}_{53} & D^{[1]}_{54} \\ D^{[1]}_{61} & D^{[1]}_{62} & D^{[1]}_{63} & D^{[1]}_{64} \\ D^{[1]}_{71} & D^{[1]}_{72} & D^{[1]}_{73} & D^{[1]}_{74} \\ D^{[1]}_{81} & D^{[1]}_{82} & D^{[1]}_{83} & D^{[1]}_{84} \\ D^{[1]}_{91} & D^{[1]}_{92} & D^{[1]}_{93} & D^{[1]}_{94} \\ D^{[1]}_{101} & D^{[1]}_{102} & D^{[1]}_{103} & D^{[1]}_{104} \\ D^{[1]}_{111} & D^{[1]}_{112} & D^{[1]}_{113} & D^{[1]}_{114} \\ D^{[1]}_{121} & D^{[1]}_{122} & D^{[1]}_{123} & D^{[1]}_{124} \\ D^{[1]}_{131} & D^{[1]}_{132} & D^{[1]}_{133} & D^{[1]}_{134} \\ D^{[1]}_{141} & D^{[1]}_{142} & D^{[1]}_{143} & D^{[1]}_{144} \\ D^{[1]}_{151} & D^{[1]}_{152} & D^{[1]}_{153} & D^{[1]}_{154} \\ D^{[1]}_{161} & D^{[1]}_{162} & D^{[1]}_{163} & D^{[1]}_{164} \end{bmatrix}. \tag{10}$$

The entries of $B^{[0]}$, $B^{[1]}$, $D^{[0]}$, and $D^{[1]}$ are listed in Appendix II in the Supplementary Material section, while the vectors $Y^{[0]}_m, Y^{[1]}_m, F^{[0]}_m, F^{[1]}_m, G^{[0]}_m, G^{[1]}_m$ are defined as follows:

$$Y^{[0]}_m = \begin{bmatrix} y_{n-t} \\ y_{n-s} \\ y_{n-r} \\ y_n \\ y'_{n-t} \\ y'_{n-s} \\ y'_{n-r} \\ y'_n \\ y''_{n-t} \\ y''_{n-s} \\ y''_{n-r} \\ y''_n \\ y'''_{n-t} \\ y'''_{n-s} \\ y'''_{n-r} \\ y'''_n \end{bmatrix},$$

$$Y^{[1]}_m = \begin{bmatrix} y_{n+r} \\ y_{n+s} \\ y_{n+t} \\ y_{n+1} \\ y'_{n+r} \\ y'_{n+s} \\ y'_{n+t} \\ y'_{n+1} \\ y''_{n+r} \\ y''_{n+s} \\ y''_{n+t} \\ y''_{n+1} \\ y'''_{n+r} \\ y'''_{n+s} \\ y'''_{n+t} \\ y'''_{n+1} \end{bmatrix},$$

$$F^{[0]}_m = \begin{bmatrix} f_{n-15} \\ f_{n-14} \\ f_{n-13} \\ f_{n-12} \\ f_{n-11} \\ f_{n-10} \\ f_{n-9} \\ f_{n-8} \\ f_{n-7} \\ f_{n-6} \\ f_{n-5} \\ f_{n-4} \\ f_{n-3} \\ f_{n-2} \\ f_{n-1} \\ f_n \end{bmatrix},$$

$$F_m^{[1]} = \begin{bmatrix} f_{n+r} \\ f_{n+s} \\ f_{n+t} \\ f_{n+1} \end{bmatrix}, \qquad G_m^{[1]} = \begin{bmatrix} g_{n+r} \\ g_{n+s} \\ g_{n+t} \\ g_{n+1} \end{bmatrix}.$$

$$G_m^{[0]} = \begin{bmatrix} g_{n-15} \\ g_{n-14} \\ g_{n-13} \\ g_{n-12} \\ g_{n-11} \\ g_{n-10} \\ g_{n-9} \\ g_{n-8} \\ g_{n-7} \\ g_{n-6} \\ g_{n-5} \\ g_{n-4} \\ g_{n-3} \\ g_{n-2} \\ g_{n-1} \\ g_n \end{bmatrix}, \tag{11}$$

3. Analysis of the Method

3.1. Zero Stability

Definition 1. The hybrid block method formula (9) is said to be zero stable if no root R_m of the first characteristic equation $\rho(R)$ has modulus greater than one; that is, $|R_m| \leqslant 1$, and if $R_m = 1$ then the multiplicity of R_m must not exceed four.

To prove that the roots of the first characteristic equation satisfy the previous definition

$$\rho(R) = \det\left[RA^{[0]} - A^{[1]}\right] = \begin{vmatrix}
R & 0 & 0 & -1 & 0 & 0 & 0 & -rh & 0 & 0 & 0 & -\frac{r^2h^2}{2} & 0 & 0 & 0 & -\frac{r^3h^3}{6} \\
0 & R & 0 & -1 & 0 & 0 & 0 & -sh & 0 & 0 & 0 & -\frac{s^2h^2}{2} & 0 & 0 & 0 & -\frac{s^3h^3}{6} \\
0 & 0 & R & -1 & 0 & 0 & 0 & -th & 0 & 0 & 0 & -\frac{t^2h^2}{2} & 0 & 0 & 0 & -\frac{t^3h^3}{6} \\
0 & 0 & 0 & R-1 & 0 & 0 & 0 & -h & 0 & 0 & 0 & -\frac{h^2}{2} & 0 & 0 & 0 & -\frac{h^3}{6} \\
0 & 0 & 0 & 0 & R & 0 & 0 & -1 & 0 & 0 & 0 & -rh & 0 & 0 & 0 & -\frac{r^2h^2}{2} \\
0 & 0 & 0 & 0 & 0 & R & 0 & -1 & 0 & 0 & 0 & -sh & 0 & 0 & 0 & -\frac{s^2h^2}{2} \\
0 & 0 & 0 & 0 & 0 & 0 & R & -1 & 0 & 0 & 0 & -th & 0 & 0 & 0 & -\frac{t^2h^2}{2} \\
0 & 0 & 0 & 0 & 0 & 0 & 0 & R-1 & 0 & 0 & 0 & -h & 0 & 0 & 0 & -\frac{h^2}{2} \\
0 & 0 & 0 & 0 & 0 & 0 & 0 & 0 & R & 0 & 0 & -1 & 0 & 0 & 0 & -rh \\
0 & 0 & 0 & 0 & 0 & 0 & 0 & 0 & 0 & R & 0 & -1 & 0 & 0 & 0 & -sh \\
0 & 0 & 0 & 0 & 0 & 0 & 0 & 0 & 0 & 0 & R & -1 & 0 & 0 & 0 & -th \\
0 & 0 & 0 & 0 & 0 & 0 & 0 & 0 & 0 & 0 & 0 & R-1 & 0 & 0 & 0 & -h \\
0 & 0 & 0 & 0 & 0 & 0 & 0 & 0 & 0 & 0 & 0 & 0 & R & 0 & 0 & -1 \\
0 & 0 & 0 & 0 & 0 & 0 & 0 & 0 & 0 & 0 & 0 & 0 & 0 & R & 0 & -1 \\
0 & 0 & 0 & 0 & 0 & 0 & 0 & 0 & 0 & 0 & 0 & 0 & 0 & 0 & R & -1 \\
0 & 0 & 0 & 0 & 0 & 0 & 0 & 0 & 0 & 0 & 0 & 0 & 0 & 0 & 0 & R-1
\end{vmatrix} = \mathbf{0}, \tag{12}$$

which imply that

$$R^{12}(R-1)^4 = 0 \Longrightarrow$$

$$R_m = \begin{cases} 0, & \text{if } m = 1\,(1)\,12, \\ 1, & \text{if } m = 13\,(1)\,16. \end{cases} \tag{13}$$

As a result, the developed method is zero stable.

3.2. Order of the Method. The linear operator Θ associated with the hybrid block methods formula (9) is defined as

$$\Theta\{y(x);h\} = A^{[0]}Y_m^{[1]} - A^{[1]}Y_m^{[0]} - \sum_{i=0}^{1} B^{[i]}F_m^{[i]} - \sum_{i=0}^{1} D^{[i]}G_m^{[i]}. \tag{14}$$

Expanding the above equation in Taylor series and combining like terms imply

$$\Theta\{y(x);h\} = \widehat{C}_0 h^0 y(x) + \widehat{C}_1 h^1 y'(x) + \widehat{C}_2 h^2 y''(x) + \cdots + \widehat{C}_{p+4} h^{p+4} y^{(p+4)}(x) + \cdots. \quad (15)$$

According to [7, 13] method (9) is said to be of order p if

$$\widehat{C}_0 = \widehat{C}_1 = \cdots = \widehat{C}_{p+3} = 0,$$

$$\widehat{C}_{p+4} \neq 0. \quad (16)$$

The term $\widehat{C}_{p+4}$ is called the error constant and the local truncation error is given by

$$t_{n+k} = \widehat{C}_{p+4} y^{(p+4)} h^{p+4} + O\left(h^{p+5}\right). \quad (17)$$

Comparing like terms of $y^{(i)}$ and h^i in (15) produces the coefficients $\widehat{C}_0 = \widehat{C}_1 = \cdots = \widehat{C}_{13} = 0$ with vector of error constants

$$\widehat{C}_{14} = \left[\widehat{C}_{14}^{[1]}\ \widehat{C}_{14}^{[2]}\ \widehat{C}_{14}^{[3]}\ \widehat{C}_{14}^{[4]}\ \widehat{C}_{14}^{[5]}\ \widehat{C}_{14}^{[6]}\ \widehat{C}_{14}^{[7]}\ \widehat{C}_{14}^{[8]}\ \widehat{C}_{14}^{[9]}\ \widehat{C}_{14}^{[10]}\ \widehat{C}_{14}^{[11]}\ \widehat{C}_{14}^{[12]}\ \widehat{C}_{14}^{[13]}\ \widehat{C}_{14}^{[14]}\ \widehat{C}_{14}^{[15]}\ \widehat{C}_{14}^{[16]}\right]^T, \quad (18)$$

where

$$\begin{aligned}\widehat{C}_{14}^{[1]} = {} & \frac{r^8}{7846046208000}\left(20r^6 - 70r^5 s - 70r^5 t - 70r^5\right. \\ & + 65r^4 s^2 + 260r^4 st + 260r^4 s + 65r^4 t^2 + 260r^4 t \\ & + 65r^4 - 260r^3 s^2 t - 260r^3 s^2 - 260r^3 st^2 \\ & - 1040r^3 st - 260r^3 s - 260r^3 t^2 - 260r^3 t \\ & + 286r^2 s^2 t^2 + 1144r^2 s^2 t + 286r^2 s^2 + 1144r^2 st^2 \\ & + 1144r^2 st + 286r^2 t^2 - 1430rs^2 t^2 - 1430rs^2 t \\ & \left. - 1430rst^2 + 2145s^2 t^2\right),\end{aligned}$$

$$\begin{aligned}\widehat{C}_{14}^{[2]} = {} & \frac{s^8}{7846046208000}\left(65r^2 s^4 - 260r^2 s^3 t - 260r^2 s^3\right. \\ & + 286r^2 s^2 t^2 + 1144r^2 s^2 t + 286r^2 s^2 - 1430r^2 st^2 \\ & - 1430r^2 st + 2145r^2 t^2 - 70rs^5 + 260rs^4 t + 260rs^4 \\ & - 260rs^3 t^2 - 1040rs^3 t - 260rs^3 + 1144rs^2 t^2 \\ & + 1144rs^2 t - 1430rst^2 + 20s^6 - 70s^5 t - 70s^5 \\ & + 65s^4 t^2 + 260s^4 t + 65s^4 - 260s^3 t^2 - 260s^3 t \\ & \left. + 286s^2 t^2\right),\end{aligned}$$

$$\begin{aligned}\widehat{C}_{14}^{[3]} = {} & \frac{t^8}{7846046208000}\left(286r^2 s^2 t^2 - 1430r^2 s^2 t\right. \\ & + 2145r^2 s^2 - 260r^2 st^3 + 1144r^2 st^2 - 1430r^2 st \\ & + 65r^2 t^4 - 260r^2 t^3 + 286r^2 t^2 - 260rs^2 t^3 \\ & + 1144rs^2 t^2 - 1430rs^2 t + 260rst^4 - 1040rst^3 \\ & + 1144rst^2 - 70rt^5 + 260rt^4 - 260rt^3 + 65s^2 t^4 \\ & - 260s^2 t^3 + 286s^2 t^2 - 70st^5 + 260st^4 - 260st^3 \\ & \left. + 20t^6 - 70t^5 + 65t^4\right),\end{aligned}$$

$$\begin{aligned}\widehat{C}_{14}^{[4]} = {} & \frac{1}{7846046208000}\left(2145r^2 s^2 t^2 - 1430r^2 s^2 t\right. \\ & + 286r^2 s^2 - 1430r^2 st^2 + 1144r^2 st - 260r^2 s \\ & + 286r^2 t^2 - 260r^2 t + 65r^2 - 1430rs^2 t^2 + 1144rs^2 t \\ & - 260rs^2 + 1144rst^2 - 1040rst + 260rs - 260rt^2 \\ & + 260rt - 70r + 286s^2 t^2 - 260s^2 t + 65s^2 - 260st^2 \\ & \left. + 260st - 70s + 65t^2 - 70t + 20\right),\end{aligned}$$

$$\begin{aligned}\widehat{C}_{14}^{[5]} = {} & \frac{r^7}{1307674368000}\left(28r^6 - 91r^5 s - 91r^5 t - 91r^5\right. \\ & + 78r^4 s^2 + 312r^4 st + 312r^4 s + 78r^4 t^2 + 312r^4 t \\ & + 78r^4 - 286r^3 s^2 t - 286r^3 s^2 - 286r^3 st^2 \\ & - 1144r^3 st - 286r^3 s - 286r^3 t^2 - 286r^3 t \\ & + 286r^2 s^2 t^2 + 1144r^2 s^2 t + 286r^2 s^2 + 1144r^2 st^2 \\ & + 1144r^2 st + 286r^2 t^2 - 1287rs^2 t^2 - 1287rs^2 t \\ & \left. - 1287rst^2 + 1716s^2 t^2\right),\end{aligned}$$

$$\begin{aligned}\widehat{C}_{14}^{[6]} = {} & \frac{s^7}{1307674368000}\left(78r^2 s^4 - 286r^2 s^3 t - 286r^2 s^3\right. \\ & + 286r^2 s^2 t^2 + 1144r^2 s^2 t + 286r^2 s^2 - 1287r^2 st^2 \\ & - 1287r^2 st + 1716r^2 t^2 - 91rs^5 + 312rs^4 t + 312rs^4 \\ & - 286rs^3 t^2 - 1144rs^3 t - 286rs^3 + 1144rs^2 t^2 \\ & + 1144rs^2 t - 1287rst^2 + 28s^6 - 91s^5 t - 91s^5 \\ & + 78s^4 t^2 + 312s^4 t + 78s^4 - 286s^3 t^2 - 286s^3 t \\ & \left. + 286s^2 t^2\right),\end{aligned}$$

$$
\begin{aligned}
\widehat{C}_{14}^{[7]} = {} & \frac{t^7}{1307674368000}\big(286r^2s^2t^2 - 1287r^2s^2t \\
& + 1716r^2s^2 - 286r^2st^3 + 1144r^2st^2 - 1287r^2st \\
& + 78r^2t^4 - 286r^2t^3 + 286r^2t^2 - 286rs^2t^3 \\
& + 1144rs^2t^2 - 1287rs^2t + 312rst^4 - 1144rst^3 \\
& + 1144rst^2 - 91rt^5 + 312rt^4 - 286rt^3 + 78s^2t^4 \\
& - 286s^2t^3 + 286s^2t^2 - 91st^5 + 312st^4 - 286st^3 \\
& + 28t^6 - 91t^5 + 78t^4\big),
\end{aligned}
$$

$$
\begin{aligned}
\widehat{C}_{14}^{[8]} = {} & \frac{1}{1307674368000}\big(1716r^2s^2t^2 - 1287r^2s^2t \\
& + 286r^2s^2 - 1287r^2st^2 + 1144r^2st - 286r^2s \\
& + 286r^2t^2 - 286r^2t + 78r^2 - 1287rs^2t^2 + 1144rs^2t \\
& - 286rs^2 + 1144rst^2 - 1144rst + 312rs - 286rt^2 \\
& + 312rt - 91r + 286s^2t^2 - 286s^2t + 78s^2 - 286st^2 \\
& + 312st - 91s + 78t^2 - 91t + 28\big),
\end{aligned}
$$

$$
\begin{aligned}
\widehat{C}_{14}^{[9]} = {} & \frac{r^6}{100590336000}\big(14r^6 - 42r^5s - 42r^5t - 42r^5 \\
& + 33r^4s^2 + 132r^4st + 132r^4s + 33r^4t^2 + 132r^4t \\
& + 33r^4 - 110r^3s^2t - 110r^3s^2 - 110r^3st^2 - 440r^3st \\
& - 110r^3s - 110r^3t^2 - 110r^3t + 99r^2s^2t^2 + 396r^2s^2t \\
& + 99r^2s^2 + 396r^2st^2 + 396r^2st + 99r^2t^2 - 396rs^2t^2 \\
& - 396rs^2t - 396rst^2 + 462s^2t^2\big),
\end{aligned}
$$

$$
\begin{aligned}
\widehat{C}_{14}^{[10]} = {} & \frac{s^6}{100590336000}\big(33r^2s^4 - 110r^2s^3t - 110r^2s^3 \\
& + 99r^2s^2t^2 + 396r^2s^2t + 99r^2s^2 - 396r^2st^2 \\
& - 396r^2st + 462r^2t^2 - 42rs^5 + 132rs^4t + 132rs^4 \\
& - 110rs^3t^2 - 440rs^3t - 110rs^3 + 396rs^2t^2 \\
& + 396rs^2t - 396rst^2 + 14s^6 - 42s^5t - 42s^5 \\
& + 33s^4t^2 + 132s^4t + 33s^4 - 110s^3t^2 - 110s^3t \\
& + 99s^2t^2\big),
\end{aligned}
$$

$$
\begin{aligned}
\widehat{C}_{14}^{[11]} = {} & \frac{t^6}{100590336000}\big(99r^2s^2t^2 - 396r^2s^2t \\
& + 462r^2s^2 - 110r^2st^3 + 396r^2st^2 - 396r^2st \\
& + 33r^2t^4 - 110r^2t^3 + 99r^2t^2 - 110rs^2t^3 + 396rs^2t^2 \\
& - 396rs^2t + 132rst^4 - 440rst^3 + 396rst^2 - 42rt^5 \\
& + 132rt^4 - 110rt^3 + 33s^2t^4 - 110s^2t^3 + 99s^2t^2 \\
& - 42st^5 + 132st^4 - 110st^3 + 14t^6 - 42t^5 + 33t^4\big),
\end{aligned}
$$

$$
\begin{aligned}
\widehat{C}_{14}^{[12]} = {} & \frac{1}{100590336000}\big(462r^2s^2t^2 - 396r^2s^2t \\
& + 99r^2s^2 - 396r^2st^2 + 396r^2st - 110r^2s + 99r^2t^2 \\
& - 110r^2t + 33r^2 - 396rs^2t^2 + 396rs^2t - 110rs^2 \\
& + 396rst^2 - 440rst + 132rs - 110rt^2 + 132rt \\
& - 42r + 99s^2t^2 - 110s^2t + 33s^2 - 110st^2 + 132st \\
& - 42s + 33t^2 - 42t + 14\big),
\end{aligned}
$$

$$
\begin{aligned}
\widehat{C}_{14}^{[13]} = {} & \frac{r^5}{50295168000}\big(28r^6 - 77r^5s - 77r^5t - 77r^5 \\
& + 55r^4s^2 + 220r^4st + 220r^4s + 55r^4t^2 + 220r^4t \\
& + 55r^4 - 165r^3s^2t - 165r^3s^2 - 165r^3st^2 - 660r^3st \\
& - 165r^3s - 165r^3t^2 - 165r^3t + 132r^2s^2t^2 \\
& + 528r^2s^2t + 132r^2s^2 + 528r^2st^2 + 528r^2st \\
& + 132r^2t^2 - 462rs^2t^2 - 462rs^2t - 462rst^2 \\
& + 462s^2t^2\big),
\end{aligned}
$$

$$
\begin{aligned}
\widehat{C}_{14}^{[14]} = {} & \frac{s^5}{50295168000}\big(55r^2s^4 - 165r^2s^3t - 165r^2s^3 \\
& + 132r^2s^2t^2 + 528r^2s^2t + 132r^2s^2 - 462r^2st^2 \\
& - 462r^2st + 462r^2t^2 - 77rs^5 + 220rs^4t + 220rs^4 \\
& - 165rs^3t^2 - 660rs^3t - 165rs^3 + 528rs^2t^2 \\
& + 528rs^2t - 462rst^2 + 28s^6 - 77s^5t - 77s^5 \\
& + 55s^4t^2 + 220s^4t + 55s^4 - 165s^3t^2 - 165s^3t \\
& + 132s^2t^2\big),
\end{aligned}
$$

$$
\begin{aligned}
\widehat{C}_{14}^{[15]} = {} & \frac{t^5}{50295168000}\big(132r^2s^2t^2 - 462r^2s^2t \\
& + 462r^2s^2 - 165r^2st^3 + 528r^2st^2 - 462r^2st \\
& + 55r^2t^4 - 165r^2t^3 + 132r^2t^2 - 165rs^2t^3 \\
& + 528rs^2t^2 - 462rs^2t + 220rst^4 - 660rst^3 \\
& + 528rst^2 - 77rt^5 + 220rt^4 - 165rt^3 + 55s^2t^4 \\
& - 165s^2t^3 + 132s^2t^2 - 77st^5 + 220st^4 - 165st^3 \\
& + 28t^6 - 77t^5 + 55t^4\big),
\end{aligned}
$$

$$\widehat{C}_{14}^{[16]} = \frac{1}{50295168000}\left(462r^2s^2t^2 - 462r^2s^2t + 132r^2s^2 - 462r^2st^2 + 528r^2st - 165r^2s + 132r^2t^2 - 165r^2t + 55r^2 - 462rs^2t^2 + 528rs^2t - 165rs^2 + 528rst^2 - 660rst + 220rs - 165rt^2 + 220rt - 77r + 132s^2t^2 - 165s^2t + 55s^2 - 165st^2 + 220st - 77s + 55t^2 - 77t + 28\right), \tag{19}$$

which concludes that the order of the developed method is $p = 10$.

3.3. Consistency

Definition 2. A block method is said to be consistent if its order p is greater than one.

Consistency property is achieved for the hybrid block method from the above analysis since the order $p = 10 \geq 1$.

3.4. Convergence

Theorem 3 (see [14]). *Consistency and zero stability are sufficient conditions for a linear multistep method to be convergent.*

The hybrid block method (9) is convergent since it fulfils both the consistency and zero stability conditions.

3.5. Region of Absolute Stability. Stability region of the hybrid block (9) is discussed in spirit of [7]. The test problem of the forms $y' = \lambda y$, $y'' = \lambda^2 y$, $y''' = \lambda^3 y$, $y^{iv} = \lambda^4 y$, and $y^v = \lambda^5 y$, where $\lambda < 0$ are substituted into the main methods of block (9), is shown below

$$\overline{A}^{[0]}\overline{Y}_{m+1} = \sum_{i=1}^{4}\overline{A}^{[i]}\overline{Y}_m^{[i]} + \sum_{i=0}^{1}\overline{B}^{[i]}\overline{F}_m^{[i]} + \sum_{i=0}^{1}\overline{D}^{[i]}\overline{G}_m^{[i]}, \tag{20}$$

where

$$\overline{A}^{[0]} = \begin{bmatrix} 1 & 0 & 0 & 0 \\ 0 & 1 & 0 & 0 \\ 0 & 0 & 1 & 0 \\ 0 & 0 & 0 & 1 \end{bmatrix},$$

$$\overline{A}^{[1]} = \begin{bmatrix} 0 & 0 & 0 & 1 \\ 0 & 0 & 0 & 1 \\ 0 & 0 & 0 & 1 \\ 0 & 0 & 0 & 1 \end{bmatrix},$$

$$\overline{A}^{[2]} = \begin{bmatrix} 0 & 0 & 0 & r \\ 0 & 0 & 0 & s \\ 0 & 0 & 0 & t \\ 0 & 0 & 0 & 1 \end{bmatrix},$$

$$\overline{A}^{[3]} = \begin{bmatrix} 0 & 0 & 0 & \frac{r^2}{2} \\ 0 & 0 & 0 & \frac{s^2}{2} \\ 0 & 0 & 0 & \frac{t^2}{2} \\ 0 & 0 & 0 & \frac{1}{2} \end{bmatrix},$$

$$\overline{A}^{[4]} = \begin{bmatrix} 0 & 0 & 0 & \frac{r^3}{6} \\ 0 & 0 & 0 & \frac{s^3}{6} \\ 0 & 0 & 0 & \frac{t^3}{6} \\ 0 & 0 & 0 & \frac{1}{6} \end{bmatrix},$$

$$\overline{B}^{[0]} = \frac{1}{h^4}\begin{bmatrix} 0 & 0 & 0 & B_{116}^{[0]} \\ 0 & 0 & 0 & B_{216}^{[0]} \\ 0 & 0 & 0 & B_{316}^{[0]} \\ 0 & 0 & 0 & B_{416}^{[0]} \end{bmatrix},$$

$$\overline{B}^{[1]} = \frac{1}{h^4}\begin{bmatrix} B_{11}^{[1]} & B_{12}^{[1]} & B_{13}^{[1]} & B_{14}^{[1]} \\ B_{21}^{[1]} & B_{22}^{[1]} & B_{23}^{[1]} & B_{24}^{[1]} \\ B_{31}^{[1]} & B_{32}^{[1]} & B_{33}^{[1]} & B_{34}^{[1]} \\ B_{41}^{[1]} & B_{42}^{[1]} & B_{43}^{[1]} & B_{44}^{[1]} \end{bmatrix},$$

$$\overline{D}^{[0]} = \frac{1}{h^5}\begin{bmatrix} 0 & 0 & 0 & D_{116}^{[0]} \\ 0 & 0 & 0 & D_{216}^{[0]} \\ 0 & 0 & 0 & D_{316}^{[0]} \\ 0 & 0 & 0 & D_{416}^{[0]} \end{bmatrix},$$

$$\overline{D}^{[1]} = \frac{1}{h^5}\begin{bmatrix} D_{11}^{[1]} & D_{12}^{[1]} & D_{13}^{[1]} & D_{14}^{[1]} \\ D_{21}^{[1]} & D_{22}^{[1]} & D_{23}^{[1]} & D_{24}^{[1]} \\ D_{31}^{[1]} & D_{32}^{[1]} & D_{33}^{[1]} & D_{34}^{[1]} \\ D_{41}^{[1]} & D_{42}^{[1]} & D_{43}^{[1]} & D_{44}^{[1]} \end{bmatrix}, \tag{21}$$

while the vectors $\overline{Y}_{m+1}, \overline{Y}_m^{[1]}, \overline{Y}_m^{[2]}, \overline{Y}_m^{[3]}, \overline{F}_m^{[0]}, \overline{F}_m^{[1]}, \overline{G}_m^{[0]}$, and $\overline{G}_m^{[1]}$ are defined as follows:

$$\overline{Y}_{m+1} = \begin{bmatrix} y_{n+r} \\ y_{n+s} \\ y_{n+t} \\ y_{n+1} \end{bmatrix},$$

$$\overline{Y}_m^{[1]} = \begin{bmatrix} y_{n-t} \\ y_{n-s} \\ y_{n-r} \\ y_n \end{bmatrix},$$

$$\overline{Y}_m^{[2]} = \begin{bmatrix} y'_{n-t} \\ y'_{n-s} \\ y'_{n-r} \\ y'_n \end{bmatrix}, \quad \overline{Y}_m^{[3]} = \begin{bmatrix} y''_{n-t} \\ y''_{n-s} \\ y''_{n-r} \\ y''_n \end{bmatrix}, \quad \overline{Y}_m^{[4]} = \begin{bmatrix} y'''_{n-t} \\ y'''_{n-s} \\ y'''_{n-r} \\ y'''_n \end{bmatrix},$$

$$\overline{F}_m^{[0]} = \begin{bmatrix} f_{n-3} \\ f_{n-2} \\ f_{n-1} \\ f_n \end{bmatrix}, \quad \overline{F}_m^{[1]} = \begin{bmatrix} f_{n+r} \\ f_{n+s} \\ f_{n+t} \\ f_{n+1} \end{bmatrix}, \quad \overline{G}_m^{[0]} = \begin{bmatrix} g_{n-3} \\ g_{n-2} \\ g_{n-1} \\ g_n \end{bmatrix}, \quad \overline{G}_m^{[1]} = \begin{bmatrix} g_{n+r} \\ g_{n+s} \\ g_{n+t} \\ g_{n+1} \end{bmatrix}, \tag{22}$$

substituting $q = \lambda h$ in (20) yields

$$\overline{Y}_{m+1} = M(q)\overline{Y}_m^{[1]}, \tag{23}$$

where the amplification matrix $M(q)$ is given by

$$M(q) = \left(\overline{A}^{[0]} - q^4\overline{B}^{[1]} - q^5\overline{D}^{[1]}\right)^{-1}\left(\overline{A}^{[1]} + q\overline{A}^{[2]} + q^2\overline{A}^{[3]} + q^3\overline{A}^{[4]} + q^4\overline{B}^{[0]} + q^5\overline{D}^{[0]}\right). \tag{24}$$

The analysis shows that the matrix $M(q)$ has eigenvalues $\{\eta_1, \eta_2, \eta_3, \eta_4\} = \{0, 0, 0, \eta_4\}$, where the dominant eigenvalue η_4 is a function of q subject to the values $\{r, s, t\} = \{1/4, 1/2, 3/4\}$ given by

$$\eta_4 = \frac{\sum_{i=0}^{20} c_i q^i}{K\sum_{j=0}^{21} c_j q^j}, \tag{25}$$

where $K = 10376293541461622784$ and the values c_i and c_j are listed in Table 1.

The region of absolute stability is depicted in the dark area in Figure 1.

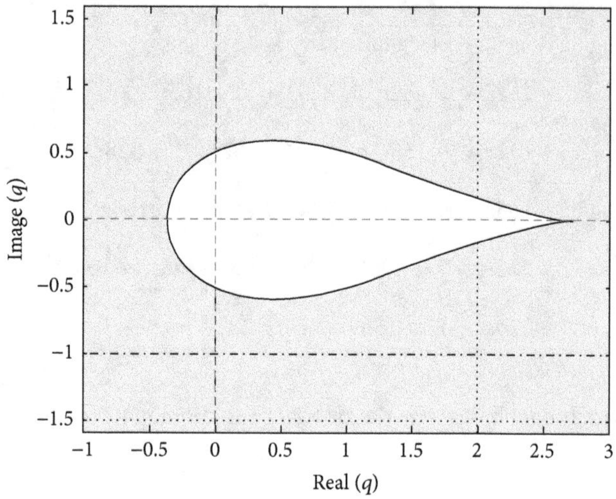

FIGURE 1: Region of absolute stability.

3.6. Pseudocode of the Developed Method. The following pseudocode illustrates how the developed method is implemented for solving fourth-order initial value problems.

Step 1. The values $y(x_0), y'(x_0), y''(x_0), y'''(x_0)$ on the interval $[a, b]$ from the problem are given.

Step 2. Set the values of the hybrid points $\{r, s, t\}$ and step size h.

Step 3. For $j = a : h : b$, set $x_{n+j} = x_n + jh$, $j = \{r, s, t, 1\}$.

Step 4. Evaluate the approximate values $y(x_{n+j})$, $y'(x_{n+j})$, $y''(x_{n+j})$, $y'''(x_{n+j})$ using the direct hybrid block method.

Step 5. Find the solution for the resultant system using the built in function fsolve in Matlab.

Step 6. Calculate the maximum error = |Exact solution − approximate solution|.

4. Numerical Examples

The general three-hybrid one-step hybrid block method (9) with order $p = 10$ in this section was employed to solve five problems from the literature. In order to show the performance of the proposed method, the problems are

Table 1

c-value	q^i coefficients	q^j coefficients
c_0	−19991259433786613699768237287038031044280320	0
c_1	−19991259433786613699768237287038031044280320	192662816967402505371648000
c_2	−99956297168933068498841186435190155221401600	0
c_3	−33318765722977689499613728811730051740467200	0
c_4	−78083522390790487043756727396404132472422400	0
c_5	−12270936643443632883203625005043197725900800	502432963738164264960000
c_6	−994813550529796720162894578620563036569600	−79502926099016318976000
c_7	59773830578760679822424316692208078028800	0
c_8	31534047193242772464338416945297159618560	0
c_9	4794949827573668397142419401567455150080	133171866570211983360
c_{10}	359049264395827172053492731789485015040	−14458011284982988800
c_{11}	−3356754007384253296917217569574748160	268452685173657600
c_{12}	−4527291309580589178991255608586403840	0
c_{13}	−632549152682975031987854772642562048	5775969118252032
c_{14}	−50983394698341254755895386446446592	−659482112163840
c_{15}	−2347009723724983385107341899462400	26223671235440
c_{16}	3581359970953878842628978864128	−529590387200
c_{17}	8897056249188627790299370721984	67083520896
c_{18}	741826115832694210392858425664	−8638140672
c_{19}	30585824514148316274163930311	506275974
c_{20}	582915589719914249658541800	−15909210
c_{21}	0	189567

solved with different values of step size h and hybrid points $\{r, s, t\}$.

Problem 1 (linear fourth-order problem).

$$\begin{aligned} y^{(iv)} &= y''' + y'' + y' + 2y, \quad 0 \le x \le 2, \\ y(0) &= 0, \\ y'(0) &= 0, \\ y''(0) &= 0, \\ y'''(0) &= 30. \end{aligned} \tag{26}$$

Exact Solution. $y = 2e^{2x} - 5e^{-x} + 3\cos(x) - 9\sin(x)$.

Source. See [15].

Problem 2 (nonlinear fourth-order problem).

$$\begin{aligned} y^{(iv)} &= (y')^2 - yy''' - 4x^2 + e^x(1 + x^2 - 4x), \quad 0 \le x \le 1, \\ y(0) &= 1, \\ y'(0) &= 1, \\ y''(0) &= 3, \\ y'''(0) &= 1. \end{aligned} \tag{27}$$

Exact Solution. $y = x^2 + e^x$ with $h = 1/10$.

Source. See [15].

Problem 3 (application to problem from ship dynamics). In this example, the derived method is implemented to solve a physical problem from ship dynamics. As stated by [3], when a sinusoidal wave of frequency passes along a ship or offshore structure, the resultant fluid actions vary with time. In a particular case study by [3], the fourth-order problem is defined as

$$\begin{aligned} y^{(iv)} + 3y'' + y(2 + \epsilon\cos(\omega(x))) &= 0, \quad x > 0, \\ y(0) &= 1, \\ y'(0) = y''(\rho) = y'''(0) &= 0. \end{aligned} \tag{28}$$

Exact Solution. $y(x) = 2\cos(x) + \cos(\sqrt{2}x)$.

Problem 4 (system of linear fourth-order IVPs).

$$\begin{aligned} y_1^{(iv)} &= y_1 + y_4 - e^{2x}, \\ y(0) &= 1, \\ y'(0) &= 1, \end{aligned}$$

$$y''(0) = 1,$$
$$y'''(0) = 1,$$
$$0 \le x \le 2,$$
$$y_2^{(iv)} = y_2 + y_3 - xe^x,$$
$$y(0) = 1,$$
$$y'(0) = -1,$$
$$y''(0) = 1,$$
$$y'''(0) = -1,$$
$$0 \le x \le 2,$$
$$y_3^{(iv)} = 4y_1 + y_3,$$
$$y(0) = 0,$$
$$y'(0) = 1,$$
$$y''(0) = 2,$$
$$y'''(0) = 3,$$
$$0 \le x \le 2,$$
$$y_4^{(iv)} = y_1 + y_2 + 16y_4 - e^x - e^{-x},$$
$$y(0) = 1,$$
$$y'(0) = 2,$$
$$y''(0) = 4,$$
$$y'''(0) = 8,$$
$$0 \le x \le 2, \tag{29}$$

Exact Solution. $y_1 = e^x$, $y_2 = e^{-x}$, $y_3 = xe^x$, and $y_4 = e^{2x}$ with $h = 1/5$.

Problem 5 (system of nonlinear fourth-order IVPs).

$$y_1^{(iv)} = y_1 + y_2 - \cos(x),$$
$$y(0) = 0,$$
$$y'(0) = 1,$$
$$y''(0) = 0,$$
$$y'''(0) = -1,$$
$$0 \le x \le \pi,$$
$$y_2^{(iv)} = y_1 + y_2 - \sin(x),$$
$$y(0) = 1,$$
$$y'(0) = 0,$$
$$y''(0) = -1,$$
$$y'''(0) = 0,$$
$$0 \le x \le \pi,$$
$$y_3^{(iv)} = 32y_1y_2,$$
$$y(0) = 0,$$
$$y'(0) = 2,$$
$$y''(0) = 0,$$
$$y'''(0) = -8,$$
$$0 \le x \le \pi,$$
$$y_4^{(iv)} = 16\left(y_2^2 - y_1^2\right),$$
$$y(0) = 1,$$
$$y'(0) = 0,$$
$$y''(0) = -2,$$
$$y'''(0) = 0,$$
$$0 \le x \le \pi, \tag{30}$$

Exact Solution. $y_1 = \sin(x)$, $y_2 = \cos(x)$, $y_3 = \sin(2x)$, and $y_4 = x^2 + \cos(2x)$, with $h = \pi/5$.

5. Conclusion

A hybrid one-step block method of order $p = 10$ with generalized three-off-step points has been derived successfully. The developed method was employed to solve general fourth-order IVPs of ODEs directly. The numerical analysis performed shows that the developed method is consistent and zero stable which leads to the convergence of the method. The numerical results were then compared with the results obtained by the existing methods in terms of error. The new method is found to have better performance over the other methods (refer to Tables 2–12).

Conflicts of Interest

The authors declare that there are no conflicts of interest regarding the publication of this paper.

TABLE 2: Comparison of the derived method with [15, 16] and Adams Bashforth-Adams Moulton method.

Step size	Method employed	Absolute error at $x = 2$
0.1	Derived method	$1.34E(-11)$
	BHCM4	$1.74E(-8)$
	Jator	$1.26E(-4)$
	Adams	$2.11E(-3)$
0.05	Derived method	$1.95E(-11)$
	BHCM4	$8.45E(-11)$
	Jator	$1.91E(-6)$
	Adams	$5.37E(-4)$

TABLE 3: Comparison of the derived method with [15, 17] and Adams Bashforth-Adams Moulton method.

Step size	Method employed	Absolute error at $x = 1$
0.2	Derived method	$2.22E(-15)$
	BHCM4	$2.38E(-12)$
	Adams	$5.01E(-7)$
	Awoyemi	$5.84E(-4)$
0.1	Derived method	$1.77E(-15)$
	BHCM4	$1.95E(-14)$
	Adams	$2.44E(-6)$
	Awoyemi	$9.26E(-5)$

TABLE 4: Comparison of the proposed method with [15] for the case $\epsilon = 0$.

Step size	Method employed	Absolute error at $x = 15$
0.25	Derived method	$1.1E(-16)$
	BHCM4	$5.2E(-7)$
	Twizell	$1.9E(-4)$
	Adams	$4.9E(-3)$
0.1	Derived method	$1.1E(-14)$
	BHCM4	$2.8E(-10)$
	Cortell	$3.7E(-5)$
	Adams	$8.4E(-5)$

TABLE 5: Exact and approximate solutions for solving y_1 using the developed method.

X-value	Exact solution	Approximate solution	Error in solving y_1
0.20	1.221402758160169900	1.221402758160169900	$0.000000E(+00)$
0.40	1.491824697641270300	1.491824697641270300	$0.000000E(+00)$
0.60	1.822118800390509100	1.822118800390508900	$2.220446E(-16)$
0.80	2.225540928492467900	2.225540928492467400	$4.440892E(-16)$
1.00	2.718281828459045500	2.718281828459045100	$4.440892E(-16)$
1.20	3.320116922736547200	3.320116922736547700	$4.440892E(-16)$
1.40	4.055199966844674500	4.055199966844674500	$0.000000E(+00)$
1.60	4.953032424395114000	4.953032424395114900	$8.881784E(-16)$
1.80	6.049647464412944800	6.049647464412945700	$8.881784E(-16)$
2.00	7.389056098930648600	7.389056098930650400	$1.776357E(-15)$

TABLE 6: Exact and approximate solutions for solving y_2 using the developed method.

X-value	Exact solution	Approximate solution	Error in solving y_2
0.20	0.818730753077981820	0.818730753077981820	0.000000E(+00)
0.40	0.670320046035639330	0.670320046035639330	0.000000E(+00)
0.60	0.548811636094026390	0.548811636094026390	0.000000E(+00)
0.80	0.449328964117221560	0.449328964117221620	5.551115E(−17)
1.00	0.367879441171442330	0.367879441171442330	0.000000E(+00)
1.20	0.301194211912202080	0.301194211912202080	0.000000E(+00)
1.40	0.246596963941606490	0.246596963941606460	2.775558E(−17)
1.60	0.201896517994655440	0.201896517994655380	5.551115E(−17)
1.80	0.165298888221586560	0.165298888221586500	5.551115E(−17)
2.00	0.135335283236612730	0.135335283236612650	8.326673E(−17)

TABLE 7: Exact and approximate solutions for solving y_3 using the developed method.

X-value	Exact solution	Approximate solution	Error in solving y_3
0.20	0.244280551632033990	0.244280551632033970	2.775558E(−17)
0.40	0.596729879056508210	0.596729879056508090	1.110223E(−16)
0.60	1.093271280234305600	1.093271280234305400	2.220446E(−16)
0.80	1.780432742793974300	1.780432742793974100	2.220446E(−16)
1.00	2.718281828459045500	2.718281828459045100	4.440892E(−16)
1.20	3.984140307283856600	3.984140307283857000	4.440892E(−16)
1.40	5.677279953582544000	5.677279953582544000	0.000000E(+00)
1.60	7.924851879032181600	7.924851879032183400	1.776357E(−15)
1.80	10.889365435943299000	10.889365435943303000	3.552714E(−15)
2.00	14.778112197861295000	14.778112197861301000	5.329071E(−15)

TABLE 8: Exact and approximate solutions for solving y_4 using the developed method.

X-value	Exact solution	Approximate solution	Error in solving y_4
0.20	1.491824697641270300	1.491824697641270300	0.000000E(+00)
0.40	2.225540928492467900	2.225540928492467400	4.440892E(−16)
0.60	3.320116922736548100	3.320116922736547700	4.440892E(−16)
0.80	4.953032424395114900	4.953032424395114900	0.000000E(+00)
1.00	7.389056098930650400	7.389056098930650400	0.000000E(+00)
1.20	11.023176380641601000	11.023176380641601000	0.000000E(+00)
1.40	16.444646771097048000	16.444646771097048000	0.000000E(+00)
1.60	24.532530197109342000	24.532530197109349000	7.105427E(−15)
1.80	36.598234443677974000	36.598234443677988000	1.421085E(−14)
2.00	54.598150033144215000	54.598150033144236000	2.131628E(−14)

TABLE 9: Exact and approximate solutions for solving y_1 using the developed method.

X-value	Exact solution	Approximate solution	Error in solving y_1
0.6283185	0.587785252292473140	0.587785252292475800	2.664535E(−15)
1.2566371	0.951056516295153530	0.951056516295303300	1.497691E(−13)
1.8849556	0.951056516295153640	0.951056516296290070	1.136424E(−12)
2.5132741	0.587785252292473250	0.587785252297075340	4.602096E(−12)
3.1415927	0.000000000000000122	0.000000000013537650	1.353753E(−11)

TABLE 10: Exact and approximate solutions for solving y_2 using the developed method.

X-value	Exact solution	Approximate solution	Error in solving y_2
0.6283185	0.809016994374947450	0.809016994374955000	$7.549517E(-15)$
1.2566371	0.309016994374947450	0.309016994375326810	$3.793632E(-13)$
1.8849556	−0.309016994374947340	−0.309016994372823590	$2.123746E(-12)$
2.5132741	−0.809016994374947340	−0.809016994368452200	$6.495138E(-12)$
3.1415927	−1.000000000000000000	−0.999999999984998780	$1.500122E(-11)$

TABLE 11: Exact and approximate solutions for solving y_3 using the developed method.

X-value	Exact solution	Approximate solution	Error in solving y_3
0.6283185	0.951056516295153530	0.951056516302223540	$7.070011E(-12)$
1.2566371	0.587785252292473250	0.587785252378536070	$8.606282E(-11)$
1.8849556	−0.587785252292473030	−0.587785251895384220	$3.970888E(-10)$
2.5132741	−0.951056516295153640	−0.951056515173170910	$1.121983E(-09)$
3.1415927	−0.000000000000000245	0.000000002364014064	$2.364014E(-09)$

TABLE 12: Exact and approximate solutions for solving y_4 using the developed method.

X-value	Exact solution	Approximate solution	Error in solving y_4
0.6283185	0.703801170418521750	0.703801170427732160	$9.210410E(-12)$
1.2566371	0.770119709799349850	0.770119709892383430	$9.303358E(-11)$
1.8849556	2.744040590017221300	2.744040590323378700	$3.061573E(-10)$
2.5132741	6.625563811072136200	6.625563811678808300	$6.066720E(-10)$
3.1415927	10.869604401089358000	10.869604401982164000	$8.928058E(-10)$

Acknowledgments

The authors thank Universiti Utara Malaysia for the financial support of the publication of this article.

References

[1] A. Malek and R. Shekari Beidokhti, "Numerical solution for high order differential equations using a hybrid neural network—optimization method," *Applied Mathematics and Computation*, vol. 183, no. 1, pp. 260–271, 2006.

[2] A. Boutayeb and A. Chetouani, "A mini-review of numerical methods for high-order problems," *International Journal of Computer Mathematics*, vol. 84, no. 4, pp. 563–579, 2007.

[3] X. J. Wu, Y. Wang, and W. G. Price, "Multiple resonances, responses, and parametric instabilities in offshore structures," *Journal of Ship Research*, vol. 32, no. 4, pp. 285–296, 1988.

[4] W. E. Milne, *Numerical Solution of Differential Equations*, John Wiley & Sons, New York, NY, USA, 1953.

[5] J. B. Rosser, "A runge-kutta for all seasons," *SIAM Review*, vol. 9, no. 3, pp. 417–452, 1967.

[6] W. B. Gragg and H. J. Stetter, "Generalized multistep predictor-corrector methods," *Journal of the ACM*, vol. 11, no. 2, pp. 188–209, 1964.

[7] J. D. Lambert, *Computational Methods in Ordinary Differential Equations*, John Wiley & Sons, New York, NY, USA, 1973.

[8] G. Dahlquist, "Convergence and stability in the numerical integration of ordinary differential equations," *Mathematica Scandinavica*, vol. 4, pp. 33–53, 1956.

[9] W. H. Enright, "Second derivative multistep methods for stiff ordinary differential equations," *SIAM Journal on Numerical Analysis*, vol. 11, no. 2, pp. 321–331, 1974.

[10] F. F. Ngwane and S. N. Jator, "Block hybrid-second derivative method for stiff systems," *International Journal of Pure and Applied Mathematics*, vol. 80, no. 4, pp. 543–559, 2012.

[11] R. K. Sahi, S. N. Jator, and N. A. Khan, "A simpson's-type second derivative method for stiff systems," *International Journal of Pure and Applied Mathematics*, vol. 81, no. 4, pp. 619–633, 2012.

[12] R. K. Sahi, S. N. Jator, and N. A. Khan, "Continuous fourth derivative method for third order boundary value problems," *International Journal of Pure and Applied Mathematics*, vol. 85, no. 5, pp. 907–923, 2013.

[13] S. O. Fatunla, *Numerical Methods for Fatunla Initial Value Problems in Ordinary Differential Equations*, Academic Press, New York, NY, USA, 1988.

[14] P. Henrici, *Discrete Variable Methods in ODEs*, John Wiley & Sons, New York, NY, USA, 1962.

[15] L. K. Yap and F. Ismail, "Block hybrid collocation method with application to fourth order differential equations," *Mathematical Problems in Engineering*, vol. 2015, Article ID 561489, 6 pages, 2015.

[16] S. N. Jator, "Numerical integrators for fourth order initial and boundary value problems," *International Journal of Pure and Applied Mathematics*, vol. 47, no. 4, pp. 563–576, 2008.

[17] D. O. Awoyemi, "Algorithmic collocation approach for direct solution of fourth-order initial-value problems of ordinary differential equations," *International Journal of Computer Mathematics*, vol. 82, no. 3, pp. 321–329, 2005.

8

First Integrals and Hamiltonians of Some Classes of ODEs of Maximal Symmetry

J. C. Ndogmo

Department of Mathematics and Applied Mathematics, University of Venda, P/B X5050, Thohoyandou, Limpopo 0950, South Africa

Correspondence should be addressed to J. C. Ndogmo; ndogmoj@yahoo.com

Academic Editor: Peter G. L. Leach

Complete sets of linearly independent first integrals are found for the most general form of linear equations of maximal symmetry algebra of order ranging from two to eight. The corresponding Hamiltonian systems are constructed and it is shown that their general solutions can also be found by a simple superposition formula from the solutions of a scalar second-order source equation.

1. Introduction

An important class of linear ordinary differential equations (LODEs) consists of those equations having a symmetry algebra of maximal dimension. This is partly due to the simple characterization of such a class of equations [1, 2], according to which they are precisely the iterative equations, and equivalently they can be reduced by a point transformation to the canonical form $y^{(n)} = 0$.

Despite the fact that linear equations are the simplest types of differential equations, they are far from being completely understood and yet they frequently appear in the study of all other types of equations and in particular as canonical or reduced form of nonlinear or partial differential equations with more complex structures.

Linear equations with maximal symmetry algebras have been studied in some recent papers [3–6], but in all such papers the emphasis has been placed more on the coefficient characterization and the elementary symmetry or the transformation properties of these equations. Nœther's theorem [7, 8] is a powerful tool that allows us to associate each variational symmetry of the equation with a conservation law, and, in the case of ordinary differential equations (ODEs), these conservation laws correspond to first integrals which in this instance are true constants of motion. The role of these first integrals is often crucial in the study of solutions and properties of differential equations, including questions related to stability [9, 10] or integrability [11–14]. The so-called first integral method for finding exact solutions of nonlinear partial differential equations has also been successfully applied to a wide range of equations modeling various phenomena in physics (see [15] and the references therein).

In this paper, we obtain a complete set of linearly independent first integrals for all linear equations of maximal symmetry of order n such that $2 \leq n \leq 8$. These first integrals are obtained for the most general form of the equations in which they depend on an arbitrary function through an application of Nœther's theorem. We also give the Hamiltonian formulation for the corresponding class of scalar equations and show amongst others that their general solution can also be obtained by a simple superposition formula from those of a scalar second-order source equation.

2. Basic Properties of Linear Iterative Equations

Let $r \neq 0$ and s be two functions of the variable x and consider the differential operator $\Psi = r(d/dx) + s$. Denote by $\Delta_n[y] = 0$ a LODE of order n in the dependent variable $y = y(x)$. Then this equation is said to be iterative if up to a normalizing factor one has $\Delta_n[y] = \Psi^n[y]$, where

$$\Psi^n[y] = \Psi^{n-1}[\Psi[y]], \quad \text{for } n \geq 1 \text{ with } \Psi^0 = I,\ \Psi^1 = \Psi \tag{1}$$

and where I is the identity operator. A LODE $\Delta_n[y] = 0$ has a maximal symmetry algebra if and only if it is iterative [1] or, equivalently [1, 2], if and only if it can be reduced by a point transformation to the canonical form $y^{(n)} = 0$. By the transformation $w = y\exp((1/n)\int_{x_0}^{x} B_{n-1}(v)dv)$, where $B_{n-1}(z)$ is the coefficient of the term of order $n-1$, any given LODE may be put into the form

$$y^{(n)} + \sum_{i=0}^{n-2} A_n^{n-i}(x)\, y^{(i)} = 0 \tag{2}$$

which is referred to as the normal form of the equation. For linear iterative equations in the latter form, n linearly independent solutions can be found in the form

$$s_k = u^{n-k-1}v^k, \quad 0 \le k \le n-1,\ n \ge 2, \tag{3}$$

for some functions u and v of the independent variable x. For a given function $F = F(t)$, we shall also use the notation F_t, F_{tt}, and so forth for $dF/dt, d^2F/dt^2$, and so forth, but where convenient we shall also denote dF/dt simply by F'.

It follows in particular from (3) that u and v are two linearly independent solutions of the second-order so-called source equation:

$$y_{xx} + \mathfrak{q}y = 0, \quad \mathfrak{q} = A_2^2. \tag{4}$$

In particular, we have

$$\mathfrak{q} = \mathcal{w}(u', v'). \tag{5}$$

Thanks to Abel's identity, the Wronskian $\mathcal{w}(u,v) = \det\begin{pmatrix} u & v \\ u' & v' \end{pmatrix}$ of two solutions u and v of a LODE in the form (2) is a constant and may be normalized to one as we shall do by convention.

By a result of Lie [16], all equations of order $n \ge 2$ reducible by a point transformation to the canonical form $y^{(n)} = 0$ have a symmetry algebra $\mathfrak{g}_n$ of maximal dimension 8 for $n = 2$ and $n+4$ for $n \ge 3$. Owing to the linear property, the $n+1$ vector fields

$$V_k = s_k\partial_y = u^{n-k-1}v^k\partial_y, \quad 0 \le k \le n-1, \tag{6a}$$

$$W_y = y\partial_y \tag{6b}$$

are symmetries of any iterative equation of the form (2). The three additional ones,

$$F_n = u^2\partial_x + (n-1)\, uu_x y\partial_y, \tag{6c}$$

$$G_n = 2uv\partial_x + (n-1)\left(uv_x + u_x v\right) y\partial_y, \tag{6d}$$

$$H_n = -v^2\partial_x - (n-1)\, vv_x y\partial_y, \tag{6e}$$

obtained in [1], are also generators of the Lie symmetry algebra $\mathfrak{g}_n$ of (2) in the case of equations of maximal symmetry. These $n+4$ symmetries form a basis of $\mathfrak{g}_n$ for $n \ge 3$. For $n = 2$, a basis of $\mathfrak{g}_n$ consists of the same $n+4$ symmetry generators together with the two non-Cartan symmetries given by

$$C_1 = yu\partial_x + y^2u'\partial_y, \qquad C_2 = yv\partial_x + y^2v'\partial_y. \tag{6f}$$

Recall that, in (6a)–(6f), u and v are assumed to be two linearly independent solutions of the source equation (4). The symmetry generators V_k in (6a) associated with each solution s_k are often called solution symmetries and generate the abelian Lie algebra $\mathscr{A}_n$, while W_y is referred to as the homogeneity symmetry. Denoting by $\mathbf{v}^Q = Q\partial_y$ the characteristic form of a vector field $\mathbf{v}$ with characteristic Q, it follows that the characteristic forms of the infinitesimal generators listed in (6a)–(6f) are given by

$$V_k^Q = V_k, \quad 0 \le k \le n-1, \qquad F_n^Q = \left[(n-1)\, uu_x y - u^2 y_x\right]\partial_y, \tag{7a}$$

$$H_n^Q = \left[(n-1)\, vv_x y + v^2 y_x\right]\partial_y, \qquad W_y^Q = W_y, \tag{7b}$$

$$C_1^Q = \left(y^2u' - uyy'\right)\partial_y, \qquad C_2^Q = \left(y^2v' - vyy'\right)\partial_y, \tag{7c}$$

$$G_n^Q = \left[(n-1)\,(uv)'\, y - 2uvy_x\right]\partial_y \tag{7d}$$

In the actual case of LODEs, the characteristic Q of $\mathbf{v}$ has the simpler expression $Q = \mathbf{v}(y) - \mathbf{v}(x)y_x$.

3. Variational Symmetries and First Integrals

Suppose that the equation $\Delta_n[y] = 0$ is the Euler-Lagrange equation of some variational problem corresponding to a Lagrangian L. Let

$$\mathbf{v} = \xi(x,y)\,\partial_x + \phi(x,y)\,\partial_y \tag{8}$$

be a symmetry generator with characteristic function Q of the equation and set

$$\mathcal{S}(\mathbf{v}) = \mathbf{v}^{[n]}(L) + L\,\text{div}\,\xi, \quad (\text{for } n \text{ even}) \qquad \mathcal{D}(\mathbf{v}) = D_\Delta^*(Q) + D_Q^*(\Delta), \tag{9}$$

where $\mathbf{v}^{[n]}$ is the nth prolongation of the vector field $\mathbf{v}$ and D_P^* is the adjoint Fréchet derivative of the differential function $P = P[y]$. The vanishing of $\mathbf{v}$ under these two operators,

$$\mathcal{S}(\mathbf{v}) = 0, \tag{10}$$

$$\mathcal{D}(\mathbf{v}) = 0, \tag{11}$$

represents some key concepts. Indeed, in the sense of [7], $\mathbf{v}$ is called a variational symmetry for $\Delta_n[y] = 0$ if it satisfies (10) and is called a divergence symmetry if it satisfies (11). The latter condition is equivalent to $\mathcal{S}(\mathbf{v}) = \operatorname{div} B$ for a certain differential function $B = B[y]$. Divergence symmetries are thus a more relaxed form of variational symmetries, and they also give rise to conservation laws and first integral as variational symmetries do by Nœther's theorem. Divergence symmetries satisfy the same condition required on generalized symmetries to be variational symmetries, and, in a more modern language [17], these divergence symmetries are simply called variational symmetries. The characteristic form of a conservation law associated with each divergence symmetry has the form $Q\Delta_n[y] = \operatorname{div} P$, for a certain differential function $P = P[y]$. In the actual case of an ODE, we have

$$Q\Delta_n[y] = DP, \quad \text{where } D = \frac{d}{dx} \tag{12}$$

is the total differential operator. Thus P is a first integral of the equation.

The general expression of iterative LODEs has been obtained in [3, 5] in explicit form in terms of the single arbitrary function $\mathfrak{q}$ upon which they depend. However, using a result of [3, 6], one can easily generate LODEs of maximal symmetry of a general order. *For even orders n, $2 \leq n \leq 8$*, the equations are as follows:

$$0 = \mathfrak{q}y + y_{xx}, \tag{13a}$$

$$0 = 10y_x\mathfrak{q}_x + 10\mathfrak{q}y_{xx} + 3y\left(3\mathfrak{q}^2 + \mathfrak{q}_{xx}\right) + y^{(4)}, \tag{13b}$$

$$\begin{aligned} 0 = {} & 7y_{xx}\left(37\mathfrak{q}^2 + 9\mathfrak{q}_{xx}\right) + 70\mathfrak{q}_x y^{(3)} + y_x\left(518\mathfrak{q}\mathfrak{q}_x + 28\mathfrak{q}^{(3)}\right) + 35\mathfrak{q}y^{(4)} + 5y\left(45\mathfrak{q}^3 + 26\mathfrak{q}_x^2 + 31\mathfrak{q}\mathfrak{q}_{xx} + \mathfrak{q}^{(4)}\right) + y^{(6)}, \end{aligned} \tag{13c}$$

$$\begin{aligned} 0 = {} & 168y^{(3)}\left(47\mathfrak{q}\mathfrak{q}_x + 2\mathfrak{q}^{(3)}\right) + 42\left(47\mathfrak{q}^2 + 9\mathfrak{q}_{xx}\right)y^{(4)} \\ & + 4y_{xx}\left(3229\mathfrak{q}^3 + 1773\mathfrak{q}\mathfrak{q}_{xx} + 45\left(33\mathfrak{q}_x^2 + \mathfrak{q}^{(4)}\right)\right) \\ & + 6y_x\left(6458\mathfrak{q}^2\mathfrak{q}_x + 524\mathfrak{q}\mathfrak{q}^{(3)} + 9\left(132\mathfrak{q}_x\mathfrak{q}_{xx} + \mathfrak{q}^{(5)}\right)\right) + 252\mathfrak{q}_x y^{(5)} + 84\mathfrak{q}y^{(6)} \\ & + 7y\left(1575\mathfrak{q}^4 + 1654\mathfrak{q}^2\mathfrak{q}_{xx} + 153\mathfrak{q}_2^2 + 226\mathfrak{q}_x\mathfrak{q}^{(3)} + 8\mathfrak{q}\left(347\mathfrak{q}_x^2 + 10\mathfrak{q}^{(4)}\right) + \mathfrak{q}^{(6)}\right) + y^{(8)}. \end{aligned} \tag{13d}$$

It can be easily verified that each of these equations is self-adjoint and thus admits a Lagrangian formulation in terms of the Euler-Lagrange operator. A Lagrangian L_n for each of the equations $\Delta_n[y] = 0$ of order n in (13a)–(13d) can be taken in the form $L_n = y\Delta_n[y]/2$, but this nth order Lagrangian can always be reduced to one of order $n/2$.

As already noted, the results of [3, 6] can be used to generate iterative ODEs of any given order and thus we omit the listing of those iterative equations of odd orders $n = 3, 5, 7$ whose first integrals are also to be found. Although *odd order* equations are not known to possess a Lagrangian formulation, in order for a symmetry $\mathbf{v}$ of any such equation to give rise to a first integral it suffices that it be a divergence symmetry, that is, that it satisfies the divergence symmetry condition (11).

On the other hand, under the assumption that u and v are linearly independent solutions of (4), one can verify that for each *even order* $n \geq 4$ the most general divergence symmetry of each of (13b)–(13d) has the form

$$\mathbf{w} \equiv \mathbf{w}^n = \mathbf{w}_1 + \mathbf{w}_2, \tag{14a}$$

$$\mathbf{w}_1 \equiv \mathbf{w}_1^n = \sum_{k=0}^{n-1} \lambda_k V_k, \quad \lambda_k \in \mathbb{R}, \ n = 4, 6, 8, \tag{14b}$$

$$\mathbf{w}_2 \equiv \mathbf{w}_2^n = \alpha F_n + \beta G_n + \gamma H_n, \quad \alpha, \beta, \gamma \in \mathbb{R}. \tag{14c}$$

In other words, for even orders $n \geq 4$, $W_y = y\partial_y$ is the only generator in (6a)–(6f) which is not a term of the divergence symmetry $\mathbf{w}$, and in particular W_y itself is not a divergence symmetry. Similarly, for $n = 2$, let $\mathbf{w}$ be an arbitrary divergence symmetry vector in the divergence symmetry algebra of (13a). Thus we have

$$\mathbf{w} = \mathbf{w}_1 + \mathbf{w}_2 + \mathbf{w}_3 + k_3 W_y, \tag{15}$$

where $\mathbf{w}_1$ and $\mathbf{w}_2$ are like in (14a)–(14c) but with $n = 2$ and

$$\mathbf{w}_3 = k_1 C_1 + k_2 C_2, \tag{16}$$

where k_j for $j = 1, 2, 3$ are arbitrary scalars and C_1 and C_2 are the non-Cartan symmetries already introduced. Then in the divergence symmetry condition $\mathcal{D}(\mathbf{w}) = 0$ expressed as a polynomial in y and its derivatives, it turns out that the coefficients of y'' and $y'y''$ are $2k_3$ and $-3(k_1 u + k_2 v)$, respectively. This shows that in the expression of $\mathbf{w}$ one must have $k_j = 0$ for $j = 1, 2, 3$. Consequently, $\mathbf{w} = \mathbf{w}_1 + \mathbf{w}_2$ in this case also, and one readily verifies that any such $\mathbf{w}$ is indeed a divergence symmetry.

Let us now denote by $\mathcal{L}_n$ the Lie subalgebra of all divergence symmetries of $\mathfrak{g}_n$ for a given order n. We have thus established the following result.

Proposition 1. *For every LODE of maximal symmetry of the most general form (13a)–(13d) and of even order n, a basis $\mathcal{B}_n$ of the Lie algebra $\mathcal{L}_n$ of its divergence symmetries is given by*

$$\mathcal{B}_n = \{F_n, G_n, H_n, V_1, \ldots, V_n\}. \tag{17}$$

In particular, $\mathcal{L}_n$ has dimension $n + 3$.

For equations of *odd orders*, the most general divergence symmetry $\mathbf{w}$ can be sought in the form

$$\mathbf{w} = \mathbf{w}_1 + \mathbf{w}_2 + k_1 W_y, \tag{18}$$

where $\mathbf{w}_1$ and $\mathbf{w}_2$ are like in (14a)–(14c) but with $n = 3, 5, 7$ and k_1 is a scalar. Then in the divergence symmetry condition $\mathcal{D}(\mathbf{w}) = 0$ expressed as a polynomial in y and its derivatives, the coefficient of $y^{(n+1)}$ is $2(\alpha u^2 + 2\beta uv + \gamma v^2)$, and owing to the linear independence of u and v the vanishing of this coefficient implies that $\alpha = \beta = \gamma = 0$. This reduces the expression of $\mathbf{w}$ to $\mathbf{w} = \mathbf{w}_1 + k_1 W_y$, and it is readily verified that the latter vector is always a divergence symmetry. We have thus established the following result.

Proposition 2. *For odd orders n, a basis $\mathscr{B}_n$ of $\mathscr{L}_n$ is given by*

$$\mathscr{B}_n = \{W_y, V_1, \ldots, V_n\}. \tag{19}$$

In particular, $\mathscr{L}_n$ has dimension $n+1$.

It follows from (12) that, for each characteristic function Q, the corresponding first integral is given by

$$P[y] = \int Q\Delta_n[y]\,dx. \tag{20}$$

However, the evaluation of this integral is not straightforward. We shall derive this expression for each of the n characteristics $Q = s_k$ $(0 \le k \le n-1)$, where s_k given in (3) are linearly independent solutions of the equation. The first integral associated with a given divergence symmetry generator $\mathbf{v}$ will be denoted by $\mathscr{F}(\mathbf{v})$ that associated with $V_k \equiv V_{n,k}$ by $\mathscr{F}_k \equiv \mathscr{F}_{n,k}$ for fixed n, and the first integral associated with the Lie algebra $\mathscr{A}_n$ generated by V_k will be denoted by $\mathscr{F}^n$.

We consider explicitly the case $n = 4$. In order to ensure that the equation reflects the fact that a given function S is a solution, we solve $\Delta_4[S] = 0$ for the coefficient A_0 of y in (13b) and substitute the resulting expression back into $Q\Delta_4[y]$. This yields

$$\begin{aligned} Q\Delta_4[y] &= S\Delta_4[y] = S\left[y^{(4)} + 10(y_x\mathfrak{q}_x + \mathfrak{q}y_{xx}) + A_0 y\right] \\ &= S\left[y^{(4)} + 10(y_x\mathfrak{q}_x + \mathfrak{q}y_{xx}) + y\frac{-10\mathfrak{q}_xS_x - 10aS_{xx} - S^{(4)}}{S}\right] \\ &= 10D(\mathfrak{q}Sy_x - \mathfrak{q}yS_x) + D\left(Sy^{(3)} - S_xy^{(2)} + S_{xx}y_x - S^{(3)}y\right). \end{aligned} \tag{21}$$

This shows that

$$P[y] = 10(\mathfrak{q}Sy_x - \mathfrak{q}S_xy) + Sy^{(3)} - S_xy^{(2)} + S_{xx}y_x - S^{(3)}y + c, \tag{22}$$

for a certain constant of integration c which will be omitted in the sequel. Letting $S = s_k$ as in (3), it follows that the most general expression of first integrals generated by solutions of $\Delta_4[y] = 0$ has the form

$$\mathscr{F}^4 = \sum_{k=0}^{3}\alpha_k\mathscr{F}_{4,k}, \quad \alpha_k \in \mathbb{R}, \tag{23a}$$

$$\mathscr{F}_{4,k} = 10(\mathfrak{q}Sy_x - \mathfrak{q}S_xy) + Sy^{(3)} - S_xy^{(2)} + S_{xx}y_x - S^{(3)}y, \tag{23b}$$

$$S = s_k = u^{3-k}v^k, \quad (0 \le k \le 3). \tag{23c}$$

By proceeding in the same manner as for $n = 4$, using identities such as

$$D(fSy_x - fyS_x) = SD(fy_x) - yD(fS_x), \tag{24a}$$

$$Sy^{(m)} + (-1)^{m-1}yS^{(m)} = D\left[\sum_{j=0}^{m-1}(-1)^j S^{(j)}y^{(m-1-j)}\right], \tag{24b}$$

where $f = f(x)$ is any given function and $m \in \mathbb{N} = \{1, 2, 3, \ldots\}$, one obtains the first integrals for all other equations of orders $n \le 8$. We first list the expressions of first integrals found starting with even order equations. Thus for equations of order $n = 2$ one has

$$\begin{aligned} \mathscr{F}^2 &= \sum_{k=0}^{1}\alpha_k\mathscr{F}_{2,k}, \quad \alpha_k \in \mathbb{R}, \\ \mathscr{F}_{2,k} &= Sy' - yS', \\ S &= s_k = u^{1-k}v^k, \quad (0 \le k \le 1). \end{aligned} \tag{25}$$

For the order $n = 6$ one has

$$\begin{aligned} \mathscr{F}^6 &= \sum_{k=0}^{5}\alpha_k\mathscr{F}_{6,k}, \quad \alpha_k \in \mathbb{R}, \\ \mathscr{F}_{6,k} &= Sy^{(5)} + y_{xx}\left(35S\mathfrak{q}_x - 35\mathfrak{q}S_x - S^{(3)}\right) + \left(35\mathfrak{q}S + S_{xx}\right)y^{(3)} + y_x\left(259\mathfrak{q}^2S + 28S\mathfrak{q}_{xx} + 35\mathfrak{q}S_{xx} + S^{(4)}\right) - S_xy^{(4)} \\ &\quad + y\left(-259\mathfrak{q}^2S_x - 28S_x\mathfrak{q}_{xx} - 35\mathfrak{q}_xS_{xx} - 35\mathfrak{q}S^{(3)} - S^{(5)}\right), \\ S &= s_k = u^{5-k}v^k, \quad (0 \le k \le 5). \end{aligned} \tag{26}$$

On the other hand, for $n = 8$, one has

$$\begin{aligned} \mathscr{F}^8 &= \sum_{k=0}^{7}\alpha_k\mathscr{F}_{8,k}, \quad \alpha_k \in \mathbb{R}, \\ \mathscr{F}_{8,k} &= y^{(3)}\left(1974\mathfrak{q}^2S - 84\mathfrak{q}_xS_x + 210S\mathfrak{q}_{xx} + 84\mathfrak{q}S_{xx} + S^{(4)}\right) + y_{xx}\left(-1974\mathfrak{q}^2S_x - 126S_x\mathfrak{q}_{xx} + 126S\mathfrak{q}^{(3)}\right. \\ &\quad \left. + 84\mathfrak{q}\left(47S\mathfrak{q}_x - S^{(3)}\right) - S^{(5)}\right) + \left(168S\mathfrak{q}_x - 84\mathfrak{q}S_x - S^{(3)}\right)y^{(4)} + \left(84\mathfrak{q}S + S_{xx}\right)y^{(5)} + y_x\left(12916\mathfrak{q}^3S\right. \\ &\quad + 1974\mathfrak{q}^2S_{xx} + 126\mathfrak{q}_{xx}S_{xx} + 84\mathfrak{q}_xS^{(3)} \end{aligned}$$

$$+ 6S\left(332\mathfrak{q}_x^2 + 9\mathfrak{q}^{(4)}\right) + 12\mathfrak{q}\left(262S\mathfrak{q}_{xx} + 7S^{(4)}\right)$$
$$+ S^{(6)}\Big) - S_x y^{(6)} + y\Big(-12916\mathfrak{q}^3 S_x - 1992\mathfrak{q}_x^2 S_x$$
$$- 3144\mathfrak{q}S_x\mathfrak{q}_{xx} - 3948\mathfrak{q}\mathfrak{q}_x S_{xx} - 126S_{xx}\mathfrak{q}^{(3)}$$
$$- 1974\mathfrak{q}^2 S^{(3)} - 210\mathfrak{q}_{xx}S^{(3)} - 54S_x\mathfrak{q}^{(4)} - 168\mathfrak{q}_x S^{(4)}$$
$$- 84\mathfrak{q}S^{(5)} - S^{(7)}\Big) + Sy^{(7)},$$
$$S = s_k = u^{7-k}v^k, \quad (0 \le k \le 7). \tag{27}$$

The algorithm for finding first integrals for iterative equations of odd order n is different from that for even order equations described above. One can always write an iterative equation $\Delta_n[y] = 0$ of the form (2) in the form $y^{(n)} + F_n(x, y^{(n-2)}) = 0$. Therefore, the characteristic form of the conservation law associated with a solution $S = s_k$ takes the form

$$S\Delta_n[y] = S\Delta_n[y] + S^{(n)}y + yF_n\left(x, S^{(m)}\right) \tag{28a}$$

$$= \left[Sy^{(n)} + S^{(n)}y\right] + \left[SF_n\left(x, y^{(n-2)}\right) + yF_n\left(x, S^{(n-2)}\right)\right]. \tag{28b}$$

Then, using the identities in (24a) and (24b), one naturally integrates each of the two expressions in square brackets appearing in (28b) to obtain the first integrals. The resulting expressions are given as follows for the order $n = 3$.

$$\mathcal{F}^3 = \sum_{k=0}^{2}\alpha_k\mathcal{F}_{3,k}, \quad \alpha_k \in \mathbb{R},$$
$$\mathcal{F}_{3,k} = 4\mathfrak{q}Sy - S_x y_x + yS_{xx} + Sy_{xx},$$
$$S = s_k = u^{2-k}v^k, \quad (0 \le k \le 2). \tag{29}$$

For $n = 5$, one has

$$\mathcal{F}^5 = \sum_{k=0}^{4}\alpha_k\mathcal{F}_{5,k}, \quad \alpha_k \in \mathbb{R},$$
$$\mathcal{F}_{5,k} = \left(20\mathfrak{q}S + S_{xx}\right)y_{xx} + y_x\left(10S\mathfrak{q}_x - 20\mathfrak{q}S_x - S^{(3)}\right)$$
$$+ y\left(64\mathfrak{q}^2 S + 10\mathfrak{q}_x S_x + 8S\mathfrak{q}_{xx} + 20\mathfrak{q}S_{xx} + S^{(4)}\right)$$
$$+ Sy^{(4)} - S_x y^{(3)},$$
$$S = s_k = u^{4-k}v^k, \quad (0 \le k \le 4). \tag{30}$$

Finally, for $n = 7$, the first integrals are given by

$$\mathcal{F}^7 = \sum_{k=0}^{6}\alpha_k\mathcal{F}_{7,k}, \quad \alpha_k \in \mathbb{R},$$
$$\mathcal{F}_{7,k} = \left(84S\mathfrak{q}_x - 56\mathfrak{q}S_x - S^{(3)}\right)y^{(3)} + y_{xx}\Big(784\mathfrak{q}^2 S$$
$$- 28\mathfrak{q}_x S_x + 84S\mathfrak{q}_{xx} + 56\mathfrak{q}S_{xx} + S^{(4)}\Big) + \Big(56\mathfrak{q}S$$
$$+ S_{xx}\Big)y^{(4)} + y_x\Big(784\mathfrak{q}S\mathfrak{q}_x - 784\mathfrak{q}^2 S_x - 56S_x\mathfrak{q}_{xx}$$
$$- 28\mathfrak{q}_x S_{xx} + 28S\mathfrak{q}^{(3)} - 56\mathfrak{q}S^{(3)} - S^{(5)}\Big) - S_x y^{(5)}$$
$$+ y\Big(2304\mathfrak{q}^3 S + 784\mathfrak{q}^2 S_{xx} + 84\mathfrak{q}_{xx}S_{xx} + 28S_x\mathfrak{q}^{(3)}$$
$$+ 84\mathfrak{q}_x S^{(3)} + 12S\left(33\mathfrak{q}_x^2 + \mathfrak{q}^{(4)}\right)$$
$$+ 8\mathfrak{q}\left(98\mathfrak{q}_x S_x + 78S\mathfrak{q}_{xx} + 7S^{(4)}\right) + S^{(6)}\Big) + Sy^{(6)},$$
$$S = s_k = u^{6-k}v^k, \quad (0 \le k \le 6). \tag{31}$$

The results obtained thus far in this section about first integrals can be summarized as follows for all orders n such that $2 \le n \le 8$.

Theorem 3. *(a) For even n, the first integrals $\mathcal{F}_{n,k}$ associated with the solution symmetries $V_k = s_k\partial_y$ are given by (23b) and (25)–(27), while they are given for odd n by (29)–(31).*

(b) For each order n, the n first integrals $\mathcal{F}_{n,k}$ $(0 \le k \le n-1)$ are linearly independent.

(c) If for each n and k we view $\mathcal{F}_{n,k} = \mathcal{F}_{n,k}[y, \mathfrak{q}]$ as a differential function of y and $\mathfrak{q}$, then $\mathcal{F}_{n,k}$ is linear in y (and its derivatives) and polynomial in $\mathfrak{q}$.

First integrals are often required to be in a form in which they depend only on the parameters of the underlying differential equation and are not in particular expressed in terms of a particular solution of the equation. The first integrals we have found can indeed be expressed solely in terms of the dependent variable y and the coefficient $\mathfrak{q}$ and their derivatives. For instance, in the case of equations of odd order n, given that $\mathcal{F}_{n,k}$ are symmetric functions of S and y and are required to be constant only on the solution space, the required type of first integrals can be obtained simply by letting $S = y$ in the expression of $\mathcal{F}_{n,k}$. Denoting by $\mathcal{F}_n^y$ the resulting first integrals for each odd order n of the equation yields the following expressions:

$$\mathcal{F}_3^y = 4\mathfrak{q}y^2 - y_x^2 + 2yy_{xx},$$
$$\mathcal{F}_5^y = 20y\mathfrak{q}_x y_x - 20\mathfrak{q}y_x^2 + 8y^2\left(8\mathfrak{q}^2 + \mathfrak{q}_{xx}\right)$$
$$+ 40\mathfrak{q}yy_{xx} + y_{xx}^2 - 2y_x y^{(3)} + 2yy^{(4)},$$

$$\begin{aligned}\mathscr{F}_7^y &= -56y_x^2\left(14\mathfrak{q}^2 + \mathfrak{q}_{xx}\right) - 56\mathfrak{q}_x y_x y_{xx} \\ &+ 56y\left(28\mathfrak{q}^2 + 3\mathfrak{q}_{xx}\right) y_{xx} + 56\mathfrak{q} y_{xx}^2 \\ &+ 56yy_x\left(28\mathfrak{q}\mathfrak{q}_x + \mathfrak{q}^{(3)}\right) + 168y\mathfrak{q}_x y^{(3)} \\ &- 112\mathfrak{q} y_x y^{(3)} - y^{(3)^2} \\ &+ 12y^2\left(192\mathfrak{q}^3 + 33\mathfrak{q}_x^2 + 52\mathfrak{q}\mathfrak{q}_{xx} + \mathfrak{q}^{(4)}\right) \\ &+ 112\mathfrak{q}yy^{(4)} + 2y_{xx}y^{(4)} - 2y_x y^{(5)} + 2yy^{(6)}.\end{aligned} \tag{32}$$

We notice that, for each n, the first integral $\mathscr{F}_n^y = \mathscr{F}_n^y[y, \mathfrak{q}]$ viewed as a differential function of y and $\mathfrak{q}$ is quadratic homogeneous in y and polynomial in $\mathfrak{q}$.

In [18], first integrals of the trivial equation $y^{(n)} = 0$ were explicitly computed for $n = 2, 3, \ldots, 6$, and they are directly expressed in terms of the linearly independent ones thus obtained. The method used in that paper for finding the first integrals F associated with each symmetry $\mathbf{v}$ of the equation is direct computation done by solving the boundary value problem

$$\mathbf{v}^{[n-1]}(F) = 0, \tag{33a}$$

$$\left.\frac{dF}{dx}\right|_{y^{(n)}=0} = 0. \tag{33b}$$

However, the calculations done in [18] are only a special case of those done in this paper for equations of maximal symmetry and low order $n \le 8$. Indeed, we have considered equations of maximal symmetry in their most general form (13a)–(13d) and not in the simpler canonical form $y^{(n)} = 0$. Of course the first integrals we have found here from a variational principle all satisfy (33a) and (33b).

It follows from Liouville's formula that if $s_1, \ldots, s_n$ are linearly independent solutions of an equation of the form (2), then n linearly independent first integrals of the equations are given by the Wronskians

$$R_j[y] = w\left(s_1, \ldots, s_{j-1}, y, s_{j+1}, \ldots, s_n\right), \quad j = 1, \ldots, n. \tag{34}$$

However, this direct calculation does not relate the first integral to any variational symmetry of the equation and thus cannot exploit the various properties of such symmetries. Moreover, it requires the knowledge of the complete set of linear independent solutions of the equation itself, while those we found involve only two linearly independent solutions of the second-order source equation. In addition, these two symbolic solutions are not required to be known explicitly but are only assumed to satisfy the second-order equation. In fact, we have ultimately expressed the first integrals found solely in terms of $\mathfrak{q}$, y and their derivatives.

4. Hamiltonian Formulation

In order to gain more insight into the properties of LODEs of maximal symmetry, we shall move one step forward in this section by finding a Hamiltonian formulation of some of these equations and investigate some of its properties. Indeed, in the case of scalar equations, the study in the previous sections was largely facilitated thanks to formula (3) which provides the general solution of any iterative LODE through a very simple superposition principle in terms of two linearly independent solutions u and v of the second-order source equation (4). Thus one of our ultimate goals in this undertaking is to obtain an extension of such a formula to Hamiltonian systems associated with equations of maximal symmetry.

For second-order equations, the transformation into a Hamiltonian system is usually achieved by the common method via the Legendre transformation by defining the conjugate momenta as $p = \partial L/\partial y$, where L is the Lagrangian. However, for higher-order equations, one can make use of the Jacobi-Ostrogradski generalized momenta [19, 20].

Indeed, consider in this more general approach a given nth order Euler-Lagrange equation $E(L) = 0$, where E is the usual Euler-Lagrange operator (see [7, Page 250]) applied to a Lagrangian $L[y] = L(x, y^{(m)})$. In the latter equality, m can be a multi-index of order $n/2$ and y a multivariable but which in our own considerations will be limited to a single dependent variable of the independent variable x. Then the generalized momenta are defined by the expressions

$$\begin{aligned} p_1 &= \frac{\partial L}{\partial y_x} - D_x\left(\frac{\partial L}{\partial y_{xx}}\right) + \cdots \\ &\quad + (-1)^{m-1} D_x^{m-1}\left(\frac{\partial L}{\partial y^{(m)}}\right) \\ p_2 &= \frac{\partial L}{\partial y_{xx}} - D_x\left(\frac{\partial L}{\partial y_{xxx}}\right) + \cdots \\ &\quad + (-1)^{m-2} D_x^{m-2}\left(\frac{\partial L}{\partial y^{(m)}}\right) \\ &\vdots \\ p_m &= \left(\frac{\partial L}{\partial y^{(m)}}\right). \end{aligned} \tag{35}$$

As for the canonical coordinates, they are given by

$$q_j = y^{(j-1)}, \quad \text{for } j = 1, \ldots, m. \tag{36}$$

The corresponding Hamiltonian function then takes the form

$$H = p_1 q_2 + p_2 q_3 + \cdots + p_{m-1} q_m + p_m y^{(m)} - L, \tag{37}$$

and in the new coordinates $(x, (q_j), (p_j))$ the Hamilton-Cartan equations are given as usual by

$$\frac{dq_j}{dx} = \frac{\partial H}{\partial p_j}, \qquad \frac{dp_j}{dx} = -\frac{\partial H}{\partial q_j}. \tag{38}$$

For each LODE $\Delta_n[y] = 0$, it is well known [7] as already noted that when the equation is of the Euler-Lagrange type, a Lagrangian L is found by setting

$$L = \frac{1}{2} y \Delta_n [y]. \tag{39}$$

Moreover, this nth order Lagrangian can always be reduced to one of order $n/2$ by eliminating some null Lagrangian terms from it. Denoting as usual by L_n the reduced Lagrangian corresponding n listed in (13a)–(13d), it is found that

$$\begin{aligned} L_2 &= \frac{1}{2}\left(\mathfrak{q} y^2 - y_x^2\right) \\ L_4 &= -5\mathfrak{q} y_x^2 + \frac{3}{2} y^2 \left(3\mathfrak{q}^2 + \mathfrak{q}_{xx}\right) + \frac{1}{2} y_{xx}^2, \end{aligned} \tag{40}$$

for the first two equations. In case $n = 2$, the canonical variables are (q, p) with $q = y$ and $p = \partial L/\partial y_x = -y_x$, and, by the Legendre transformation, the Hamiltonian in this case has the expression

$$H_2 = -(y_x p - L) = -\frac{1}{2}\left(p^2 + \mathfrak{q} q^2\right) \tag{41}$$

with corresponding Hamiltonian system

$$\dot{q} = -p, \qquad \dot{p} = \mathfrak{q} q, \tag{42}$$

where we have $\dot{F} = dF/dx$, for any function $F = F(x)$. It turns out that (42) is linear as was to be expected given that H_2 is quadratic, but also (42) coincides with the standard representation of the second-order source equation (4) as a first-order system. Thus two linearly independent solutions of (42) are $(u, -u_x)$ and $(v, -v_x)$, with the usual notations.

For $n = 4$, the canonical coordinates (q_1, q_2, p_1, p_2) are obtained more explicitly from (35) and (36) as

$$\begin{aligned} q_1 &= y, \\ q_2 &= y_x, \\ p_1 &= \frac{\partial L_4}{\partial y_x} - \frac{d}{dx}\left(\frac{\partial L_4}{y_{xx}}\right) = -10\mathfrak{q} y_x - y^{(3)}, \\ p_2 &= \frac{\partial L_4}{\partial y_{xx}} = y_{xx}, \end{aligned} \tag{43}$$

and the quadratic Hamiltonian H_4 has the expression

$$\begin{aligned} H_4 &= p_1 y_x + p_2 y_{xx} - L_4 \\ &= \frac{1}{2}\left(p_2^2 - 9\mathfrak{q}^2 q_1^2 + 2 p_1 q_2 + 10\mathfrak{q} q_2^2 - 3 q_1^2 \ddot{\mathfrak{q}}\right), \end{aligned} \tag{44}$$

whereas the corresponding linear Hamiltonian system takes the form

$$\begin{aligned} \dot{q}_1 &= q_2, \\ \dot{q}_2 &= p_2, \\ \dot{p}_1 &= 3 q_1 \ddot{\mathfrak{q}} + 9\mathfrak{q}^2 q_1, \\ \dot{p}_2 &= -p_1 - 10\mathfrak{q} q_2. \end{aligned} \tag{45}$$

A question which is worth considering at this point is whether solutions to the Hamiltonian system (45) can also be obtained through a simple superposition formula from the solutions of the second-order system (42), as in the case of scalar equations of maximal symmetry. By construction of (45), this question is already positively answered by formulas (43) which provide an explicit expression for the unknown variables q_j and p_j directly in terms of the solutions of the scalar source equation (4). Indeed, if $s = s(x)$ is any given function of x, then (43) shows that $S = \{s, s_x, -10\mathfrak{q} s_x - s^{(3)}, s_{xx}\}$ is a solution to (45) if and only if s is a solution to the corresponding fourth-order scalar equation (13b). Therefore, a complete set of linearly independent solutions for (45) is given by

$$\begin{aligned} S_k &= \left\{s_k, (s_k)_x, -10\mathfrak{q}\,(s_k)_x - s_k^{(3)}, (s_k)_{xx}\right\}, \\ s_k &= u^{3-k} v^k, \quad k = 0, \ldots, 3, \end{aligned} \tag{46}$$

where as usual u and v are two linearly independent solutions of (4), using also the fact that $\mathfrak{q} = \varpi(u_x, v_x)$ by (5). Note that, for any choice of s in the expression of S, all equations in (45) are always satisfied, except for the third one which reduces exactly to (13b).

More generally, it follows from (35) and (36) that the general solution to every Hamiltonian system derived from an equation of maximal symmetry as those given in (13a)–(13d) can be found by a mere superposition formula from the solutions of the second-order source equation (4). It will therefore be useful to obtain a characterization of the equivalence class under point transformations of these nontrivial families of Hamiltonian systems depending on the arbitrary function $\mathfrak{q}$, as well as its symmetry properties.

Restricting our attention for now to the fourth-order system (45), the existence of the superposition formula (46) indicates that this system of equations can be reduced in order, and we know this is possible in particular if the system has a two-parameter abelian symmetry subgroup. As is well known for Hamiltonian systems, symmetries are essentially in one-one correspondence with first integrals F of the system, the determining equation of which is given by

$$\frac{\partial F}{\partial x} + \{F, H\} = 0 \Longleftrightarrow \tag{47a}$$

$$\frac{\partial F}{\partial x} + \widehat{\mathbf{v}}_H (F) = 0. \tag{47b}$$

In (47a) and (47b), $\{,\}$ is the associated Poisson bracket for the canonical coordinates and $\widehat{\mathbf{v}}_H$ is the Hamiltonian vector field associated with the Hamiltonian function H. In the case of (45), we have

$$\begin{aligned}\hat{\mathbf{v}}_H &= -q_2\partial_{q_1} - p_2\partial_{q_2} - 3q_1\left(3\mathsf{q}^2 + \ddot{\mathsf{q}}\right)\partial_{p_1}\\ &\quad + \left(p_1 + 10\mathsf{q}q_2\right)\partial_{p_2}.\end{aligned} \tag{48}$$

Due to the arbitrariness of the function $\mathsf{q} = \mathsf{q}(x)$ in the expression of $\hat{\mathbf{v}}_H$, it does not seem to be possible to solve directly the system corresponding to (48). However, it should be possible to find these first integrals as in the scalar case by exploiting the fact that u and v are solutions of (4). Here, we look at what happens when $\mathsf{q} = 0$. In this case, we find the four functionally independent first integrals

$$\begin{aligned} f_1 &= \frac{1}{6}\left(-x^3 p_1 + 3x^2 p_2 + 6q_1 + 6xq_2\right),\\ f_2 &= \frac{1}{2}\left(-x^2 p_1 + 2xp_2 + 2q_2\right),\\ f_3 &= -xp_1 + p_2,\\ f_4 &= p_1. \end{aligned} \tag{49}$$

We clearly have, for instance, $\{f_1, f_2\} = 0$ and thus $\hat{\mathbf{v}}_{f_1}$ and $\hat{\mathbf{v}}_{f_2}$ generate an abelian Hamiltonian symmetry group of order two of the equation. Therefore, system (45) can be reduced in order by four and hence its general solution can be found by a mere superposition formula from those of a second-order reduced system, that is, in particular from the solutions of the source equation (4). Moreover, the corresponding system with two degrees of freedom (45) is maximally superintegrable in the sense of [21, 22]. It is our belief that all Hamiltonian systems derived from equations of maximal symmetry (13a)–(13d) should satisfy similar properties in their most general form. But this investigation is postponed to a future research project.

5. Concluding Remarks

We would like to comment on two other possible methods to find the first integrals obtained in this paper and more generally the first integrals of any system of differential equations. The first method is due to two important results recently obtained in [6]. The first of these two results is an efficient algorithm for generating equations of maximal symmetry of arbitrary orders and of the most general form solely in terms of the coefficient q and its derivatives. The second result is the point transformation

$$\begin{aligned}\zeta &= \frac{v}{u},\\ \theta &= k_0 u^{1-n} y,\end{aligned} \tag{50}$$

where k_0 is an arbitrary constant, which maps the canonical form $\theta^{(n)} = 0$ to the most general form $\Delta_n[y] = \Delta_n[y, \mathsf{q}]$ of a LODE of maximal symmetry. Thanks to these two results, to find first integrals of linear equations of maximal symmetry in their most general form, it suffices to find that for the canonical form of the corresponding equation. Indeed, the first integral of the transformed equation under (50) is just the transformed version under (50) of the first integral of the original equation. For instance, the most general first integral $\mathcal{F}$ associated with solution symmetries of $\theta^{(2)} = 0$ is easily found to be given by $\mathcal{F} = -a_1\theta + (a_0 + a_1\zeta)\theta'$, for some arbitrary constants a_0 and a_1. Under (50), $\mathcal{F}$ is simply transformed into (25). A research project is currently in progress to use this method to find first integrals for the most general forms of linear and nonlinear ODEs of maximal symmetry and of a general order.

Most often, to find first integrals of an nth order differential equation $\Delta_n[y] = 0$, one looks for an integrating factor $\Pi[y]$ as a differential function of y of order at most $n-1$ such that the expression $\Pi[y]\Delta_n[y]$ is a null Lagrangian, that is, such that $E(\Pi[y]\Delta_n[y]) = 0$, where E is the Euler operator. Indeed, the latter equality holds if and only if $\Pi[y]\Delta_n[y] = \operatorname{div} P$ is a total divergence for a certain differential function $P = P[y]$. Thus, in the case of ODEs, when $\Pi[y]$ is a finite expression, $P[y]$ is a first integral and may be expressed as the integral of $\Pi[y]\Delta_n[y]$. The first problem with this method is to find the integrating factor $\Pi[y]$. This can be achieved by an application of Nœther's theorem as we did in this paper. There are several other more or less similar methods based on adjoint symmetries or other types of symmetries [23–25] or on more direct methods [26, 27] achieved by solving equations similar to $E(\Pi[y]\Delta_n[y]) = 0$. The second problem with these rather common methods is that once an integrating factor has been found the components of the vector $P[y]$ are obtained as integrals of generally complicated expressions, and thus evaluating such integrals remains usually challenging as it was the case in this paper.

Competing Interests

The author declares that there are no competing interests regarding the publication of this article.

References

[1] J. Krause and L. Michel, "Equations différentielles linéaires d'ordre n > 2 ayant une algèbre de Lie de symétrie de dimension n + 4," *Comptes Rendus de l'Académie des Sciences. Paris*, vol. 307, Article ID 905910, pp. 905–910, 1988.

[2] S. Lie, "Classification und Integration von gewöhnlichen Differentialgleichungen zwischenxy, die eine Gruppe von Transformationen gestatten," *Mathematische Annalen*, vol. 32, no. 2, pp. 213–281, 1888.

[3] J.-C. Ndogmo and F. M. Mahomed, "On certain properties of linear iterative equations," *Central European Journal of Mathematics*, vol. 12, no. 4, pp. 648–657, 2014.

[4] J. C. Ndogmo, "Coefficient characterization of linear differential equations with maximal symmetries," *Communications in Nonlinear Science and Numerical Simulation*, vol. 19, no. 5, pp. 1270–1278, 2014.

[5] F. M. Mahomed and P. G. Leach, "Symmetry Lie algebras of nth order ordinary differential equations," *Journal of Mathematical Analysis and Applications*, vol. 151, no. 1, pp. 80–107, 1990.

[6] J. C. Ndogmo, "Generation and identification of ordinary differential equations of maximal symmetry algebra," *Abstract and Applied Analysis*, vol. 2016, Article ID 1796316, 9 pages, 2016.

[7] P. J. Olver, *Applications of Lie groups to Differential Equations*, vol. 107 of *Graduate Texts in Mathematics*, Springer-Verlag, New York, NY, USA, 1986.

[8] E. Noether, "Invariante variationsprobleme," *Math-Phys Kl Heft*, pp. 235–257, 1918.

[9] C. Nitsch, "Some stability properties involving first integrals," *Rendiconti dell'Accademia delle Scienze Fisiche e Matemàtiche di Napoli*, vol. 72, pp. 29–40, 2005.

[10] Z. Rzesztko, "Strong stability of linear symplectic actions and the orbit method," *Buletinul Academiei de Stiinte a Republicii Moldova. Matematica*, no. 2, pp. 99–103, 2005.

[11] M. F. Singer, "Liouvillian first integrals of differential equations," *Transactions of the American Mathematical Society*, vol. 333, no. 2, pp. 673–688, 1992.

[12] J. Chavarriga, H. Giacomini, J. Giné, and J. Llibre, "Darboux integrability and the inverse integrating factor," *Journal of Differential Equations*, vol. 194, no. 1, pp. 116–139, 2003.

[13] J. Giné and M. Grau, "Weierstrass integrability of differential equations," *Applied Mathematics Letters*, vol. 23, no. 5, pp. 523-526, 2010.

[14] J. Giné, "Reduction of integrable planar polynomial differential systems," *Applied Mathematics Letters. An International Journal of Rapid Publication*, vol. 25, no. 11, pp. 1862–1865, 2012.

[15] A. R. Ünal, "Exact solutions of the Zakharov equations by using the first integral method," *Communications de la Faculté des Sciences de l'Université d'Ankara. Séries A1. Mathematics and Statistics*, vol. 61, pp. 9–16, 2012.

[16] S. Lie, *Vorlesungüber Continuerliche Gruppen Mit Geometrischen und Anderen Anwendungen Bearbeitet und Herausgegeben von Dr. G. Scheffers*, Teubner, Liepzig, Germany, 1893.

[17] C. Muriel, J. L. Romero, and P. J. Olver, "Variational C^{∞}-symmetries and Euler–Lagrange equation," *Journal of Differential Equations*, vol. 222, no. 1, pp. 164–184, 2006.

[18] G. P. Flessas, K. S. Govinder, and P. G. L. Leach, "Characterisation of the algebraic properties of first integrals of scalar ordinary differential equations of maximal symmetry," *Journal of Mathematical Analysis and Applications*, vol. 212, no. 2, pp. 349–374, 1997.

[19] P. Guha, A. Ghose Choudhury, and A. S. Fokas, "Hamiltonians and conjugate Hamiltonians of some fourth-order nonlinear ODEs," *Nonlinear Analysis: Theory, Methods & Applications*, vol. 75, no. 4, pp. 2126–2138, 2012.

[20] M. C. Nucci and P. G. Leach, "The method of Ostrogradsky, quantization, and a move toward a ghost-free future," *Journal of Mathematical Physics*, vol. 50, no. 11, 2009.

[21] P. Winternitz and I. Yurdusen, "Integrable and superintegrable systems with spin in three-dimensional Euclidean space," *Journal of Physics A*, vol. 42, no. 38, Article ID 385203, 20 pages, 2009.

[22] R. G. Smirnov and P. Winternitz, "A class of superintegrable systems of Calogero type," *Journal of Mathematical Physics*, vol. 47, no. 9, Article ID 093505, 8 pages, 2006.

[23] S. C. Anco and G. Bluman, "Integrating factors and first integrals for ordinary differential equations," *European Journal of Applied Mathematics*, vol. 9, no. 3, pp. 245–259, 1998.

[24] S. V. Meleshko, S. Moyo, C. Muriel, J. L. Romero, P. Guha, and A. G. Choudhury, "On first integrals of second-order ordinary differential equations," *Journal of Engineering Mathematics*, vol. 82, pp. 17–30, 2013.

[25] R. Naz, F. M. Mahomed, and D. P. Mason, "Comparison of different approaches to conservation laws for some partial differential equations in fluid mechanics," *Applied Mathematics and Computation*, vol. 205, no. 1, pp. 212–230, 2008.

[26] S. C. Anco and G. Bluman, "Direct construction method for conservation laws of partial differential equations. Part II: general treatment," *European Journal of Applied Mathematics*, vol. 13, no. 5, pp. 567–585, 2002.

[27] T. Wolf, "A comparison of four approaches to the calculation of conservation laws," *European Journal of Applied Mathematics*, vol. 13, no. 2, pp. 129–152, 2002.

9

Axioms for Consensus Functions on the n-Cube

C. Garcia-Martinez,[1] F. R. McMorris,[2,3] O. Ortega,[4] and R. C. Powers[3]

[1]*Departamento de Ciencias Básicas, Universidad Autóma Metropolitana Unidad Azcapotzalco, Av. San Pablo 180, Col. Reynosa Tamaulipas, C.P. 02200 Ciudad de México, Mexico*
[2]*Department of Applied Mathematics, Illinois Institute of Technology, Chicago, IL 60616, USA*
[3]*Department of Mathematics, University of Louisville, Louisville, KY 40292, USA*
[4]*Department of Mathematics, Harold Washington College, Chicago, IL 60601, USA*

Correspondence should be addressed to F. R. McMorris; mcmorris@iit.edu

Academic Editor: Dimitris Fotakis

A p value of a sequence $\pi = (x_1, x_2, \ldots, x_k)$ of elements of a finite metric space (X, d) is an element x for which $\sum_{i=1}^{k} d^p(x, x_i)$ is minimum. The ℓ_p-function with domain the set of all finite sequences on X and defined by $\ell_p(\pi) = \{x\colon x \text{ is a } p \text{ value of } \pi\}$ is called the ℓ_p-function on (X, d). The ℓ_1 and ℓ_2 functions are the well-studied median and mean functions, respectively. In this note, simple characterizations of the ℓ_p-functions on the n-cube are given. In addition, the center function (using the minimax criterion) is characterized as well as new results proved for the median and antimedian functions.

1. Introduction

A *consensus function* (aka *location function*) on a finite connected graph $G = (X, E)$ is a mapping $L : X^* \to 2^X \setminus \{\emptyset\}$, where 2^X denotes the set of all subsets of X, and $X^* = \bigcup_{k \geq 1} X^k$ with $X^k = \overbrace{X \times \cdots \times X}^{k \text{ times}}$. The elements of X^* are called *profiles* and a generic one of *length* k is denoted by $\pi = (x_1, x_2, \ldots, x_k)$. Let d denote the usual geodesic distance, where $d(x, y)$ is the length of a minimum length path joining vertices x and y. Suppose the graph $G = (X, E)$ represents the totality of possible locations. Then a profile $\pi = (x_1, \ldots, x_k)$ is formed where x_i represents the best location from the point-of-view of client (voter, customer, and user) i. A typical approach in location theory is to find those vertices (locations) in X that are "closest" to the profile π. There has been much work in this area of research, ranging from practical computational methods to more theoretical aspects. Since Holzman's paper in 1990 [1], there have been many axiomatic studies of the procedures themselves which resulted in a much better understanding of the process of location (for a small sample, see [2–4] and references within). Now suppose the vertex set X is the set of all linear orders (preference ranking) on a given set of alternatives. In this consensus situation, a profile $\pi = (x_1, \ldots, x_k)$ could represent the collection of ballots of the voters labeled by the set $\{1, \ldots, k\}$; that is, x_i is the preferred ranking of alternatives by voter i. Here a closest vertex to π would represent the entire group's preferred consensus ranking. Many references for this classical situation can be found in [5] and other books on voting theory. Another classic situation, and one pertinent to our study, is the process of selecting a committee from a slate of n candidates. Here each of k voters is to nominate a subset of candidates, so a ballot is simply a profile $\pi = (x_1, \ldots, x_k)$ where each x_i is a subset of the candidates [6, 7]. The vertices of the graph G are the subsets of candidates and the committee consensus function will return one or more subsets closest to the profile.

Four popular measures of the closeness, or remoteness, of a vertex x to a profile $\pi = (x_1, \ldots, x_k)$ are as follows:

(1) The *eccentricity* of x, $e(x, \pi) = \max\{d(x, x_1) d(x, x_2), \ldots, d(x, x_k)\}$

(2) The *status* of x, $S_\pi(x) = \sum_{i=1}^{k} d(x, x_i)$

(3) The *square status* of x, $SS_\pi(x) = \sum_{i=1}^{k} d^2(x, x_i)$

(4) The ℓ_p *status* of x, $\ell_p S_\pi(x) = \sum_{i=1}^{k} d^p(x, x_i)$

The consensus functions based on the these measures of remoteness have been defined as follows:

(a) The *center function*, denoted by Cen, is defined by

$$\text{Cen}(\pi) = \{x \in X : e(x,\pi) \text{ is minimum}\}. \quad (1)$$

(b) The *median function*, denoted by Med, is defined by

$$\text{Med}(\pi) = \{x \in X : S_\pi(x) \text{ is minimum}\}. \quad (2)$$

(c) The *mean function*, denoted by Mean, is defined by

$$\text{Mean}(\pi) = \{x \in X : SS_\pi(x) \text{ is minimum}\}. \quad (3)$$

(d) The ℓ_p-*function*, denoted by ℓ_p, is defined by

$$\ell_p(\pi) = \{x \in X : \ell_p S_\pi(x) \text{ is minimum}\}. \quad (4)$$

The median and mean functions are special cases of the ℓ_p-function, but earlier work [8–10] shows a striking difference between the case of $p = 1$ and $p > 1$.

In this paper we focus on consensus functions on the n-*dimensional hypercube* $Q_n = (X, E)$ whose vertex set is $X = \{(w_1, \ldots, w_n) : w_i \in \{0,1\}\}$. Of course the natural realization of Q_n is the set of all subsets of an n-element set. Recall that, for $u = (u_1, \ldots, u_n)$ and $v = (v_1, \ldots, v_n)$ vertices in Q_n, uv is an edge of Q_n if and only if $\sum_{i=1}^{n} |u_i - v_i| = 1$. We set $u \le v$ if and only if $u_i \le v_i$ for all i. Let d be the usual Hamming distance, where $d(u,v) = \sum_{i=1}^{n} |u_i - v_i|$, so that uv is an edge if and only if $d(u,v) = 1$. Let $\oplus$ denote the addition *modulo* 2, and define $u \oplus v = (u_1 \oplus v_1, \ldots, u_n \oplus v_n)$. For a profile $\pi = (x_1, \ldots, x_k)$ and $u \in Q_n$ let $\pi \oplus u = (x_1 \oplus u, \ldots, x_k \oplus u)$. Let $\mathbf{0} = (0, \ldots, 0)$ and $\mathbf{1} = (1, \ldots, 1)$. Note that $x \oplus x = \mathbf{0}$ for all $x \in Q_n$. Also it is easy to see that, for x, y and z vertices in Q_n, $d(x,y) = d(x \oplus z, y \oplus z)$. We set $e_j \in Q_n$ to be the vertex with 0's everywhere except 1 in the jth coordinate. So, for example, in Q_5

$$\begin{aligned}
(0,0,1,1,0) &= e_3 \oplus e_4, \\
(0,0,1,0,1) \oplus e_3 &= (0,0,0,0,1) \\
(0,1,0,1,1) \oplus (0,0,1,1,0) &= (0,1,1,0,1) \\
&= e_2 \oplus e_3 \oplus e_5.
\end{aligned} \quad (5)$$

Let $\langle \pi \rangle$ denote the subgraph induced by the vertices comprising π. Note that $\langle \pi \oplus v \rangle$ is isomorphic to $\langle \pi \rangle$ for all $v \in Q_n$, and so intuitively $\langle \pi \oplus v \rangle$ is simply a "translation" of $\langle \pi \rangle$ to another position within Q_n. Our goal is to use the particular structure of Q_n to present a very simple unifying approach to give axiomatic characterizations of the consensus functions Cen, Med, and ℓ_p on these graphs. Mulder and Novick [10, 11] have given an elegant set of axioms characterizing the function Med on all median graphs (of which Q_n is a special case) whereas our axioms are essentially straightforward properties that follow from the definitions. At present the most general graph for which characterizations exist for Cen, Mean, and ℓ_p is a tree [9, 12–14]. An interesting weighted version of Cen on Q_n is studied in [6].

We mention that the following results can be framed in the more abstract context of finite Boolean algebras, as it is done in [15–17]. We prefer to work in the more specific situation of the n-cube where properties become quite easy to visualize, and yet we are working without loss of generality because every finite Boolean algebra is isomorphic to an n-cube.

2. The Axioms and Characterizations of Cen, Med, and ℓ_p-Function

In this section we give two very simple properties that will allow us to establish a general result that can be used to give a new way to view Cen, Med, and ℓ_p defined on Q_n. Let $f : X^* \to 2^X \setminus \{\emptyset\}$ be a consensus function on $Q_n = (X, E)$. Our key axiom for a consensus function f is the following.

Translation (T). For any profile π and vertices u and v of Q_n,

$$\begin{gathered} u \in f(\pi) \\ \text{implies that } u \oplus v \in f(\pi \oplus v). \end{gathered} \quad (6)$$

Note that this is equivalent to $u \in f(\pi)$ if and only if $u \oplus v \in f(\pi \oplus v)$.

Now let f and g be consensus functions on Q_n and let x_0 be a vertex. We say f and g *agree* at x_0 if for any profile π

$$x_0 \in f(\pi) \quad \text{iff } x_0 \in g(\pi). \quad (7)$$

Theorem 1. *If the consensus functions f and g on Q_n both satisfy (T) and agree at a vertex x_0, then $f = g$.*

Proof. Let π be a profile and $v \in X$. Then there exists $v' \in X$ such that $v \oplus v' = x_0$. Since f satisfies (T), we have

$$v \in f(\pi) \quad \text{iff } v \oplus v' = x_0 \in f(\pi \oplus v'). \quad (8)$$

Because f and g agree at x_0,

$$x_0 \in f(\pi \oplus v') \quad \text{iff } x_0 \in g(\pi \oplus v'). \quad (9)$$

Since g satisfies (T),

$$v \oplus v' = x_0 \in g(\pi \oplus v') \quad \text{iff } v \in g(\pi). \quad (10)$$

Hence $v \in f(\pi)$ if and only if $v \in g(\pi)$. □

Theorem 1 implies that if f and g are consensus functions on Q_n and both satisfy (T); then $f = g$ if the conditions placing $\mathbf{0}$ in $f(\pi)$ are the same as the conditions placing $\mathbf{0}$ in $g(\pi)$.

As observed before, $d(x,y) = d(x \oplus z, y \oplus z)$ for x, y, and z vertices in Q_n. Using this and the definitions it is easy to see that Cen, Med, and ℓ_p all satisfy (T). Therefore, characterizations will follow once the conditions are obtained for when $\mathbf{0} \in \text{Cen}(\pi)$, $\mathbf{0} \in \text{Med}(\pi)$, and $\mathbf{0} \in \ell_p(\pi)$. We present these results in a series of lemmas and corollaries.

Let $u \in Q_n$ and set $\|u\| = d(\mathbf{0}, u)$, that is, the number of ones that appear in the representation u as a vertex of Q_n. Let $\pi = (x_1, x_2, \ldots, x_k)$ be a profile on Q_n. Then $\|\pi\|$ is defined to be

$$\|\pi\| = \max\{\|x_1\|, \|x_2\|, \ldots, \|x_k\|\}. \quad (11)$$

Lemma 2. *Let Cen be the center function on Q_n and π a profile. Then*

$$\mathbf{0} \in Cen(\pi) \quad \text{iff } \|\pi\| \le \|\pi \oplus u\|, \ \forall u \in Q_n. \quad (12)$$

Proof. The result is clear because $d(x, y) = d(x \oplus z, y \oplus z)$ in Q_n, and $e(\mathbf{0}, \pi) = \|\pi\|$ for any profile π. □

Corollary 3. *Let f be a consensus function on Q_n. Then $f = Cen$ if and only f satisfies (T) and for every profile π and $u \in Q_n$*

$$\mathbf{0} \in f(\pi) \quad \text{iff } \|\pi\| \le \|\pi \oplus u\|. \quad (13)$$

Mulder and Novick [10] give an elegant characterization of Med on Q_n, which was extended to all median graphs in [11]. We will give another characterization using the approach given by Theorem 1. For a profile $\pi = (x_1, \ldots, x_k)$ let $x_i = (x_1^i, \ldots, x_n^i)$. The next result has been noted in [10].

Lemma 4. *Let Med be the median function on Q_n and $\pi = (x_1, \ldots, x_k)$ a profile. Then*

$$\mathbf{0} \in Med(\pi) \quad \text{iff } \sum_{j=1}^{k} x_i^j \le \frac{k}{2} \ \forall i. \quad (14)$$

Corollary 5. *Let f be a consensus function on Q_n. Then $f = Med$ if and only f satisfies (T) and for any profile $\pi = (x_1, \ldots, x_k)$,*

$$\mathbf{0} \in f(\pi) \quad \text{iff } \sum_{j=1}^{k} x_i^j \le \frac{k}{2} \ \forall i. \quad (15)$$

For the function ℓ_p it is easy to see from the definitions that, for any profile π and a in Q_n,

$$\mathbf{0} \in \ell_p(\pi) \quad \text{iff } a = \mathbf{0} \oplus a \in \ell_p(\pi \oplus a). \quad (16)$$

As in [17] we consider the *p-characteristic* of a profile $\pi = (x_1, x_2, \ldots, x_k)$ to be the number

$$\mathrm{Char}_p(\pi) = \sum_{i=1}^{n} \|x_i\|^p. \quad (17)$$

Lemma 3.12 in [17] gives the following result.

Lemma 6. *Consider the function ℓ_p on Q_n, and let $\pi = (x_1, \ldots, x_k)$ be a profile. Then*

$$\mathbf{0} \in \ell_p(\pi) \quad \text{iff } Char_p(\pi) \le Char_p(\pi \oplus a) \text{ for every } a \text{ in } Q_n. \quad (18)$$

Corollary 7. *Let f be a consensus function on Q_n. Then $f = \ell_p$ if and only f satisfies (T) and for any profile $\pi = (x_1, \ldots, x_k)$,*

$$\mathbf{0} \in f(\pi) \quad \text{iff } Char_p(\pi) \le Char_p(\pi \oplus a) \text{ for every vertex } a \text{ in } Q_n. \quad (19)$$

Here are three other examples of consensus functions that satisfy the Translation property. However it is clear that these functions would not be useful in committee elections or as location functions, for instance.

Example 1. Let f_1 be the consensus function on Q_n defined by $f(\pi) = \{x_1\}$ for any profile $\pi = (x_1, \ldots, x_k)$. That is, f_1 is a standard projection function. Then clearly f_1 satisfies (T).

Example 2. Let f_2 be the consensus function on Q_n defined by $f_2(\pi) = X$ for all profiles π. That is, f_2 is the constant function with ouput being the entire vertex set X. Then f_2 satisfies (T), and moreover it can be easily shown that it is the only constant function that satisfies (T).

Example 3. Let f_3 be the consensus function on Q_n defined by $f_3(\pi) = \{\pi\}$ for all π where $\{\pi\}$ is the set of vertices appearing in the profile π. Then clearly f_3 satisfies (T).

The function f_2 allows us to see some of the implications of imposing (T). First we need to recall one of the crucial axioms for the characterization of the consensus function Med [10, 11, 18].

Consistency (C). The consensus function f satisfies (C) if, for profiles π_1 and π_2,

$$f(\pi_1) \cap f(\pi_2) \neq \emptyset \quad \text{implies } f(\pi_1\pi_2) = f(\pi_1) \cap f(\pi_2). \quad (20)$$

Proposition 8. *A consensus function f on Q_n satisfies (T), (C), and*

$$\bigcap_{x \in X} f(x) \neq \emptyset \quad (21)$$

if and only if $f = f_2$.

Proof. Clearly f_2 satisfies the conditions, so now let f be a consensus function that satisfies (T), (C), and the intersection condition. Let $v \in f(x)$ for all $x \in X$. Then since f satisfies (T) we have $v \oplus x \in f(x \oplus x) = f(\mathbf{0})$ for all $x \in$ X. Now let w be an arbitrary vertex. Then $w = v \oplus (v \oplus w) \in f(\mathbf{0})$ and thus $f(\mathbf{0}) = X$. So if z is any vertex in X, $z \oplus x \in f(\mathbf{0})$ and since f satisfies (T) we have

$$z = (z \oplus x) \oplus x \in f(\mathbf{0} \oplus x) = f(x). \quad (22)$$

Therefore $f(x) = X$ for all $x \in X$, which means that $f(\pi) = X$ for all profiles π of length 1. Using (C) and induction we conclude that $f(\pi) = X$ for all profiles π, that is, $f = f_2$. □

3. Alternative Characterizations of the Median and Antimedian Functions on Q_n

For any profile $\pi = (v_1, \dots, v_k)$ such that

$$v_i = \left(x_1^i, \dots, x_n^i\right) \in \{0,1\}^n \tag{23}$$

for $i = 1, \dots k$, let $\text{Maj}(\pi) = (w_1, \dots, w_n)$ be the vertex in X such that

$$w_i = 1 \quad \text{iff} \sum_{j=1}^{k} x_i^j > \frac{k}{2} \tag{24}$$

for $i = 1, \dots, n$. We will say that a location function f satisfies the condition (Maj) if

$$\text{Maj}(\pi) \in f(\pi) \tag{25}$$

for any profile π. We have previously noted that the median function satisfies (T) and we will show below that, as expected, Med satisfies (Maj). However, there are other location functions that satisfy these two conditions, such as f_2, for example. But, arguably f_2 is not a very reasonable method of consensus or location. So our next step is to invoke a condition that restricts the range of a location function.

For any profile $\pi = (v_1, \dots, v_k)$ such that

$$v_i = \left(x_1^i, \dots, x_n^i\right) \in \{0,1\}^n \tag{26}$$

for $i = 1, \dots, k$ define the *Condorcet score* of π to be

$$\text{Cs}(\pi) = \left|\left\{ i : \sum_{j=1}^{k} x_i^j = \frac{k}{2} \right\}\right|. \tag{27}$$

Observe that if the profile length k is odd, then $\text{Cs}(\pi) = 0$. A location function f satisfies *Restricted Range* (RR) if

$$|f(\pi)| \le 2^{\text{Cs}(\pi)} \tag{28}$$

for any profile π.

We can now give a completely different characterization of Med from that found in [10].

Theorem 9. *Let f be a location function on Q_n. Then f = Med if and only if f satisfies (T), (Maj), and (RR).*

Proof. Assume f = Med. We already know that f satisfies (T), so we only need to show that Med satisfies (Maj) and (RR).

We will follow the notation given above. Let $\pi = (v_1, \dots, v_k)$ be a profile such that

$$v_i = \left(x_1^i, \dots, x_n^i\right) \in \{0,1\}^n \tag{29}$$

for $i = 1, \dots, k$ and let $\text{Maj}(\pi) = (w_1, \dots, w_n) = w$. Now let $a = (y_1, \dots, y_n) \ne w$ be such that $y_m \ne w_m$ for some m. First note that, for every j, because w_j and x_j^i are equal for at least $k/2$ of the i's,

$$\sum_{i=1}^{k} \left|y_j - x_j^i\right| \ge \sum_{i=1}^{k} \left|w_j - x_j^i\right|. \tag{30}$$

Since

$$S_\pi(a) = \sum_{i=1}^{k} d(a, v_i) \quad \text{where } d(a, v_i) = \sum_{j=1}^{n} \left|y_j - x_j^i\right| \tag{31}$$

we have

$$\begin{aligned} S_\pi(a) &= \sum_{i=1}^{k}\sum_{j=1}^{n} \left|y_j - x_j^i\right| = \sum_{j=1}^{n}\sum_{i=1}^{k} \left|y_j - x_j^i\right| \\ &\ge \sum_{j=1}^{n}\sum_{i=1}^{k} \left|w_j - x_j^i\right| = S_\pi(w). \end{aligned} \tag{32}$$

Therefore $w \in \text{Med}(\pi)$ and f satisfies (Maj).

Let $u = (u_1, \dots, u_n)$ be the vertex in X such that

$$u_i = 1 \quad \text{iff} \sum_{j=1}^{k} x_i^j \ge \frac{k}{2} \tag{33}$$

for $i = 1, \dots, n$. For any vertex $a = (y_1, \dots, y_n)$ such that $w \le a \le u$ and for any $i \in \{1, \dots, n\}$ such that $\sum_{j=1}^{k} x_i^j = k/2$ we get that $w_i = 0$, $u_i = 1$, and of course $y_i \in \{0,1\}$. Observe that

$$\sum_{i=1}^{k} \left|y_j - x_j^i\right| = \frac{k}{2} = \sum_{i=1}^{k} \left|w_j - x_j^i\right|. \tag{34}$$

Since $y_i = w_i$ whenever $\sum_{j=1}^{k} x_i^j \ne k/2$ it follows that $S_\pi(a) = S_\pi(w)$ and so $a \in \text{Med}(\pi)$. Moreover, if $b = (z_1, \dots, z_n)$ is vertex in X such that $z_m \ne w_m$ for some $m \in \{1, \dots, n\}$ where $\sum_{j=1}^{k} x_m^j \ne k/2$, then

$$\sum_{i=1}^{k} \left|z_m - x_m^i\right| > \frac{k}{2} > \sum_{i=1}^{k} \left|w_m - x_m^i\right|. \tag{35}$$

In this case, $S_\pi(b) > S_\pi(w)$ and so $b \notin \text{Med}(\pi)$. It now follows that

$$\text{Med}(\pi) = \left\{ \text{Maj}(\pi) \oplus \sum_{\alpha \in A} i_\alpha : A \subseteq S \right\}, \tag{36}$$

where

$$S = \left\{ \alpha \in \{1, \dots, n\} : \sum_{j=1}^{k} x_\alpha^j = \frac{k}{2} \right\}. \tag{37}$$

Therefore, $|\text{Med}(\pi)| = 2^{|S|} = 2^{Cs(\pi)}$ and hence Med satisfies (RR).

For the converse, assume that f satisfies (T), (Maj), and (RR). We will show that f = Med. Let $\pi = (v_1, \dots, v_k)$ be a profile. Then, using Theorem 9,

$$v \in \text{Med}(\pi) \text{ iff } \mathbf{0} \in \text{Med}(\pi \oplus v) \text{ iff } \sum_{j=1}^{k} y_i^j \le \frac{k}{2} \quad \forall i, \tag{38}$$

where $v_j \oplus v = (y_1^j, \ldots, y_n^j)$ for $j = 1, \ldots, k$. Observe that $\mathrm{Maj}(\pi \oplus v) = \mathbf{0}$, and since f satisfies (Maj) it follows that $\mathbf{0} \in f(\pi \oplus v)$. Since f satisfies (T) we get

$$v = \mathbf{0} \oplus v \in f(\pi). \quad (39)$$

It now follows that $\mathrm{Med}(\pi) \subseteq f(\pi)$ for any profile π. Therefore,

$$|\mathrm{Med}(\pi)| \leq |f(\pi)| \quad (40)$$

for any profile π. We know that $\mathrm{Med}(\pi) = 2^{\mathrm{Cs}(\pi)}$ and, by (RR), that $|f(\pi)| \leq 2^{\mathrm{Cs}(\pi)}$ for any profile π. Hence $f(\pi) = \mathrm{Med}(\pi)$ for any profile π and we are done. □

The three consensus functions we have considered all minimize a criterion in order to produce vertices that are close to a given profile of vertices, and as such are useful in location theory. When finding locations to place noxious entities, it is more appropriate to maximize rather than minimize these objective functions, and the resulting "anti"-functions have also been well-studied. Because we have proved Theorem 9 about the median function, we mention the *antimedian function*, denoted by AM, defined by

$$\mathrm{AM}(\pi) = \{x \in X : S_\pi(x) \text{ is maximum}\}. \quad (41)$$

AM has been characterized on Q_n in [19], but we will give an alternate characterization as a corollary to Theorem 9. As before $\pi = (v_1, \ldots, v_k)$ is a profile such that

$$v_i = (x_1^i, \ldots, x_n^i) \in \{0,1\}^n \quad (42)$$

for $i = 1, \ldots, k$. Let $\mathrm{Min}(\pi) = (m_1, \ldots, m_n)$ be the vertex in X such that

$$m_i = 1 \quad \text{iff} \sum_{j=1}^{k} x_i^j < \frac{k}{2} \quad (43)$$

for $i = 1, \ldots, n$. We will say that a location function f satisfies condition (Min) if

$$\mathrm{Min}(\pi) \in f(\pi) \quad (44)$$

for any profile π. Corollary 3 now follows from the proof of Theorem 9 in the obvious way by reversing the inequalities.

Corollary 10. *Let f be a location function on Q_n. Then f = AM if and only if f satisfies (T), (Min), and (RR).*

Competing Interests

The authors declare that there is no conflict of interests regarding the publication of this paper.

References

[1] R. Holzman, "An axiomatic approach to location on networks," *Mathematics of Operations Research*, vol. 15, no. 3, pp. 553–563, 1990.

[2] P. Hansen and F. S. Roberts, "An impossibility result in axiomatic location theory," *Mathematics of Operations Research*, vol. 21, no. 1, pp. 195–208, 1996.

[3] F. R. McMorris, H. M. Mulder, and R. V. Vohra, "Axiomatic characterization of location functions," in *Advances in Interdisciplinary Applied Discrete Mathematics*, H. Kaul and H. M. Mulder, Eds., pp. 71–91, World Scientific, Singapore, 2011.

[4] P. B. Mirchandani and R. L. Francis, Eds., *Discrete Location Theory*, John Wiley & Sons, New York, NY, USA, 1990.

[5] W. H. E. Day and F. R. McMorris, *Axiomatic consensus theory in group choice and biomathematics*, vol. 29 of *Frontiers in Applied Mathematics*, Society for Industrial and Applied Mathematics (SIAM), Philadelphia, Pa, USA, 2003.

[6] S. J. Brams, D. M. Kilgour, and M. R. Sanver, "A minimax procedure for electing committees," *Public Choice*, vol. 132, no. 3, pp. 401–420, 2007.

[7] D. M. Kilgour, S. J. Brams, and M. R. Sanver, "How to elect a representative committee using approval balloting," in *Mathematics and Democracy*, B. Simeone and F. Pukelsheim, Eds., pp. 83–95, Springer, Berlin, Germany, 2006.

[8] O. Ortega and G. Kriston, "The median function on trees," *Discrete Mathematics, Algorithms and Applications*, vol. 5, no. 4, Article ID 1350033, 14 pages, 2013.

[9] F. R. McMorris, H. M. Mulder, and O. Ortega, "The ℓ_p-function on trees," *Networks*, vol. 60, no. 2, pp. 94–102, 2012.

[10] H. M. Mulder and B. Novick, "An axiomatization of the median procedure on the n-cube," *Discrete Applied Mathematics*, vol. 159, no. 9, pp. 939–944, 2011.

[11] H. M. Mulder and B. Novick, "A tight axiomatization of the median procedure on median graphs," *Discrete Applied Mathematics*, vol. 161, no. 6, pp. 838–846, 2013.

[12] F. R. McMorris, F. S. Roberts, and C. Wang, "The center function on trees," *Networks*, vol. 38, no. 2, pp. 84–87, 2001.

[13] F. R. McMorris, H. M. Mulder, and O. Ortega, "Axiomatic characterization of the mean function on trees," *Discrete Mathematics, Algorithms and Applications*, vol. 2, no. 3, pp. 313–329, 2010.

[14] H. M. Mulder, M. J. Pelsmajer, and K. B. Reid, "Axiomization of the center function on trees," *The Australasian Journal of Combinatorics*, vol. 41, pp. 223–226, 2008.

[15] A. Kezdy and R. C. Powers, *The Center Function on Boolean Algebras*, Department of Mathematics, University of Louisville, 2001.

[16] O. Ortega and C. Garcia-Martinez, "The median function on Boolean lattices," *Discrete Mathematics, Algorithms and Applications*, vol. 6, no. 4, Article ID 1450056, 21 pages, 2014.

[17] O. Ortega, C. Garcia-Martinez, and K. Adamski, "The ℓ_p function on finite Boolean lattices," *Discrete Mathematics, Algorithms and Applications*, vol. 8, Article ID 1650044, 15 pages, 2016.

[18] F. R. McMorris, H. M. Mulder, and F. S. Roberts, "The median procedure on median graphs," *Discrete Applied Mathematics*, vol. 84, no. 1–3, pp. 165–181, 1998.

[19] K. Balakrishnan, M. Changat, H. M. Mulder, and A. R. Subhamathi, "Axiomatic characterization of the antimedian function on paths and hypercubes," *Discrete Mathematics, Algorithms and Applications*, vol. 4, no. 4, 1250054, 20 pages, 2012.

10

Determination of the Creep Parameters of Linear Viscoelastic Materials

Alibay Iskakbayev,[1] Bagdat Teltayev,[2] and Sergei Alexandrov[3]

[1] *Department of Mechanics, Al-Farabi Kazakh National University, 71 Al-Farabi, Almaty 050040, Kazakhstan*
[2] *Kazakhstan Highway Research Institute, 2A Nurpeisov, Almaty 050061, Kazakhstan*
[3] *Institute for Problems in Mechanics, Russian Academy of Sciences, 101 Prospect Vernadskogo, Moscow 117526, Russia*

Correspondence should be addressed to Bagdat Teltayev; bagdatbt@yahoo.com

Academic Editor: M. Shariyat

Creep process of linear viscoelastic materials is described by the integral equation of Boltzmann-Volterra in which creep kernel is approximated by Rabotnov's fractional exponential function. The creep equation contains four unknown parameters: α, singularity parameter; β, fading parameter; λ, rheological parameter; and ε_0, conditionally instantaneous strain. Two-stage determination method of creep parameters is offered. At the first stage, taking into account weak singularity properties of Abel's function at the initial moment of loading, parameters ε_0 and α are determined. At the second stage, using already known parameters ε_0 and α, parameters β and λ are determined. Analytical expressions for calculating these parameters are obtained. An accuracy evaluation of the offered method with using experimentally determined creep strains of material Nylon 6 and asphalt concrete showed its high accuracy.

1. Introduction

Many materials used in technique and production (rocks, soils, metals, concretes, polymers, composites, etc.) under loading show viscoelastic properties. Currently a number of methods for quantitative evaluation of viscoelastic properties of materials are known [1–4]. One of the most frequently used methods is creep testing under uniaxial stress [5]. In the mathematical description of creep process of linear viscoelastic material the integral equation of Boltzmann-Volterra with corresponding creep kernel is widely used. A large number of analytical expressions for the creep kernel are known [1, 4, 5].

In testing samples of viscoelastic materials on creep, usually it is difficult to measure instantaneous elastic strain $\varepsilon_0 = \varepsilon\ (t = 0)$. This is due to the fact that in the initial time after application of load to sample the rate of creep is extremely high. Its accurate measurement is practically impossible due to the dynamic effects caused by inertia of loading devices and measuring equipment.

Creep process of many materials particularly at small loads and low temperatures proceeds for a very long time. At high times of loading the rate of creep starts to fade and strain slowly approaching asymptotic value.

Thus, we can assume that creep kernel of integral equation must have the following two properties: firstly, having a weak singularity at the initial time point and, secondly, having a property of exponential function at high times of loading.

The most universal description of viscoelastic properties of materials satisfying the above requirements is a kernel in the form of fractional exponential function of Rabotnov [4–7]. Fractional exponential function is well studied and for simplification of calculations by using it, a special table has been developed [8]. Another advantage of the fractional exponential function is that for known creep kernel parameters constructed with using the fractional exponential function a kernel of relaxation becomes definite [9, 10].

Some methods for determining the parameters of Rabotnov's creep kernel are known: using Mittag-Leffler's function [11, 12], using Laplace-Carson's transform [9, 12–14], and a method of direct approximation [9, 12–15].

In this paper we offer a new double stage method for determining the creep kernel parameters of linear viscoelastic

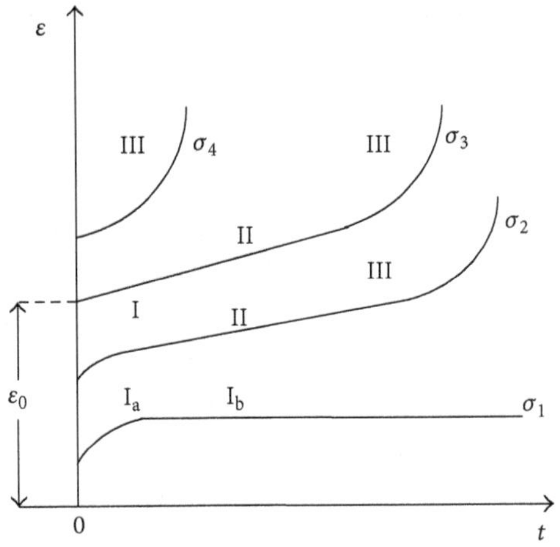

FIGURE 1: Various types of creep curves.

materials constructed with using Rabotnov's fractional exponential function. At the first stage, taking into account weak singularity properties of Abel's function at the initial moment of loading, parameters ε_0 and α are determined. At the second stage, using already known parameters ε_0 and α, parameters β and λ are determined.

Fractional derivative constitutive models for finite deformation of viscoelastic materials developed with using a kernel close to Abel's kernel are given in [16]. A new model for the linear theory of generalized thermoviscoelasticity using a relaxation kernel close to Abel's kernel is demonstrated in [17].

2. Creep Equation and Kernel

As is known generally creep curve of the viscoelastic material has three characteristic sites: I, site with a decreasing rate of creep (unstabilized creep), II, site with a constant (minimum) rate of creep (stabilized creep), and III, site of accelerating creep, preceding failure (Figure 1).

In the present work sites I and II of creep curves of linear viscoelastic materials under uniaxial tension are described by the integral equation of Boltzmann-Volterra:

$$\varepsilon(t) = \frac{1}{E}\left[\sigma(t) + \int_0^t K(t-\tau)\,\sigma(\tau)\,d\tau\right], \tag{1}$$

where $\varepsilon(t)$ is strain at time point t, $\sigma(t)$, $\sigma(\tau)$ are stresses at time points t and τ, $K(t-\tau)$ is creep kernel, E is instant modulus of elasticity, t is observation time, and τ is time, preceding observation time.

Taking into account that for creep process $\sigma_0 = \text{const}$, from (1) we will find

$$\varepsilon(t) = \varepsilon_0\left[1 + \int_0^t K(t-\tau)\,d\tau\right], \tag{2}$$

where ε_0 is conditionally instantaneous strain.

In (2) it was accounted that on Hooke's law $\sigma_0 = E\cdot\varepsilon_0$.

As mentioned above, creep kernel recorded using Rabotnov's fractional exponential function in the following form [6]:

$$\begin{aligned} K(t-\tau) &= \lambda \ni_{-\alpha}(-\beta, t-\tau) \\ &= \lambda (t-\tau)^{-\alpha} \sum_{n=0}^{\infty} \frac{(-\beta)^n (t-\tau)^{(1-\alpha)n}}{\Gamma[(1-\alpha)(1+n)]}, \end{aligned} \tag{3}$$

where $\ni_{-\alpha}(-\beta, t-\tau)$ is Rabotnov's fractional exponential function, α, β, and λ are creep kernel parameters ($\lambda > 0$, $0 < \alpha < 1$, and $\beta > 0$) and Γ is gamma-function.

Substituting expression (3) for creep kernel in the creep equation (2) we obtain the following equation:

$$\varepsilon(t) = \varepsilon_0\left[1 + \lambda \sum_{n=0}^{\infty} \frac{(-\beta)^n t^{(1-\alpha)(1+n)}}{\Gamma[(1-\alpha)(1+n)+1]}\right]. \tag{4}$$

Equation (4) represents creep equation recorded using Rabotnov's fractional exponential function. It contains four unknown parameters: α is the singularity parameter; β is the attenuation parameter; λ is the rheological parameter; and ε_0 is conditionally instantaneous strain.

3. Method of Determining the Creep Parameters

3.1. Parameters ε_0 and α. Taking $n = 0$ from series (4) we find the first term:

$$\varepsilon(t) = \varepsilon_0\left[1 + \frac{\delta}{1-\alpha} t^{1-\alpha}\right]. \tag{5}$$

As one can see, obtained expression contains well-known Abel's function with parameter $\delta > 0$. Abel's function at $t = 0$ has a singularity at order α. Based on this property of Abel's function, unknown parameters ε_0 and δ are determined by use of (5).

Knowing that parameter α takes value from interval $(0, 1)$ we will consider as unknown only parameters ε_0 and δ.

According to the least squares method the best values of parameters ε_0 and δ are those for which the following condition is met:

$$S(\varepsilon_0, \delta) = \sum_{i=1}^{m}\left[\varepsilon_0\left(1 + \frac{\delta}{1-\alpha} t_i^{1-\alpha}\right) - \varepsilon_{\text{ei}}\right]^2 \longrightarrow \min, \tag{6}$$

where $S(\varepsilon_0, \delta)$ is sum of squares of deviations, ε_{ei} is values of creep strain determined experimentally, and m is number of creep strains.

From two equations based on expressions $\partial S(\varepsilon_0, \delta)/\partial\varepsilon_0 = 0$ and $\partial S(\varepsilon_0, \delta)/\partial\delta = 0$, we find expressions for determining the parameters ε_0 and δ:

$$\varepsilon_0 = \frac{\sum_{i=1}^m \varepsilon_i \sum_{i=1}^m t_i^{2(1-\alpha)} - \sum_{i=1}^m t_i^{(1-\alpha)} \sum_{i=1}^m \varepsilon_1 t_i^{(1-\alpha)}}{m\sum_{i=1}^m t_i^{2(1-\alpha)} - \left[\sum_{i=1}^m t_i^{(1-\alpha)}\right]^2}, \tag{7}$$

$$\delta = \frac{\sum_{i=1}^m (\varepsilon_i/\varepsilon_0 - 1)\, t_i^{(1-\alpha)}}{(1/(1-\alpha))\sum_{i=1}^m t_i^{2(1-\alpha)}}. \tag{8}$$

Setting values of parameter α from interval (0, 1) with a certain step from expression (7) we will find values of parameter $\varepsilon_0 = \varepsilon_0(\alpha)$. Substituting the found values of parameter ε_0 and the corresponding values of singularity parameter α in expression (8) values of parameter $\delta = \delta(\varepsilon_0, \alpha)$ are determined.

Further, sequentially substituting values of singularity parameter α and calculated corresponding values of parameters ε_0 and δ in expression (5), values of creep strain $\varepsilon(t_i) = \varepsilon(t_i, \alpha, \varepsilon_0, \delta)$ are calculated.

If we designate an average deviation of calculated values of creep strains from experimental values through $\Delta\varepsilon_m(\lambda, \varepsilon_0, \delta)$, as a selecting criterion of the best values of parameters α, ε_0, and δ the following condition can be accepted:

$$\Delta\varepsilon_{\mathrm{mi}}\left(\alpha_i, \varepsilon_{0i}, \delta_i\right) \longrightarrow \min. \tag{9}$$

3.2. Parameters β and λ. Rewrite equation of creep (4) in the following form:

$$\varepsilon(t) = \varepsilon_0\left(1 + \lambda t^{1-\alpha} F_{2i}\right), \tag{10}$$

$$\text{where } F_{2i} = \sum_{n=0}^{\infty} \frac{(-\beta)^n t^{(1-\alpha)n}}{\Gamma\left[(1-\alpha)(1+n)+1\right]}. \tag{11}$$

According to the least squares method analogically to condition (6) we write an extremum condition using the equation of creep (10):

$$S(\beta, \lambda) = \sum_{i=1}^{m}\left[\varepsilon_0\left(1 + \lambda t_i^{1-\alpha} F_{2i}\right) - \varepsilon_{\mathrm{ei}}\right]^2 \longrightarrow \min. \tag{12}$$

From two equations based on the expressions $\partial S(\beta, \lambda)/\partial\beta = 0$ and $\partial S(\beta, \lambda)/\partial\lambda = 0$, expressions for determining the parameters β and λ are found:

$$\begin{aligned}&\sum_{i=1}^{m}\left[\left(1 - \frac{\varepsilon_i}{\varepsilon_0}\right)\sum_{i=1}^{m}\left(t_i^{1-\alpha}F_{2i}\right)^2\right.\\&\left.\quad + t_i^{1-\alpha}F_{2i}\sum_{i=1}^{m}\left(\frac{\varepsilon_i}{\varepsilon_0} - 1\right)t_i^{1-\alpha}F_{2i}\right]\left(t_i^{1-\alpha}\right)^2 F_{3i} = 0,\end{aligned} \tag{13}$$

$$\lambda = \frac{\sum_{i=1}^{m}\left(\varepsilon_i/\varepsilon_0 - 1\right)t_i^{1-\alpha}F_{2i}}{\sum_{i=1}^{m}\left(t_i^{1-\alpha}F_{2i}\right)^2}, \tag{14}$$

$$\text{where } F_{3i} = \sum_{n=1}^{\infty}(-1)^n \frac{n(\beta)^{n-1} t_i^{(1-\alpha)(n-1)}}{\Gamma\left[(1-\alpha)(n+1)+1\right]}, \tag{15}$$

but series F_{21} is determined by expression (11).

Value of parameter of fading β is determined from (13) by substituting trial values. If (13) has a unique solution, it will be desired value of the parameter $\beta > 0$. Obviously, for determining the values of parameters β and λ from expressions (13) and (14) earlier calculated values of parameters ε_0 and α are used.

TABLE 1: Strains in site I of material Nylon 6 creep curve.

Time t, h	Creep strain (%) calculated by expression		Deviation, %
	(16)	(17)	
0,01	0,2919	0,2957	+1,30
0,03	0,3184	0,3186	+0,06
0,05	0,3315	0,3306	−0,27
0,07	0,3404	0,3389	−0,44
0,09	0,3472	0,3455	−0,49
0,10	0,3501	0,3483	−0,51
0,30	0,3819	0,3806	−0,34
0,50	0,3976	0,3975	−0,01
0,70	0,4083	0,4094	+0,27
0,90	0,4165	0,4187	+0,53

4. Experimental Approbation

4.1. Material Nylon 6. In work [9] the creep curve of Nylon 6 at a stress of $\sigma_0 = 5$ MPa was approximated by expression

$$\varepsilon_e(t) = 0{,}42t^{0{,}079}, \tag{16}$$

where ε [%] is strain and t [h] is time.

Firstly, for determination of parameters ε_0 and α we calculate some strain values in time interval $t = 0{,}01 \cdots 0{,}9$ h. Time values and corresponding creep strain values are given in the first and the second columns of Table 1.

Substituting values of time t, strain ε, and various values of singularity parameter α from interval $0{,}05 \cdots 0{,}95$ with step of 0,05 in expressions (7) and (8) and taking into account condition (9) it has been found that $\alpha = 0{,}85$; $\varepsilon_0 = 0{,}1682$; $\delta = 0{,}2269$.

With consideration of found values of parameters α, ε_0, and δ, (5) constructed using Abel's function for describing strain in site I of creep curve takes the form

$$\varepsilon(t) = 0{,}1682\left(1 + 1{,}5127t^{0{,}15}\right). \tag{17}$$

Calculated by (17) values of creep strain and corresponding approximation errors are presented in the third and the fourth columns of Table 1.

As can be seen from Table 1, (17) with parameters ε_0 and α found by the offered method has high accuracy of approximation.

Since the parameter β characterizes rate of material creep strain at long times, for definition it and parameter λ consider creep process of material Nylon 6 in the time interval $t = 0{,}01 \cdots 100$ h. Time values and corresponding values of creep strain calculated according to expression (16) are given in the first and the second columns of Table 2.

Using values of time t, strain ε, parameters $\alpha = 0{,}85$, $\varepsilon_0 = 0{,}1682$, and trial values of parameter $\beta > 0$, from expression (13) it has been found that $\beta = 0{,}18$. Further from expression (14) we will define $\lambda = 1{,}6682$.

TABLE 2: Strains in sites I_a and I_b of material Nylon 6 creep curve.

Time t, h	Creep strain (%) calculated by expression (16)	(18)	Deviation, %
0,01	0,2919	0,3060	+4,83
0,03	0,3184	0,3282	+3,08
0,05	0,3315	0,3396	+2,44
0,07	0,3404	0,3474	+2,06
0,09	0,3472	0,3535	+1,82
0,1	0,3501	0,3561	+1,72
0,3	0,3819	0,3852	+0,86
0,5	0,3976	0,3999	+0,58
0,7	0,4083	0,4102	+0,46
0,9	0,4165	0,4180	+0,36
1	0,4200	0,4214	+0,33
3	0,4581	0,4584	+0,07
5	0,4769	0,4770	+0,02
7	0,4898	0,4898	0,00
9	0,4996	0,4996	0,00
10	0,5038	0,5038	0,00
20	0,5321	0,5321	0,00
30	0,5495	0,5495	0,00
40	0,5621	0,5621	0,00
50	0,5721	0,5721	0,00
60	0,5804	0,5804	0,00
70	0,5875	0,5875	0,00
80	0,5937	0,5937	0,00
90	0,5993	0,5992	0,00
100	0,6043	0,6042	0,00

After substitution values of the found parameters ε_0, α, β, and λ (10) and auxiliary function F_{2i} (11) will have the next form:

$$\varepsilon(t) = 0{,}1682\left(1 + 1{,}6682t^{0{,}15}F_{2i}\right), \tag{18}$$

$$F_{2i} = \sum_{n=0}^{\infty} \frac{(-0{,}18)^n t^{0{,}15n}}{\Gamma[0{,}15(1+n)+1]}. \tag{19}$$

Calculated by (18) values of creep strain and corresponding errors of approximation are given in the last two columns of Table 2.

As can be seen from Table 2 creep (18) with parameters ε_0, α, β, and λ defined by the offered method has high accuracy of approximation.

4.2. Asphalt Concrete. The strains calculated by approximation formula (16) were taken for determination of the above creep parameters of material Nylon 6 instead of strains obtained experimentally. The reason for this was the fact that in work [9] the creep strains were given only graphically in the form of a creep curve and numerical values of the strain are not shown. It appeared almost impossible to find the strain

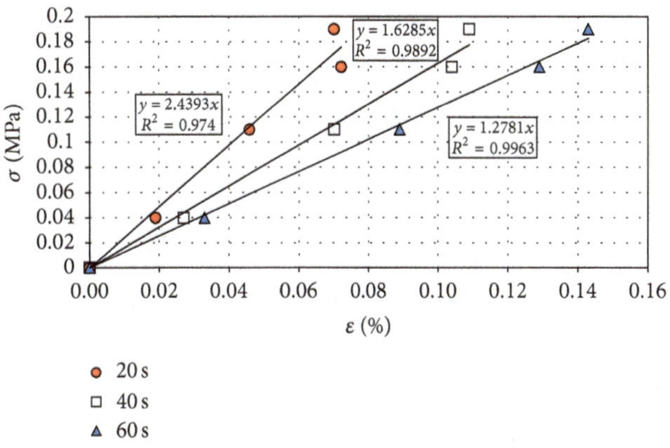

FIGURE 2: Isochronous lines of deformation of asphalt concrete.

values from the graphical creep curve with sufficient accuracy especially at short times of loading.

Possibly, in order to reduce the length of material in published works containing an approximation of creep curves of various materials, as a rule, information on creep strain is given graphically, for example, in works [9, 10, 12–14, 18–20].

Further, we will show determination of creep parameters ε_0 and α by the above offered method (expressions (7) and (8)) using experimentally determined values of strain at site I of creep of hot fine-grained asphalt concrete. This type of asphalt is used for application on top layer of road surface in many countries, including Kazakhstan. Asphalt concrete mix that meets the requirements of the Kazakhstan standard ST RK 1225 [21] was prepared with use of viscous road bitumen of grade BND 100/130 from Pavlodar petrochemical plant, which meets the standard requirements of ST RK 1373 [22]. Samples of asphalt concrete with rectangular cross-section dimensions of 5 × 5 × 15 cm were prepared by means of Cooper compactor (UK, model CRT-RC2S) in accordance with European Union standard EN 12697-33 [23]. Detailed information on preparation of asphalt mixtures and samples, as well as standard properties of bitumen and asphalt concrete, is given in [24]. Samples of asphalt concrete were tested for creep according to the scheme of uniaxial tension at a temperature of 20 ± 2°C and different loads.

Isochronous lines of deformation for the asphalt concrete constructed using creep curves within site I at durations of loadings 20, 40, and 60 s are shown in Figure 2. As it can be seen dependence between stress and strain is linear; that is, asphalt concrete under these conditions is linearly deformable material and for description of its creep process the equation of Boltzmann-Volterra is applicable.

Figures 3 and 4 show site I of creep curves of asphalt concrete at stresses of 0.16 and 0.19 MPa, respectively. In these figures points designate the creep strain defined experimentally and lines designate its approximation by Abel's kernel (5). It is seen that compliance of approximating curves to the experimental data is quite good. Average deviations of calculated strains from experimental creep curves at 0,16 and 0,19 MPa are 1.5% and 0.5%, respectively. The values of parameters ε_0 and α found by the offered method are equal to

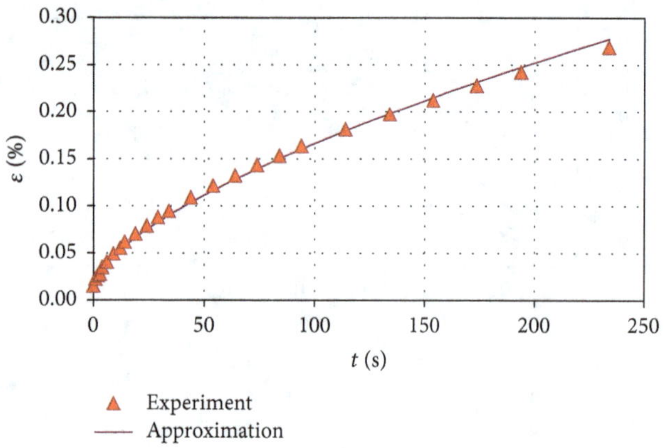

FIGURE 3: Creep curve of asphalt concrete at stress σ_0 = 0,16 MPa.

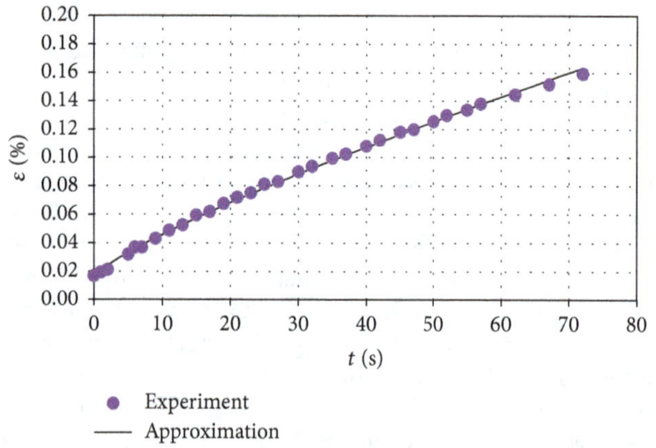

FIGURE 4: Creep curve of asphalt concrete at stress σ_0 = 0,19 MPa.

ε_0 = 0,015217% and α = 0,35 at σ_0 = 0,16 MPa and ε_0 = 0,016871% and α = 0,18 at σ_0 = 0,19 MPa.

Thus, the integral equation of Boltzmann-Volterra with creep kernel constructed using Rabotnov's fractional exponential function with parameters defined by the offered method describes the creep process of linear viscoelastic materials with high accuracy.

5. Conclusions

(1) Describing the creep process of linear viscoelastic materials by the integral equation of Boltzmann-Volterra with creep kernel which is constructed using Rabotnov's fractional exponential function is suggested.

(2) A new method of determining the creep kernel parameters of linear viscoelastic materials constructed using Rabotnov's fractional exponential function is offered. It is implemented in two stages. At the first stage based on the property of singularity of Abel's function at the initial moment of time parameters ε_0 and α are defined. At the second stage using already known parameters ε_0 and α parameters β and λ are defined. Analytical expressions for calculating these parameters are obtained.

(3) Applicability of the offered method is evaluated using data on creep strain of material Nylon 6 and hot fine-grained asphalt concrete obtained experimentally. The results showed high accuracy of the offered method.

Competing Interests

The authors declare that they have no competing interests.

References

[1] R. M. Cristensen, *Theory of Viscoelasticity: An Introduction*, Academic Press, New York, NY, USA, 1971.

[2] N. W. Tschoegl, *The Phenomenological Theory of Linear Viscoelastic Behavior: An Introduction*, Springer, Berlin, Germany, 1989.

[3] J. D. Ferry, *Viscoelastic Properties of Polymers*, John Wiley & Sons, New York, NY, USA, 3rd edition, 1980.

[4] Yu. N. Rabotnov, *Mechanics of Deformed Solid Body*, Nauka, Moscow, Russia, 1988.

[5] Yu. N. Rabotnov, *Creep of Structure Elements*, Nauka, Moscow, Russia, 1966.

[6] Y. N. Rabotnov, *Elements of Hereditary Mechanics of Solids*, Nauka, Moscow, Russia, 1977.

[7] Yu. N. Rabotnov, "Balance of the elastic medium with a aftereffect," *Applied Mathematics and Mechanics*, vol. 12, pp. 53–62, 1948.

[8] Y. N. Rabotnov, L. Kh. Papernik, and E. N. Zvonov, *Tables of a Fractional Exponential Function of Negative Parameter and Its Integral*, Nauka, Moscow, Russia, 1969.

[9] Yu. V. Suvorova, "About Yu.N. Rabotnov's nonlinear hereditary equation and its applications," *News of the Russian Academy of Sciences, Mechanics of solids*, no. 1, pp. 174–181, 2004.

[10] Y. V. Suvorova, "Master equations for viscoelastic materials taking into account the influence of temperature and humidity," *Strength of Materials*, vol. 32, no. 5, pp. 490–496, 2000.

[11] M. I. Rozovskii, "Some features of elasticity hereditary media," *News of the USSR Academy of Sciences*, no. 2, pp. 30–36, 1961.

[12] V. P. Golub, P. V. Fernati, and Y. G. Lyashenko, "Determining the parameters of the fractional exponential heredity kernels of linear viscoelastic materials," *International Applied Mechanics*, vol. 44, no. 9, pp. 963–974, 2008.

[13] E. N. Zvonov, N. I. Malinin, L. Kh. Papernik, and B. M. Tseitlin, "Determination of the creep characteristics of linear elastic hereditary materials by using a digital computer," *Mechanics of Solid*, no. 5, pp. 76–82, 1968.

[14] Yu. V. Suvorova and A. V. Mosin, "Determination of parameters of the Rabotnov's fractional exponential function with use of integral transform and modern software," *Problems of Mechanical Engineering and Automation*, vol. 4, pp. 54–56, 2002.

[15] D. A. Gavrilov and V. A. Markov, "Numerical method of determining the rheologic parameters of composites from test results," *Mechanics of Composite Materials*, vol. 22, no. 4, pp. 417–421, 1987.

[16] M. Fukunaga and N. Shimizu, "Fractional derivative constitutive models for finite deformation of viscoelastic materials,"

Journal of Computational and Nonlinear Dynamics, vol. 10, no. 6, Article ID 061002, 8 pages, 2015.

[17] M. A. Ezzat, A. S. El-Karamany, and A. A. El-Bary, "Thermo-viscoelastic materials with fractional relaxation operators," *Applied Mathematical Modelling*, vol. 39, no. 23-24, pp. 7499–7512, 2015.

[18] R. D. Maksimov, L. Jirgens, J. Jansons, and E. Plume, "Mechanical properties of polyester polymer-concrete," *Mechanics of Composite Materials*, vol. 35, no. 2, pp. 99–110, 1999.

[19] R. D. Maksimov and E. Plume, "Long-term creep of hybrid aramid/glass-fiber-reinforced plastics," *Mechanics of Composite Materials*, vol. 37, no. 4, pp. 271–280, 2001.

[20] E. A. Sokolov and R. D. Maksimov, "Possibilities of predicting the creep of a polymeric fiber-reinforced plastic from the properties of the components," *Polymer Mechanics*, vol. 14, no. 6, pp. 807–813, 1978.

[21] ST RK 1225-2003, "Asphalt concrete mixes for road, airfield and asphalt concrete," Technical specifications, Astana, Kazakhstan, 2003.

[22] "Bitumen and bituminous binders. Oil road viscous bitumens," ST RK 1373-2005, Technical Specifications, Astana, Kazakhstan, 2005.

[23] European Committee for Standardization, "Bituminous mixtures. Test methods for hot mix asphalt —part 33: specimen prepared by roller compactor," Tech. Rep. EN 12697-33, 2003.

[24] B. Teltayev, "A new failure criterion for asphalt mixtures under fatigue loading," *International Journal of Pavement Research and Technology*, vol. 8, no. 4, pp. 276–282, 2015.

11

Chemical Entropy Generation and MHD Effects on the Unsteady Heat and Fluid Flow through a Porous Medium

Gamal M. Abdel-Rahman Rashed

Department of Mathematics, Faculty of Science, Benha University, Benha 13518, Egypt

Correspondence should be addressed to Gamal M. Abdel-Rahman Rashed; gamalm60@yahoo.com

Academic Editor: Oluwole D. Makinde

Chemical entropy generation and magnetohydrodynamic effects on the unsteady heat and fluid flow through a porous medium have been numerically investigated. The entropy generation due to the use of a magnetic field and porous medium effects on heat transfer, fluid friction, and mass transfer have been analyzed numerically. Using a similarity transformation, the governing equations of continuity, momentum, and energy and concentration equations, of nonlinear system, were reduced to a set of ordinary differential equations and solved numerically. The effects of unsteadiness parameter, magnetic field parameter, porosity parameter, heat generation/absorption parameter, Lewis number, chemical reaction parameter, and Brinkman number parameter on the velocity, the temperature, the concentration, and the entropy generation rates profiles were investigated and the results were presented graphically.

1. Introduction

Industries developing technology related to heat transfer are more concerned with the design of new thermal systems; thus research is in progress to investigate the hydrodynamic and heat transfer behavior of new forms of heat transfer fluid. Recently, entropy generation (or production) has been used to gauge the significance of irreversibility related to heat transfer, friction, and other nonideal processes within thermal system by Bejan [1]. Entropy generation and its minimization have been considered as an effective tool to improve the performance of any heat transfer process. Entropy generation minimization of diabetic distillation column with trays has been investigated using a new approach by Spasojević et al. [2] in which the exchanged heat has been considered as a control variable instead of temperature.

Andersson et al. [3] analyzed the momentum and heat transfer in a laminar liquid film on a horizontal stretching sheet governed by time-dependent boundary layer equations. Tsai et al. [4] studied the nonuniform heat source/sink effect on the flow and heat transfer from an unsteady stretching sheet through a quiescent fluid medium extending to infinity. Elbashbeshy and Bazid [5] presented similarity solutions of the boundary layer equations, which describe the unsteady flow and heat transfer over an unsteady stretching sheet. Shateyi and Motsa [6] investigated thermal radiation effects on heat and mass transfer over an unsteady stretching surface. Bouabid et al. [7] studied analysis of the magnetic field effect on entropy generation at thermosolutal convection in a square cavity. Oliveski et al. [8] proposed an entropy generation and natural convection in rectangular cavities. Achintya [9] Analyzed the entropy generation due to natural convection in square enclosures with multiple discrete heat sources. Magherbi et al. [10] investigated second law analysis in convective heat and mass transfer. El Jery et al. [11] studied effect of external oriented magnetic field on entropy generation in natural convection and Abd El-Aziz [12] studied radiation effect on the flow and heat transfer over an unsteady stretching sheet.

This study is a complementary study to the work of [6] to include the thermal radiation effects on heat and mass transfer over an unsteady stretching surface. This also presents chemical entropy generation and magnetohydrodynamic effects on the unsteady heat and fluid flow through a porous medium which are numerically investigated. The entropy generation due to the use of a magnetic field and porous medium effects on heat transfer, fluid friction, and mass transfer have been analyzed numerically. The effects of unsteadiness parameter, magnetic field parameter, porosity parameter, heat generation/absorption parameter, Lewis number, chemical reaction parameter, and Brinkman number parameter on the velocity, the temperature, the concentration, and the entropy generation rates profiles were investigated and the results were presented graphically.

2. Problems Development

The flow, assumed to be unsteady, laminar, and incompressible fluid on a horizontal sheet with chemical entropy generation and magnetohydrodynamic effects through a porous medium, has been considered. The viscous dissipation effect is also taken into consideration. The radiative heat flux in the x-direction is negligible in comparison with that in the y-direction. The fluid motion arises due to the stretching of the elastic sheet. The continuous sheet aligned with the x-axis at $y = 0$ moves in its own plane with a surface velocity $U_w(x,t)$, the surface temperature $T_w(x,t)$, and the surface concentration $C_w(x,t)$ varying both along the sheet and with time. The magnetic Reynolds number of the flow is taken to be small enough that the induced magnetic field is assumed to be negligible in comparison with the applied magnetic field, that is, $B = (0, B_0, 0)$, where B_0 is the uniform magnetic field acting normal to the plate. u and v are the velocity of x and y component, and T and C are the temperature and concentration, respectively. The governing boundary layer equations of continuity, momentum, energy, and concentration equations under Boussinesq approximations could be written as follows.

The continuity equation:

$$\frac{\partial u}{\partial x} + \frac{\partial v}{\partial y} = 0. \tag{1}$$

The momentum equations:

$$\frac{\partial u}{\partial t} + u\frac{\partial u}{\partial x} + v\frac{\partial u}{\partial y} = \nu\frac{\partial^2 u}{\partial y^2} - \frac{\sigma B^2(x,t)}{\rho}u - S_1(x,t)\,u. \tag{2}$$

The energy equation:

$$\frac{\partial T}{\partial t} + u\frac{\partial T}{\partial x} + v\frac{\partial T}{\partial y} = \alpha\frac{\partial^2 T}{\partial y^2} - \frac{1}{\rho C_p}\frac{\partial q_r}{\partial y} + \frac{Q_1(x,t)}{\rho C_p}(T - T_\infty). \tag{3}$$

The concentration equations:

$$\frac{\partial C}{\partial t} + u\frac{\partial C}{\partial x} + v\frac{\partial C}{\partial y} = D\frac{\partial^2 C}{\partial y^2} - k_1(x,t)(C - C_\infty). \tag{4}$$

The boundary conditions are

$$\begin{gathered} u(x,0) = U_w(x,t), \\ v(x,0) = 0, \\ T(x,0) = T_w(x,t), \\ C(x,0) = C_w(x,t), \\ u(x,\infty) \longrightarrow 0, \\ T(x,\infty) \longrightarrow T_\infty, \\ C(x,\infty) \longrightarrow C_\infty, \end{gathered} \tag{5}$$

where σ is the electrical conductivity, ρ is the density of the fluid, μ is the viscosity of the fluid, $\alpha = k/\rho C_p$ is the thermal diffusivity of the fluid, C_p is the heat capacity at constant pressure, k_1 is the thermal conductivity, Q_1 is heat generation/absorption rate (is positive in the case of the sheet's generation of heat and is negative in the case of the sheet's absorption of heat from the fluid flow), D is the mass diffusivity, T_∞ is temperature of the fluid at infinity, and q_r is the radiative heat flux in the y-direction. Using the Rosseland approximation (Sparrow and Cess [13] and Moradi et al. [14]), the radiative heat flux q_r is given by

$$q_r = -\frac{4\sigma^*}{3k^*}\frac{\partial T^4}{\partial y}, \tag{6}$$

where σ^* is the Stefan-Boltzmann constant and k^* is the mean absorption coefficient. Assuming that the temperature difference within the flow is sufficiently small such that T^4 could be approached as the linear function of temperature:

$$T^4 \cong 4T_\infty^3 T - 3T_\infty^4. \tag{7}$$

Following Andersson et al. [15], the surface velocity $U_w(x,t)$ is assumed to be $U_w(x,t) = bx/(1 - at)$, where both a and b are positive constants with dimension reciprocal time. We have b as the initial stretching rate and $b/(1-at)$ is increasing with time. In the context of polymer extrusion, the material properties particularly the elasticity of the extruded sheet may vary with time even though the sheet is being pulled by a constant force. With unsteady stretching, however, a^{-1} becomes the representative time scale of the resulting unsteady boundary layer problem. We assume that all of the surface temperature $T_w(x,t)$, the surface concentration $C_w(x,t)$, the applied transverse magnetic field $B(x,t)$, the volumetric heat generation/absorption rate $Q_1(x,t)$, the thermal conductivity $k_1(x,t)$, and the porous medium $S_1(x,t)$ are on a stretching sheet to vary with the distance x along the sheet and time in the following forms:

$$\begin{aligned}
B(x,t) &= B_0(1-at)^{-1/2},\\
Q_1(x,t) &= Q_0(1-at)^{-1},\\
T_w(x,t) &= T_\infty + T_0\left(\frac{bx^2}{2\nu}\right)(1-at)^{-3/2},\\
C_w(x,t) &= C_\infty + C_0\left(\frac{bx^2}{2\nu}\right)(1-at)^{-3/2},\\
S_1(x,t) &= \frac{\nu}{\kappa} = S_0(1-at)^{-1},\\
k_1(x,t) &= k_0(1-at)^{-1}.
\end{aligned} \tag{8}$$

We introduce the similarity transformations

$$\begin{aligned}
\psi &= \sqrt{\nu b}(1-at)^{-1/2}xf(\eta),\\
T &= T_\infty + T_0\left(\frac{bx^2}{2\nu}\right)(1-at)^{-3/2}\theta(\eta),\\
C &= C_\infty + C_0\left(\frac{bx^2}{2\nu}\right)(1-at)^{-3/2}\varphi(\eta),\\
\eta &= \sqrt{\frac{b}{\nu}}(1-at)^{-1/2}y.
\end{aligned} \tag{9}$$

Equation (1) is satisfied automatically and so are governing equations (2)–(5); we have

$$f''' - A\left(\frac{\eta f''}{2} + f'\right) + ff'' - f'^2 - (M+S)f' = 0,$$

$$\begin{aligned}
&(3R+4)\theta''\\
&\quad + 3RP_r\left(f\theta' - 2f'\theta - \left(\frac{A}{2}\right)(3\theta + \eta\theta') + H\theta\right)\\
&= 0,
\end{aligned} \tag{10}$$

$$\begin{aligned}
&\varphi'' + P_rL_e\left(f\varphi' - 2f'\varphi - \left(\frac{A}{2}\right)(3\varphi + \eta\varphi') - \gamma\varphi\right)\\
&= 0,
\end{aligned}$$

with the boundary conditions

$$\begin{aligned}
f(0) &= 0,\\
f'(0) &= 1,\\
\theta(0) &= 1,\\
\varphi(0) &= 1,\\
f'(\infty) &= 0,\\
\theta(\infty) &= 0,\\
\varphi(\infty) &= 0.
\end{aligned} \tag{11}$$

Here the prime denotes a partial differentiation with respect to η, κ is the permeability of the porous medium, $A = a/b$ is the unsteadiness parameter, $M = \sigma B_0^2/b\rho$ is the magnetic field parameter, $S = S_0/b$ is the porosity parameter, $P_r = \mu C_p \nu/k$ is the Prandtl number, $R = k^*k/4\sigma^* T_\infty^3$ is the thermal radiation parameter, $H = Q_0/b\rho C_p$ is the heat generation/absorption parameter, $L_e = k/D\rho C_p$ is the Lewis number, and $\gamma = k_0/b$ is the chemical reaction parameter.

3. Entropy Generation

In the present problem, the volumetric entropy generation is therefore the sum of irreversibilities due to heat transfer, fluid friction, mass transfer by pure concentration gradients, and mass transfer by mixed product of concentration and thermal gradients with magnetic field and porous medium effects which is given by

$$\begin{aligned}
S_G &= \frac{k}{T_\infty^2}\left(\frac{\partial T}{\partial y}\right)^2 + \frac{\mu}{T_\infty}\left(\frac{\partial u}{\partial y}\right)^2 + \frac{\mu}{C_\infty}\left(\frac{\partial C}{\partial y}\right)^2\\
&\quad + \frac{\mu}{C_\infty}\left(\frac{\partial T}{\partial y}\right)\left(\frac{\partial C}{\partial y}\right) + \frac{\sigma B_0^2}{T_\infty}u^2 + \frac{\mu}{\kappa T_\infty}u^2.
\end{aligned} \tag{12}$$

Using the nondimensional quantities, we obtain the local entropy generation rates in nondimensional form:

$$\begin{aligned}
N_s &= \lambda_1\frac{k}{T_\infty^2}\left(\frac{\partial T}{\partial y}\right)^2 + \lambda_2\frac{\mu}{T_\infty}\left(\frac{\partial u}{\partial y}\right)^2\\
&\quad + \lambda_3\frac{\mu}{C_\infty}\left(\frac{\partial C}{\partial y}\right)^2 + \lambda_4\frac{\mu}{C_\infty}\left(\frac{\partial T}{\partial y}\right)\left(\frac{\partial C}{\partial y}\right)\\
&\quad + \lambda_5\frac{\sigma B_0^2}{T_\infty}u^2 + \lambda_6\frac{\mu}{\kappa T_\infty}u^2,
\end{aligned} \tag{13}$$

where $\lambda_1 = x^2T_\infty^2(1-at)^4/k(\Delta T)^2$, $\lambda_2 = x^2T_\infty^2(1-at)^3/k(\Delta T)^2$, $\lambda_3 = T_\infty(\Delta C/\Delta T)^2(1-at)^4/k$, $\lambda_4 = T_\infty(\Delta C/\Delta T)(1-at)^4/k$, $\lambda_5 = x^2T_\infty(1-at)^2/k(\Delta T)^2$, and $\lambda_6 = x^2T_\infty(1-at)^3/k(\Delta T)^2$ are the characteristic entropy generation rate.

The total dimensionless entropy generation is obtained by

$$\begin{aligned}
N_s &= B_rR_e\left[\theta'^2 + f''^2 + \frac{\Omega_1}{\Omega}\varphi'^2 + \Omega_1\theta'\varphi'\right.\\
&\quad \left. + \frac{1}{\Omega}(M+S)f'^2\right],
\end{aligned} \tag{14}$$

where $B_r = \mu T_\infty b^2x^2/k(\Delta T)$ is the Brinkman number, $R_e = bx^2/\nu$ is local Reynolds number, $\Omega = \Delta T/T_\infty$ is the dimensionless temperature ratio, and $\Omega_1 = \Delta C/C_\infty$ is the dimensionless concentration ratio.

4. Result and Discussion

The system of nonlinear ordinary differential equations (10) together with the boundary conditions (11) is locally similar and solved numerically by using the Control Volume Finite-Element Method. Numerical values of the velocity, the temperature, and the concentration profiles have been used to

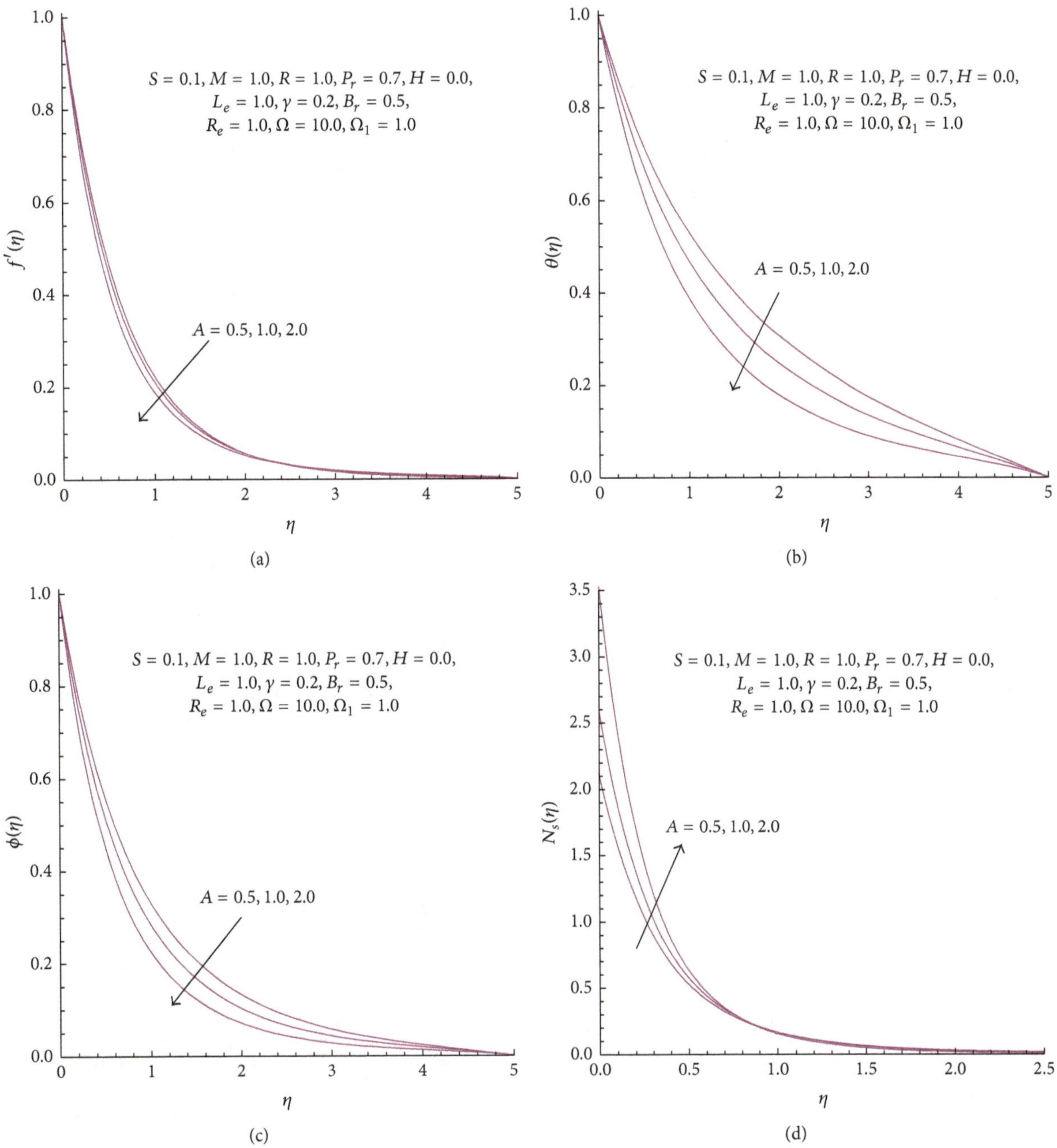

FIGURE 1: Influence of unsteadiness parameter on (a) the velocity profile, (b) the temperature profile, (c) the concentration profile, and (d) the entropy generation rates profile.

compute the entropy generation rates, found for different values of the various parameters occurring in the problem with unsteadiness parameter, magnetic field parameter, porosity parameter, heat generation/absorption parameter, Lewis number, chemical reaction parameter, and Brinkman number parameter; the results are displayed in Figures 1–7, for the velocity, the temperature, the concentration, and the entropy generation rates profiles. In order to verify the accuracy of our present method, we have compared our results with those of Shateyi and Motsa [6] and Abd El-Aziz [12]. Table 1 shows the values of $-\theta'(0)$ for several of A and P_r. The comparisons in all the above cases are found to agree with each other excellently. Also the results are found to be similar to those by Shateyi and Motsa [6] and Abd El-Aziz [12]. So it is good.

In Figures 1(a), 1(b), 1(c), and 1(d), respectively, the influence of the unsteadiness parameter on the velocity profile, the temperature profile, the concentration profile, and the entropy generation rates profile is shown. It is observed that the velocity profile, the temperature profile, and the concentration profile decrease while the entropy generation rates profile increases with the increase of the unsteadiness

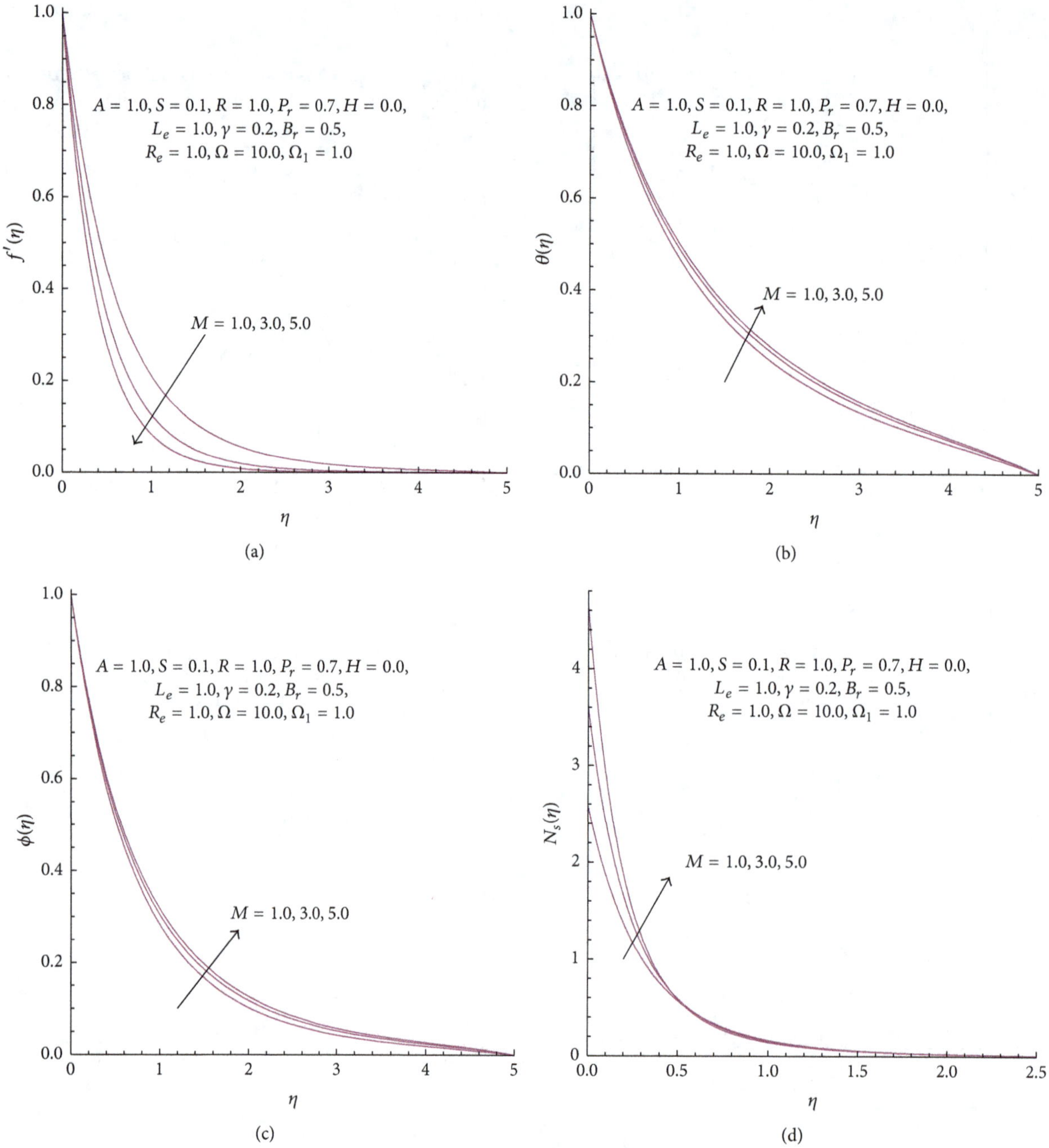

FIGURE 2: Influence of the magnetic field parameter on (a) the velocity profile, (b) the temperature profile, (c) the concentration profile, and (d) the entropy generation rates profile.

parameter. When $A = 0$, we have a steady state flow and for $A > 0$, we have an unsteady flow.

In absence of both magnetic field parameter and porosity parameter effects on the velocity profile, the temperature profile, the concentration profile, and the entropy generation rates profile are illustrated in Figures 2(a), 2(b), 2(c), 2(d), 3(a), 3(b), 3(c), and 3(d), respectively; also, we have found that the velocity profile decreases while the temperature profile, the concentration profile, and the entropy generation rates profile increase with the increase of each of the magnetic field parameter and porosity parameter, and this is due to the fact that the thermal boundary layer increases with magnetic field parameter and porosity parameter. The presence of each of the magnetic field and porosity creates additional entropy.

The effects of heat generation/absorption parameter on the temperature and the entropy generation rates profiles are presented in Figures 4(a) and 4(b), respectively. It is observed that the temperature profile increases while the entropy generation rates profile decreases with increase of heat generation/absorption parameter. This is due to the fact

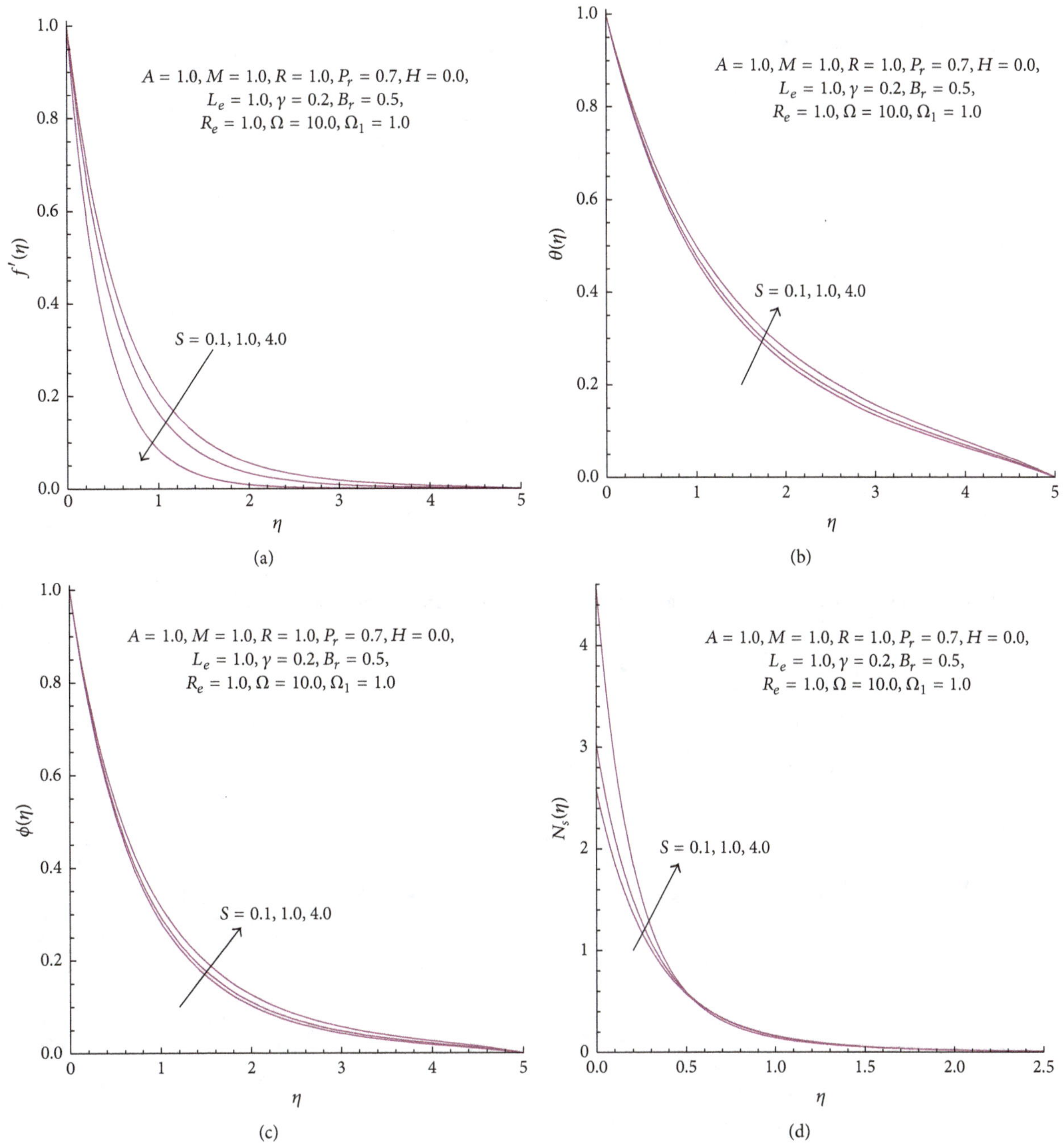

FIGURE 3: Influence of the porosity parameter on (a) the velocity profile, (b) the temperature profile, (c) the concentration profile, and (d) the entropy generation rates profile.

that the increase of the heat source/sink parameter means an increase of the heat generated inside the boundary layer leading to higher temperature profile.

The effect of the Lewis number parameter on the concentration and the entropy generation rates profiles is shown on Figures 5(a) and 5(b), respectively, and we have found that the concentration profile decreases while entropy generation rates profile increases with increase of the values of the Lewis number parameter.

The concentration profile for different values of the chemical reaction parameter is plotted in Figure 6; we observe that the concentration profile decreases with the increase of the chemical reaction parameter. The influence of the Brinkman number parameter on the entropy generation rates profile is shown in Figure 7; we have found that the entropy generation rates profile increases with the increase of the Brinkman number parameter.

5. Conclusions

The entropy generation due to the use of a magnetic field and porous medium effects on heat transfer, fluid friction,

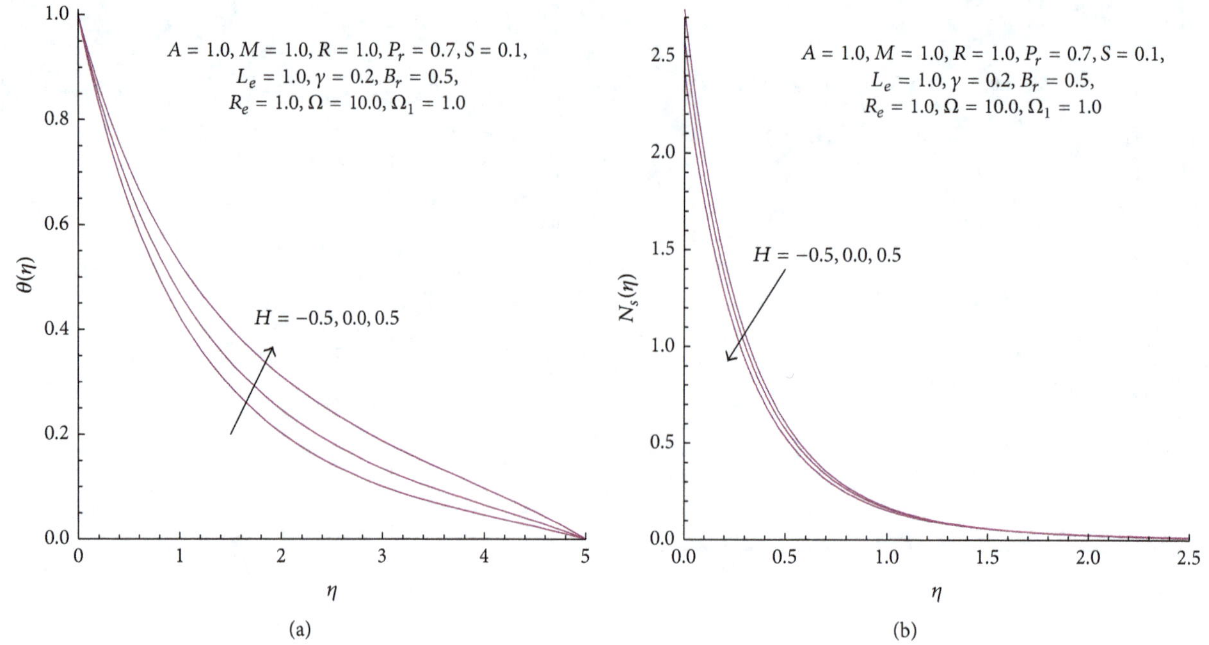

FIGURE 4: Influence of the heat generation/absorption parameter on (a) the temperature profile and (b) the entropy generation rates profile.

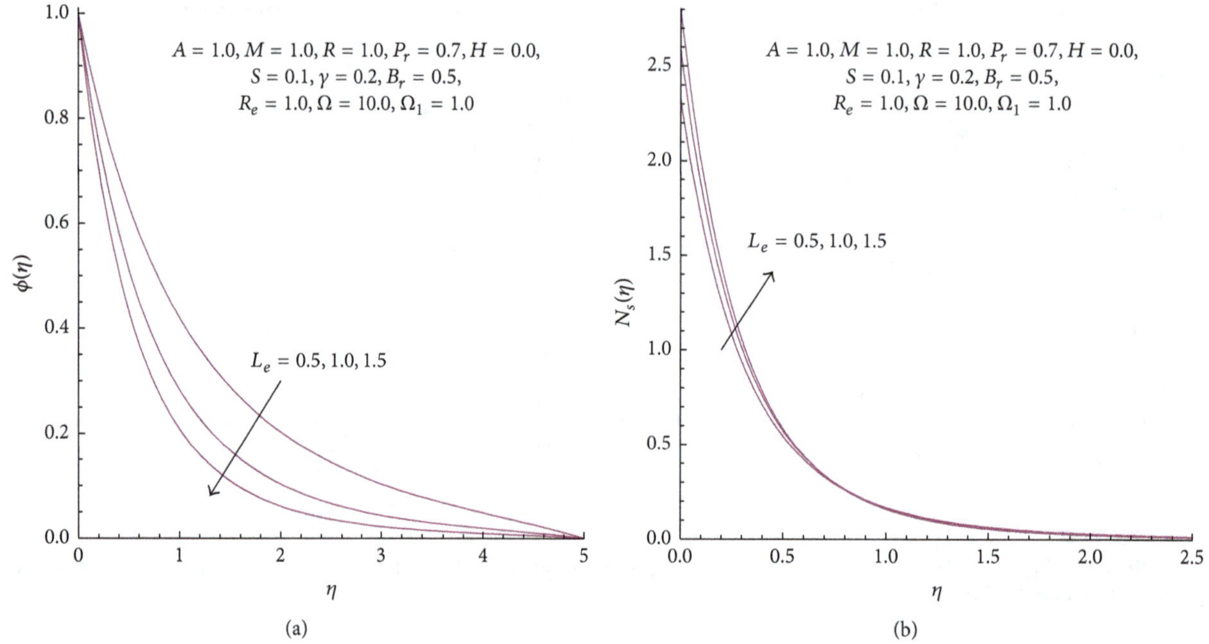

FIGURE 5: Influence of the Lewis number parameter on (a) the concentration profile and (b) the entropy generation rates profile.

and mass transfer have been analyzed numerically. Numerical values of the velocity, the temperature, and the concentration profiles have been used to compute the entropy generation rates, found for different values of the various parameters occurring in the problem. The effects of unsteadiness parameter, magnetic field parameter, porosity parameter, heat generation/absorption parameter, Lewis number, chemical reaction parameter, and Brinkman number parameter on the velocity, the temperature, the concentration, and the entropy generation rates profiles are discussed graphically. The main conclusions derived from this study are given below:

(1) The increasing values of porosity parameter and magnetic field parameter lead to increase in the entropy generation rates profile, and we also observed that the temperature profile and the concentration profile increase, while the velocity profile decreases.

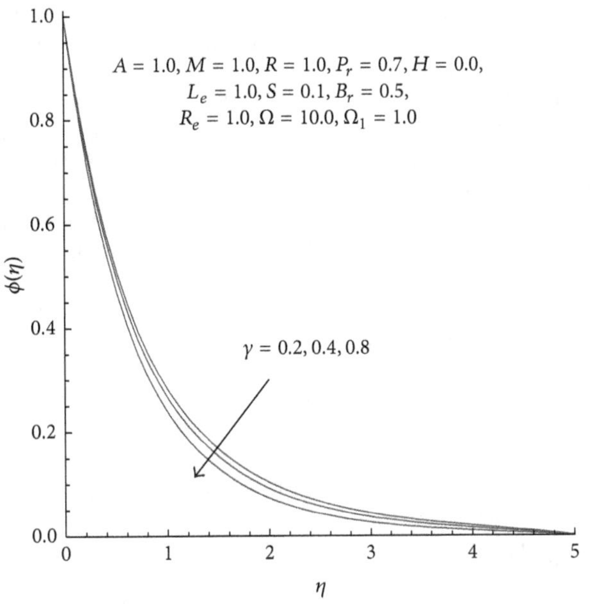

FIGURE 6: Influence of the chemical reaction parameter on the concentration profile.

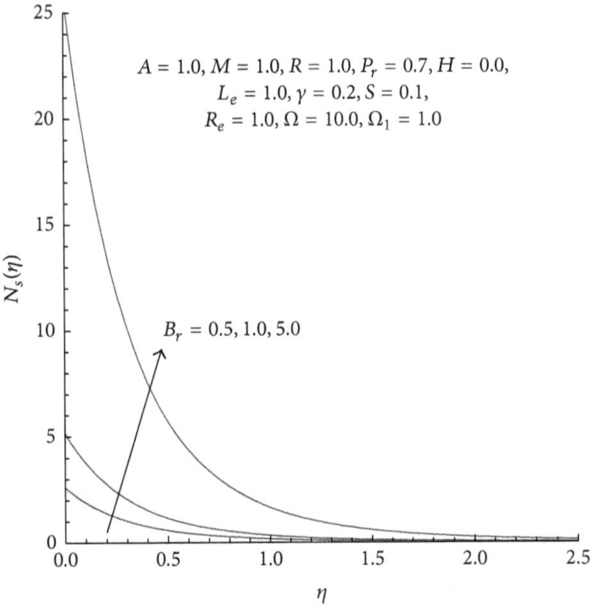

FIGURE 7: Influence of the Brinkman number parameter on the entropy generation rates profile.

(2) The entropy generation rates profile decreases with the increase of the heat generation/absorption parameter. And also, the temperature profile increases.

(3) As the Lewis number parameter increases, we found that the entropy generation rates profile increases, while the concentration profile decreases.

TABLE 1: Comparison of the value $-\theta'(0)$ for several of A and P_r with $M = 0.0$, $S = 0.0$, $H = 0.0$, $L_e = 0.0$, $\gamma = 0.0$, and $B_r = 0.0$.

A	P_r	Abd El-Aziz [12]	Shateyi and Motsa [6]	Present study
	0.1	0.4517	0.45149	0.4518
0.8	1.0	1.6728	1.67285	1.67275
	10.0	5.70503	5.70598	5.70573
	0.1	0.5087	0.50850	0.50841
1.2	1.0	1.818	1.81801	1.81793
	10.0	6.12067	6.12102	6.12012
	0.1	0.606013	0.60352	0.60341
2.0	1.0	2.07841	2.07841	2.07830
	10.0	6.88506	6.88615	6.88586

(4) The entropy generation rates profile decreases with the increase of the Brinkman number parameter.

(5) The unsteadiness parameter increases with the increase of the entropy generation rates profile, and also, the velocity profile, the temperature profile, and the concentration profile decrease.

Nomenclature

A: Unsteadiness parameter ($= a/b$)
a: Positive constant
B: Magnetic field
B_0: Uniform transverse magnetic field
B_r: Brinkman number ($= \mu T_\infty b^2 x^2 / k(\Delta T)$)
b: Positive constant
C: Concentration profile
C_p: Heat capacity at constant pressure
C_w: Surface concentration
D: Mass diffusivity
H: Heat generation/absorption parameter ($= Q_0/b\rho C_p$)
k: Thermal conductivity
k^*: Mean absorption coefficient
L_e: Lewis number ($= k/D\rho C_p$)
M: Magnetic field parameter ($= \sigma B_0^2/b\rho$)
N_s: Entropy generation rates
Q_1: Volumetric heat generation/absorption rate
Q_0: Heating source
q_r: Radiative heat flux
T: Temperature profile
T_w: Surface temperature
T_∞: Temperature of the fluid at infinity
S: Porosity parameter ($= S_0/b$)
S_G: Volumetric entropy generation
P_r: Prandtl number ($= \mu C_p \nu/k$)
R: Thermal radiation parameter ($= k^* k/4\sigma^* T_\infty^3$)
R_e: Local Reynolds number ($= bx^2/\nu$)
u: Velocity in the x-direction
U_w: Surface velocity
v: Velocity in the y-direction
x: Horizontal distance
y: Vertical distance.

Greek Symbols

$\lambda_1, \lambda_2, \lambda_3, \lambda_4, \lambda_5$: Characteristic entropy generation rate
α: Thermal diffusivity of the fluid $(= k/\rho C_p)$
γ: Chemical reaction parameter $(= k_0/b)$
Ω: Dimensionless temperature ratio $(= \Delta T/T_\infty)$
Ω_1: Dimensionless concentration ratio $(= \Delta C/C_\infty)$
η: Similarity variable
θ: Dimensionless temperature distribution
ψ: Stream function
μ: Viscosity of the fluid
ρ: Density of the fluid
ν: Kinematic viscosity of the fluid
κ: Permeability of the porous medium
σ: Electrical conductivity
σ^*: Stephan-Boltzmann constant.

Subscripts

w, ∞: Conditions at the surface and in the free stream.

Conflict of Interests

The author declares that there is no conflict of interests regarding the publication of this paper.

References

[1] A. Bejan, *Entropy Generation through Heat and Fluid Flow*, John Wiley & Sons, New York, NY, USA, 1982.

[2] M. D. Spasojević, M. R. Janković, and D. D. Djaković, "A new approach to entropy production minimization in diabatic distillation column with trays," *Thermal Science*, vol. 14, no. 2, pp. 317–328, 2010.

[3] H. I. Andersson, J. B. Aarseth, and B. S. Dandapat, "Heat transfer in a liquid film on an unsteady stretching surface," *International Journal of Heat and Mass Transfer*, vol. 43, no. 1, pp. 69–74, 2000.

[4] R. Tsai, K. H. Huang, and J. S. Huang, "Flow and heat transfer over an unsteady stretching surface with non-uniform heat source," *International Communications in Heat and Mass Transfer*, vol. 35, no. 10, pp. 1340–1343, 2008.

[5] E. M. Elbashbeshy and M. A. Bazid, "Heat transfer over an unsteady stretching surface with internal heat generation," *Applied Mathematics and Computation*, vol. 138, no. 2-3, pp. 239–245, 2003.

[6] S. Shateyi and S. S. Motsa, "Thermal radiation effects on heat and mass transfer over an unsteady stretching surface," *Mathematical Problems in Engineering*, vol. 2009, Article ID 965603, 13 pages, 2009.

[7] M. Bouabid, N. Hidouri, M. Magherbi, and A. Ben Brahim, "Analysis of the magnetic field effect on entropy generation at thermosolutal convection in a square cavity," *Entropy*, vol. 13, no. 5, pp. 1034–1054, 2011.

[8] R. D. C. Oliveski, M. H. Macagnan, and J. B. Copetti, "Entropy generation and natural convection in rectangular cavities," *Applied Thermal Engineering*, vol. 29, no. 8-9, pp. 1417–1425, 2009.

[9] M. Achintya, "Analysis of entropy generation due to natural convection in square enclosures with multiple discrete heat sources," *International Communications in Heat and Mass Transfer*, vol. 37, no. 7, pp. 867–872, 2010.

[10] M. Magherbi, H. Abbassi, N. Hidouri, and A. Ben Brahim, "Second law analysis in convective heat and mass transfer," *Entropy*, vol. 8, no. 1, pp. 1–17, 2006.

[11] A. El Jery, N. Hidouri, M. Magherbi, and A. B. Ben Brahim, "Effect of external oriented magnetic field on entropy generation in natural convection," *Entropy*, vol. 12, no. 6, pp. 1391–1417, 2010.

[12] M. Abd El-Aziz, "Radiation effect on the flow and heat transfer over an unsteady stretching sheet," *International Communications in Heat and Mass Transfer*, vol. 36, no. 5, pp. 521–524, 2009.

[13] E. M. Sparrow and R. D. Cess, *Radiation Heat Transfer*, Brooks/Cole, Belmont, Calif, USA, 1970.

[14] A. Moradi, H. Ahmadikia, T. Hayat, and A. Alsaedi, "On mixed convection-radiation interaction about an inclined plate through a porous medium," *International Journal of Thermal Sciences*, vol. 64, pp. 129–136, 2013.

[15] H. I. Andersson, J. B. Aarseth, N. Braud, and B. S. Dandapat, "Flow of a power-law fluid film on an unsteady stretching surface," *Journal of Non-Newtonian Fluid Mechanics*, vol. 62, no. 1, pp. 1–8, 1996.

12

Implicit One-Step Block Hybrid Third-Derivative Method for the Direct Solution of Initial Value Problems of Second-Order Ordinary Differential Equations

Mohammad Alkasassbeh and Zurni Omar

Department of Mathematics, School of Quantitative Sciences, College of Arts and Sciences, Universiti Utara Malaysia, Sintok, Malaysia

Correspondence should be addressed to Mohammad Alkasassbeh; mfkasassbeh@yahoo.com

Academic Editor: Mehmet Sezer

A new one-step block method with generalized three hybrid points for solving initial value problems of second-order ordinary differential equations directly is proposed. In deriving this method, a power series approximate function is interpolated at $\{x_n, x_{n+r}\}$ while its second and third derivatives are collocated at all points $\{x_n, x_{n+r}, x_{n+s}, x_{n+t}, x_{n+1}\}$ in the given interval. The proposed method is then tested on initial value problems of second-order ordinary differential equations solved by other methods previously. The numerical results confirm the superiority of the new method to the existing methods in terms of accuracy.

1. Introduction

Numerous problems such as chemical kinetics, orbital dynamics, circuit and control theory, and Newton's second law applications involve second-order ordinary differential equations (ODEs). Normally, those equations have no analytical solutions. To approximate the solution of such problems several numerical methods were developed on the hands of many scholars such as [1–3].

Block methods for solving ODEs were first proposed by Milne ([4]). Later [5] adopted Milne's methods to provide starting values for predictor-corrector scheme. However, the block methods have some drawbacks and this led to the introduction of hybrid methods. According to [6], hybrid methods were initially introduced to overcome zero-stability barrier that occurred in block methods in Dahlquists ([7]). Besides the ability to change step size, the other benefit of these methods is utilizing data off-step points which contribute to the accuracy of the methods.

To increase the accuracy of the numerical methods further, researchers such as [8, 9] proposed high method derivative to overcome stiffness in ODEs. The former presented another type of hybrid methods called second-derivative methods, while the later proposed a Simpson's type second-derivative method for the solution of a stiff system of first-order IVPs. These scholars motivated us to develop a new generalized three-hybrid one-step third-derivative implicit method for solving second-order ODEs directly using the approach of interpolation and collocation for the general use to improve the efficiency of the approximate solution.

This article is organized as follows: in the coming section we demonstrate the derivation of the method, where we consider three off-step points through the approach of interpolation and collocation. The details of the analysis of the method are discussed in Section 3 which include zero stability, order, consistency, and convergence. In Section 4 some numerical problems are solved and the performance of the developed method is compared with other methods mentioned in literature. Finally, the conclusion is discussed in Section 5.

2. Development of the Method

An approximate power series basis function taking the form

$$p(x) = \sum_{j=0}^{2v+u-1} a_j \left(\frac{x - x_n}{h}\right)^j, \tag{1}$$

where $u = 2$ and $v = 5$ are the number of interpolation and collocation points, respectively, is considered to be a solution to the following ODE:

$$\begin{aligned} y'' &= f\left(x, y, y'\right), \\ y(a) &= p_0, \\ y_0'(a) &= p_0', \\ & \qquad x \in [a, b]. \end{aligned} \tag{2}$$

On derivation of (1) twice and thrice we obtain

$$\begin{aligned} p''(x) &= \sum_{j=2}^{2v+u-1} \frac{a_j j!}{h^2 (j-2)!} \left(\frac{x - x_n}{h}\right)^{j-2} \\ &= f\left(x, y, y'\right), \\ p'''(x) &= \sum_{j=3}^{2v+u-1} \frac{a_j j!}{h^3 (j-3)!} \left(\frac{x - x_n}{h}\right)^{j-3} \\ &= g\left(x, y, y'\right). \end{aligned} \tag{3}$$

Interpolating (1) at $x_{n+\hat{u}} = x_n + \hat{u}h, \hat{u} = \{0, r\}$ and collocating (3) at all points $x_{n+\hat{v}} = x_n + \hat{v}h, \hat{v} = \{0, r, s, t, 1\}$, where $\{r, s, t\} \in (0, 1)$, a system of equations in matrix form is produced as below:

$$AX = U, \tag{4}$$

where

$$A = \begin{bmatrix} a_0 \\ a_1 \\ a_2 \\ a_3 \\ a_4 \\ a_5 \\ a_6 \\ a_7 \\ a_8 \\ a_9 \\ a_{10} \\ a_{11} \end{bmatrix},$$

$$X = \begin{bmatrix} 1 & 0 & 0 & 0 & 0 & \cdots & 0 \\ 1 & r & r^2 & r^3 & r^4 & \cdots & r^{(2v+u-1)} \\ 0 & 0 & \frac{2!}{0!h^2} & 0 & 0 & \cdots & 0 \\ 0 & 0 & \frac{2!}{0!h^2} & \frac{3!r}{1!h^2} & \frac{4!r^2}{2!h^2} & \cdots & \frac{(2v+u-1)!r^{(2v+u-3)}}{(2v+u-3)!h^2} \\ 0 & 0 & \frac{2!}{0!h^2} & \frac{3!s}{1!h^2} & \frac{4!s^2}{2!h^2} & \cdots & \frac{(2v+u-1)!s^{(2v+u-3)}}{(2v+u-3)!h^2} \\ 0 & 0 & \frac{2!}{0!h^2} & \frac{3!t}{1!h^2} & \frac{4!t^2}{2!h^2} & \cdots & \frac{(2v+u-1)!t^{(2v+u-3)}}{(2v+u-3)!h^2} \\ 0 & 0 & \frac{2!}{0!h^2} & \frac{3!}{1!h^2} & \frac{4!}{2!h^2} & \cdots & \frac{(2v+u-1)!}{(2v+u-3)!h^2} \\ 0 & 0 & 0 & \frac{3!}{0!h^3} & 0 & \cdots & 0 \\ 0 & 0 & 0 & \frac{3!r}{0!h^3} & \frac{4!r^2}{1!h^3} & \cdots & \frac{(2v+u-1)!r^{(2v+u-4)}}{(2v+u-4)!h^3} \\ 0 & 0 & 0 & \frac{3!s}{0!h^3} & \frac{4!s^2}{1!h^3} & \cdots & \frac{(2v+u-1)!s^{(2v+u-4)}}{(2v+u-4)!h^3} \\ 0 & 0 & 0 & \frac{3!t}{0!h^3} & \frac{4!t^2}{1!h^3} & \cdots & \frac{(2v+u-1)!t^{(2v+u-4)}}{(2v+u-4)!h^3} \\ 0 & 0 & 0 & \frac{3!}{0!h^3} & \frac{4!}{1!h^3} & \cdots & \frac{(2v+u-1)!}{(2v+u-4)!h^3} \end{bmatrix},$$

$$U = \begin{bmatrix} y_n \\ y_{n+r} \\ f_n \\ f_{n+r} \\ f_{n+s} \\ f_{n+t} \\ f_{n+1} \\ g_n \\ g_{n+r} \\ g_{n+s} \\ g_{n+t} \\ g_{n+1} \end{bmatrix}. \tag{5}$$

Using matrix manipulation to solve (4) for the unknown coefficients $a_j^{\prime}$s and then substituting them back into (1) yield

$$\begin{aligned} p(x) &= \sum_{i=0,r} \alpha_i y_{n+i} + h^2 \left[\sum_{i=r,s,t} \beta_i f_{n+i} + \sum_{i=0}^{1} \beta_i f_{n+i} \right] \\ &\quad + h^3 \left[\sum_{i=r,s,t} \gamma_i g_{n+i} + \sum_{i=0}^{1} \gamma_i g_{n+i} \right], \end{aligned} \tag{6}$$

where $n = 0, 1, 2, \ldots, N-1$, $h = x_n - x_{n-1}$ is the constant step size for the partition π_N of the interval $[a, b]$ which is given by $\pi_N = [a = x_0 < x_1 < \cdots < x_{N-1} < x_N = b]$, α_i, β_i, and γ_i are undetermined constants listed in Appendix I in Supplementary Material available online at https://doi.org/10.1155/2017/8510948, $f_{n+i} = f(x + ih)$, and $g_{n+i} = df(x_{n+i}, y_{n+i}, y'_{n+i})/dx$ whose first partial derivative is

$$p'(x) = \frac{1}{h}\sum_{i=0,r}\alpha_i' y_{n+i} + h\left[\sum_{i=r,s,t}\beta_i' f_{n+i} + \sum_{i=0}^{1}\beta_i' f_{n+i}\right] + h^2\left[\sum_{i=r,s,t}\gamma_i' g_{n+i} + \sum_{i=0}^{1}\gamma_i' g_{n+i}\right]. \tag{7}$$

Evaluating (6) at the noninterpolating points $\{x_{n+s}, x_{n+t}, x_{n+1}\}$ and (7) at all points x_{n+i}, $i = \{0, r, s, t, 1\}$, produces the following general equations in block form:

$$A^{[0]}Y_m^{[1]} = A^{[1]}Y_m^{[0]} + \sum_{i=0}^{1}B^{[i]}F_m^{[i]} + \sum_{i=0}^{1}D^{[i]}G_m^{[i]}, \tag{8}$$

where $A^{[0]}$ is an 8×8 identity matrix and

$$A^{[1]} = \begin{bmatrix} 0&0&0&1&0&0&0&rh \\ 0&0&0&1&0&0&0&sh \\ 0&0&0&1&0&0&0&th \\ 0&0&0&1&0&0&0&h \\ 0&0&0&0&0&0&0&1 \\ 0&0&0&0&0&0&0&1 \\ 0&0&0&0&0&0&0&1 \\ 0&0&0&0&0&0&0&1 \end{bmatrix},$$

$$B^{[0]} = \begin{bmatrix} 0&0&0&0&0&0&0&B_{18}^{[0]} \\ 0&0&0&0&0&0&0&B_{28}^{[0]} \\ 0&0&0&0&0&0&0&B_{38}^{[0]} \\ 0&0&0&0&0&0&0&B_{48}^{[0]} \\ 0&0&0&0&0&0&0&B_{58}^{[0]} \\ 0&0&0&0&0&0&0&B_{68}^{[0]} \\ 0&0&0&0&0&0&0&B_{78}^{[0]} \\ 0&0&0&0&0&0&0&B_{88}^{[0]} \end{bmatrix},$$

$$B^{[1]} = \begin{bmatrix} B_{11}^{[1]} & B_{12}^{[1]} & B_{13}^{[1]} & B_{14}^{[1]} \\ B_{21}^{[1]} & B_{22}^{[1]} & B_{23}^{[1]} & B_{24}^{[1]} \\ B_{31}^{[1]} & B_{32}^{[1]} & B_{33}^{[1]} & B_{34}^{[1]} \\ B_{41}^{[1]} & B_{42}^{[1]} & B_{43}^{[1]} & B_{44}^{[1]} \\ B_{51}^{[1]} & B_{52}^{[1]} & B_{53}^{[1]} & B_{54}^{[1]} \\ B_{61}^{[1]} & B_{62}^{[1]} & B_{63}^{[1]} & B_{64}^{[1]} \\ B_{71}^{[1]} & B_{72}^{[1]} & B_{73}^{[1]} & B_{74}^{[1]} \\ B_{81}^{[1]} & B_{82}^{[1]} & B_{83}^{[1]} & B_{84}^{[1]} \end{bmatrix},$$

$$D^{[0]} = \begin{bmatrix} 0&0&0&0&0&0&0&D_{18}^{[0]} \\ 0&0&0&0&0&0&0&D_{28}^{[0]} \\ 0&0&0&0&0&0&0&D_{38}^{[0]} \\ 0&0&0&0&0&0&0&D_{48}^{[0]} \\ 0&0&0&0&0&0&0&D_{58}^{[0]} \\ 0&0&0&0&0&0&0&D_{68}^{[0]} \\ 0&0&0&0&0&0&0&D_{78}^{[0]} \\ 0&0&0&0&0&0&0&D_{88}^{[0]} \end{bmatrix},$$

$$D^{[1]} = \begin{bmatrix} D_{11}^{[1]} & D_{12}^{[1]} & D_{13}^{[1]} & D_{14}^{[1]} \\ D_{21}^{[1]} & D_{22}^{[1]} & D_{23}^{[1]} & D_{24}^{[1]} \\ D_{31}^{[1]} & D_{32}^{[1]} & D_{33}^{[1]} & D_{34}^{[1]} \\ D_{41}^{[1]} & D_{42}^{[1]} & D_{43}^{[1]} & D_{44}^{[1]} \\ D_{51}^{[1]} & D_{52}^{[1]} & D_{53}^{[1]} & D_{54}^{[1]} \\ D_{61}^{[1]} & D_{62}^{[1]} & D_{63}^{[1]} & D_{64}^{[1]} \\ D_{71}^{[1]} & D_{72}^{[1]} & D_{73}^{[1]} & D_{74}^{[1]} \\ D_{81}^{[1]} & D_{82}^{[1]} & D_{83}^{[1]} & D_{84}^{[1]} \end{bmatrix}. \tag{9}$$

Their entries are listed in Appendix II in Supplementary Material, while the vectors $Y_m^{[0]}, Y_m^{[1]}, F_m^{[0]}, F_m^{[1]}, G_m^{[0]}, G_m^{[1]}$ are defined as follows:

$$Y_m^{[0]} = \begin{bmatrix} y_{n-t} \\ y_{n-s} \\ y_{n-r} \\ y_n \\ y'_{n-t} \\ y'_{n-s} \\ y'_{n-r} \\ y'_n \end{bmatrix},$$

$$Y_m^{[1]} = \begin{bmatrix} y_{n+r} \\ y_{n+s} \\ y_{n+t} \\ y_{n+1} \\ y'_{n+r} \\ y'_{n+s} \\ y'_{n+t} \\ y'_{n+1} \end{bmatrix}, \quad F_m^{[0]} = \begin{bmatrix} f_{n-7} \\ f_{n-6} \\ f_{n-5} \\ f_{n-4} \\ f_{n-3} \\ f_{n-2} \\ f_{n-1} \\ f_n \end{bmatrix}, \quad F_m^{[1]} = \begin{bmatrix} f_{n+r} \\ f_{n+s} \\ f_{n+t} \\ f_{n+1} \end{bmatrix}, \quad G_m^{[0]} = \begin{bmatrix} g_{n-7} \\ g_{n-6} \\ g_{n-5} \\ g_{n-4} \\ g_{n-3} \\ g_{n-2} \\ g_{n-1} \\ g_n \end{bmatrix}, \quad G_m^{[1]} = \begin{bmatrix} g_{n+r} \\ g_{n+s} \\ g_{n+t} \\ g_{n+1} \end{bmatrix}. \tag{10}$$

3. Analysis of the Method

3.1. Zero Stability

Definition 1. The hybrid block method formula (8) is said to be zero stable if no root of the first characteristic equation $\rho(R)$ has modulus greater than one; that is, $|R_m| \leqslant 1$ and if $R_m = 1$ then the multiplicity of R_m must not exceed two.

To illustrate that the root of the first characteristic equation satisfies the prior definition we assume that $\{r, s, t\} \in (0, 1)$ and hence

$$\rho(R) = \det\left[RA^{[0]} - A^{[1]}\right] = \begin{vmatrix} R & 0 & 0 & -1 & 0 & 0 & 0 & -rh \\ 0 & R & 0 & -1 & 0 & 0 & 0 & -sh \\ 0 & 0 & R & -1 & 0 & 0 & 0 & -th \\ 0 & 0 & 0 & R-1 & 0 & 0 & 0 & -h \\ 0 & 0 & 0 & 0 & R & 0 & 0 & -1 \\ 0 & 0 & 0 & 0 & 0 & R & 0 & -1 \\ 0 & 0 & 0 & 0 & 0 & 0 & R & -1 \\ 0 & 0 & 0 & 0 & 0 & 0 & 0 & R-1 \end{vmatrix} = 0 \tag{11}$$

which imply that

$$R^6 (R-1)^2 = 0 \Longrightarrow R_i = \begin{cases} 0, & \text{if } i = 1\,(1)\,6 \\ 1, & \text{if } i = 7, 8. \end{cases} \tag{12}$$

As a result, the developed method is zero stable.

3.2. Order of the Method. The linear operator Δ associated with the hybrid block methods formula (8) is defined as

$$\Delta\{y(x); h\} = A^{[0]} Y_m^{[1]} - A^{[1]} Y_m^{[0]} - \sum_{i=0}^{1} B^{[i]} F_m^{[i]} - \sum_{i=0}^{1} D^{[i]} G_m^{[i]}. \tag{13}$$

Expanding the above equation in Taylor series and combining like terms we wind up with

$$\Delta\{y(x); h\} = \widehat{C}_0 h^0 y(x) + \widehat{C}_1 h^1 y'(x) + \widehat{C}_2 h^2 y''(x) + \cdots + \widehat{C}_{p+2} h^{p+2} y^{(p+2)}(x) + \cdots \tag{14}$$

According to [6, 10] method (8) is said to be of order p if

$$\widehat{C}_0 = \widehat{C}_1 = \cdots = \widehat{C}_{p+1} = 0, \quad \widehat{C}_{p+2} \neq 0. \tag{15}$$

The term $\widehat{C}_{p+2}$ is called the error constant and the local truncation error is given by

$$t_{n+k} = \widehat{C}_{p+2} y^{(p+2)} h^{p+2} + O\left(h^{p+3}\right). \tag{16}$$

Comparing like terms of $y^{(i)}$ and h^i in (14) produces the coefficients $\widehat{C}_0 = \widehat{C}_1 = \cdots = \widehat{C}_{11} = 0$ with vector of error constants

$$\widehat{C}_{12} = \left[\widehat{C}_{12}^{[1]}\ \widehat{C}_{12}^{[2]}\ \widehat{C}_{12}^{[3]}\ \widehat{C}_{12}^{[4]}\ \widehat{C}_{12}^{[5]}\ \widehat{C}_{12}^{[6]}\ \widehat{C}_{12}^{[7]}\ \widehat{C}_{12}^{[8]}\right]^T, \tag{17}$$

where

$$\begin{aligned}
\widehat{C}_{12}^{[1]} &= \frac{h^{12}r^{6}y^{12}}{100590336000}\Big(\big(14r^6 - 42r^5s - 42r^5t - 42r^5 \\
&\quad + 33r^4s^2 + 132r^4st + 132r^4s + 33r^4t^2 + 132r^4t \\
&\quad + 33r^4 - 110r^3s^2t - 110r^3s^2 - 110r^3st^2 - 440r^3st \\
&\quad - 110r^3s - 110r^3t^2 - 110r^3t + 99r^2s^2t^2 + 396r^2s^2t \\
&\quad + 99r^2s^2 + 396r^2st^2 + 396r^2st + 99r^2t^2 - 396rs^2t^2 \\
&\quad - 396rs^2t - 396rst^2 + 462s^2t^2\big)\Big), \\
\widehat{C}_{12}^{[2]} &= \frac{h^{12}s^{6}y^{12}}{100590336000}\Big(\big(33r^2s^4 - 110r^2s^3t - 110r^2s^3 \\
&\quad + 99r^2s^2t^2 + 396r^2s^2t + 99r^2s^2 - 396r^2st^2 \\
&\quad - 396r^2st + 462r^2t^2 - 42rs^5 + 132rs^4t + 132rs^4 \\
&\quad - 110rs^3t^2 - 440rs^3t - 110rs^3 + 396rs^2t^2 \\
&\quad + 396rs^2t - 396rst^2 + 14s^6 - 42s^5t - 42s^5 \\
&\quad + 33s^4t^2 + 132s^4t + 33s^4 - 110s^3t^2 - 110s^3t \\
&\quad + 99s^2t^2\big)\Big), \\
\widehat{C}_{12}^{[3]} &= \frac{h^{12}t^{6}y^{12}}{100590336000}\Big(\big(99r^2s^2t^2 - 396r^2s^2t \\
&\quad + 462r^2s^2 - 110r^2st^3 + 396r^2st^2 - 396r^2st \\
&\quad + 33r^2t^4 - 110r^2t^3 + 99r^2t^2 - 110rs^2t^3 + 396rs^2t^2 \\
&\quad - 396rs^2t + 132rst^4 - 440rst^3 + 396rst^2 - 42rt^5 \\
&\quad + 132rt^4 - 110rt^3 + 33s^2t^4 - 110s^2t^3 + 99s^2t^2 \\
&\quad - 42st^5 + 132st^4 - 110st^3 + 14t^6 - 42t^5 + 33t^4\big)\Big), \\
\widehat{C}_{12}^{[4]} &= \frac{h^{12}y^{12}}{100590336000}\Big(\big(462r^2s^2t^2 - 396r^2s^2t \\
&\quad + 99r^2s^2 - 396r^2st^2 + 396r^2st - 110r^2s + 99r^2t^2 \\
&\quad - 110r^2t + 33r^2 - 396rs^2t^2 + 396rs^2t - 110rs^2 \\
&\quad + 396rst^2 - 440rst + 132rs - 110rt^2 + 132rt \\
&\quad - 42r + 99s^2t^2 - 110s^2t + 33s^2 - 110st^2 + 132st \\
&\quad - 42s + 33t^2 - 42t + 14\big)\Big), \\
\widehat{C}_{12}^{[5]} &= \frac{h^{11}r^{5}y^{12}}{50295168000}\Big(\big(28r^6 - 77r^5s - 77r^5t - 77r^5 \\
&\quad + 55r^4s^2 + 220r^4st + 220r^4s + 55r^4t^2 + 220r^4t \\
&\quad + 55r^4 - 165r^3s^2t - 165r^3s^2 - 165r^3st^2 - 660r^3st \\
&\quad - 165r^3s - 165r^3t^2 - 165r^3t + 132r^2s^2t^2 \\
&\quad + 528r^2s^2t + 132r^2s^2 + 528r^2st^2 + 528r^2st \\
&\quad + 132r^2t^2 - 462rs^2t^2 - 462rs^2t - 462rst^2 \\
&\quad + 462s^2t^2\big)\Big), \\
\widehat{C}_{12}^{[6]} &= \frac{h^{11}s^{5}y^{12}}{50295168000}\Big(\big(55r^2s^4 - 165r^2s^3t - 165r^2s^3 \\
&\quad + 132r^2s^2t^2 + 528r^2s^2t + 132r^2s^2 - 462r^2st^2 \\
&\quad - 462r^2st + 462r^2t^2 - 77rs^5 + 220rs^4t + 220rs^4 \\
&\quad - 165rs^3t^2 - 660rs^3t - 165rs^3 + 528rs^2t^2 \\
&\quad + 528rs^2t - 462rst^2 + 28s^6 - 77s^5t - 77s^5 \\
&\quad + 55s^4t^2 + 220s^4t + 55s^4 - 165s^3t^2 - 165s^3t \\
&\quad + 132s^2t^2\big)\Big), \\
\widehat{C}_{12}^{[7]} &= \frac{h^{11}t^{5}y^{12}}{50295168000}\Big(\big(132r^2s^2t^2 - 462r^2s^2t \\
&\quad + 462r^2s^2 - 165r^2st^3 + 528r^2st^2 - 462r^2st \\
&\quad + 55r^2t^4 - 165r^2t^3 + 132r^2t^2 - 165rs^2t^3 \\
&\quad + 528rs^2t^2 - 462rs^2t + 220rst^4 - 660rst^3 \\
&\quad + 528rst^2 - 77rt^5 + 220rt^4 - 165rt^3 + 55s^2t^4 \\
&\quad - 165s^2t^3 + 132s^2t^2 - 77st^5 + 220st^4 - 165st^3 \\
&\quad + 28t^6 - 77t^5 + 55t^4\big)\Big), \\
\widehat{C}_{12}^{[8]} &= \frac{h^{11}y^{12}}{50295168000}\Big(\big(462r^2s^2t^2 - 462r^2s^2t \\
&\quad + 132r^2s^2 - 462r^2st^2 + 528r^2st - 165r^2s \\
&\quad + 132r^2t^2 - 165r^2t + 55r^2 - 462rs^2t^2 + 528rs^2t \\
&\quad - 165rs^2 + 528rst^2 - 660rst + 220rs - 165rt^2 \\
&\quad + 220rt - 77r + 132s^2t^2 - 165s^2t + 55s^2 - 165st^2 \\
&\quad + 220st - 77s + 55t^2 - 77t + 28\big)\Big),
\end{aligned} \tag{18}$$

which conclude that the order of the developed method is $p = 10$.

3.3. Consistency

Definition 2. A block method is said to be consistent if its order p is greater than one.

TABLE 1: Comparison of the proposed method with [11].

X value	Exact solution	Computed solution for $r = 1/2$; $s = 2/3$; $t = 3/4$	Error in new method	Error for [11]
0.100	0.904837418035959520	0.904837418035948970	1.054712$E(-14)$	2:05$E(-11)$
0.200	0.818730753077981820	0.818730753077964060	1.776357$E(-14)$	4:39$E(-11)$
0.300	0.740818220681717770	0.740818220681694340	2.342571$E(-14)$	6:55$E(-11)$
0.400	0.670320046035639330	0.670320046035611350	2.797762$E(-14)$	8:38$E(-11)$
0.500	0.606530659712633420	0.606530659712602120	3.130829$E(-14)$	9:86$E(-11)$
0.600	0.548811636094026500	0.548811636093992530	3.397282$E(-14)$	1:10$E(-10)$
0.700	0.496585303791409530	0.496585303791373890	3.563816$E(-14)$	1:19$E(-10)$
0.800	0.449328964117221620	0.449328964117184870	3.674838$E(-14)$	1:24$E(-10)$
0.900	0.406569659740599170	0.406569659740561860	3.730349$E(-14)$	1:28$E(-10)$
1.000	0.367879441171442330	0.367879441171404920	3.741452$E(-14)$	1:30$E(-10)$

TABLE 2: Comparison of the proposed method with [12].

X value	Exact solution	Computed solution for $r = 1/4$; $s = 1/2$; $t = 3/4$	Error in new method	Error for [12]
0.100	1.050041729278491400	1.050041729278491200	2.220446$E(-16)$	1.1102$E(-15)$
0.200	1.100335347731075600	1.100335347731075300	2.220446$E(-16)$	5.9952$E(-15)$
0.300	1.151140435936466800	1.151140435936466100	6.661338$E(-16)$	2.5535$E(-14)$
0.400	1.202732554054082100	1.202732554054081000	1.110223$E(-15)$	7.1054$E(-14)$
0.500	1.255412811882995200	1.255412811882994800	4.440892$E(-16)$	1.1568$E(-13)$
0.600	1.309519604203111900	1.309519604203112800	8.881784$E(-16)$	1.1990$E(-13)$
0.700	1.365443754271396400	1.365443754271398000	1.554312$E(-15)$	6.8567$E(-13)$
0.800	1.423648930193601900	1.423648930193606400	4.440892$E(-15)$	3.4754$E(-12)$
0.900	1.484700278594052000	1.484700278594060600	8.659740$E(-16)$	1.2219$E(-11)$
1.000	1.549306144334055000	1.549306144334067700	1.265654$E(-14)$	3.7282$E(-11)$

Consistency property is achieved for the hybrid block method from the above analysis since the order $p = 10 \geq 1$.

3.4. Convergence

Theorem 3 (see [16]). *Consistency and zero stability are sufficient conditions for a linear multistep method to be convergent*

The hybrid block method equation (8) is convergent since it fulfills both the consistency and zero-stability conditions.

4. Numerical Examples

In this section, the efficiency and the performance of the general three-hybrid one-step implicit hybrid block method (8) with order $p = 10$ is investigated on five test problems. The first example is highly stiff linear IVP problem with step size $h = 1/10$, the second is nonlinear IVP with $h = 5/100$, the third is linear with $h = 1/100$, the fourth is a nonlinear system with $h = 1/10$, and finally the fifth is a nonlinear undamped Duffing equation with $h = \pi/5$. It is worth mentioning that this method works even for large interval and different values of step size. The values mentioned in this article are chosen just for the sake of comparison with the existing methods only.

Problem 4. $f(x, y, y') = -1001y' - 1000y$, $y(0) = 1$, $y'(0) = -1$, $0 \leq x \leq 1$.

Exact Solution. $y = e^{-x}$ with $h = 1/10$.

Source (see [11]). See Table 1.

Problem 5. $f(x, y, y') = x(y')^2$, $y(0) = 1$, $y'(0) = 1/2$, $0 \leq x \leq 1$.

Exact Solution. $y = 1 + \ln((2 + x)/(2 - x))$ with $h = 5/100$.

Source (see [12]). See Table 2.

Problem 6. $f(x, y, y') = y'$, $y(0) = 0$, $y'(1) = -1$, $0 \leq x \leq 1$.

Exact Solution. $y = 1 - e^{x}$ with $h = 1/100$.

Source (see [13]). See Table 3.

Problem 7 (a system of nonlinear IVP). $y_1'' = 1 - \cos(x) + \sin(y_2') + \cos(y_2')$, $y_1(0) = 0$, $y_1'(0) = 0$, $0 \leq x \leq 1$.

TABLE 3: Comparison of the proposed method with [13].

X value	Exact solution	Computed solution for $r = 1/5;\ s = 1/3;\ t = 1/2$	Error in new method	Error for [13]
0.100	−0.105170918075647710	−0.105170918075647660	5.551115E(−17)	2.2360E(−13)
0.200	−0.221402758160169850	−0.221402758160169990	1.387779E(−16)	1.6425E(−12)
0.300	−0.349858807576003180	−0.349858807576003520	3.330669E(−16)	3.4625E(−12)
0.400	−0.491824697641270570	−0.491824697641271070	4.996004E(−16)	8.6628E(−12)
0.500	−0.648721270700128640	−0.648721270700129420	7.771561E(−16)	1.1338E(−11)
0.600	−0.822118800390509550	−0.822118800390510880	1.332268E(−15)	2.0317E(−11)
0.700	−1.013752707470477500	−1.013752707470479300	1.776357E(−15)	3.2476E(−11)
0.800	−1.225540928492468800	−1.225540928492471600	2.886580E(−15)	4.5463E(−11)
0.900	−1.459603111156951200	−1.459603111156955000	3.774758E(−15)	6.1781E(−11)
1.000	−1.718281828459047300	−1.718281828459052400	5.107026E(−15)	8.2113E(−11)

TABLE 4: Exact and approximate solutions for solving y_1 using the developed method.

X value	Exact solution	Approximate solution	Error in developed method	Error in [14]
0.200	0.980066577841241630	0.980066580773730210	2.93E(−09)	3.331E(−08)
0.400	0.921060994002885100	0.921060786088783830	2.08E(−07)	7.27E(−07)
0.600	0.825335614909678330	0.825334546366523920	1.07E(−06)	3.09E(−06)
0.800	0.696706709347165500	0.696703745729634890	2.96E(−06)	8.01E(−06)
1.000	0.540302305868139770	0.540296131232123940	6.17E(−06)	1.60E(−05)

TABLE 5: Exact and approximate solutions for solving y_2 using the developed method.

X value	Exact solution	Approximate solution	Error in developed method	Error in [14]
0.200	0.408407044966673130	0.628318530716756810	1.20E(−12)	5.74E(−12)
0.400	1.036725575684631900	1.256637061492416500	5.65E(−10)	3.40E(−10)
0.600	1.665044106402590500	1.884955593121642800	9.68E(−10)	3.48E(−09)
0.800	2.293362637120548900	2.513274128242366100	5.37E(−09)	1.66E(−08)
1.000	2.921681167838507500	3.141592671767204200	1.82E(−08)	5.24E(−08)

TABLE 6: Comparison of the proposed method with [15].

X value	Exact solution	Approximate solution	Error in developed method	Error in [15]
π	−0.200326851873144250	−0.200326851873131260	1.30E(−14)	2.77E(−6)
2π	0.200027330586423440	0.200027330586373150	5.03E(−14)	2.97E(−8)
4π	0.198830853474466220	0.198830853474288970	1.77E(−13)	1.16E(−7)
6π	0.196842430954904000	0.196842430954587670	3.16E(−13)	2.53E(−7)
8π	0.194070581011836470	0.194070581011452750	3.84E(−13)	4.28E(−7)
10π	0.190527147620306290	0.190527147620002170	3.04E(−13)	6.29E(−7)

$y_2'' = 1/(4 + y_1^2) - 1/(5 - (\sin(x))^2)$, $y_2(0) = 0$, $y_2'(0) = \pi$, $0 \le x \le 1$.

Exact Solutions. $y_1 = \cos(x)$, $y_2 = \pi x$, with $h = (1/10)(r = 1/10, s = 3/10, t = 3/5)$; see Tables 4 and 5.

Problem 8 (the nonlinear undamped Duffing equation). $y'' + y + y^3 = Q\cos(\omega x)$, $y(0) = k_0$, $y'(0) = 0$, $0 \le x \le 10\pi$.

Exact Solutions. $y(x) = k_1\cos(\omega x) + k_2\cos(3\omega x) + k_3\cos(5\omega x) + k_4\cos(7\omega x)$, where $\omega = 1.01$, $Q = 0.002$, $k_0 = 0.200426728069$, $k_1 = 0.200179477536$, $k_2 = 0.246946143e(-3)$, $k_3 = 0.304016e(-6)$, $k_4 = 0.374e(-9)$, and $h = \pi/5$. $(r = 1/4, s = 1/2, t = 3/4)$; see Table 6.

5. Conclusion

A general three-hybrid one-step block method of order 10 has been proposed for the direct solution of general second-order ODEs. The developed method is tested on five different problems. Numerical analysis shows that the developed method is consistent and zero stable which conclude its convergence. The computed results are then compared with the results of existing methods in terms of error by considering different values of r, s, and t. The new method is found to have superiority over them as shown in Tables 1–6.

Competing Interests

The authors declare that there is no conflict of interests regarding the publication of this paper.

Acknowledgments

The authors thank Universiti Utara Malaysia for the financial support of the publication of this article.

References

[1] M. M. Chawla and S. R. Sharma, "Families of three-stage third order Runge-Kutta-Nystrom methods for $y'' = f(x, y, y')$," *The Journal of the Australian Mathematical Society. Series B. Applied Mathematics*, vol. 26, no. 3, pp. 375–386, 1985.

[2] E. Hairer and G. Wanner, "A theory for Nyström methods," *Numerische Mathematik*, vol. 25, no. 4, pp. 383–400, 1976.

[3] J. Vigo-Aguiar and H. Ramos, "Variable stepsize implementation of multistep methods for y'' = f(x,y,y')," *Journal of Computational and Applied Mathematics*, vol. 192, no. 1, pp. 114–131, 2006.

[4] W. E. Milne, *Numerical Solution of Differential Equations*, John Wiley & Sons, New York, NY, USA, 1953.

[5] R. J. Barkley, "A Runge-kutta for all seasons," *SIAM Review*, vol. 9, no. 3, pp. 417–452, 1967.

[6] J. D. Lambert, *Computational Methods in Ordinary Differential Equations*, John Wiley & Sons, New York, NY, USA, 1973.

[7] G. Dahlquist, "Convergence and stability in the numerical integration of ordinary differential equations," *Mathematica Scandinavica*, vol. 4, pp. 33–53, 1956.

[8] F. F. Ngwane and S. N. Jator, "Block hybrid-second derivative method for stiff systems," *International Journal of Pure and Applied Mathematics*, vol. 80, no. 4, pp. 543–559, 2012.

[9] R. K. Sahi, S. N. Jator, and N. A. Khan, "A Simpson's-type second derivative method for stiff systems," *International Journal of Pure and Applied Mathematics*, vol. 81, no. 4, pp. 619–633, 2012.

[10] S. O. Fatunla, *Numerical Methods for Fatunla Initial Value Problems in Ordinary Differential Equations*, Academic Press, New York, NY, USA, 1988.

[11] O. Adeniran and B. Ogundare, "An efficient hybrid numerical scheme for solving general second order initial value problems (IVPs)," *International Journal of Applied Mathematical Research*, vol. 4, no. 2, pp. 411–419, 2015.

[12] A. A. James, A. O. Adesanya, and S. Joshua, "Continuous block method for the solution of second order initial value problems of ordinary differential equation," *International Journal of Pure and Applied Mathematics*, vol. 83, no. 3, pp. 405–416, 2013.

[13] O. O. Olanegan, D. O. Awoyemi, B. G. Ogunware, and F. O. Obarhua, "Continuous double-hybrid point method for the solution of second order ordinary differential equations," *International Journal of Advanced Scientific and Technical Research*, vol. 5, no. 2, pp. 549–562, 2015.

[14] R. F. Abdelrahim and Z. Omar, "Direct solution of second-order ordinary differential equation using a single-step hybrid block method of order five," *Mathematical and Computational Applications*, vol. 21, no. 2, pp. 1–12, 2016.

[15] S. N. Jator, "Block third derivative method based on trigonometric polynomials for periodic initial-value problems," *Afrika Matematika*, vol. 27, no. 3-4, pp. 365–377, 2016.

[16] P. Henrici, *Discrete Variable Methods in ODEs*, John Wiley & Sons, New York, NY, USA, 1962.

13

The Fixed Point Theory and the Existence of the Periodic Solution on a Nonlinear Differential Equation

Ni Hua

Faculty of Science, Jiangsu University, 301 Xuefu Road, Zhenjiang, Jiangsu 212013, China

Correspondence should be addressed to Ni Hua; nihua979@126.com

Academic Editor: Mehmet Sezer

This paper deals with a nonlinear differential equation, by using the fixed point theory. The existence of the periodic solution of the nonlinear differential equation is obtained; these results are new.

1. Introduction

The nonlinear Abel-type first-order differential equation

$$\frac{dx}{dt} = a(t)x^3 + b(t)x^2 + c(t)x + d(t) \tag{1}$$

plays an important role in many physical and technical applications; because of its importance, many scholars have studied it [1–7].

Recently, A. Cima, A. Gasull, and F. Manosas [8] gave the maximum number of polynomial solutions of some integrable polynomial Abel differential equations; Jaume Giné Claudia and Valls [9] studied the center problem for Abel polynomial differential equations of second kind; Jianfeng Huang and Haihua Liang [10] were devoted to the investigation of Abel equation by means of Lagrange interpolation formula; they gave a criterion to estimate the number of limit cycles of the Abel's equations; Berna Bülbül and Mehmet Sezer [11] introduced a numerical power series algorithm which is based on the improved Taylor matrix method for the approximate solution of Abel-type differential equations; Ni et al. [12] discussed the existence and stability of the periodic solutions of (1) and obtained the sufficient conditions which guaranteed the existence and stability of the periodic solutions for (1) from a particular one.

In this paper, we consider the following more general nonlinear differential equation:

$$\frac{dx}{dt} = f(t,x)x + g(t,x); \tag{2}$$

firstly, we give two results about the existence and uniqueness of the periodic solution of (2) by using the fixed point theory; then, we use Lyapunov function method and discuss the stability of the periodic solution; further, we discuss (1) and get the existence of the periodic solution of (1); some new results are obtained.

2. Preliminaries

Lemma 1 (see [12]). *Consider the following linear differential equation:*

$$\frac{dx}{dt} = a(t)x + b(t), \tag{3}$$

where $a(t), b(t)$ are ω−periodic continuous functions; if $\int_0^\omega a(t)dt \neq 0$, then for (3) there exists a unique ω−periodic continuous solution $\eta(t)$, and $\eta(t)$ can be written as follows:

$$\eta(t) = \begin{cases} \int_{-\infty}^{t} e^{\int_s^t a(\tau)d\tau} b(s)\,ds, & \int_0^\omega a(t)\,dt < 0 \\ -\int_{t}^{+\infty} e^{\int_s^t a(\tau)d\tau} b(s)\,ds, & \int_0^\omega a(t)\,dt > 0 \end{cases} \tag{4}$$

For the sake of convenience, suppose that $f(t)$ is an ω-periodic continuous function on R; we denote

$$f_M = \sup_{t\in[0,\omega]} f(t), \qquad f_L = \inf_{t\in[0,\omega]} f(t). \tag{5}$$

The rest of the paper is arranged as follows. We will study the existence and stability property of the periodic solution of system (2) in the next section and discuss the existence of the periodic solution of system (1) on Section 4. We end this paper with two examples.

3. Periodic Solution on a Nonlinear Differential Equation

Theorem 2. *Consider (2); $f(t,x)$, $g(t,x)$ are both ω–periodic continuous functions on $t \in R$ and uniformly with respect to $x \in R$; suppose that the following conditions hold:*

$$(H_1)\quad f(t,x) \le -\alpha < 0;$$

$$(H_2)\quad |f(t,x) - f(t,y)| \le L_1 |x-y|, \qquad |g(t,x) - g(t,y)| \le L_2 |x-y|, \quad L_2 < \alpha; \tag{6}$$

$$(H_3)\quad \frac{\alpha L_2 + L_1 \sup_{t\in[0,\omega]} |g(t,0)| - L_2^2}{\alpha(\alpha - L_2)} < 1;$$

then

(1) *for (2) there exists a unique ω-periodic continuous solution $\gamma(t)$, and*

$$|\gamma(t)| \le \frac{1}{\alpha - L_2} \sup_{t\in[0,\omega]} |g(t,0)|, \tag{7}$$

(2) *$\gamma(t)$ is uniformly asymptotic stable.*

Proof. (1) Let

$$B = \left\{ \varphi(t) \mid \varphi \in C(R,R),\ \varphi(t+\omega) = \varphi(t),\ |\varphi(t)| \le \frac{1}{\alpha - L_2} \sup_{t\in[0,\omega]} |g(t,0)| \right\}. \tag{8}$$

Given any $\varphi(t) \in B$, the norm is defined as follows:

$$\|\varphi\| = \sup_{t\in[0,\omega]} |\varphi(t)|; \tag{9}$$

thus $(B, \|\cdot\|)$ is a Banach space. Consider the following equation:

$$\frac{dx}{dt} = f[t,\varphi(t)]x + g[t,\varphi(t)]; \tag{10}$$

by (H_1), according to Lemma 1, (10) has a unique ω–periodic continuous solution as follows:

$$\eta(t) = \int_{-\infty}^{t} e^{\int_s^t f[\tau,\varphi(\tau)]d\tau} g[s,\varphi(s)]\,ds \tag{11}$$

and

$$\begin{aligned} |\eta(t)| &\le \int_{-\infty}^{t} e^{-\alpha(t-s)} |g[s,\varphi(s)]|\,ds \\ &\le \int_{-\infty}^{t} e^{-\alpha(t-s)} (L_2|\varphi(s)| + |g(s,0)|)\,ds \\ &\le \frac{L_2|\varphi|_M + \sup_{t\in[0,\omega]}|g(t,0)|}{\alpha} \\ &= \frac{L_2(1/(\alpha - L_2)) \sup_{t\in[0,\omega]} |g(t,0)| + \sup_{t\in[0,\omega]} |g(t,0)|}{\alpha} \\ &= \frac{1}{\alpha - L_2} \sup_{t\in[0,\omega]} |g(t,0)| \end{aligned} \tag{12}$$

Define a map as follows:

$$(T\varphi)(t) = \eta(t) = \int_{-\infty}^{t} e^{\int_s^t f[\tau,\varphi(\tau)]d\tau} g[s,\varphi(s)]\,ds; \tag{13}$$

thus if given any $\varphi(t) \in B$, then $(T\varphi)(t) \in B$; hence $T: B \longrightarrow B$; given any $\varphi(t), \psi(t) \in B$, we have

$$\begin{aligned} |(T\varphi)(t) - (T\psi)(t)| &= \left| \int_{-\infty}^{t} e^{\int_s^t f[\tau,\varphi(\tau)]d\tau} g[s,\varphi(s)]\,ds - \int_{-\infty}^{t} e^{\int_s^t f[\tau,\psi(\tau)]d\tau} g[s,\psi(s)]\,ds \right| \\ &= \left| \int_{-\infty}^{t} e^{\int_s^t f[\tau,\varphi(\tau)]d\tau} (g(s,\varphi(s)) - g(s,\psi(s)))\,ds + \int_{-\infty}^{t} \left(e^{\int_s^t f[\tau,\varphi(\tau)]d\tau} - e^{\int_s^t f[\tau,\psi(\tau)]d\tau} \right) g(s,\psi(s))\,ds \right| \\ &= \left| \int_{-\infty}^{t} e^{\int_s^t f[\tau,\varphi(\tau)]d\tau} (g(s,\varphi(s)) - g(s,\psi(s)))\,ds + \int_{-\infty}^{t} e^{\xi} \left(\int_s^t f[\tau,\varphi(\tau)]\,d\tau - \int_s^t f[\tau,\psi(\tau)]\,d\tau \right) \cdot g(s,\psi(s))\,ds \right|; \end{aligned} \tag{14}$$

here, ξ is between $\int_s^t f[\tau,\varphi(\tau)]d\tau$ and $\int_s^t f[\tau,\psi(\tau)]d\tau$; thus $\xi \le -\alpha(t-s)$, so we have

$$\begin{aligned} |(T\varphi)(t) - (T\psi)(t)| &\le \int_{-\infty}^{t} e^{\int_s^t f[\tau,\varphi(\tau)]d\tau} |g(s,\varphi(s)) - g(s,\psi(s))|\,ds \\ &\quad + \int_{-\infty}^{t} e^{\xi} \left| \int_s^t f[\tau,\varphi(\tau)]\,d\tau - \int_s^t f[\tau,\psi(\tau)]\,d\tau \right| |g(s,\psi(s))|\,ds\ | \\ &\le \int_{-\infty}^{t} e^{\int_s^t f[\tau,\varphi(\tau)]d\tau} |g(s,\varphi(s)) - g(s,\psi(s))|\,ds + \int_{-\infty}^{t} e^{\xi} \int_s^t |f[\tau,\varphi(\tau)] - f[\tau,\psi(\tau)]|\,d\tau\, |g(s,\psi(s))|\,ds\ | \end{aligned}$$

$$\begin{aligned}
&\le L_2 \int_{-\infty}^{t} e^{-\alpha(t-s)} \left|\varphi(s) - \psi(s)\right| ds + L_1 \int_{-\infty}^{t} e^{-\alpha(t-s)} \int_{s}^{t} \left|\varphi(\tau) - \psi(\tau)\right| d\tau \left|g(s, \psi(s))\right| ds \mid \\
&\le L_2 \int_{-\infty}^{t} e^{-\alpha(t-s)} \left|\varphi(s) - \psi(s)\right| ds + L_1 \int_{-\infty}^{t} e^{-\alpha(t-s)} \int_{s}^{t} \left|\varphi(\tau) - \psi(\tau)\right| d\tau \left(L_2 \left|\psi(s)\right| + \left|g(s, 0)\right|\right) ds \\
&= \left(L_2 \int_{-\infty}^{t} e^{-\alpha(t-s)} ds + L_1 \int_{-\infty}^{t} e^{-\alpha(t-s)} (t-s) \left(L_2 \left|\psi(s)\right| + \left|g(s, 0)\right|\right) ds\right) \|\varphi - \psi\| \\
&\le \left(\frac{L_2}{\alpha} + \frac{L_1 \left(L_2 |\psi|_M + \sup_{t\in[0,\omega]} |g(t, 0)|\right)}{\alpha^2}\right) \|\varphi - \psi\| \\
&= \left(\frac{L_2}{\alpha} + \frac{L_1 \left(L_2 \left(1/(\alpha - L_2)\right) \sup_{t\in[0,\omega]} |g(t, 0)| + \sup_{t\in[0,\omega]} |g(t, 0)|\right)}{\alpha^2}\right) \|\varphi - \psi\| \\
&= \left(\frac{L_2}{\alpha} + \frac{\alpha L_1 \sup_{t\in[0,\omega]} |g(t, 0)| / (\alpha - L_2)}{\alpha^2}\right) \|\varphi - \psi\| = \frac{\alpha L_2 + L_1 \sup_{t\in[0,\omega]} |g(t, 0)| - L_2^2}{\alpha(\alpha - L_2)} \|\varphi - \psi\|
\end{aligned} \tag{15}$$

By (H_2), (H_3), it follows that

$$0 < \frac{\alpha L_2 + L_1 \sup_{t\in[0,\omega]} |g(t, 0)| - L_2^2}{\alpha(\alpha - L_2)} < 1; \tag{16}$$

by (16), T is a compression mapping; according to the compression mapping principle, T has a fixed point on B, the fixed point is the unique ω–periodic continuous solution $\gamma(t)$ of (2), and $|\gamma(t)| \le (1/(\alpha - L_2))\sup_{t\in[0,\omega]}|g(t, 0)|$.

(2) Define a Lyapunov function as follows:

$$V(t, x(t)) = (x(t) - \gamma(t))^2, \tag{17}$$

where $x(t)$ is the unique solution with the initial value $x(t_0) = x_0$ of (2); differentiating both sides of (17) along the solution of (2), we get

$$\begin{aligned}
\frac{dV(t, x(t))}{dt} &= 2(x(t) - \gamma(t)) \left[\frac{dx(t)}{dt} - \frac{d\gamma(t)}{dt}\right] \\
&= 2(x(t) - \gamma(t)) [f(t, x(t)) x(t) \\
&\quad - f(t, \gamma(t)) \gamma(t) + (g(t, x(t)) - g(t, \gamma(t)))] \\
&= 2(x(t) - \gamma(t)) [f(t, x(t)) (x(t) - \gamma(t)) \\
&\quad + (f(t, x(t)) - f(t, \gamma(t))) \gamma(t) \\
&\quad + (g(t, x(t)) - g(t, \gamma(t)))] = 2f(t, x) (x(t) \\
&\quad - \gamma(t))^2 + 2(x(t) - \gamma(t)) (f(t, x(t)) \\
&\quad - f(t, \gamma(t))) \gamma(t) + 2(x(t) - \gamma(t)) (g(t, x(t)) \\
&\quad - g(t, \gamma(t))) \le 2f(t, x) (x(t) - \gamma(t))^2 + 2|x(t) \\
&\quad - \gamma(t)| |f(t, x(t)) - f(t, \gamma(t))| |\gamma(t)| + 2|x(t) \\
&\quad - \gamma(t)| |g(t, x(t)) - g(t, \gamma(t))| \le 2(x(t) - \gamma(t)) \\
&\quad \cdot [f(t, x(t)) (x(t) - \gamma(t)) \\
&\quad + L_1 (x(t) - \gamma(t)) |\gamma(t)| + L_2 (x(t) - \gamma(t))] \\
&= 2(x(t) - \gamma(t))^2 [f(t, x(t)) + L_1 |\gamma(t)| + L_2] \\
&\le 2(x(t) - \gamma(t))^2 \left[-\alpha + \frac{L_1}{\alpha - L_2} \sup_{t\in[0,\omega]} |g(t, 0)| \right. \\
&\quad \left. + L_2\right] = \frac{-\alpha^2 + 2\alpha L_2 - L_2^2 + L_1 \sup_{t\in[0,\omega]} |g(t, 0)|}{\alpha - L_2} \\
&\quad \cdot 2(x(t) - \gamma(t))^2
\end{aligned} \tag{18}$$

By (H_2) and (H_3), we have $(-\alpha^2 + 2\alpha L_2 - L_2^2 + L_1\sup_{t\in[0,\omega]}|g(t, 0)|)/(\alpha - L_2) < 0$; there is a positive number δ such that $(-\alpha^2 + 2\alpha L_2 - L_2^2 + L_1\sup_{t\in[0,\omega]}|g(t, 0)|)/(\alpha - L_2) \le -\delta < 0$, and hence we have

$$\frac{dV(t, x(t))}{dt} \le -2\delta (x(t) - \gamma(t))^2. \tag{19}$$

Therefore, the periodic solution $\gamma(t)$ is uniformly asymptotic stable. □

Theorem 3. *Consider (2); $f(t, x)$, $g(t, x)$ are both ω–periodic continuous functions on $t \in R$ and uniformly, with respect to $x \in R$, suppose that the following conditions hold:*

(H_1) $f(t, x) \ge \alpha > 0$;

(H_2)
$$\begin{aligned}
&|f(t, x) - f(t, y)| \le L_1 |x - y|, \\
&|g(t, x) - g(t, y)| \le L_2 |x - y|, \quad L_2 < \alpha;
\end{aligned} \tag{20}$$

(H_3) $\dfrac{\alpha L_2 + L_1 \sup_{t\in[0,\omega]} |g(t, 0)| - L_2^2}{\alpha(\alpha - L_2)} < 1;$

then

(1) *for (2) there exists a unique ω-periodic continuous solution γ(t), and*

$$\left|\gamma(t)\right| \le \frac{1}{\alpha - L_2}\sup_{t\in[0,\omega]}\left|g(t,0)\right|, \tag{21}$$

(2) *γ(t) is unstable.*

Proof. (1) Let

$$B = \left\{\varphi(t) \mid \varphi \in C(R,R),\ \varphi(t+\omega) = \varphi(t),\ \left|\varphi(t)\right| \le \frac{1}{\alpha - L_2}\sup_{t\in[0,\omega]}\left|g(t,0)\right|\right\}. \tag{22}$$

Given any $\varphi(t) \in B$, the norm is defined as follows:

$$\|\varphi\| = \sup_{t\in[0,\omega]}\left|\varphi(t)\right|; \tag{23}$$

thus $(B, \|\cdot\|)$ is a Banach space. Consider the following equation:

$$\frac{dx}{dt} = f[t,\varphi(t)]x + g[t,\varphi(t)]; \tag{24}$$

by (H_1), according to Lemma 1, (24) has a unique ω–periodic continuous solution as follows:

$$\eta(t) = -\int_t^{+\infty} e^{\int_s^t f[\tau,\varphi(\tau)]d\tau} g[s,\varphi(s)]\,ds \tag{25}$$

and

$$\begin{aligned}
\left|\eta(t)\right| &\le \int_t^{+\infty} e^{-\alpha(s-t)}\left|g[s,\varphi(s)]\right| ds \\
&\le \int_t^{+\infty} e^{-\alpha(s-t)}\left(L_2\left|\varphi(s)\right| + \left|g(s,0)\right|\right) ds \\
&\le \frac{L_2\left|\varphi\right|_M + \sup_{t\in[0,\omega]}\left|g(t,0)\right|}{\alpha} \\
&= \frac{1}{\alpha - L_2}\sup_{t\in[0,\omega]}\left|g(t,0)\right|
\end{aligned} \tag{26}$$

Define a map as follows:

$$(T\varphi)(t) = \eta(t) = -\int_t^{+\infty} e^{\int_s^t f[\tau,\varphi(\tau)]d\tau} g[s,\varphi(s)]\,ds; \tag{27}$$

thus if given any $\varphi(t) \in B$, then $(T\varphi)(t) \in B$; hence $T : B \longrightarrow B$; given any $\varphi(t), \psi(t) \in B$, we have

$$\begin{aligned}
&\left|(T\varphi)(t) - (T\psi)(t)\right| \\
&= \left|-\int_t^{+\infty} e^{\int_s^t f[\tau,\varphi(\tau)]d\tau} g[s,\varphi(s)]\,ds + \int_t^{+\infty} e^{\int_s^t f[\tau,\psi(\tau)]d\tau} g[s,\psi(s)]\,ds\right| \\
&= \left|\int_t^{+\infty} e^{\int_s^t f[\tau,\varphi(\tau)]d\tau}\left(g(s,\varphi(s)) - g(s,\psi(s))\right) ds + \int_t^{+\infty}\left(e^{\int_s^t f[\tau,\varphi(\tau)]d\tau} - e^{\int_s^t f[\tau,\psi(\tau)]d\tau}\right) \cdot g(s,\psi(s))\,ds\right| \\
&= \left|\int_t^{+\infty} e^{\int_s^t f[\tau,\varphi(\tau)]d\tau}\left(g(s,\varphi(s)) - g(s,\psi(s))\right) ds + \int_t^{+\infty} e^{\xi}\left(\int_s^t f[\tau,\varphi(\tau)]\,d\tau - \int_s^t f[\tau,\psi(\tau)]\,d\tau\right) \cdot g(s,\psi(s))\,ds\right|;
\end{aligned} \tag{28}$$

here, ξ is between $\int_s^t f[\tau,\varphi(\tau)]d\tau$ and $\int_s^t f[\tau,\psi(\tau)]d\tau$; thus $\xi \le -\alpha(s-t)$, so we have

$$\begin{aligned}
\left|(T\varphi)(t) - (T\psi)(t)\right| &\le L_2\int_t^{+\infty} e^{-\alpha(t-s)}\left|\varphi(s) - \psi(s)\right| ds + L_1\int_t^{+\infty} e^{-\alpha(t-s)}\int_s^t\left|\varphi(\tau) - \psi(\tau)\right| d\tau\left|g(s,\psi(s))\right| ds\ | \\
&\le L_2\int_t^{+\infty} e^{-\alpha(t-s)]d\tau}\left|\varphi(s) - \psi(s)\right| ds \\
&\quad + L_1\int_t^{+\infty} e^{-\alpha(t-s)}\int_s^t\left|\varphi(\tau) - \psi(\tau)\right| d\tau\left(L_2\left|\psi(s)\right| + \left|g(s,0)\right|\right) ds \\
&= \left(L_2\int_t^{+\infty} e^{-\alpha(t-s)}ds + L_1\int_t^{+\infty} e^{-\alpha(t-s)}(t-s)\left(L_2\left|\psi(s)\right| + \left|g(s,0)\right|\right) ds\right)\|\varphi - \psi\| \\
&\le \left(\frac{L_2}{\alpha} + \frac{L_1\left(L_2\left|\psi\right|_M + \sup_{t\in[0,\omega]}\left|g(t,0)\right|\right)}{\alpha^2}\right)\|\varphi - \psi\|
\end{aligned}$$

$$= \left(\frac{L_2}{\alpha} + \frac{L_1 \left(L_2 \left(1/\left(\alpha - L_2\right)\right) \sup_{t\in[0,\omega]} |g(t,0)| + \sup_{t\in[0,\omega]} |g(t,0)| \right)}{\alpha^2} \right) \|\varphi - \psi\|$$

$$= \left(\frac{L_2}{\alpha} + \frac{\alpha L_1 \sup_{t\in[0,\omega]} |g(t,0)| / \left(\alpha - L_2\right)}{\alpha^2} \right) \|\varphi - \psi\| = \frac{\alpha L_2 + L_1 \sup_{t\in[0,\omega]} |g(t,0)| - L_2^2}{\alpha \left(\alpha - L_2\right)} \|\varphi - \psi\| \tag{29}$$

By (H_2), (H_3), it follows that

$$0 < \frac{\alpha L_2 + L_1 \sup_{t\in[0,\omega]} |g(t,0)| - L_2^2}{\alpha \left(\alpha - L_2\right)} < 1 \tag{30}$$

by (30), T is a compression mapping; according to the compression mapping principle, T has a fixed point on B, the fixed point is the unique ω–periodic continuous solution $\gamma(t)$ of (2), and $|\gamma(t)| \le (1/(\alpha - L_2))\sup_{t\in[0,\omega]}|g(t,0)|$.

(2) Define a Lyapunov function as follows:

$$V(t, x(t)) = (x(t) - \gamma(t))^2, \tag{31}$$

where $x(t)$ is the unique solution with the initial value $x(t_0) = x_0$ of (2); differentiating both sides of (31) along the solution of (2), we get

$$\begin{aligned}
\frac{dV(t,x(t))}{dt} &= 2(x(t) - \gamma(t)) \left[\frac{dx(t)}{dt} - \frac{d\gamma(t)}{dt} \right] \\
&= 2(x(t) - \gamma(t)) [f(t,x(t)) x(t) \\
&\quad - f(t,\gamma(t)) \gamma(t) + (g(t,x(t)) - g(t,\gamma(t)))] \\
&= 2(x(t) - \gamma(t)) [f(t,x(t)) (x(t) - \gamma(t)) \\
&\quad + (f(t,x(t)) - f(t,\gamma(t))) \gamma(t) \\
&\quad + (g(t,x(t)) - g(t,\gamma(t)))] = 2f(t,x)(x(t) \\
&\quad - \gamma(t))^2 + 2(x(t) - \gamma(t))(f(t,x(t)) \\
&\quad - f(t,\gamma(t))) \gamma(t) + 2(x(t) - \gamma(t))(g(t,x(t)) \\
&\quad - g(t,\gamma(t))) \ge 2f(t,x(t))(x(t) - \gamma(t))^2 \\
&\quad - 2|x(t) - \gamma(t)| |f(t,x(t)) - f(t,\gamma(t))| |\gamma(t)| \\
&\quad - 2|x(t) - \gamma(t)| |g(t,x(t)) - g(t,\gamma(t))| \\
&\ge 2(x(t) - \gamma(t)) [f(t,x(t))(x(t) - \gamma(t)) \\
&\quad - L_1(x(t) - \gamma(t)) |\gamma(t)| - L_2(x(t) - \gamma(t))] \\
&= 2(x(t) - \gamma(t))^2 [f(t,x(t)) - L_1 |\gamma(t)| - L_2] \\
&\ge 2(x(t) - \gamma(t))^2 \left[\alpha - \frac{L_1}{\alpha - L_2} \sup_{t\in[0,\omega]} |g(t,0)| \right. \\
&\quad \left. - L_2 \right] = \frac{\alpha^2 - 2\alpha L_2 - L_1 \sup_{t\in[0,\omega]} |g(t,0)| + L_2^2}{\alpha - L_2} \\
&\quad \cdot 2(x(t) - \gamma(t))^2;
\end{aligned} \tag{32}$$

By (H_2) and (H_3), we have $(\alpha^2 - 2\alpha L_2 - L_1 \sup_{t\in[0,\omega]}|g(t,0)| + L_2^2)/(\alpha - L_2) > 0$; there is a positive number δ such that $(\alpha^2 - 2\alpha L_2 - L_1 \sup_{t\in[0,\omega]}|g(t,0)| + L_2^2)/(\alpha - L_2) \ge \delta > 0$, and hence we have

$$\frac{dV(t,x(t))}{dt} \ge 2\delta (x(t) - \gamma(t))^2 \tag{33}$$

Therefore, the periodic solution $\gamma(t)$ is unstable.

Consider the following nonlinear differential equation:

$$\frac{dx}{dt} = f(t,x) x + g(t); \tag{34}$$

it is easy for us to draw the two corollaries about the existence and stability of the periodic solution on (34). □

Corollary 4. *Consider (34); $f(t,x)$ is ω–periodic continuous functions on $t \in R$ and uniformly with respect to $x \in R$, $g(t)$ is an ω–periodic continuous function on $t \in R$; suppose that the following conditions hold:*

$$\begin{aligned}
&(H_1) \quad f(t,x) \le -\alpha < 0; \\
&(H_2) \quad |f(t,x) - f(t,y)| \le L|x - y|; \\
&(H_3) \quad L|g|_M < \alpha^2;
\end{aligned} \tag{35}$$

then

(1) *for (34) there exists a unique ω-periodic continuous solution $\gamma(t)$, and*

$$|\gamma(t)| \le \frac{|g|_M}{\alpha}, \tag{36}$$

(2) *$\gamma(t)$ is uniformly asymptotic stable.*

Corollary 5. *Consider (34); $f(t,x)$ is ω–periodic continuous functions on $t \in R$ and uniformly with respect to $x \in R$, $g(t)$ is* *an ω–periodic continuous function on $t \in R$, suppose that the following conditions hold:*

$$\begin{aligned} &(H_1) \quad f(t,x) \geq \alpha > 0; \\ &(H_2) \quad |f(t,x) - f(t,y)| \leq L|x-y|; \\ &(H_3) \quad L|g|_M < \alpha^2; \end{aligned} \tag{37}$$

then

(1) *for (34) there exists a unique ω-periodic continuous solution $\gamma(t)$, and*

$$|\gamma(t)| \leq \frac{|g|_M}{\alpha}, \tag{38}$$

(2) *$\gamma(t)$ is unstable.*

4. Periodic Solution on Abel's Differential Equation

Theorem 6. *Consider (1); $a(t), b(t), c(t), d(t)$ are all ω–periodic continuous functions; suppose that the following conditions hold:*

$$\begin{aligned} &(H_1) \quad a(t) < 0; \\ &(H_2) \quad b^2(t) - 3a(t)c(t) < 0; \\ &(H_3) \quad -2a_L|d|_M^2 + \alpha|b|_M|d|_M - \alpha^3 < 0; \end{aligned} \tag{39}$$

here, $\alpha = ((b^2 - 4ac)/4a)_L$; then

(1) *for (1) there exists a unique ω-periodic continuous solution $\gamma(t)$, and*

$$|\gamma(t)| \leq \frac{|d|_M}{\alpha}, \tag{40}$$

(2) *$\gamma(t)$ is uniformly asymptotic stable.*

Proof. (1) Define a set

$$B = \left\{\varphi(t) \mid \varphi \in C(R,R),\ \varphi(t+\omega) = \varphi(t),\ |\varphi(t)| \leq \frac{|d|_M}{\alpha}\right\}; \tag{41}$$

given any $\varphi(t) \in B$, the norm is defined as follows:

$$\|\varphi\| = \sup_{t\in[0,\omega]} |\varphi(t)|; \tag{42}$$

thus $(B, \|\cdot\|)$ is a Banach space. Consider the following equation:

$$\frac{dx}{dt} = \left[a(t)\varphi^2(t) + b(t)\varphi(t) + c(t)\right]x + d(t) \tag{43}$$

By (H_1) and (H_2), we get that

$$b^2(t) - 4a(t)c(t) < 0 \tag{44}$$

and

$$\begin{aligned} a(t)\varphi^2(t) + b(t)\varphi(t) + c(t) &\leq \frac{4a(t)c(t) - b^2(t)}{4a(t)} \\ &\leq \left(\frac{4ac-b^2}{4a}\right)_M = -\left(\frac{b^2-4ac}{4a}\right)_L = -\alpha < 0; \end{aligned} \tag{45}$$

by Lemma 1, (43) has a unique continuous ω–periodic solution as follows:

$$\eta(t) = \int_{-\infty}^{t} e^{\int_s^t [a(\tau)\varphi^2(\tau)+b(\tau)\varphi(\tau)+c(\tau)]d\tau} d(s)\,ds; \tag{46}$$

by (46), we get

$$|\eta(t)| \leq \frac{|d|_M}{\alpha}. \tag{47}$$

Define a map as follows:

$$\begin{aligned} (T\varphi)(t) &= \eta(t) \\ &= \int_{-\infty}^{t} e^{\int_s^t [a(\tau)\varphi^2(\tau)+b(\tau)\varphi(\tau)+c(\tau)]d\tau} d(s)\,ds; \end{aligned} \tag{48}$$

thus if given any $\varphi(t) \in B$, then $(T\varphi)(t) \in B$; hence $T : B \longrightarrow B$; given any $\varphi(t), \psi(t) \in B$, we have

$$\begin{aligned} |(T\varphi)(t) - (T\psi)(t)| &= \left|\int_{-\infty}^{t} \left\{e^{\int_s^t [a(\tau)\varphi^2(\tau)+b(\tau)\varphi(\tau)+c(\tau)]d\tau} - e^{\int_s^t [a(\tau)\psi^2(\tau)+b(\tau)\psi(\tau)+c(\tau)]d\tau}\right\} d(s)\,ds\right| \\ &= \left|\int_{-\infty}^{t} e^{\xi} \int_s^t \left\{a(\tau)\left[\varphi^2(\tau) - \psi^2(\tau)\right] + b(\tau)\left[\varphi(\tau) - \psi(\tau)\right]\right\} d\tau\, d(s)\,ds\right| \\ &= \left|\int_{-\infty}^{t} e^{\xi} \int_s^t \left[a(\tau)(\varphi(\tau) + \psi(\tau)) + b(\tau)\right]\left[\varphi(\tau) - \psi(\tau)\right] d\tau\, d(s)\,ds\right|; \end{aligned} \tag{49}$$

here, ξ is between $\int_s^t [a(\tau)\varphi^2(\tau) + b(\tau)\varphi(\tau) + c(\tau)]d\tau$ and $\int_s^t [a(\tau)\psi^2(\tau) + b(\tau)\psi(\tau) + c(\tau)]d\tau$; thus $\xi \leq -\alpha(t-s)$, so we have

$$\begin{aligned} &|(T\varphi)(t) - (T\psi)(t)| \\ &\leq \int_{-\infty}^t e^{-\alpha(t-s)} \int_s^t |a(\tau)(\varphi(\tau) + \psi(\tau)) + b(\tau)| d\tau \\ &\cdot |d(s)| ds \|\varphi - \psi\| \\ &\leq \int_{-\infty}^t e^{-\alpha(t-s)} \int_s^t (2|a|_M |\varphi|_M + |b|_M) d\tau \\ &\cdot |d|_M ds \|\varphi - \psi\| \\ &= \int_{-\infty}^t e^{-\alpha(t-s)} (2|a|_M |\varphi|_M + |b|_M)(t-s) \\ &\cdot |d|_M ds \|\varphi - \psi\| \\ &= \frac{(2|a|_M |\varphi|_M + |b|_M)|d|_M}{\alpha^2} \|\varphi - \psi\| \\ &= \frac{(-2a_L |\varphi|_M + |b|_M)|d|_M}{\alpha^2} \|\varphi - \psi\| \\ &= \frac{(-2a_L(|d|_M/\alpha) + |b|_M)|d|_M}{\alpha^2} \|\varphi - \psi\|; \end{aligned} \tag{50}$$

therefore,

$$\|T\varphi - T\psi\| \leq \frac{(-2a_L|d|_M + \alpha|b|_M)|d|_M}{\alpha^3} \|\varphi - \psi\|. \tag{51}$$

By (H_3), it follows that

$$0 < \frac{(-2a_L|d|_M + \alpha|b|_M)|d|_M}{\alpha^3} < 1 \tag{52}$$

by (51) and (52), T is a compression mapping; according to the compression mapping principle, T has a fixed point on B, the fixed point is the unique ω–periodic continuous solution $\gamma(t)$ of (1), and $|\gamma(t)| \leq |d|_M/\alpha$.

(2) Define a Lyapunov function as follows:

$$V(t, x(t)) = (x(t) - \gamma(t))^2; \tag{53}$$

where $x(t)$ is the unique solution with the initial value $x(t_0) = x_0$ of (1); differentiating both sides of (53) along the solution of (1), we get

$$\begin{aligned} \frac{dV(t,x(t))}{dt} &= 2(x(t) - \gamma(t))\left[\frac{dx(t)}{dt} - \frac{d\gamma(t)}{dt}\right] \\ &= 2(x(t) - \gamma(t))\left[a(t)(x^3(t) - \gamma^3(t)) + b(t)(x^2(t) - \gamma^2(t)) + c(t)(x(t) - \gamma(t))\right] \\ &= 2(x(t) - \gamma(t))^2\left[a(t)(x^2(t) + x(t)\gamma(t) + \gamma^2(t)) + b(t)(x(t) + \gamma(t)) + c(t)\right] \\ &= 2(x(t) - \gamma(t))^2\left[a(t)x^2(t) + (a(t)\gamma(t) + b(t))x(t) + a(t)\gamma^2(t) + b(t)\gamma(t) + c(t)\right] \\ &\leq 2(x(t) - \gamma(t))^2\left[\frac{4a(t)(a(t)\gamma^2(t) + b(t)\gamma(t) + c(t)) - (a(t)\gamma(t) + b(t))^2}{4a(t)}\right] \\ &= 2(x(t) - \gamma(t))^2\left[\frac{3a^2(t)\gamma^2(t) + 2a(t)b(t)\gamma(t) + 4a(t)c(t) - b^2(t)}{4a(t)}\right] \\ &\leq \frac{(x(t) - \gamma(t))^2}{2a(t)}\left[\frac{12a^2(t)(4a(t)c(t) - b^2(t)) - 4a^2(t)b^2(t)}{12a^2(t)}\right] = \frac{2(3a(t)c(t) - b^2(t)))}{3a(t)}(x(t) - \gamma(t))^2 \end{aligned} \tag{54}$$

By (H_2), there is a positive number δ such that $3a(t)c(t) - b^2(t) \geq \delta > 0$, and hence we have

$$\frac{dV(t,x(t))}{dt} \leq \frac{2\delta}{3a_L}(x(t) - \gamma(t))^2. \tag{55}$$

Therefore, the periodic solution $\gamma(t)$ of (1) is uniformly asymptotic stable. □

Theorem 7. *Consider (1); $a(t), b(t), c(t), d(t)$ are all ω–periodic continuous functions; suppose that the following conditions hold:*

$$\begin{aligned} &(H_1) \quad a(t) > 0; \\ &(H_2) \quad b^2(t) - 3a(t)c(t) < 0; \\ &(H_3) \quad 2a_M|d|_M^2 + \alpha|b|_M|d|_M - \alpha^3 < 0, \end{aligned} \tag{56}$$

where $\alpha = ((4ac - b^2)/4a)_L$; then

(1) *for (1) there exists a unique ω-periodic continuous solution $\gamma(t)$, and*

$$|\gamma(t)| \le \frac{|d|_M}{\alpha}, \quad (57)$$

(2) *$\gamma(t)$ is unstable.*

Proof. (1) Define a set

$$B = \left\{\varphi(t) \mid \varphi \in C(R,R),\ \varphi(t+\omega) = \varphi(t),\ |\varphi(t)| \le \frac{|d|_M}{\alpha}\right\}; \quad (58)$$

given any $\varphi(t) \in B$, the norm is defined as follows:

$$\|\varphi\| = \sup_{t\in[0,\omega]} |\varphi(t)|; \quad (59)$$

thus $(B, \|\cdot\|)$ is a Banach space. Consider the following equation:

$$\frac{dx}{dt} = \left[a(t)\varphi^2(t) + b(t)\varphi(t) + c(t)\right]x + d(t) \quad (60)$$

By (H_1) and (H_2), we get that

$$b^2(t) - 4a(t)c(t) < 0 \quad (61)$$

and

$$a(t)\varphi^2(t) + b(t)\varphi(t) + c(t) \ge \frac{4a(t)c(t) - b^2(t)}{4a(t)} \ge \left(\frac{4ac-b^2}{4a}\right)_L = \alpha > 0. \quad (62)$$

by Lemma 1, (60) has a unique ω–periodic continuous solution as follows:

$$\eta(t) = -\int_t^{+\infty} e^{\int_s^t [a(\tau)\varphi^2(\tau)+b(\tau)\varphi(\tau)+c(\tau)]d\tau} d(s)\,ds \quad (63)$$

and

$$|\eta(t)| \le \int_t^{+\infty} e^{-\alpha(s-t)} |d(s)|\,ds \le \int_t^{+\infty} e^{-\alpha(s-t)} |d|_M\,ds = \frac{|d|_M}{\alpha}. \quad (64)$$

Define a map as follows:

$$(T\varphi)(t) = \eta(t) = -\int_t^{+\infty} e^{\int_s^t [a(\tau)\varphi^2(\tau)+b(\tau)\varphi(\tau)+c(\tau)]d\tau} d(s)\,ds; \quad (65)$$

thus if given any $\varphi(t) \in B$, then $(T\varphi)(t) \in B$; hence $T : B \longrightarrow B$; given any $\varphi(t), \psi(t) \in B$, we have

$$\begin{aligned}
|(T\varphi)(t) - (T\psi)(t)| &= \left|\int_t^{+\infty} \left\{e^{\int_s^t [a(\tau)\psi^2(\tau)+b(\tau)\psi(\tau)+c(\tau)]d\tau} - e^{\int_s^t [a(\tau)\varphi^2(\tau)+b(\tau)\varphi(\tau)+c(\tau)]d\tau}\right\} d(s)\,ds\right| \\
&= \left|\int_t^{+\infty} e^{\xi} \int_s^t \left\{a(\tau)\left[\psi^2(\tau) - \varphi^2(\tau)\right] + b(\tau)\left[\psi(\tau) - \varphi(\tau)\right]\right\} d\tau\, d(s)\,ds\right| \\
&= \left|\int_t^{+\infty} e^{\xi} \int_s^t \left[a(\tau)(\varphi(\tau) + \psi(\tau)) + b(\tau)\right]\left[\psi(\tau) - \varphi(\tau)\right] d\tau\, d(s)\,ds\right|,
\end{aligned} \quad (66)$$

where ξ is between $\int_s^t [a(\tau)\varphi^2(\tau) + b(\tau)\varphi(\tau) + c(\tau)]d\tau$ and $\int_s^t [a(\tau)\psi^2(\tau) + b(\tau)\psi(\tau) + c(\tau)]d\tau$; thus $\xi \le -\alpha(s-t)$, so we have

$$\begin{aligned}
|(T\varphi)(t) - (T\psi)(t)| &\le \int_t^{+\infty} e^{-\alpha(s-t)} \cdot \int_s^t |a(\tau)(\varphi(\tau) + \psi(\tau)) + b(\tau)|\, d\tau\, |d(s)|\, ds \cdot \|\varphi - \psi\| \\
&\le \int_t^{+\infty} e^{-\alpha(s-t)} \cdot \int_s^t (2|a|_M|\varphi|_M + |b|_M)\, d\tau\, |d|_M\, ds\, \|\varphi - \psi\| \\
&= \int_t^{+\infty} e^{-\alpha(s-t)} (2a_M|\varphi|_M + |b|_M)(t-s)|d|_M\, ds \cdot \|\varphi - \psi\| = \frac{(2a_M|\varphi|_M + |b|_M)|d|_M}{\alpha^2}\|\varphi - \psi\| \\
&= \frac{(2a_M(|d|_M/\alpha) + |b|_M)|d|_M}{\alpha^2}\|\varphi - \psi\|;
\end{aligned} \quad (67)$$

therefore,

$$\|T\varphi - T\psi\| \le \frac{(2a_M(|d|_M/\alpha) + |b|_M)|d|_M}{\alpha^2}\|\varphi - \psi\|. \quad (68)$$

By (H_3), it follows that

$$0 < \frac{(2a_M|d|_M + \alpha|b|_M)|d|_M}{\alpha^3} < 1 \quad (69)$$

by (68) and (69), T is a compression mapping; according to the compression mapping principle, T has a fixed point on B, the fixed point is the unique ω−periodic continuous solution $\gamma(t)$ of (1), and $|\gamma(t)| \le |d|_M/\alpha$.

(2) Define a Lyapunov function as follows:

$$V(t, x(t)) = (x(t) - \gamma(t))^2; \tag{70}$$

where $x(t)$ is the unique solution with the initial value $x(t_0) = x_0$ of (1); differentiating both sides of (70) along the solution of (1), we get

$$\begin{aligned}
\frac{dV(t,x(t))}{dt} &= 2(x(t)-\gamma(t))\left[\frac{dx(t)}{dt} - \frac{d\gamma(t)}{dt}\right] \\
&= 2(x(t)-\gamma(t))\left[a(t)\left(x^3(t)-\gamma^3(t)\right) + b(t)\left(x^2(t)-\gamma^2(t)\right) + c(t)(x(t)-\gamma(t))\right] \\
&= 2(x(t)-\gamma(t))^2\left[a(t)\left(x^2(t)+x(t)\gamma(t)+\gamma^2(t)\right) + b(t)(x(t)+\gamma(t)) + c(t)\right] \\
&= 2(x(t)-\gamma(t))^2\left[a(t)x^2(t) + (a(t)\gamma(t)+b(t))x(t) + a(t)\gamma^2(t) + b(t)\gamma(t) + c(t)\right] \\
&\ge 2(x(t)-\gamma(t))^2\left[\frac{4a(t)\left(a(t)\gamma^2(t)+b(t)\gamma(t)+c(t)\right) - (a(t)\gamma(t)+b(t))^2}{4a(t)}\right] \\
&= 2(x(t)-\gamma(t))^2\left[\frac{3a^2(t)\gamma^2(t) + 2a(t)b(t)\gamma(t) + 4a(t)c(t) - b^2(t)}{4a(t)}\right] \\
&\ge \frac{(x(t)-\gamma(t))^2}{2a(t)}\left[\frac{12a^2(t)\left(4a(t)c(t)-b^2(t)\right) - 4a^2(t)b^2(t)}{12a^2(t)}\right] = \frac{2\left(3a(t)c(t)-b^2(t)\right)}{3a(t)}(x(t)-\gamma(t))^2
\end{aligned} \tag{71}$$

By (H_2), there is a positive number δ such that $3a(t)c(t) - b^2(t) \ge \delta > 0$, and hence we have

$$\frac{dV(t,x(t))}{dt} \ge \frac{2\delta}{3a_M}(x(t)-\gamma(t))^2. \tag{72}$$

Therefore, the periodic solution $\gamma(t)$ of (1) is unstable. □

5. Examples

The following examples show the feasibility of our main results.

Example 1. Consider the following equation:

$$\frac{dx}{dt} = (-10 + 2\sin x + \sin t)x + 4 + 2\cos x + \cos t \tag{73}$$

It is easy to calculate that $\alpha = 7, L_1 = 2, L_2 = 2, \sup_{t\in[0,2\pi]} g(t,0) = 7$, and

$$\begin{aligned}
|f(t,x) - f(t,y)| &\le 2|x-y|, \\
|g(t,x) - g(t,y)| &\le 2|x-y|, \\
L_2 &= 2 < \alpha;
\end{aligned} \tag{74}$$

$$\frac{\alpha L_2 + L_1 \sup_{t\in[0,\omega]}|g(t,0)| - L_2^2}{\alpha(\alpha - L_2)} = \frac{24}{35} < 1.$$

Clearly, conditions $(H_1), (H_2), (H_3)$ of Theorem 2 are satisfied. It follows from Theorem 2 that system (73) has a unique 2π−periodic solution $\gamma(t)$, which is uniformly asymptotic stable. Clearly, by the graph of solution curve, given any initial value $x(0) = x_0$(e.g., $x(0) = 0$), the solution curve of (73) tends to the curve of the periodic solution $\gamma(t)$ (see Figure 1).

Example 2. Consider the following equation:

$$\begin{aligned}
\frac{dx}{dt} &= (-2 + \sin t)x^3 + (1 + \cos t)x^2 - (8 + \cos t)x \\
&\quad + 2 + \sin t
\end{aligned} \tag{75}$$

It is easy to calculate that $b^2(t) - 3a(t)c(t) < 0, \alpha = ((b^2 - 4ac)/4a)_L = 7, a_L = -3, |d|_M = 3, |b|_M = 2$, and

$$-2a_L|d|_M^2 + \alpha|b|_M|d|_M - \alpha^3 = -247 < 0. \tag{76}$$

Clearly, conditions $(H_1), (H_2), (H_3)$ of Theorem 6 are satisfied. It follows from Theorem 6 that (75) has a unique 2π−periodic solution $\gamma(t)$, which is uniformly asymptotic stable.

Clearly, by the graph of solution curve, given any initial value $x(0) = x_0$(e.g., $x(0) = 0$), the solution curve of (75) tends to the curve of the periodic solution $\gamma(t)$ (see Figure 2).

6. Conclusion

Abel's differential equation plays an important role in many physical and technical applications; because of this, the paper also deals with Abel's differential equation; we first give two results about the existence and uniqueness of the periodic

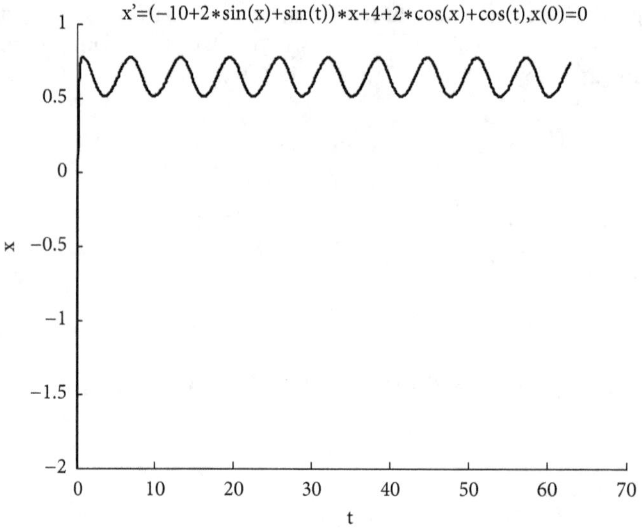

FIGURE 1: The solution curve of (73) with initial value $x(0) = 0$.

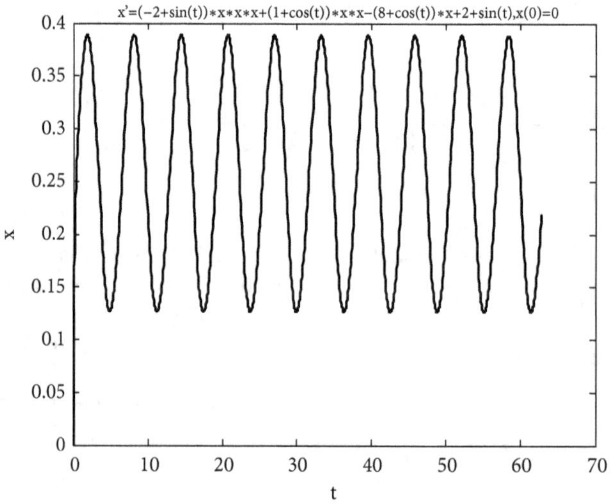

FIGURE 2: The solution curve of (75) with initial value $x(0) = 0$.

solution on a nonlinear differential equation by using the fixed point theory; then, we get the existence of the periodic solution of Abel's differential equation, the results having great application value.

Conflicts of Interest

The author declares that there are no conflicts of interest.

Acknowledgments

This work is supported by special project supported by senior personnel of Jiangsu University (14JDG176).

References

[1] J. M. Lebrun, *On Two Coupled Abel-Type Differential Equations Arising in A Magnetostatic Problem*, vol. 103, II Nuovo Cimento, 1990.

[2] M. K. Mak and T. Harko, "Exact causal viscous cosmologies," *General Relativity and Gravitation*, vol. 30, no. 8, pp. 1171–1186, 1998.

[3] Y. Matsuno, "Two-dimensional dynamical system associated with Abel's nonlinear differential equation," *Journal of Mathematical Physics*, vol. 33, no. 1, pp. 412–421, 1992.

[4] G. L. Strobel and J. L. Reid, "Nonlinear superposition rule for Abel's equation," *Physics Letters A*, vol. 91, no. 5, pp. 209-210, 1982.

[5] J. L. Reid and G. L. Strobel, "The nonlinear superposition theorem of Lie and Abel's differential equations," *Lettere al Nuovo Cimento. Rivista Internazionale della Società Italiana di Fisica. Serie 2*, vol. 38, no. 13, pp. 448–452, 1983.

[6] M. K. Mak, H. W. Chan, and T. Harko, "Solutions generating technique for Abel-type nonlinear ordinary differential equations," *Computers & Mathematics with Applications*, vol. 41, no. 10-11, pp. 1395–1401, 2001.

[7] M. K. Mak and T. Harko, "New method for generating general solution of Abel differential equation," *Computers & Mathematics with Applications*, vol. 43, no. 1-2, pp. 91–94, 2002.

[8] A. Cima, A. Gasull, and F. Mañosas, "On the number of polynomial solutions of Bernoulli and Abel polynomial differential equations," *Journal of Differential Equations*, vol. 263, no. 11, pp. 7099–7122, 2017.

[9] J. Ginï and C. Valls, "Nondegenerate centers for Abel polynomial differential equations of second kind," *Journal of Computational and Applied Mathematics*, vol. 321, pp. 469–477, 2017.

[10] J. Huang and H. Liang, "Estimate for the number of limit cycles of Abel equation via a geometric criterion on three curves," *Nonlinear Differential Equations and Applications NoDEA*, vol. 24, no. 4, Art. 47, 31 pages, 2017.

[11] B. Bïlbïl and M. Sezer, "A numerical approach for solving generalized Abel-type nonlinear differential equations," *Applied Mathematics and Computation*, vol. 262, no. 1, pp. 169–177, 2015.

[12] H. Ni, L. Tian, and H. Zhang, "The existence and stability of the periodic solutions on Abel differential equation," *Mathematica Applicata*, vol. 25, no. 4, pp. 854–862, 2012.

14

Periodic Travelling Wave Solutions of Discrete Nonlinear Schrödinger Equations

Dirk Hennig

Erich-Steinfurth-Str. 10/11, 16227 Eberswalde, Germany

Correspondence should be addressed to Dirk Hennig; hennigd@physik.hu-berlin.de

Academic Editor: Erik Van Vleck

The existence of nonzero periodic travelling wave solutions for a general discrete nonlinear Schrödinger equation (DNLS) on one-dimensional lattices is proved. The DNLS features a general nonlinear term and variable range of interactions going beyond the usual nearest-neighbour interaction. The problem of the existence of travelling wave solutions is converted into a fixed point problem for an operator on some appropriate function space which is solved by means of Schauder's Fixed Point Theorem.

1. Introduction

Coherent structures arising in the form of travelling waves, solitons, and breathers in systems of coupled oscillators have attracted considerable interest not least due to the important role they play for applications in physics, biology, and chemistry (for reviews see [1–3]). In this context a variety of nonlinear lattice systems has been studied including Fermi-Pasta-Ulam systems, discrete nonlinear Klein-Gordon systems, phase oscillator lattices, Josephson junction systems, reaction-diffusion systems, and the discrete nonlinear Schrödinger equation. Some exact results concerning the existence, stability, and uniqueness of coherent structures in the above-mentioned systems have been obtained; see, for example, [4–27].

In particular with regard to the existence of periodic travelling waves (TWs) in nonlinear lattice systems various methods have been used. For instance, the existence of small amplitude waves in nonlinear discrete Klein-Gordon systems was proved with the usage of spatial dynamics and centre manifold reduction [15, 21]. For the Frenkel-Kontorova model the existence of TWs was shown by means of fixed point methods in [16]. Utilising a modified Lyapunov-Schmidt technique, existence of periodic TWs in Newton's cradle was proved in [25]. Systems with nonlocal interactions were considered in [19, 24, 28].

In [24] existence and bifurcation results for periodic TWs of a general infinite DNLS system as given in (1) in the next section were derived using variational methods. With the current study we present a (less involved) proof of the existence of periodic TWs for the same system but on finite lattices. To obtain our existence result, some appropriate function space is introduced on which the original problem is formulated as a fixed point problem for a corresponding operator. By exploiting Schauder's Fixed Point Theorem the existence of periodic TWs is established. The main advantage of considering finite lattices is that for periodic TW solutions the associated conserved norm (power), given in terms of a sum over the squares of the amplitudes at the lattice sites, is of finite value. Hereby the coercivity of the power can be used to define suitable subsets in function space in order to apply Schauder's Fixed Point Theorem. Nevertheless, imposing periodic boundary conditions an infinite lattice is constructed. The presented method can readily be applied to prove the existence of periodic TWs not only in general DNLS systems of higher dimensions but also in various other nonlinear lattice systems.

2. General DNLS Systems

In the current study we are interested in the existence of periodic TW solutions of the following general discrete nonlinear Schrödinger equation on finite one-dimensional lattices:

$$i\frac{d\psi_n}{dt} = \sum_{j=1}^{N_c} \kappa_j \left[\psi_{n+j} - 2\psi_n + \psi_{n-j}\right] + F\left(|\psi_n|^2\right)\psi_n, \quad 1 \le n \le N, \tag{1}$$

with $\psi_n \in \mathbb{C}$.

The solutions satisfy periodicity conditions:

$$\psi_{N+m}(t) = \psi_m(t), \tag{2}$$

for $m \in \mathbb{Z}$; namely, we consider the DNLS on rings. Moreover, by means of the periodic boundary conditions an infinite one-dimensional lattice can be obtained. Each unit interacts with its N_c neighbouring oscillators to the left and right, respectively. $N_c = 1, \dots, [(N-1)/2]$ determines the interaction radius, which ranges from nearest-neighbour interaction obtained for $N_c = 1$ to global coupling when $N_c = (N-1)/2$ and N is odd.

Assume the following condition on $F(|\psi_n|^2)$ holds.

(A) $F \in C(\mathbb{R}_+, \mathbb{R})$ for $\mathbb{R}_+ = [0, \infty)$, $F(0) = 0$. There are constants $a > 0$, $b > 0$ such that

$$|F(x)| < a\left(1 + x^b\right), \tag{3}$$

for any $x \ge 0$.

The standard DNLS, arising for $F(|\psi_n|^2) = |\psi_n|^2$ and $\kappa_1 \ne 0$, $\kappa_{j\ne 1} = 0$ in (1), is known to support periodic travelling wave solutions (see, e.g., [29]). As mentioned above the existence of periodic TWs in system (1) on the lattice $\mathbb{Z}$ was given in [24] using variational methods. Here we present a proof of the existence of periodic travelling wave solutions of (1) on finite lattices. Nevertheless, by imposing periodic boundary conditions an infinite lattice is constructed. For the existence statement we introduce some appropriate function space on which the original problem is converted into a fixed point problem for a corresponding operator. By means of Schauder's Fixed Point Theorem the existence of periodic TWs is established.

System (1) possesses two conserved quantities: the energy

$$\mathcal{H} = \sum_{n=1}^{N}\left[G\left(|\psi_n|^2\right) - \sum_{m\ne n}\kappa_m |\psi_m - \psi_n|^2\right], \qquad G(x) = \int_0^x f(x)\,dx, \tag{4}$$

and the power

$$\mathcal{P} = \sum_{n=1}^{N} |\psi_n|^2. \tag{5}$$

We consider travelling wave solutions of the form

$$\psi_n(t) = \Psi(kn - \omega t), \tag{6}$$

with a 2π-periodic function $\Psi(u)$, $u = kn - \omega t$, where $k \in (-\pi, \pi)$ and $\omega \in \mathbb{R} \setminus \{0\}$ are the wave parameters.

In order for a travelling wave solution to satisfy the periodicity conditions in (2) we adopt the lattice size accordingly. This means that for a given wavenumber $|k| = \pi q$ with rational $q = r/s$ and two relatively prime integers $r, s \in \mathbb{Z}_+ \setminus \{0\}$, $r < s$, the number of sites of the lattice, N, is supposed to be an appropriate multiple of the minimal spatial period of the associated periodic travelling wave, determined by $L = s/r$, so that the periodicity conditions in (2) are fulfilled.

3. Statement of the Existence Problem

Regarding the existence of periodic travelling wave solutions we state the following.

Theorem 1. *Let* (A) *hold. Then for any rational number* $q \in \mathbb{Q} \cap (0,1)$ *there exists nonzero periodic travelling wave solution* $\psi_n(t) = \Psi(kn - \omega t) \equiv \Psi(u)$ *of (1) with* $\Psi \in C^1(\mathbb{R}, \mathbb{C})$, *such that*

$$\Psi(u + 2\pi) = \Psi(u), \quad \forall u \in \mathbb{R}, \tag{7}$$

provided that

$$|\omega| \ge \mathcal{R}\left(1 + p\frac{a\left(1 + \mathcal{P}^b\right)}{\mathcal{R}(1+q) + 4\bar{\kappa} + a\left(1 + \mathcal{P}^b\right)}\right), \tag{8}$$

where

$$\bar{\kappa} = \sum_{j=1}^{N_c} \kappa_j, \tag{9}$$

$$p = \tilde{q} + \frac{1}{1-q}, \tag{10}$$

$$\tilde{q} = \begin{cases} \dfrac{1}{q} & \text{for } 0 < q < \dfrac{1}{2} \\ \dfrac{1}{1-q} & \text{for } \dfrac{1}{2} \le q < 1, \end{cases} \tag{11}$$

and $\mathcal{R}$ *determines the range* $[-\mathcal{R}, \mathcal{R}]$ *of the function* $g \in C(\mathbb{R} \setminus \{0\}, \mathbb{R})$ *given by*

$$g(x) = \frac{2}{x}\sum_{j=1}^{N_c} \kappa_j \sin^2(jx). \tag{12}$$

In the following we reformulate the original problem as a fixed point problem in a Banach space in a similar vein to the approach in [16].

4. Proof of the Existence Theorem

To prove the assertions of the theorem we utilise Schauder's Fixed Point Theorem (see, e.g., in [30]): let M be a closed convex subset of a Banach space X. Suppose $T: M \to M$ is continuous mapping such that $T(M)$ is a relatively compact subset of M. Then T has a fixed point.

Proof. Travelling wave solutions Ψ satisfy the advance-delay equation

$$-i\omega\Psi'(u) = \sum_{j=1}^{N_c} \kappa_j \Delta_j \Psi(u) + F\left(|\Psi(u)|^2\right)\Psi(u), \tag{13}$$

where $\Delta_j\Psi(u) = \Psi(u+j) - 2\Psi(u) + \Psi(u-j)$ and $\Psi(u+2\pi) = \Psi(u)$, $\forall u \in \mathbb{R}$, so that according to the Bloch-Floquet Theorem a solution must be of the form

$$\Psi(u) = \exp(iqu)\,\Phi(u), \tag{14}$$

where $q \in \mathbb{Q} \cap (0,1)$ and

$$\Phi(u+2\pi) = \Phi(u), \quad \forall u \in \mathbb{R}. \tag{15}$$

Substituting (14) into (13) one obtains

$$\begin{aligned}-i\omega\Phi'(u) + \omega q\Phi(u) &= \sum_{j=1}^{N_c} \kappa_j \tilde{\Delta}_j \Phi(u) \\ &\quad + F\left(|\Phi(u)|^2\right)\Phi(u),\end{aligned} \tag{16}$$

with

$$\begin{aligned}\tilde{\Delta}_j\Phi(u) &= \Phi(u+j)\exp(iqj) - 2\Phi(u) \\ &\quad + \Phi(u-j)\exp(-iqj).\end{aligned} \tag{17}$$

Thus, the task amounts to finding 2π-periodic functions $\Phi \in C^1(\mathbb{R}, \mathbb{C})$ satisfying (16).

For the forthcoming discussion (16) is suitably rearranged as follows:

$$\begin{aligned}-i\omega\Phi'(u) + \omega q\Phi(u) - \sum_{j=1}^{N_c} \kappa_j \tilde{\Delta}_j \Phi(u) \\ = F\left(|\Phi(u)|^2\right)\Phi(u).\end{aligned} \tag{18}$$

Note that terms nonlinear in Φ feature only on the r.h.s. of (18).

Let $q \in \mathbb{Q} \cap (0,1)$ be fixed. We identify $\mathbb{C}$ with $\mathbb{R}^2$. Denote by X_q^h the real Banach spaces

$$X_q^h = \left\{\Theta \in C_{2\pi}^h(\mathbb{R}, \mathbb{C})\right\}, \quad h = 0, 1, \tag{19}$$

where $C_\pi^h(\mathbb{R}, \mathbb{C})$ is the Banach space of 2π-periodic and C^h functions $\Theta : \mathbb{R} \to \mathbb{C}$ equipped with norms given by

$$\begin{aligned}\|\Theta\|_{C_{2\pi}^0} &= \max_{u\in[0,2\pi]} |\Theta(u)|, \quad \Theta \in C_{2\pi}^0(\mathbb{R}, \mathbb{C}), \\ \|\Theta\|_{C_{2\pi}^1} &= \max_{u\in[0,2\pi]} |\Theta(u)| + \max_{u\in(0,2\pi)} \left|\Theta'(u)\right|, \\ &\qquad \Theta \in C_{2\pi}^1(\mathbb{R}, \mathbb{C}),\end{aligned} \tag{20}$$

respectively. X_q^1 is compactly embedded in X_q^0 ($X_q^1 \Subset X_q^0$).

We decompose functions $\Theta \in X_q^1$ in a Fourier series

$$\Theta(u) = \sum_{l\in\mathbb{Z}} \Theta_l \exp(ilu). \tag{21}$$

Related to the l.h.s. of (18) we consider the linear mapping: $M_q : X_q^1 \to X_q^0$:

$$M_q(\Theta) = -i\omega\Theta'(u) + \omega q\Theta(u) - \sum_{j=1}^{N_c} \kappa_j \tilde{\Delta}_j \Theta(u). \tag{22}$$

We demonstrate that this mapping is invertible and get an upper bound for the norm of its inverse.

Applying the operator M_q to the Fourier elements $\exp(ilu)$ in (21) results in

$$M_q \exp(ilu) = \nu_l(q)\exp(ilu), \tag{23}$$

where

$$\nu_l(q) = \omega(q+l) + 4\sum_{j=1}^{N_c} \kappa_j \sin^2\left(\frac{q+l}{2}j\right). \tag{24}$$

By the assumption (8) one has $\nu_l(q) \neq 0$, $\forall l \in \mathbb{Z}$, so that the mapping M_q possesses an inverse obeying $M_q^{-1}\exp(ilu) = (1/\nu_l)\exp(ilu)$. For the bounded linear operator $M_q^{-1} : X_q^0 \to X_q^1$ one derives:

$$\begin{aligned}\left\|M_q^{-1}\right\|_{X_q^0,X_q^1} &\equiv \left\|M_q^{-1}\right\| = \sup_{0\neq\Theta\in X_q^0} \frac{\left\|M_q^{-1}\Theta\right\|_{X_q^1}}{\|\Theta\|_{X_q^0}} = \sup_{0\neq\Theta\in X_q^0} \frac{\left\|\sum_{l\in\mathbb{Z}}(1/\nu_l)\Theta_l\exp(ilu)\right\|_{X_q^1}}{\|\Theta\|_{X_q^0}} \\ &= \sup_{0\neq\Theta\in X_q^0} \frac{\left(\sup_{u\in[0,2\pi]}\left|\sum_{l\in\mathbb{Z}}(1/\nu_l)\Theta_l\exp(ilu)\right| + \sup_{u\in[0,2\pi]}\left|\left(\sum_{l\in\mathbb{Z}}(1/\nu_l)\Theta_l\exp(ilu)\right)'\right|\right)}{\|\Theta\|_{X_q^0}} \\ &\le \sup_{l\in\mathbb{Z}} \frac{1+|l|}{|\nu_l|} \sup_{0\neq\Theta\in X_q^0} \frac{\sup_{u\in[0,2\pi]}\left|\sum_{l\in\mathbb{Z}}\Theta_l\exp(ilu)\right|}{\|\Theta\|_{X_q^0}} = \sup_{l\in\mathbb{Z}} \frac{1+|l|}{|\nu_l|}\end{aligned}$$

$$= \sup_{l\in\mathbb{Z}} \frac{1+|l|}{\left|(q+l)\left(\omega+(4/(q+l))\sum_{j=1}^{N_c}\kappa_j \sin^2(((q+l)/2)j)\right)\right|} \leq \left(\tilde{q}+\frac{1}{1-q}\right)\frac{1}{|\omega|-\mathcal{R}}$$
$$\leq \frac{(1+q)\mathcal{R}+4\overline{\kappa}+a\left(1+\mathcal{P}^b\right)}{a\left(1+\mathcal{P}^b\right)\mathcal{R}}, \tag{25}$$

where $\tilde{q}$ is given in (11).

For periodic travelling wave solutions $\Phi \in C^1_{2\pi}(\mathbb{R},\mathbb{C})$ one derives, using (3), (5), (8), and (16), the bounds

$$\max_{u\in[0,2\pi]} |\Phi(u)| \leq \mathcal{P}^{1/2},$$
$$\max_{u\in(0,2\pi)} |\Phi'(u)| \leq \left(q+\frac{4\overline{\kappa}+a\left(1+\mathcal{P}^b\right)}{\mathcal{R}}\right)\mathcal{P}^{1/2}. \tag{26}$$

We consider then the closed and convex subsets of X^0_q and X^1_q determined by

$$Y^0_q = \left\{\Theta \in X^0_q : \|\Theta\|_{C^0_{2\pi}} \leq \mathcal{P}^{1/2}\right\},$$
$$Y^1_q = \left\{\Theta \in X^1_q : \|\Theta\|_{C^1_{2\pi}} \leq \left(1+q+\frac{4\overline{\kappa}+a\left(1+\mathcal{P}^b\right)}{\mathcal{R}}\right)\mathcal{P}^{1/2}\right\}, \tag{27}$$

respectively. Y^1_q is compactly embedded in Y^0_q ($Y^1_q \Subset Y^0_q$).

Furthermore associated with the r.h.s. of (18) we introduce the nonlinear operator $N_q : Y^0_q \to Y^0_q$, as

$$N_q(\Theta) = F\left(|\Theta|^2\right)\Theta. \tag{28}$$

Clearly, the operator N_q is uniformly continuous on Y^0_q. The range is contained in a bounded ball in Y^0_q, since

$$\left\|N_q(\Theta)\right\|_{Y^0_q} = \left\|F\left(|\Theta|^2\right)\Theta\right\|_{C^0_{2\pi}}$$
$$= \max_{u\in[0,2\pi]} \left|F\left(|\Theta(u)|^2\right)\Theta(u)\right| \tag{29}$$
$$\leq a\left(1+\mathcal{P}^b\right)\mathcal{P}^{1/2}.$$

Finally, we express problem (18) as a fixed point equation in terms of a mapping $Y^0_q \to Y^1_q \Subset Y^0_q$:

$$\Phi = M_q^{-1} \circ N_q(\Phi) \equiv T_q(\Phi). \tag{30}$$

We get

$$\left\|T_q(\Phi)\right\|_{Y^1_q} = \left\|M_q^{-1}\left(N_q(\Phi)\right)\right\| \leq \left\|M_q^{-1}\right\|\left\|N_q(\Phi)\right\|_{Y^0_q}$$
$$\leq \left(1+q+\frac{4\overline{\kappa}+a\left(1+\mathcal{P}^b\right)}{\mathcal{R}}\right)\mathcal{P}^{1/2}, \tag{31}$$

verifying that indeed

$$T_q\left(Y^0_q\right) \subseteq Y^1_q. \tag{32}$$

Hence T_q maps bounded subsets Y^0_q of X^0_q into relatively compact subsets Y^1_q of Y^0_q.

It remains to prove that T_q is continuous on Y^0_q. As N is uniformly continuous on Y^0_q, one has $\forall t \in [0,2\pi]$ and $\forall \Phi_1,\Phi_2 \in Y^0_q$ that for a fixed arbitrary $\epsilon > 0$ there exists $\delta > 0$ such that

$$\left\|N_q(\Phi_1)(t)-N_q(\Phi_2)(t)\right\|_{Y^0_q}$$
$$< \frac{a\left(1+\mathcal{P}^b\right)\mathcal{R}}{(1+q)\mathcal{R}+4\overline{\kappa}+a\left(1+\mathcal{P}^b\right)}\epsilon \tag{33}$$

if $\|\Phi_1-\Phi_2\|_{Y^0_q} < \delta$. Hence, for arbitrary $\Phi_1,\Phi_2 \in Y^0_q$, we have

$$\left\|M_q^{-1}\left(N_q(\Phi_1)\right)-M_q^{-1}\left(N_q(\Phi_2)\right)\right\|_{Y^1_q\Subset Y^0_q}$$
$$\leq \left\|M_q^{-1}\right\|\left\|N_q(\Phi_1)-N_q(\Phi_2)\right\|_{Y^0_q} < \epsilon, \tag{34}$$

verifying that $T_q(\Phi)$ is continuous on Y^0_q. Schauder's Fixed Point Theorem implies then that the fixed point equation $\Phi = T_q(\Phi)$ has at least one solution.

Furthermore, the spectrum of linear plane wave solutions (phonons) arising for zero nonlinear term, determined by the r.h.s. of system (18), forms a continuous band with values in the interval $[-\mathcal{R},\mathcal{R}]$. However, since by hypothesis (8) the values of the frequency of oscillations ω lie outside the range of the linear (phonon) band, the corresponding orbits are anharmonic. This necessitates amplitude-depending tuning of the frequency so that the latter comes to lie outside of the phonon spectrum. Thus it must hold that $\|N_q(\Phi)\|_{Y^0_q} = \|F(|\Phi|^2)\Phi\|_{C^0_{2\pi}} \not\equiv 0$ which is fulfilled only if $\Phi \not\equiv 0$. That is, the fixed point equation (30) possesses only nonzero solutions and the proof is finished. □

5. Summary

To summarise, we have proven the existence of nonzero periodic travelling wave solutions for a general DNLS (including as a special case the standard DNLS) on finite one-dimensional lattices. To this end the existence problem has been reformulated as a fixed point problem for an operator on a function space which is solved with the help of Schauder's Fixed Point Theorem. Our method can be straightforwardly extended not only to treating the general DNLS on lattices of higher dimension but also to other types of nonlinear lattice systems such as nonlinear discrete Klein-Gordon systems.

Conflicts of Interest

The author declares that there are no conflicts of interest.

References

[1] O. M. Braun and Y. S. Kivshar, *The Frenkel-Kontorova Model: Concepts, Methods, and Applications*, Theoretical and Mathematical Physics, Springer, 2004.

[2] M. Remoissenet, *Waves Called Solitons*, Advanced Texts in Physics, Springer, Berlin, Germany, 1999.

[3] T. Dauxois and M. Peyrard, *Physics of Solitons*, Cambridge University Press, Cambridge, UK, 2006.

[4] B. Zinner, G. Harris, and W. Hudson, "Traveling wavefronts for the discrete Fisher's equation," *Journal of Differential Equations*, vol. 105, no. 1, pp. 46–62, 1993.

[5] N. Flytzanis, B. A. Malomed, and J. A. D. Wattis, "Analysis of stability of solitons in one-dimensional lattices," *Physics Letters A*, vol. 180, no. 1-2, pp. 107–112, 1993.

[6] G. Friesecke and J. A. Wattis, "Existence theorem for solitary waves on lattices," *Communications in Mathematical Physics*, vol. 161, no. 2, pp. 391–418, 1994.

[7] D. Smets and M. Willem, "Solitary waves with prescribed speed on infinite lattices," *Journal of Functional Analysis*, vol. 149, no. 1, pp. 266–275, 1997.

[8] S.-N. Chow, J. Mallet-Paret, and W. Shen, "Traveling waves in lattice dynamical systems," *Journal of Differential Equations*, vol. 149, no. 2, pp. 248–291, 1998.

[9] A. Pankov, *Travelling Waves and Periodic Oscillations in Fermi-Pasta-Ulam Lattices*, Imperial College Press, London, UK, 2005.

[10] D. B. Duncan, J. C. Eilbeck, H. Feddersen, and J. A. D. Wattis, "Solitons on lattices," *Physica D: Nonlinear Phenomena*, vol. 68, no. 1, pp. 1–11, 1993.

[11] J. Mallet-Paret, "The global structure of traveling waves in spatially discrete dynamical systems," *Journal of Dynamics and Differential Equations*, vol. 11, no. 1, pp. 49–127, 1999.

[12] G. Friesecke and R. L. Pego, "Solitary waves on FPU lattices. I. Qualitative properties, renormalization and continuum limit," *Nonlinearity*, vol. 12, no. 6, pp. 1601–1627, 1999.

[13] A. A. Pankov and K. Pflüger, "Travelling waves in lattice dynamical systems," *Mathematical Methods in the Applied Sciences*, vol. 23, no. 14, pp. 1223–1235, 2000.

[14] A. Carpio, S. J. Chapman, S. Hastings, and J. . McLeod, "Wave solutions for a discrete reaction-diffusion equation," *European Journal of Applied Mathematics*, vol. 11, no. 4, pp. 399–412, 2000.

[15] G. Iooss and K. Kirchgässner, "Travelling waves in a chain of coupled nonlinear oscillators," *Communications in Mathematical Physics*, vol. 211, no. 2, pp. 439–464, 2000.

[16] R. Mirollo and N. Rosen, "Existence, uniqueness, and nonuniqueness of single-wave-form solutions to Josephson junction systems," *SIAM Journal on Applied Mathematics*, vol. 60, no. 5, pp. 1471–1501, 2000.

[17] P. W. Bates, X. Chen, and A. J. Chmaj, "Traveling waves of bistable dynamics on a lattice," *SIAM Journal on Mathematical Analysis*, vol. 35, no. 2, pp. 520–546, 2003.

[18] D. E. Pelinovsky and V. M. Rothos, "Bifurcations of travelling wave solutions in the discrete NLS equations," *Physica D: Nonlinear Phenomena*, vol. 202, no. 1-2, pp. 16–36, 2005.

[19] P. W. Bates and C. Zhang, "Traveling pulses for the Klein-Gordon equation on a lattice or continuum with long-range interaction," *Discrete and Continuous Dynamical Systems - Series A*, vol. 16, no. 1, pp. 235–252, 2006.

[20] T. R. Melvin, A. R. Champneys, P. G. Kevrekidis, and J. Cuevas, "Travelling solitary waves in the discrete Schrödinger equation with saturable nonlinearity: existence, stability and dynamics," *Physica D: Nonlinear Phenomena*, vol. 237, no. 4, pp. 551–567, 2008.

[21] G. James and Y. Sire, "Center Manifold Theory in the Context of Infinite One-Dimensional Lattices," in *The Fermi-Pasta-Ulam Problem*, vol. 728 of *Lecture Notes in Physics*, pp. 207–238, 2008.

[22] X. Chen, J.-S. Guo, and C.-C. Wu, "Traveling waves in discrete periodic media for bistable dynamics," *Archive for Rational Mechanics and Analysis*, vol. 189, no. 2, pp. 189–236, 2008.

[23] J. Cuevas, N. I. Karachalios, and F. Palmero, "Energy thresholds for the existence of breather solutions and travelling waves on lattices," *Applicable Analysis: An International Journal*, vol. 89, no. 9, pp. 1351–1385, 2010.

[24] M. Fečkan and V. M. Rothos, "Travelling waves of discrete nonlinear Schrödinger equations with nonlocal interactions," *Applicable Analysis: An International Journal*, vol. 89, no. 9, pp. 1387–1411, 2010.

[25] G. James, "Nonlinear waves in Newton's cradle and the discrete p-Schrödinger equation," *Mathematical Models and Methods in Applied Sciences*, vol. 21, no. 11, pp. 2335–2377, 2011.

[26] M. Syafwan, H. Susanto, S. M. Cox, and B. A. Malomed, "Variational approximations for traveling solitons in a discrete nonlinear Schrödinger equation," *Journal of Physics A: Mathematical and Theoretical*, vol. 45, no. 7, Article ID 075207, 2012.

[27] J.-S. Guo and Y.-C. Lin, "Traveling wave solution for a lattice dynamical system with convolution type nonlinearity," *Discrete and Continuous Dynamical Systems - Series A*, vol. 32, no. 1, pp. 101–124, 2012.

[28] M. Fečkan, M. Pospíšil, V. M. Rothos, and H. Susanto, "Periodic Travelling Waves of Forced FPU Lattices," *Journal of Dynamics and Differential Equations*, vol. 25, no. 3, pp. 795–820, 2013.

[29] J. C. Eilbeck and M. Johansson, "The discrete nonlinear Schrödinger equation—20 years on," in *Proceedings of the Third Conference on Localization and Energy Transfer in Nonlinear Systems*, L. Vazquez, R. S. MacKay, and M.-P. Zorzano, Eds., pp. 44–67, San Lorenzo de El Escorial Madrid, Spain, June 2002.

[30] E. Zeidler, *Nonlinear Functional Analysis and Its Applications I: Fixed Point Theorems*, Nonlinear Functional Analysis & Its Applications, Springer, New York, NY, USA, 1998.

15

The Equivalent Linearization Method with a Weighted Averaging for Solving Undamped Nonlinear Oscillators

D. V. Hieu,[1] **N. Q. Hai,**[2] **and D. T. Hung**[1]

[1] *Thai Nguyen University of Technology, Thai Nguyen, Vietnam*
[2] *Hanoi Architectural University, Ha Noi, Vietnam*

Correspondence should be addressed to D. V. Hieu; hieudv@tnut.edu.vn

Academic Editor: Peter G. L. Leach

The Equivalent Linearization Method (ELM) with a weighted averaging is applied to analyze five undamped oscillator systems with nonlinearities. The results obtained via this method are compared with the ones achieved by Parameterized Perturbation Method (PPM), Min-Max Approach (MMA), Variational Iteration Method (VIM), Homotopy Perturbation Method (HPM), Energy Balance Method (EBM), Harmonic Balance Method (HBM), 4th-Order Runge-Kutta Method, and the exact ones. The obtained results demonstrate that this method is very convenient for solving nonlinear equations and also can be successfully applied to a lot of practical engineering and physical problems.

1. Introduction

Nonlinear oscillation problem is very important in the physical science, mechanical structures, and other kinds of mathematical sciences. Most of real systems are modeled by nonlinear differential equations which are important issues in mechanical structures, mathematical physics, and engineering. In most cases, it is difficult to solve such equations, especially analytically; and in addition, the most important information, such as the natural circular frequency of a nonlinear oscillation which depends on the initial conditions, will be lost during the procedure of numerical simulation.

Recently, nonlinear oscillator models have been widely considered in physics and engineering. It is obvious that there are many nonlinear equations in the study of different branches of science which do not have analytical solutions. Due to the limitation of existing exact solutions, many analytical approaches have been investigated. Many researchers have been working on various analytical methods for solving nonlinear oscillation systems in the last decades, such as Homotopy Perturbation Method (HPM) [1–5], Max–Min Approach (MMA) [5–10], Variational Iteration Method (VIM) [2, 5], Energy Balance Method (EBM) [5, 11–16], Amplitude-Frequency Formulation (AFF) [5–7, 15, 17, 18], Improved Amplitude-Frequency Formulation (IAFF) [5], Parameter Expansion Method (PEM) [5, 7, 18–20], Homotopy Analysis Method (HAM) [5, 21], Modified Homotopy Perturbation Method (MHPM) [5, 6], Modified Lindstedt-Poincare Method [22], Harmonic Balance Method [23, 24], and combined Newton's Method with the Harmonic Balance Method [25].

The Equivalent Linearization Method is one of the common approaches to approximate analysis of dynamical systems. The original linearization for deterministic systems was proposed by Krylov and Bogoliuboff [26]. Then Caughey expanded the method for stochastic systems [27]. Thenceforward, there have been some extended versions of the Equivalent Linearization Method [28–32]. It has been shown that the Equivalent Linearization Method is presently the simplest tool widely used for analyzing nonlinear stochastic problems. Nevertheless, the accuracy of the Equivalent Linearization Method with conventional averaging normally reduces for middle or strong nonlinear systems. A reason is that some terms will vanish in the averaging process; for example, the averaging value of the functions $\sin(t)$ and $\cos(t)$ over one period is equal to zero. Recently, Anh proposed a new way for determining averaging values: instead of using conventional averaging process the author introduced weighted coefficient functions $h(t)$ [31]; by this manner, the averaging value is calculated in a new way called the weighted averaging value.

This proposed method has been applied effectively to analyze some strongly nonlinear oscillations such as the nonlinear Duffing oscillator with third, fifth, and seventh powers of the amplitude, the strongly nonlinear oscillators in forms $(1 + \varepsilon u^2)\ddot{u} + u = 0$ and $\ddot{u} + u^3/(1 + u^2) = 0$, and the cubic Duffing with discontinuity [33].

In this paper, the Equivalent Linearization Method with a weighted averaging continues to be applied to analyze five undamped nonlinear oscillators: the Duffing oscillator with cubic nonlinearity (Example 1); the motion of simple pendulum attached to a rotating rigid frame (Example 2); the motion of a mass attached to the center of a stretched elastic wire (Example 3); the Duffing oscillator with discontinuity (Example 4); and the nonlinear oscillator with fractional elastic force (Example 5). It should be emphasized that the generalized forms of the Duffing equation (in Example 1) and the oscillator (in Example 3) have been solved by Energy Balance Method [12, 16], and the generalized form of the nonlinear oscillator with fractional elastic force (in Example 5) has been solved by Energy Balance Method and Variational Approach [14]; these solutions are achieved by Younesian et al. [12, 14, 16]. The frequency-amplitude relationship is analytically obtained by using this proposed method. Accuracy and validity of results are then presented by comparing the results with the ones obtained by the other well-known techniques and the exact and numerical ones.

2. The Equivalent Linearization Method with a Weighted Averaging

2.1. The Equivalent Linearization Method. In order to introduce the general idea of the Equivalent Linearization Method, we consider a nonlinear oscillator governed by the following equation [33]:

$$\ddot{X} + 2h\dot{X} + \omega_0^2 X + g(\dot{X}, X) = 0, \tag{1}$$

where $g(\dot{X}, X)$ is a nonlinear function only depending on two variables of velocity $\dot{X}(t)$ and displacement $X(t)$ and h and ω_0 are constants. The equivalent linear oscillator is described by the equation as follows:

$$\ddot{X} + (2h + \mu)\dot{X} + (\omega_0^2 + \lambda)X = 0. \tag{2}$$

The equation error between the two oscillators is taken as

$$e(\dot{X}, X) = g(\dot{X}, X) - \mu\dot{X} - \lambda X. \tag{3}$$

The coefficients of linearization in the linearized equation (2) are found from an optimal criterion. There are some criteria for determining these coefficients. The most common criterion is the mean square error criterion which requires that the mean square of equation error be minimum:

$$\langle e^2(\dot{X}, X)\rangle = \left\langle \left(g(\dot{X}, X) - \mu\dot{X} - \lambda X\right)^2 \right\rangle \longrightarrow \min_{\mu,\lambda}. \tag{4}$$

Thus,

$$\frac{\partial}{\partial\lambda}\langle e^2(\dot{X}, X)\rangle = 0$$
$$\frac{\partial}{\partial\mu}\langle e^2(\dot{X}, X)\rangle = 0 \tag{5}$$

yields

$$\lambda = \frac{\langle gX\rangle\langle \dot{X}^2\rangle - \langle g\dot{X}\rangle\langle X\dot{X}\rangle}{\langle X^2\rangle\langle \dot{X}^2\rangle - \langle X\dot{X}\rangle^2} \tag{6a}$$

$$\mu = \frac{\langle g\dot{X}\rangle\langle X^2\rangle - \langle gX\rangle\langle X\dot{X}\rangle}{\langle X^2\rangle\langle \dot{X}^2\rangle - \langle X\dot{X}\rangle^2}. \tag{6b}$$

In the formulations in (4), (6a), and (6b), the symbol $\langle\bullet\rangle$ denotes the time-averaging operator in classical meaning [26]:

$$\langle f(t)\rangle = \lim_{T\to+\infty}\frac{1}{T}\int_0^T f(t)\,dt. \tag{7}$$

For a ω-frequency function $f(\omega t)$, the averaging process is taken during one period T; that is,

$$\langle f(\omega t)\rangle = \frac{1}{T}\int_0^T f(\omega t)\,dt = \frac{1}{2\pi}\int_0^{2\pi} f(\tau)\,d\tau, \qquad \tau = \omega t. \tag{8}$$

In this technique, the importance of the attended terms is considered to be the same on time scale. In fact, their roles generally differ from time to time. That may be one of the reasons causing the classical equivalent replacement to be effective only for oscillators with weak nonlinearity but normally not being good for ones with strong nonlinearity. In order to improve this shortcoming, the averaging operation with weighting functions is proposed as in the next section. This idea is introduced by Anh [31].

2.2. The Weighted Averaging. The conventional averaging value of an integrable deterministic function $x(t)$ on a domain D: $(0, d)$ is a constant value defined by

$$\langle x(t)\rangle = \frac{1}{d}\int_0^d x(t)\,dt. \tag{9}$$

In many cases when the function $x(\omega t)$ is periodic with period $2\pi/\omega$, the value d is taken as $2\pi/\omega$ and it leads to the averaging value of $x(t)$ over one period:

$$\langle x(\omega t)\rangle = \frac{\omega}{2\pi}\int_0^{2\pi/\omega} x(\omega t)\,dt = \frac{1}{2\pi}\int_0^{2\pi} x(\tau)\,d\tau, \tag{10}$$

where $\tau = \omega t$. Definition (9) has some deficiencies; for example, if (9) or (10) are equal to zero, the information about $x(t)$ will be lost. For all harmonic functions such as $\cos(n\omega t)$ and $\sin(n\omega t)$, this observation is true. In [31], Anh suggested replacing the constant coefficient $1/d$ in (9) by a weighted coefficient function $h(t)$. Thus one gets the so-called weighted average value:

$$\langle x(t)\rangle = \int_0^d h(t)\,x(t)\,dt \tag{11}$$

with the condition

$$\int_0^d h(t)\,dt = 1. \quad (12)$$

There are three basic weighted coefficients:

(i) *Basic Optimistic Weighted Coefficients.* They are increasing functions of t and denoted as $O(t)$. Examples are αt^{β} and $\alpha e^{\beta t}$, $\alpha, \beta > 0$.

(ii) *Basic Pessimistic Weighted Coefficients.* They are decreasing functions of t and denoted as $P(t)$. Examples are αt^{β} and $\alpha e^{\beta t}$, $\alpha < 0,\ \beta > 0$; or $\alpha > 0,\ \beta < 0$.

(iii) *Neutral Weighted Coefficients.* They are denoted as $N(t)$ and are constants.

An arbitrary weighted coefficient $h(t)$ can be obtained as summation and/or product of basic weighted coefficients. Example is

$$h(t) = \sum_{i=1}^{n} A_i O_i(t) + B_i P_i(t) + C_i O_i(t) P_i(t) + N(t), \quad (13)$$

where A_i, B_i, C_i are constant.

In this paper, a special form of weighted coefficient is introduced as follows [31, 33]:

$$h(t) = s^2\omega^2 t e^{-s\omega t}, \quad s > 0, \quad (14)$$

where s is constant.

The properties of the weighted function $h(t)$ in (14) can be viewed in [31, 33].

Based on the weighted coefficient (14), a new weighted averaging value is proposed:

$$\begin{aligned} \langle x(\omega t)\rangle &= \int_0^{\infty} s^2\omega^2 t e^{-s\omega t} x(\omega t)\,dt \\ &= \int_0^{\infty} s^2 \tau e^{-s\tau} x(\tau)\,d\tau \end{aligned} \quad (15)$$

which is a linear operator. From Laplace transform, we get, for example,

$$\begin{aligned} \langle \cos(n\omega t)\rangle &= \int_0^{\infty} s^2\omega^2 t e^{-s\omega t}\cos(n\omega t)\,dt \\ &= \int_0^{\infty} s^2 \tau e^{-s\tau}\cos(n\tau)\,d\tau = s^2\frac{s^2-n^2}{(s^2+n^2)^2} \\ \langle \sin(n\omega t)\rangle &= \int_0^{\infty} s^2\omega^2 t e^{-s\omega t}\sin(n\omega t)\,dt \\ &= \int_0^{\infty} s^2 \tau e^{-s\tau}\sin(n\tau)\,d\tau = s^2\frac{2sn}{(s^2+n^2)^2}. \end{aligned} \quad (16)$$

As ω-periodic functions $x(\omega t)$ can be expanded into Fourier series; hence we can easily calculate (15) by using (16).

In this paper, for the sake of comparison, the parameter s is chosen equal to 2. In the next section, some practical examples for some strongly nonlinear vibration systems are illustrated to show the applicability, accuracy, and effectiveness of the present method.

3. Some Examples and Discussions

3.1. Example 1. Consider the following Duffing equation with cubic nonlinearity:

$$\ddot{u} + \alpha u + \beta u^3 = 0, \quad u(0) = A,\ \dot{u}(0) = 0. \quad (17)$$

The linearized equation of (17) is

$$\ddot{u} + \omega^2 u = 0. \quad (18)$$

The equation error between (17) and (18) is

$$e(u) = \alpha u + \beta u^3 - \omega^2 u, \quad (19)$$

where ω^2 is determined by using the mean square criterion, as follows:

$$\omega^2 = \frac{\alpha\langle u^2\rangle + \beta\langle u^4\rangle}{\langle u^2\rangle}. \quad (20)$$

The periodic solution of the linearized equation (18) is

$$u(t) = A\cos(\omega t). \quad (21)$$

Using definition (11) and weighted coefficient (14), we can calculate averaging operators in (20):

$$\begin{aligned} \langle u^2\rangle &= \langle A^2\cos^2\omega t\rangle \\ &= \int_0^{+\infty} A^2 s^2\omega^2 t e^{-s\omega t}\cos^2(\omega t)\,dt \\ &= \int_0^{+\infty} A^2 s^2 \tau e^{-s\tau}\cos^2(\tau)\,d\tau = A^2\frac{s^4+2s^2+8}{(s^2+4)^2} \\ \langle u^4\rangle &= \langle A^4\cos^4\omega t\rangle \\ &= \int_0^{+\infty} A^4 s^2\omega^2 t e^{-s\omega t}\cos^4(\omega t)\,dt \\ &= \int_0^{+\infty} A^4 s^2 \tau e^{-s\tau}\cos^4(\tau)\,d\tau \\ &= A^4\frac{248s^4+416s^2+1536+28s^6+s^8}{(s^2+4)^2(s^2+16)^2}. \end{aligned} \quad (22)$$

Substituting (22) into (20), we get the approximate frequency of this oscillator:

$$\omega = \sqrt{\alpha + \beta A^2\frac{248s^4+416s^2+1536+28s^6+s^8}{(s^4+2s^2+8)(s^2+16)^2}}. \quad (23)$$

With s chosen equal to 2, we have

$$\omega = \sqrt{\alpha + \frac{9216}{12800}\beta A^2} = \sqrt{\alpha + 0.72\beta A^2}. \quad (24)$$

From (21), the approximate solution of this oscillator is

$$u(t) = A\cos\left(\sqrt{\alpha + 0.72\beta A^2 t}\right). \quad (25)$$

Accuracy of this method for this example is shown in Table 1 and Figures 1 and 2. Table 1 shows the comparison between the approximate frequencies ω_{present} obtained by this method, the ones obtained by Parameterized Perturbation

Method ω_{PPM} (PPM) [5], and the exact ones ω_e; the relative error is again obtained for small and large values of the oscillation amplitude A. Table 1 shows that the maximum relative error is less than 0.4733 for this method and 2.2135 for Parameterized Perturbation Method. The numerical results obtained by three different methods are illustrated in Figures 1 and 2. Numerical results validate again accuracy of this method. As it is shown in Figures 3 and 4, validity of the solution technique is guaranteed for stronger nonlinearities. Relative error versus the oscillation amplitude is illustrated in Figure 3.

The approximate frequency obtained by PPM [5] is the same as that obtained by EMB [12], as follows:

$$\omega_{\text{PPM}} = \omega_{\text{EBM}} = \sqrt{\alpha + \frac{3}{4}\beta A^2}. \tag{26}$$

And the exact frequency of this oscillator is [5]

$$\omega_e = \frac{2\pi}{4\sqrt{2}\int_0^{\pi/2} \dfrac{dt}{\sqrt{\beta A^2 \cos^2(t) + \beta A^2 + 2\alpha}}}. \tag{27}$$

3.2. Example 2. The motion of a simple pendulum attached to a rotating rigid frame that is shown in Figure 4 has following nonlinear differential equation [6, 14]:

$$\ddot{\theta} + [1 - \Lambda\cos(\theta)]\sin(\theta) = 0 \tag{28}$$

with the initial conditions:

$$\begin{aligned}\theta(0) &= A,\\ \dot{\theta}(0) &= 0,\end{aligned} \tag{29}$$

in which θ and t are generalized dimensionless displacements and time variables, and $\Lambda = \Omega^2 r/g$.

Equation (28) can be rewritten as follows:

$$\ddot{\theta} + \sin(\theta) - \frac{\Lambda}{2}\sin(2\theta) = 0. \tag{30}$$

Based on Taylor-Maclaurin expansion, the approximations of $\sin(\theta)$ and $\sin(2\theta)$ are considered:

$$\begin{aligned}\sin(\theta) &\approx \theta - \frac{\theta^3}{6} + \frac{\theta^5}{120},\\ \sin(2\theta) &\approx 2\theta - \frac{4\theta^3}{3} + \frac{4\theta^5}{15}.\end{aligned} \tag{31}$$

Substituting (31) into (30) yields

$$\ddot{\theta} + (1-\Lambda)\theta + \left(-\frac{1}{6} + \frac{2\Lambda}{3}\right)\theta^3 + \left(\frac{1}{120} - \frac{2\Lambda}{15}\right)\theta^5 = 0. \tag{32}$$

The linearized equation of (32) is

$$\ddot{\theta} + (1 - \Lambda + \alpha)\theta = 0. \tag{33}$$

The equation error between (32) and (33) is

$$e(\theta) = \left(-\frac{1}{6} + \frac{2\Lambda}{3}\right)\theta^3 + \left(\frac{1}{120} - \frac{2\Lambda}{15}\right)\theta^5 - \alpha\theta. \tag{34}$$

The unknown coefficient α is determined from the mean square error criterion:

$$\frac{\partial}{\partial\alpha}\left\langle e^2(\theta)\right\rangle = 0. \tag{35}$$

It yields

$$\alpha = \frac{(-1/6 + 2\Lambda/3)\left\langle\theta^4\right\rangle + (1/120 - 2\Lambda/15)\left\langle\theta^6\right\rangle}{\left\langle\theta^2\right\rangle}. \tag{36}$$

The periodic solution and the frequency of (33) are

$$\theta(t) = A\cos(\omega t), \quad \omega = \sqrt{1 - \Lambda + \alpha}. \tag{37}$$

Similarly to Example 1, we calculate averaging operators $\langle\theta^2\rangle$, $\langle\theta^4\rangle$, and $\langle\theta^6\rangle$:

$$\begin{aligned}\left\langle\theta^2\right\rangle &= \left\langle A^2\cos^2(\omega t)\right\rangle = A^2\frac{s^4 + 2s^2 + 8}{(s^2+4)^2}\\ \left\langle\theta^4\right\rangle &= \left\langle A^4\cos^4(\omega t)\right\rangle = A^2\frac{248s^4 + 416s^2 + 1536 + 28s^6 + s^8}{(s^2+4)^2(s^2+16)^2}\\ \left\langle\theta^6\right\rangle &= \left\langle A^6\cos^6(\omega t)\right\rangle = A^6\frac{1658880 + 440064s^2 + 282496s^4 + 45712s^6 + 3168s^8 + 94s^{10} + s^{12}}{(s^2+4)^2(s^2+16)^2(s^2+36)^2}.\end{aligned} \tag{38}$$

Substituting averaging operators in (38) into (36), and then substituting (36) into (37), we get the approximate frequency (with s chosen equal to 2):

$$\omega = \sqrt{1 - \Lambda + 0.72\left(-\frac{1}{6} + \frac{2\Lambda}{3}\right)A^2 + 0.575\left(\frac{1}{120} - \frac{2\Lambda}{15}\right)A^4}, \tag{39}$$

TABLE 1: Comparison of the approximate frequencies with the exact frequency, Example 1.

A	α	β	ω_e	ω_{PPM}	R. error (%)	$\omega_{Present}$	R. error (%)
0.1	0.5	0.1	0.7076	0.7076	0.0000	0.7076	0.0000
0.5	0.1	2	0.6800	0.6892	1.3501	0.6782	0.2647
1	2	0.5	1.5403	1.5411	0.0520	1.5362	0.2662
2	5	2	3.2958	3.3166	0.6313	3.2802	0.4733
5	2	5	9.5818	9.7852	2.1228	9.5917	0.1033
10	1	0.5	6.0722	6.2048	2.0994	6.0828	0.1746
15	0.5	2	17.9866	18.3848	2.2135	18.0139	0.1518
20	5	1	17.0977	17.4642	2.1436	17.1172	0.1141

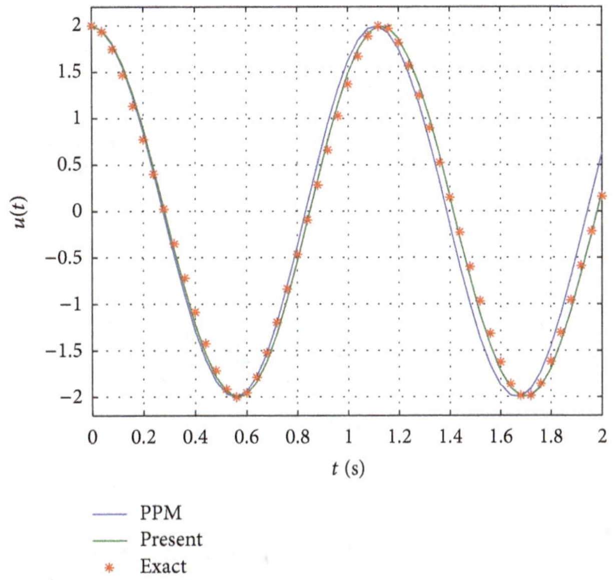

FIGURE 1: Comparison of time history diagram of displacement between the present, PPM, and exact solutions at $\alpha = 2$, $\beta = 10$, $A = 2$, Example 1.

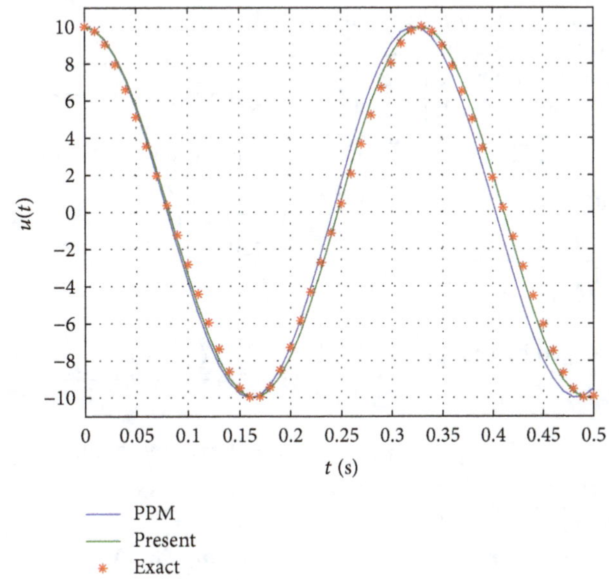

FIGURE 2: Comparison of time history diagram of displacement between the present, PPM, and exact solutions at $\alpha = 5$, $\beta = 5$, $A = 10$, Example 1.

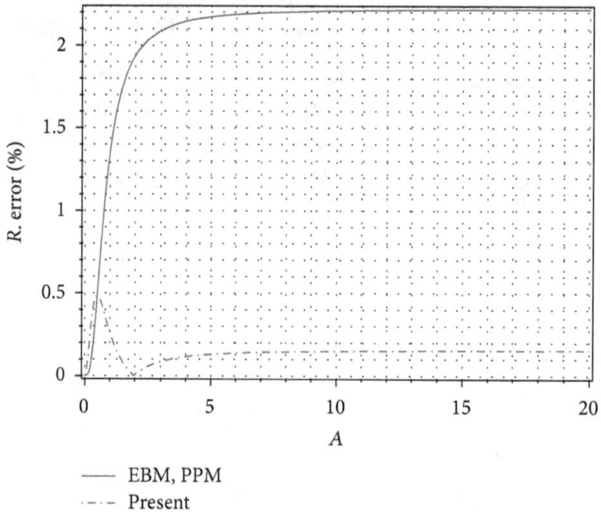

FIGURE 3: Comparison of relative error between the present and PPM solutions with $\alpha = 2$ and $\beta = 5$, Example 1.

and the approximate solution:

$$\theta(t) = A\cos\left[\sqrt{1-\Lambda+0.72\left(-\frac{1}{6}+\frac{2\Lambda}{3}\right)A^2+0.575\left(\frac{1}{120}-\frac{2\Lambda}{15}\right)A^4}\,t\right]. \tag{40}$$

Comparisons between the results of the present method and the ones achieved by Jouybari and Ramzani [6] using Min–Max Approach (MMA), by Younesian et al. [14] using EBM, and using accurate numerical method are demonstrated in Tables 2 and 3, for two cases of $\Lambda = 0.75$ and $\Lambda = 1$. A very interesting agreement between the results is observed.

To illustrate and verify the accuracy of the current analytical approach, the Equivalent Linearization Method with a weighted averaging, in comparison with other well-known techniques, we performed a comparison between the results obtained by this method and outcomes achieved by Jouybari and Ramzani [6], by Younesian et al. [15], and by using the Runge-Kutta 4th-Order Method. Variations of $\theta(t)$ for different Λ are illustrated in Figures 5–7.

The approximate frequency obtained by using MMA [6] is

$$\omega_{\mathrm{MMA}} = \sqrt{1-\frac{1}{8}A^2-\frac{1}{12}\Lambda A^4+\frac{1}{2}\Lambda A^2-\Lambda+\frac{1}{192}A^4}. \tag{41}$$

The approximate frequency obtained by using EBM [15] is

$$\omega_{\mathrm{EBM}} = \frac{2}{A} \cdot \sqrt{\cos\left(\frac{A}{\sqrt{2}}\right)-\cos(A)-\frac{\Lambda}{4}\cos\left(\sqrt{2}A\right)+\frac{\Lambda}{4}\cos(2A)}. \tag{42}$$

3.3. Example 3. The equation of motion of a mass attached to the center of a stretched elastic wire in dimensionless form is as follows [5, 16, 18]:

$$\ddot{u}+u-\frac{\eta u}{\sqrt{1+u^2}} = 0, \quad u(0) = A, \; \dot{u}(0) = 0, \tag{43}$$

where η is a parameter, $0 < \eta \le 1$.

Equation (43) can be written as follows:

$$\sqrt{1+u^2}\ddot{u}+\sqrt{1+u^2}u-\eta u = 0. \tag{44}$$

The linearized equation of (44) is

$$\ddot{u}+\omega^2 u = 0. \tag{45}$$

The equation error between (44) and (45) is

$$e(u) = \sqrt{1+\varepsilon u^2}\ddot{u}+\left(1+\varepsilon u^2\right)u-\eta u-\ddot{u}-\omega^2 u, \tag{46}$$

where ω^2 is determined by using the mean square criterion, as follows:

$$\omega^2 = \frac{\left\langle\sqrt{1+u^2}\ddot{u}u\right\rangle+\left\langle u^2\sqrt{1+u^2}\right\rangle-\eta\left\langle u^2\right\rangle-\langle u\ddot{u}\rangle}{\left\langle u^2\right\rangle}. \tag{47}$$

The periodic solution of the linearized equation (45) is

$$u(t) = A\cos(\omega t). \tag{48}$$

Using (48), we have

$$u\ddot{u} = -A\cos(\omega t)A\omega^2\cos(\omega t) = -A^2\omega^2\cos^2(\omega t) = -\omega^2 u^2. \tag{49}$$

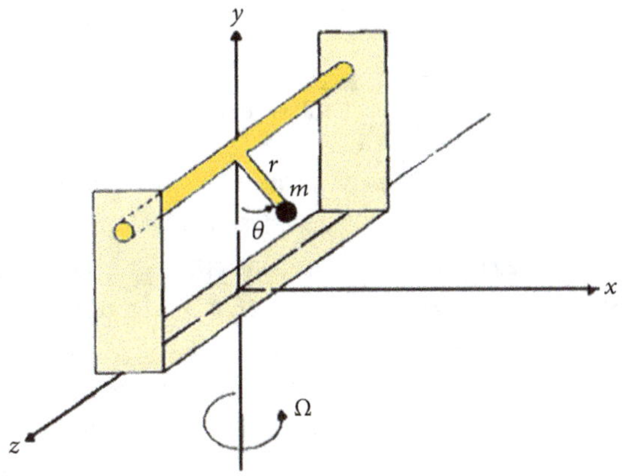

FIGURE 4: A simple pendulum attached to a rotating rigid frame.

TABLE 2: The results of present, approximate, obtained by Jouybari & Ramzani and by Yousesian et al., and numerical solution for $\theta(t)$ when $\Lambda = 0.75$ and $A = 1$, Example 2.

t	Numerical solution [12]	MMA solution [6]	R. error (%)	EBM solution [14]	R. error (%)	Present solution	R. error (%)
0	1.000000000	1.000000000	0.000000000	1.000000000	0.000000000	1.000000000	0.000000000
0.5	0.939545503	0.945169971	0.598637105	0.944189348	0.494265045	0.945834709	0.669388122
1	0.769687769	0.786692549	2.20930885	0.782987048	1.727879737	0.789206596	2.535940908
1.5	0.520367974	0.541946378	4.146758655	0.534386712	2.694004762	0.547083274	5.13392471
2	0.226896964	0.237770336	4.792206916	0.226137434	0.334746656	0.245694103	8.284438306
2.5	−0.082424976	−0.092479615	12.19853434	−0.107353589	30.24400395	−0.082311253	0.137971529
3	−0.386261801	−0.412588246	6.815699852	−0.428861683	11.0287587	−0.401399783	3.919098902
3.5	−0.660436345	−0.687452426	4.090641165	−0.702499667	6.369019864	−0.677004442	2.50865918
4	−0.871990011	−0.886930535	1.713382471	−0.897723721	2.951147338	−0.879268816	0.834734906
4.5	−0.986529936	−0.989147789	0.265359712	−0.992742682	0.629757473	−0.986281489	0.025183929
5	−0.982752773	−0.982895042	0.014476581	−0.970950401	1.200950262	−0.986449715	0.376182302
5.5	−0.861382758	−0.868857967	0.867815025	−0.852109657	1.07653664	−0.879755271	2.132909305
6	−0.644890958	−0.659541879	2.271844692	−0.632155313	1.974852468	−0.677756428	5.09628327
6.5	−0.368002797	−0.377900390	2.689542873	−0.341638967	7.164029789	−0.402335838	9.329559797
7	−0.063204827	−0.054818323	13.26877139	−0.012988435	79.45024832	−0.083329972	31.84115194
7.5	0.245785975	0.274275125	11.59103972	0.317111883	29.01951912	0.244703077	0.440585758
8	0.537446333	0.573291546	6.669542762	0.611815759	13.8375539	0.546227301	1.633831596
8.5	0.78289188	0.809440784	3.391132885	0.838227903	7.068156461	0.788578403	0.726348445
9	0.946701653	0.956826699	1.069507586	0.971676667	2.638108207	0.945502349	0.12668236
9.5	0.999764591	0.999286944	0.047775947	0.995531392	0.423419577	0.999999478	0.023494231
10	0.931963943	0.932165326	0.021608454	0.908764206	2.489338474	0.946166081	1.523893505

Thus

$$\left\langle \sqrt{1+u^2}\ddot{u}u \right\rangle = -\omega^2 \left\langle u^2\sqrt{1+u^2} \right\rangle, \quad \langle u\ddot{u}\rangle = -\omega^2 \left\langle u^2 \right\rangle. \tag{50}$$

Substituting (49) and (50) into (47), we get

$$\omega^2 = \frac{\left(1-\omega^2\right)\left\langle u^2\sqrt{1+u^2}\right\rangle}{\left\langle u^2\right\rangle} - \eta + \omega^2. \tag{51}$$

Using definition (11) and weighted coefficient (14), we calculate averaging operators in (51):

$$\left\langle u^2\right\rangle = \left\langle A^2\cos^2\omega t\right\rangle = \int_0^{+\infty} A^2 s^2\omega^2 t e^{-s\omega t}\cos^2(\omega t)\,dt = \int_0^{+\infty} A^2 s^2 \tau e^{-s\tau}\cos^2(\tau)\,d\tau = A^2\frac{s^4+2s^2+8}{\left(s^2+4\right)^2}$$

$$\left\langle u^2\sqrt{1+u^2}\right\rangle = \left\langle A^2\cos^2(\omega t)\sqrt{1+A^2\cos^2(\omega t)}\right\rangle$$

TABLE 3: The results of present, approximate obtained by Jouybari & Ramzani, and numerical solution for $\theta(t)$ when $\Lambda = 1$ and $A = 1$, Example 2.

t	Numerical solution [2]	MMA solution [6]	R. error (%)	EBM solution [14]	R. error (%)	Present solution	R. error (%)
0	1.000000000	1.000000000	0.000000000	1.000000000	0.000000000	1.000000000	0.000000000
0.5	0.953968489	0.963119576	0.959265123	0.961786066	0.819479583	0.964200044	1.072525468
1	0.825442672	0.855198634	3.604849011	0.850064873	2.982908666	0.859363450	4.109404463
1.5	0.638888746	0.684197515	7.091808908	0.673375035	5.397855138	0.692996509	8.469043059
2	0.422287711	0.462729406	9.576810773	0.445220578	5.430626183	0.477011079	12.95878771
2.5	0.195637881	0.207129984	5.874170657	0.183038862	6.439969057	0.226871698	15.96511721
3	−0.032545617	−0.063747522	95.87129659	−0.093132124	186.1587291	−0.039511677	21.40398813
3.5	−0.260619945	−0.329922957	26.5915995	−0.362185221	38.97064593	−0.303066019	16.28657929
4	−0.485739543	−0.571762994	17.70978959	−0.603557273	24.25533019	−0.544920861	12.18375544
4.5	−0.696309233	−0.771429307	10.78832083	−0.79880073	14.71925003	−0.747759417	7.388984888
5	−0.869266232	−0.914194339	5.16850941	−0.93299355	7.331162267	−0.897058465	3.197206101
5.5	−0.976272484	−0.989527622	1.357729345	−0.995879662	2.008371466	−0.982128207	0.599804163
6	−0.996192580	−0.991872507	0.43365842	−0.982652815	1.35915136	−0.996877655	0.068769334
6.5	−0.924854298	−0.921056034	0.410687825	−0.894323908	3.301102678	−0.940250751	1.664743629
7	−0.776876524	−0.782301685	0.698329893	−0.737643731	5.050068034	−0.816301976	5.074867213
7.5	−0.579081909	−0.585844101	1.167743612	−0.524587017	9.410567167	−0.633906052	9.467424582
8	−0.358097172	−0.346174158	3.329547098	−0.271437236	24.20011739	−0.406122511	13.41125894
8.5	−0.130576292	−0.080970115	37.99018661	0.002457915	101.8823593	−0.149260634	14.30913814
9	0.097635936	0.190206352	94.8118283	0.276165213	182.8520157	0.118288292	21.15241257
9.5	0.325420035	0.447353037	37.46942071	0.528765792	62.48716586	0.377367786	15.96329218
10	0.548087739	0.671502582	22.51735155	0.740953929	35.1889262	0.609427780	11.19164627

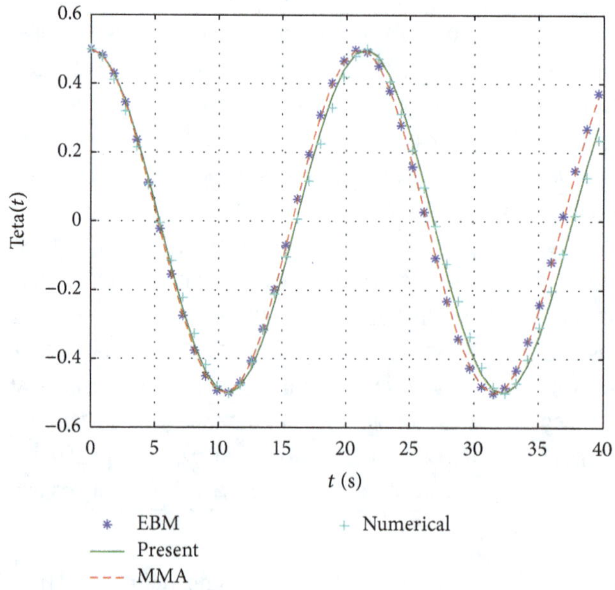

FIGURE 5: A comparison between the approximate and numerical solutions with $\Lambda = 1$ and $A = 0.5$, Example 2.

$$= \int_0^{+\infty} A^2 s^2 \omega^2 t e^{-s\omega t} \cos^2(\omega t) \sqrt{1 + A^2 \cos^2(\omega t)} dt$$

$$= \int_0^{+\infty} A^2 s^2 \tau e^{-s\tau} \cos^2(\tau) \sqrt{1 + A^2 \cos^2(\tau)} d\tau. \quad (52)$$

With s chosen equal to 2, substituting (52) into (51), we get the approximate frequency of this oscillator:

$$\omega = \sqrt{1 - \frac{\eta A^2}{2I}} \quad (53)$$

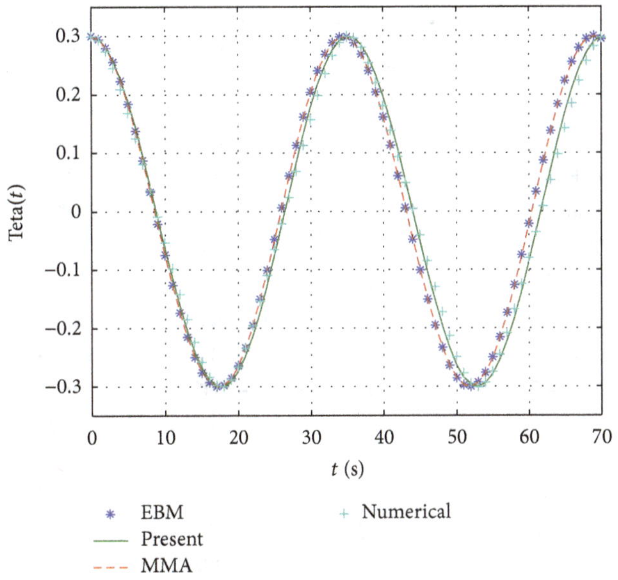

FIGURE 6: A comparison between the approximate and numerical solutions with $\Lambda = 1$ and $A = 0.3$, Example 2.

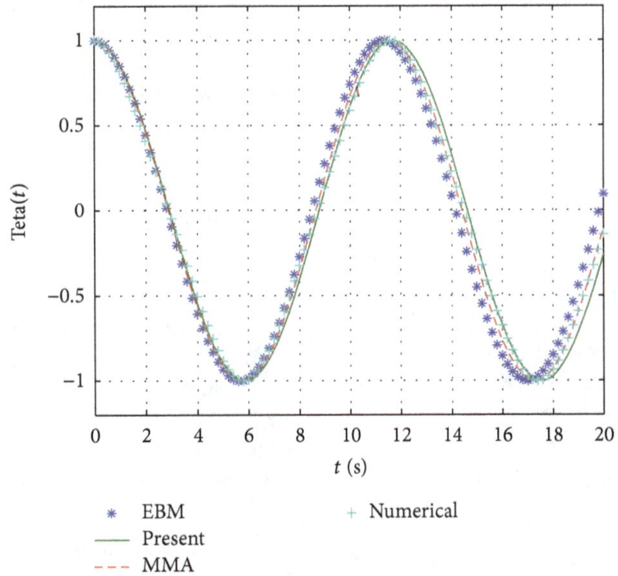

FIGURE 7: A comparison between the approximate and numerical solutions with $\Lambda = 1$ and $A = 1$, Example 2.

and the approximate solution of this oscillator:

$$u(t) = A\cos\left(\sqrt{1 - \frac{\eta A^2}{2I}}t\right), \qquad (54)$$

where

$$I = \int_0^{+\infty} 4A^2 \tau e^{-2\tau} \cos^2(\tau)\sqrt{1 + A^2\cos^2(\tau)}d\tau. \qquad (55)$$

Thus, the approximate period of this oscillator is

$$T = \frac{2\pi}{\omega} = \frac{2\pi}{\sqrt{1 - \eta A^2/2I}}. \qquad (56)$$

For $\eta = 0.1$, $\eta = 0.5$, $\eta = 0.75$, and $\eta = 0.95$, comparison of the approximate periods T in (56), the approximate Variational Iteration Method (VIM) periods T_{VIM} [5] in (57), and the Energy Balance Method (EBM) periods T_{EBM} [16] in (58) with exact periods T_{ex} in (59) is tabulated in Table 4. From Table 4, the accuracy of this proposed method can be observed.

The approximate period obtained by Variational Iteration Method is as follows [5]:

$$T_{\text{VIM}} = \frac{2\pi\sqrt{1 + A^2}}{\sqrt{1 + A^2 - \sqrt{1 + A^2}\eta}}. \qquad (57)$$

TABLE 4: Comparison of the approximate periods with the exact periods, Example 3.

A	η	T_{ex} [18]	T_{VIM} [5]	R. error (%)	$T_{Present}$	R. error (%)	T_{EBM} [16]	R. error (%)
0.1	0.1	6.62168	6.621237	0.00669	6.62174	0.00091	6.62169	0.00011
1	0.1	6.537508	6.517854	0.30063	6.53806	0.00844	6.53573	0.02723
10	0.1	6.322938	6.314678	0.13064	6.32097	0.03112	6.32006	0.04558
100	0.1	6.287182	6.286329	0.01357	6.28696	0.00353	6.28687	0.00498
0.1	0.5	8.869257	8.863794	0.06159	8.86990	0.00725	8.86925	0.0001
1	0.5	7.992133	7.814722	2.21982	7.99408	0.02436	7.97282	0.2416
10	0.5	6.490208	6.445572	0.68774	6.47925	0.16884	6.47431	0.24498
100	0.5	6.303280	6.298952	0.06866	6.30214	0.01809	6.30167	0.02557
0.1	0.75	12.49673	12.47385	0.18309	12.49943	0.02161	12.4967	0.00041
1	0.75	9.625404	9.168186	4.75012	9.61936	0.06279	9.56406	0.63733
10	0.75	6.602092	6.531632	1.06724	6.58442	0.26767	6.57665	0.38536
100	0.75	6.313404	6.306879	0.10335	6.31168	0.02731	6.31097	0.03854
0.1	0.95	27.15679	26.86136	1.08787	27.18975	0.12137	27.1540	0.01043
1	0.95	12.075277	10.96676	9.18005	12.00887	0.54994	11.8737	1.66932
10	0.95	6.696117	6.60302	1.39031	6.67235	0.35494	6.66211	0.50781
100	0.95	6.321539	6.313243	0.13123	6.31934	0.03479	6.31844	0.04897

The approximate period obtained by Energy Balance Method is as follows [16]:

$$T_{EBM} = \frac{2\pi}{(2/A)\sqrt{A^2/4 - \eta\left(-\sqrt{A^2/2+1} + \sqrt{A^2+1}\right)}}. \quad (58)$$

The exact period of this oscillator is as follows [18]:

$$T_{ex} = 4\int_0^{\pi/2}\left[1 - \frac{2\eta}{\sqrt{1+A^2\sin^2\theta} + \sqrt{1+A^2}}\right]^{-1/2} d\theta. \quad (59)$$

Comparisons of the approximate solutions obtained by two analytical methods with the exact ones is presented in Figures 8 and 9. And the relative error versus the oscillation amplitude is illustrated in Figure 10.

3.4. Example 4. Consider the Duffing oscillator with discontinuity [3]:

$$\ddot{u} + \alpha u + \beta u|u| = 0, \quad u(0) = A, \ \dot{u}(0) = 0. \quad (60)$$

The linearized equation of (60) is

$$\ddot{u} + \omega^2 u = 0. \quad (61)$$

The equation error between (60) and (61) is

$$e(u) = \alpha u + \beta u|u| - \omega^2 u. \quad (62)$$

The unknown coefficient ω^2 is determined from the mean square error criterion, as follows:

$$\omega^2 = \alpha + \frac{\beta\left\langle u^2|u|\right\rangle}{\langle u^2\rangle}. \quad (63)$$

The periodic solution of (61) is

$$u(t) = A\cos(\omega t). \quad (64)$$

Similar to Examples 1–3, we calculate averaging operators $\langle u^2\rangle$ and $\langle u^2|u|\rangle$; then, substituting these operators into (63) with note that the parameter s is chosen equal to 2, we have the approximate frequency of this oscillator:

$$\omega = \sqrt{\alpha + 0.8324\beta A}. \quad (65)$$

Thus, the approximate solution is

$$u(t) = A\cos\left(\sqrt{\alpha + 0.8324\varepsilon A}t\right). \quad (66)$$

Accuracy of the approach for this example is shown in Figures 11 and 12. We performed a comparison between the results obtained by this method, the ones obtained by He [3] using the Homotopy Perturbation Method, and outcomes achieved using the Runge-Kutta 4th-Order Method for different values of A, α, and β.

3.5. Example 5. Consider the nonlinear oscillator equation of the form [23, 24, 34]

$$\ddot{u} + u^{1/(2n+1)} = 0, \quad u(0) = A, \ \dot{u}(0) = 0, \quad (67)$$

where n is a positive integer.

The linearized equation of (67) is

$$\ddot{u} + \omega^2 u = 0. \quad (68)$$

The equation error between (67) and (68) is

$$e(u) = u^{1/(2n+1)} - \omega^2 u. \quad (69)$$

The unknown coefficient ω^2 is determined from the mean square error criterion, as follows:

$$\omega^2 = \frac{\left\langle uu^{1/(2n+1)}\right\rangle}{\langle u^2\rangle}. \quad (70)$$

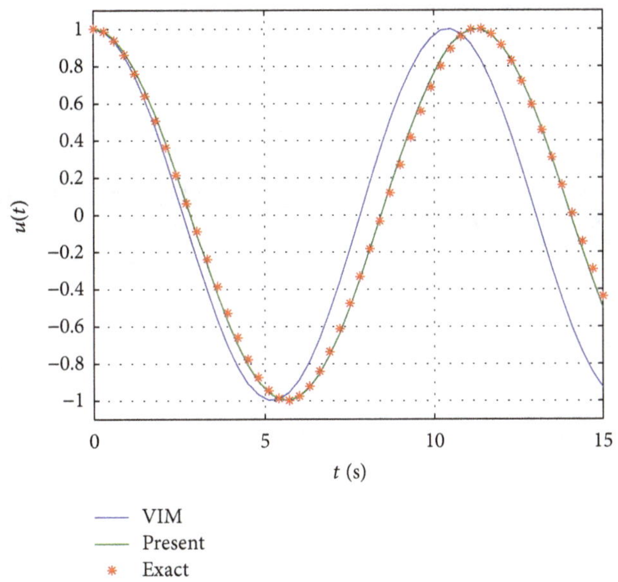

FIGURE 8: Comparison of time history diagram of displacement between the present, VIM, and exact solutions at $\eta = 0.9$, $A = 1$, Example 3.

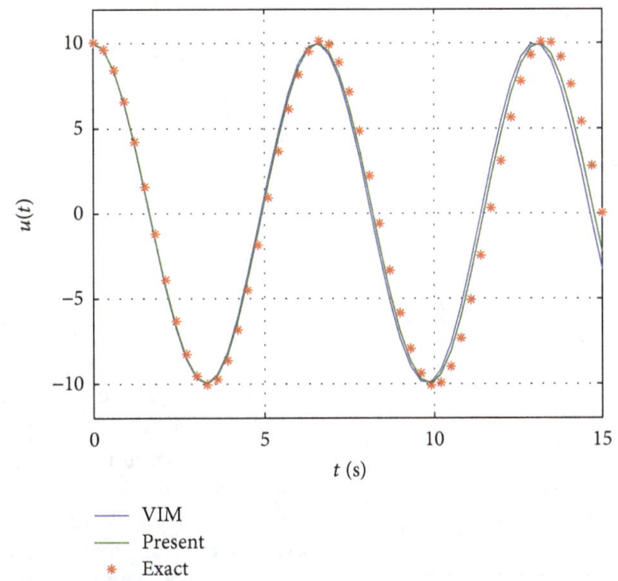

FIGURE 9: Comparison of time history diagram of displacement between the present, VIM, and exact solutions at $\eta = 0.7$, $A = 10$, Example 3.

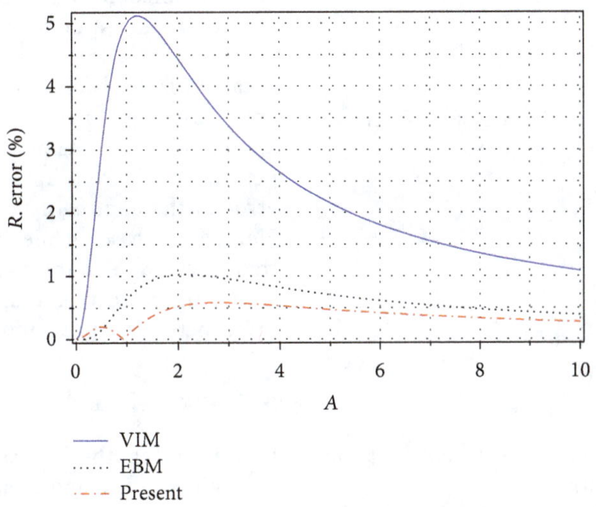

FIGURE 10: Comparison of relative error between the present, VIM, and EBM solutions with $\eta = 0.75$, Example 3.

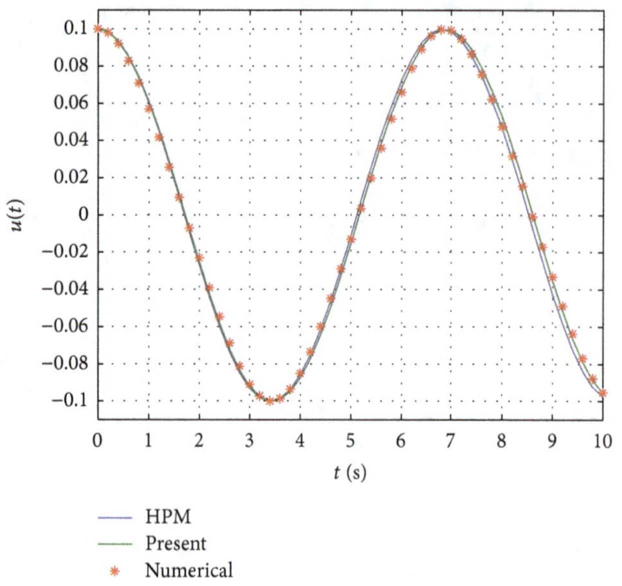

FIGURE 11: Comparison of time history diagram of displacement between the present, HPM, and numerical solutions at $\alpha = 0$, $\beta = 10$, $A = 0.1$, Example 4.

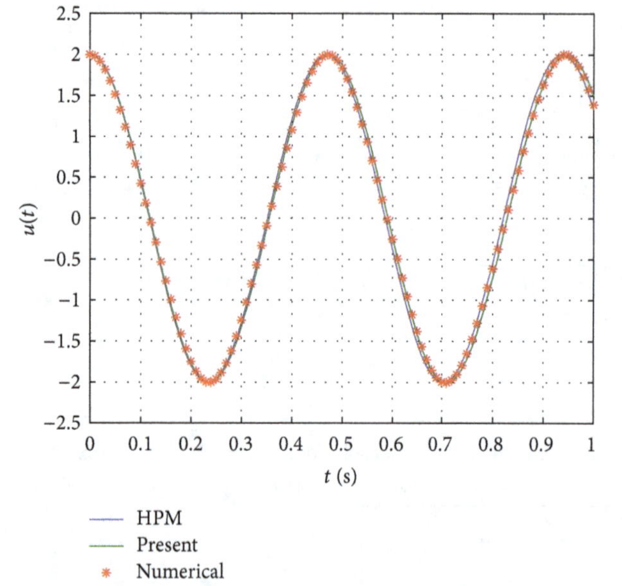

FIGURE 12: Comparison of time history diagram of displacement between the Present, HPM, and numerical solutions at $\alpha = 10$, $\beta = 100$, $A = 2$, Example 4.

The periodic solution of (68) is

$$u(t) = A\cos(\omega t). \tag{71}$$

Using definition (11) and weighted coefficient (14), we calculate averaging operators in (70):

$$\begin{aligned}\left\langle u^2\right\rangle &= \left\langle A^2\cos^2\omega t\right\rangle = \int_0^{+\infty} A^2 s^2\omega^2 t e^{-s\omega t}\cos^2(\omega t)\,dt \\ &= \int_0^{+\infty} A^2 s^2 \tau e^{-s\tau}\cos^2(\tau)\,d\tau = A^2\frac{s^4+2s^2+8}{\left(s^2+4\right)^2}\end{aligned} \tag{72}$$

$$\begin{aligned}\left\langle uu^{1/(2n+1)}\right\rangle &= \left\langle A\cos(\omega t)\, A^{1/(2n+1)}(\cos(\omega t))^{1/(2n+1)}\right\rangle \\ &= \int_0^{+\infty} AA^{1/(2n+1)} s^2\omega^2 t e^{-s\omega t}\cos(\omega t) \\ &\cdot (\cos(\omega t))^{1/(2n+1)}\,dt = \int_0^{+\infty} AA^{1/(2n+1)} s^2 \tau e^{-s\tau} \\ &\cdot \cos(\tau)(\cos(\tau))^{1/(2n+1)}\,d\tau.\end{aligned} \tag{73}$$

With s chosen equal to 2, from (72) and (73), we have

$$\left\langle u^2\right\rangle = \frac{A^2}{2} \tag{74}$$

$$\begin{aligned}\left\langle uu^{1/(2n+1)}\right\rangle &= \int_0^{+\infty} A^{(2n+2)/(2n+1)} 4\tau e^{-2\tau}\cos(\tau) \\ &\cdot (\cos(\tau))^{1/(2n+1)}\,d\tau.\end{aligned} \tag{75}$$

Substituting (71) and (72) into (70), we get the approximate frequency of this oscillator:

$$\omega = A^{-n/(2n+1)}\sqrt{2I_n}, \tag{76}$$

where

$$I_n = \int_0^{+\infty} 4\tau e^{-2\tau}\cos(\tau)(\cos(\tau))^{1/(2n+1)}\,d\tau. \tag{77}$$

For $n = 0$ (the harmonic oscillator case) it follows from (77) and (76) that ω is equal to the well-known value 1. And, for $n \to \infty$, ω is equal to $0.97979A^{-1/2}$. For other values of n, the integral in (77) can be calculated numerically by using the formula manipulation package Maple. For some values of n approximations of the frequency are given in Table 5. Table 5 presents the comparison of the approximate frequencies ω_{present} in (76), the approximate frequencies ω_{app} obtained by van Horssen [34], and the approximate frequencies ω_{HBM} obtained by Mickens [23]. As can be seen from this table the fractional errors of the approximations obtained in [34] for $n \geq 1$ range from approximately 2 to approximately 11 percent, while the fractional errors of the approximations obtained in this paper range from 2 to 10 percent. And, from Table 5, when n increases, the relative error of the approximate frequencies ω_{app} obtained by van Horssen increases, while the relative error of the approximate frequencies ω_{app} in this paper decreases.

Comparison of the approximate solutions obtained by three methods is present in Figures 13 and 14.

4. Conclusions

In this paper, the Equivalent Linearization Method with a weighted averaging is applied to analyze five undamped

TABLE 5: Comparison of the approximate frequencies, Example 5.

n	ω_{HBM} [23]	ω_{app} [34]	R. error (%)	$\omega_{present}$	R. error (%)
0	1.00000	1.000000	0.00000	1.000000	0.00000
1	$A^{-1/3}1.04912$	$A^{-1/3}1.07045$	2.03313	$A^{-1/3}0.94307$	10.10847
2	$A^{-2/5}1.04812$	$A^{-2/5}1.08613$	3.62649	$A^{-2/5}0.95714$	8.680304
3	$A^{-3/7}1.04405$	$A^{-3/7}1.09302$	4.69039	$A^{-3/7}0.96342$	7.72281
4	$A^{-4/9}1.04017$	$A^{-4/9}1.09689$	5.45295	$A^{-4/9}0.96697$	7.037311
5	$A^{-5/11}1.03684$	$A^{-5/11}1.09938$	6.03179	$A^{-5/11}0.96926$	6.517881
6	$A^{-6/13}1.03403$	$A^{-6/13}1.10110$	6.48627	$A^{-6/13}0.97085$	6.110074
7	$A^{-7/15}1.03164$	$A^{-7/15}1.10238$	6.85704	$A^{-7/15}0.97202$	5.779148
8	$A^{-8/17}1.02960$	$A^{-8/17}1.10335$	7.16298	$A^{-8/17}0.97293$	5.504079
9	$A^{-9/19}1.02783$	$A^{-9/19}1.10412$	7.42243	$A^{-9/19}0.97364$	5.272273
10	$A^{-10/21}1.02628$	$A^{-10/21}1.10474$	7.64509	$A^{-10/21}0.97422$	5.07269
50	$A^{-50/101}1.00919$	$A^{-50/101}1.10947$	9.93668	$A^{-50/101}0.97863$	3.028171
500	$A^{-500/1001}1.00149$	$A^{-500/1001}1.11059$	10.8938	$A^{-500/1001}0.97968$	2.17776
∞	$A^{-1/2}1.00000$	$A^{-1/2}\left(\pi/2\sqrt{2}\right)$	11.07200	$A^{-1/2}0.97979$	2.02100

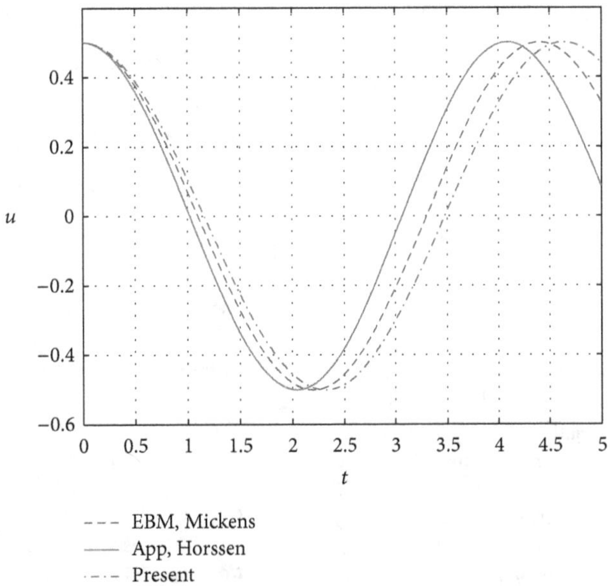

FIGURE 13: Comparison of time history diagram of displacement between the present, HBM [23], and App [34] solutions with $n = 10$, Example 5.

nonlinear oscillation systems. This method is proposed by Anh in 2015. Instead of using the classical averaging process in (9) and (10), a special form of weighted coefficient $h(t)$ is introduced in (14), whereby a new weighted average value is computed by (15). The proposed weighted function is a combination of two basic coefficient functions (the optimistic weighted coefficient t and the pessimistic weighted coefficient $e^{-s\omega t}$). The frequency-amplitude relationship was analytically obtained by this method. The frequency depends not only on the initial amplitude of the oscillator but also on the parameter s in the expression of weighted coefficient $h(t)$. Accuracy of this method is investigated by five nonlinear oscillation systems. The results show that this method is useful in obtaining analytical solutions for oscillators and vibration problems with nonlinearities. The approximate solutions obtained by this method are valid not only for small parameters but also for very large parameters. And the results indicate that the solution procedure is easy and provides a remarkable accuracy. In this paper, s is chosen equal to 2 for comparison and convenient calculating purposes. However, the value of the parameter s in the expression of the weighted coefficient $h(t)$ should be chosen to give better results and the best solution still requires further investigation.

Conflicts of Interest

The authors declare that no conflicts of interest took place during the preparation of the manuscript.

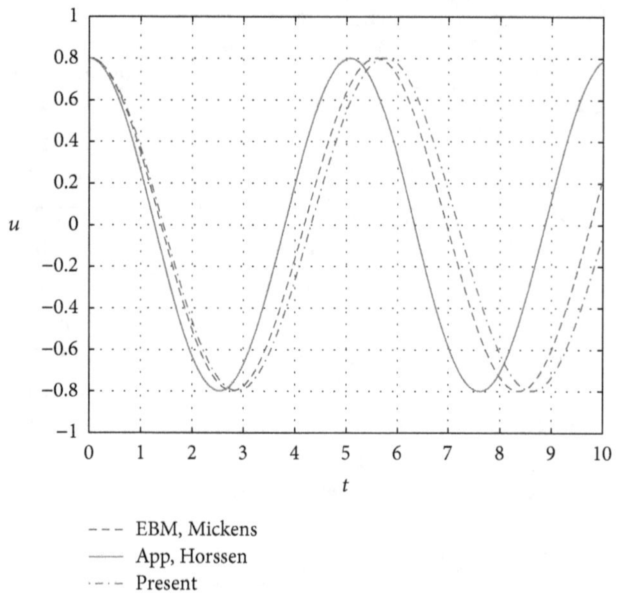

Figure 14: Comparison of time history diagram of displacement between the Present, HBM [23] and App [34] solutions with $n = 50$, Example 5.

Acknowledgments

The research is supported by Thai Nguyen University of Technology grant for a scientific project (no. "*T2018-B27*").

References

[1] T. Öziş and A. Yildirim, "A note on He's homotopy perturbation method for van der Pol oscillator with very strong nonlinearity," *Chaos, Solitons & Fractals*, vol. 34, no. 3, pp. 989–991, 2007.

[2] S. Ghafoori, M. Motevalli, M. G. Nejad, F. Shakeri, D. D. Ganji, and M. Jalaal, "Efficiency of differential transformation method for nonlinear oscillation: Comparison with HPM and VIM," *Current Applied Physics*, vol. 11, no. 4, pp. 965–971, 2011.

[3] J.-H. He, "The homotopy perturbation method nonlinear oscillators with discontinuities," *Applied Mathematics and Computation*, vol. 151, no. 1, pp. 287–292, 2004.

[4] J.-H. He, "Homotopy perturbation technique," *Computer Methods Applied Mechanics and Engineering*, vol. 178, no. 3-4, pp. 257–262, 1999.

[5] M. Bayat, I. Pakar, and G. Domairry, "Recent developments of some asymptotic methods and their applications for nonlinear vibration equations in engineering problems: A review," *Latin American Journal of Solids and Structures*, vol. 9, no. 2, pp. 145–234, 2012.

[6] A. K. Jouybari and M. Ramzani, "Analytical methods for solving nonlinear motion of simple pendulum attached to a rotating rigid frame," *International Journal of Mechatronics, Electrical and Computer Technology*, vol. 4, pp. 11–22, 2014.

[7] S. Durmaz, S. A. Demirbağ, and M. O. Kaya, "Approximate solutions for nonlinear oscillation of a mass attached to a stretched elastic wire," *Computers and Mathematics with Applications*, vol. 61, no. 3, pp. 578–585, 2011.

[8] M. K. Yazdi, H. Ahmadian, A. Mirzabeigy, and A. Yildirim, "Dynamic analysis of vibrating systems with nonlinearities," *Communications in Theoretical Physics*, vol. 57, no. 2, pp. 183–187, 2012.

[9] J. H. He, "Max-min approach to nonlinear oscillators," *International Journal of Nonlinear Sciences and Numerical Simulation*, vol. 9, no. 2, pp. 207–210, 2008.

[10] S. S. Ganji, D. D. Ganji, A. G. Davodi, and S. Karimpour, "Analytical solution to nonlinear oscillation system of the motion of a rigid rod rocking back using max-min approach," *Applied Mathematical Modelling: Simulation and Computation for Engineering and Environmental Systems*, vol. 34, no. 9, pp. 2676–2684, 2010.

[11] H. Ebrahimi Khah and D. D. Ganji, "A study on the motion of a rigid rod rocking back and cubic-quintic duffing oscillators by using He's energy balance method," *International Journal of Nonlinear Science*, vol. 10, no. 4, pp. 447–451, 2010.

[12] D. Younesian, H. Askari, Z. Saadatnia, and M. KalamiYazdi, "Frequency analysis of strongly nonlinear generalized Duffing oscillators using HE's frequency-amplitude formulation and HE's energy balance method," *Computers & Mathematics with Applications. An International Journal*, vol. 59, no. 9, pp. 3222–3228, 2010.

[13] S. S. Ganji, D. D. Ganji, Z. Z. Ganji, and S. Karimpour, "Periodic solution for strongly nonlinear vibration systems by He's energy balance method," *Acta Applicandae Mathematicae*, vol. 106, no. 1, pp. 79–92, 2009.

[14] D. Younesian, H. Askari, Z. Saadatnia, and A. Yildirim, "Periodic solutions for the generalized nonlinear oscillators containing fraction order elastic force," *International Journal of Nonlinear Sciences and Numerical Simulation*, vol. 11, no. 12, pp. 1027–1032, 2010.

[15] D. Younesian, H. Askari, Z. Saadatnia, and M. KalamiYazdi, "Frequency analysis of strongly nonlinear generalized Duffing oscillators using HE's frequency-amplitude formulation and HE's energy balance method," *Computers & Mathematics with Applications*, vol. 59, no. 9, pp. 3222–3228, 2010.

[16] D. Younesian, H. Askari, Z. Saadatnia, and M. KalamiYazdi, "Analytical approximate solutions for the generalized nonlinear oscillator," *Applicable Analysis: An International Journal*, vol. 91, no. 5, pp. 965–977, 2012.

[17] G.-h. Chen, Z.-L. Tao, and J.-Z. Min, "Notes on a conservative nonlinear oscillator," *Computers & Mathematics with Applications*, vol. 61, no. 8, pp. 2120–2122, 2011.

[18] L. Xu, "Application of He's parameter-expansion method to an oscillation of a mass attached to a stretched elastic wire," *Physics Letters A*, vol. 368, no. 3-4, pp. 259–262, 2007.

[19] M. O. Kaya and S. Altay Demirbağ, "Application of parameter expansion method to the generalized nonlinear discontinuity equation," *Chaos, Solitons & Fractals*, vol. 42, no. 4, pp. 1967–1973, 2009.

[20] M. T. Darvishi, A. Karami, and B.-C. Shin, "Application of He's parameter-expansion method for oscillators with smooth odd nonlinearities," *Physics Letters A*, vol. 372, no. 33, pp. 5381–5384, 2008.

[21] J.-H. He, "Comparison of homotopy perturbation method and homotopy analysis method," *Applied Mathematics and Computation*, vol. 156, no. 2, pp. 527–539, 2004.

[22] J.-H. He, "Modified Lindstedt-Poincare methods for some strongly non-linear oscillations," *International Journal of Non-Linear Mechanics*, vol. 37, no. 2, pp. 309–314, 2002.

[23] R. E. Mickens, "Oscillations in an $x^{4/3}$ potential," *Journal of Sound and Vibration*, vol. 246, no. 2, pp. 375–378, 2001.

[24] K. Cooper and R. E. Mickens, "Generalized harmonic balance/ numerical method for determining analytical approximations to the periodic solutions of the $x^{4/3}$ potential," *Journal of Sound and Vibration*, vol. 250, no. 5, pp. 951–954, 2002.

[25] C. W. Lim, B. S. Wu, and W. P. Sun, "Higher accuracy analytical approximations to the Duffing-harmonic oscillator," *Journal of Sound and Vibration*, vol. 296, no. 4-5, pp. 1039–1045, 2006.

[26] N. Krylov and N. Bogoliuboff, *Introduction to nonlinear mechanics*, Princenton University Press, New York, NY, USA, 1943.

[27] T. K. Caughey, "Equivalent linearization techniques," *The Journal of the Acoustical Society of America*, vol. 35, pp. 1706–1711, 1959.

[28] R. N. Iyengar, "Higher order linearization in nonlinear random vibration," *International Journal of Non-Linear Mechanics*, vol. 23, no. 5-6, pp. 385–391, 1988.

[29] N. D. Anh and W. Schiehlen, "New criterion for Gaussian equivalent linearization," *European Journal of Mechanics. A Solids*, vol. 16, no. 6, pp. 1025–1039, 1997.

[30] N. D. Anh and N. N. Linh, "Some extensions of Gaussian equivalent linearization," in *Proceedings of the International Conference on Nonlinear Stochastic Dynamics*, vol. 36, Hanoi, Vietnam, 1995.

[31] N. D. Anh, "Short Communication Dual approach to averaged values of functions: a form for weighting coefficient," *Vietnam Journal of Mechanics*, vol. 37, no. 2, pp. 145–150, 2015.

[32] I. Elishakoff, L. Andriamasy, and M. Dolley, "Application and extension of the stochastic linearization by Anh and Di Paola," *Acta Mechanica*, vol. 204, no. 1-2, pp. 89–98, 2009.

[33] N. D. Anh, N. Q. Hai, and D. V. Hieu, "The equivalent linearization method with a weighted averaging for analyzing of nonlinear vibrating systems," *Latin American Journal of Solids and Structures*, vol. 14, no. 9, pp. 1723–1740, 2017.

[34] W. T. van Horssen, "On the periods of the periodicsolutions of the non-linear oscillator equation," *Journal of Sound and Vibration 260*, pp. 961–964, 2003.

16

The Dynamics of Epidemic Model with Two Types of Infectious Diseases and Vertical Transmission

Raid Kamel Naji and Reem Mudar Hussien

Department of Mathematics, College of Science, University of Baghdad, Baghdad, Iraq

Correspondence should be addressed to Reem Mudar Hussien; reem.m.hussien@gmail.com

Academic Editor: Zhen Jin

An epidemic model that describes the dynamics of the spread of infectious diseases is proposed. Two different types of infectious diseases that spread through both horizontal and vertical transmission in the host population are considered. The basic reproduction number R_0 is determined. The local and the global stability of all possible equilibrium points are achieved. The local bifurcation analysis and Hopf bifurcation analysis for the four-dimensional epidemic model are studied. Numerical simulations are used to confirm our obtained analytical results.

1. Introduction

Mathematical models can be defined as a method of emulating real life situations with mathematical equations to expect their future behavior. In epidemiology, mathematical models play role as a tool in analyzing the spread and control of infectious diseases. Although one of the most famous principles of ecology is the competitive exclusion principle that stipulates "two species competing for the same resources cannot coexist indefinitely with the same ecological niche" [1, 2], Volterra was the first scientist who used the mathematical modeling and showed that the indefinite coexistence of two or more species limited by the same resource is impossible [3]. Moreover, Ackleh and Allen [4] were the first who used the competitive exclusion principle of the infectious disease with different levels in single host population.

It is well known that one of the most useful parameters concerning infectious diseases is called basic reproduction number. It can be specific to each strain of an epidemic model. In fact the basic reproduction number of the model is defined as the maximum reproduction numbers of other strains [5–7]. Diekmann et al. [8] had studied epidemic models with one strain, while Martcheva in [9] studied the SIS-type of disease with multistrain. However, Ackleh and Allen [10] studied SIR-type of disease with **n** strain and vertical transmission.

Keeping the above in view, in our proposed model two strains with two different types of infectious diseases are considered. Accordingly two different reproduction numbers are obtained and then competitive exclusion principle is presented. It is assumed that two different types of diseases transmission, say horizontal and vertical transmission, are used too. The horizontal transmission occurs by direct contact between infected and susceptible individuals, while vertical transmission occurs when the parasite is transmitted from parent to offspring [11–13]. The incidence of an epidemiological model is defined as the rate at which susceptible becomes infectious. Different types of incidence rates are introduced into literatures [14–17]. Finally two types of incidence rates, say bilinear mass action and nonlinear type, are used with the horizontal and vertical transmission, respectively. The local and global stability for all possible equilibria are carried out with the help of Lyapunov function and LaSalle's invariant principle [18]. An application of Sotomayor theorem [19, 20] for local bifurcations is used to study the occurrence of local bifurcations near the equilibria. The Hopf bifurcation [21, 22] conditions are derived. Finally, numerical simulations are used to confirm our obtained analytical results and specify the control set of parameters.

2. Model Formulation

Consider a real world system consisting of a host population $N(t)$ that is divided into four compartments: $S(t)$ which represents the number of susceptible individuals at time t; $I_1(t)$ and $I_2(t)$ that represent the number of infected individuals at time t for *SIRS*-type of disease and *SIS*-type of disease, respectively; finally $R(t)$ that represents the number of recovered individuals at time t, thus $N(t) = S(t) + I_1(t) + I_2(t) + R(t)$. Now in order to formulate the dynamics of the above system mathematically, the following assumptions have been adopted:

(1) There is a constant number of the host populations entering to the system with recruitment rate $\Lambda > 0$.

(2) There is a vertical transmission of both of the diseases; that is, the infectious host gives birth to a new infected host of rates $0 \le p_1 \le 1$ and $0 \le p_2 \le 1$ for the diseases I_1 and I_2, respectively. Consequently $p_1 I_1$ and $p_2 I_2$ individuals enter into infected compartments I_1 and I_2, respectively, and the same quantities are disappearing from recruitment in the susceptible compartment.

(3) The diseases are transmitted by contact, according to the mass action law, between the individuals in the *S*-compartment and those in I_i $(i = 1, 2)$ compartments with nonlinear incidence rate for I_1 that is given by $\beta_1 S I_1/(1 + I_1)$, in which $\beta_1 > 0$ represents the infection force rate while $1/(1 + I_1)$ represents the inhibition effect of the crowding effect of the infected individuals, and linear incidence rate for I_2 that is given by $\beta_2 S I_2$, where $\beta_2 > 0$ represents the infection rate.

(4) The individuals in the I_1 compartment are facing death due to the disease with infection death rate $\alpha_1 \ge 0$. They recover from disease and get immunity with a recovery rate $\delta > 0$.

(5) The individuals in the I_2 compartment are facing death due to the disease with infection death rate $\alpha_2 \ge 0$. They also recover from the disease but return back to be susceptible with recovery rate $\gamma > 0$.

(6) The individuals in the R compartment are losing the immunity from the I_1 disease and return back to be susceptible again with losing immunity rate $0 \le \eta < 1$.

(7) There is a natural death rate $\mu > 0$ for the individuals in the host population. Finally, it is assumed that both the diseases cannot be transmitted to the same individual simultaneously.

According to these assumptions the dynamics of the above real world system can be represented mathematically by the following set of differential equations:

$$
\begin{aligned}
\frac{dS}{dt} &= \Lambda - \left(\frac{\beta_1 I_1}{1 + I_1} + \beta_2 I_2\right) S + (\gamma - p_2) I_2 - \mu S - p_1 I_1 \\
&\quad + \eta R, \\
\frac{dI_1}{dt} &= \frac{\beta_1 S I_1}{1 + I_1} - (\mu + \alpha_1 + \delta - p_1) I_1, \\
\frac{dI_2}{dt} &= \beta_2 S I_2 - (\mu + \alpha_2 + \gamma - p_2) I_2, \\
\frac{dR}{dt} &= \delta I_1 - (\eta + \mu) R
\end{aligned}
\tag{1}
$$

with the initial condition $S(0) > 0$, $I_1(0) > 0$, $I_2(0) > 0$, and $R(0) > 0$. Moreover to insure that the recruitment Λ in the susceptible compartment is always positive the following hypotheses are assumed to be holding always:

$$
\begin{aligned}
\delta &\ge p_1, \\
\gamma &\ge p_2.
\end{aligned}
\tag{2}
$$

Theorem 1. *The closed set $\Omega = \{(S, I_1, I_2, R) \in \mathbb{R}_+^4 : N \le \Lambda/\mu\}$ is positively invariant and attracting with respect to model (1).*

Proof. Let $(S(t), I_1(t), I_2(t), R(t))$ be any solution of system (1) with any given initial condition. Then by adding all the equations in system (1) we obtain that

$$
\begin{aligned}
\frac{dN}{dt} &= \Lambda - \mu S - (\mu + \alpha_1) I_1 - (\mu + \alpha_2) I_2 - \mu R \\
&\le \Lambda - \mu N.
\end{aligned}
\tag{3}
$$

Thus, from standard comparison theorem [20], we obtain

$$
N(t) \le N(0) e^{-\mu t} + \frac{\Lambda}{\mu}\left(1 - e^{-\mu t}\right). \tag{4}
$$

Consequently it is easy to verify that

$$
N(t) \le \frac{\Lambda}{\mu}, \quad \text{when } (0) \le \frac{\Lambda}{\mu}. \tag{5}
$$

Thus, Ω is positively invariant. Further, when $N(0) > \Lambda/\mu$, then either the solution enters Ω in finite time, or $N(t)$ approaches Λ/μ as $t \to \infty$. Hence, Ω is attracting (i.e., all solutions in $\mathbb{R}_+^4$ eventually approach, enter, or stay in Ω). □

Therefore, the system of equations given in model (1) is mathematically well-posed and epidemiologically reasonable, since all the variables remain nonnegative $\forall t \ge 0$. Further since the equations of model (1) are continuous and have continuously partial derivatives then they are Lipschitzian. In addition to that from Theorem 1, model (1) is uniformly bounded. Therefore the solution of it exists and is unique. Hence, from now onward it is sufficient to consider the dynamics of model (1) in Ω.

3. Equilibrium Points and Basic Reproduction Number

Model (1) has four equilibrium points that are obtained by setting the right hand sides of this model equal to zero. The first equilibrium point is the disease-free equilibrium (DFE) point that is denoted by $E_0 = (S_0, 0, 0, 0)$ with $S_0 = \Lambda/\mu$. Moreover the basic reproduction number of model (1), which is denoted by R_0, is the maximum eigenvalue of the next generation matrix (i.e., the maximum of the reproduction numbers, those computed of each disease). That is,

$$R_0 = \max\{R_1, R_2\}. \quad (6)$$

Here $R_1 = (\beta_1 S_0 + p_1)/(\mu + \alpha_1 + \delta)$ and $R_2 = (\beta_2 S_0 + p_2)/(\mu + \alpha_2 + \gamma)$.

The other three equilibrium points can be described as follows.

The first disease-free equilibrium point, which is located in the boundary SI_2-plane, is denoted by $E_1 = (\bar{S}, 0, \bar{I}_2, 0)$ where

$$\bar{S} = \frac{S_0}{T_2}, \qquad \bar{I}_2 = \frac{\Lambda}{(\mu + \alpha_2)}\left(1 - \frac{1}{T_2}\right), \quad (7)$$

and here $T_2 = \beta_2 S_0/(\mu+\alpha_2+\gamma-p_2)$. Clearly E_1 exists uniquely in the interior of SI_2-plane provided that

$$T_2 > 1. \quad (8)$$

The second disease-free equilibrium point that is located in the boundary SI_1R-space is given by $E_2 = (\tilde{S}, \tilde{I}_1, 0, \tilde{R})$ where

$$\tilde{S} = \frac{S_0\left(1 + \tilde{I}_1\right)}{T_1}; \qquad \tilde{I}_1 = \frac{\Lambda\left(1 - 1/T_1\right)}{(\mu + \alpha_1 + \delta) - \eta\delta/(\eta + \mu) + \Lambda/T_1}, \qquad \tilde{R} = \frac{\tilde{I}_1}{\eta + \mu}, \quad (9)$$

and here $T_1 = \beta_1 S_0/(\mu + \alpha_1 + \delta - p_1)$. Obviously E_2 exists uniquely in the interior of positive octant of SI_1R-space provided that

$$T_1 > 1. \quad (10)$$

Finally, the endemic equilibrium point, which is denoted by $E_3 = (S^*, I_1^*, I_2^*, R^*)$, where

$$S^* = \frac{S_0}{T_2}; \qquad I_1^* = \left(\frac{T_1}{T_2} - 1\right); \qquad R^* = \frac{\delta\left((T_1/T_2) - 1\right)}{\eta + \mu},$$

$$I_2^* = \frac{\Lambda}{(\mu + \alpha_2)}\left(1 - \frac{1}{T_2}\right) - \frac{\left((T_1/T_2) - 1\right)}{(\mu + \alpha_2)}\left[(\mu + \alpha_1 + \delta) - \frac{\eta\delta}{\eta + \mu}\right], \quad (11)$$

exists uniquely in the interior of Ω provided that the following conditions hold:

$$T_1 > T_2 > 1, \qquad \Lambda > \frac{\Lambda}{T_2} + I_1^*\left((\mu + \alpha_1 + \delta) - \frac{\eta\delta}{\eta + \mu}\right). \quad (12)$$

Keeping the above in view, it is easy to verify with the help of condition (2) that

$$T_i > 1 \ (T_i < 1) \Longleftrightarrow R_i > 1 \ (R_i < 1), \quad i = 1, 2. \quad (13)$$

Then directly we obtain $T_i > 1$ $(T_i < 1) \Leftrightarrow R_0 > 1$ $(R_0 < 1)$. Consequently, T_i represent the threshold parameters for the existence of the last three equilibrium points of model (1). Moreover, it is well known that the basic reproduction number (R_0) is representing the average number of secondary infections that occur from one infected individual in contact with susceptible individuals. Therefore if $R_0 < 1$, then each infected individual in the entire period of infectivity will produce less than one infected individual on average, which shows the disease will be wiped out of the population. However, if $R_0 > 1$, then each infected individual in the entire infection period having contact with susceptible individuals will produce more than one infected individual; this leads to the disease invading the susceptible population.

4. Local Stability Analysis

In this section, the local stability analyses of all possible equilibrium points of model (1) are discussed by determining the Jacobian matrix with their eigenvalues. Now the general Jacobian matrix of model (1) can be written:

$$J = \begin{bmatrix} -\frac{\beta_1 I_1}{1 + I_1} - \beta_2 I_2 - \mu & -\frac{\beta_1 S}{(1 + I_1)^2} - p_1 & -\beta_2 S + (\gamma - p_2) & \eta \\ \frac{\beta_1 I_1}{1 + I_1} & -\frac{\beta_1 S I_1}{(1 + I_1)^2} + f(S, I_1) & 0 & 0 \\ \beta_2 I_2 & 0 & \beta_2 S - (\mu + \alpha_2 + \gamma - p_2) & 0 \\ 0 & \delta & 0 & -(\eta + \mu) \end{bmatrix}, \quad (14)$$

where $f(S,I_1) = \beta_1 S/(1+I_1) - (\mu+\alpha_1+\delta-p_1)$. Therefore the local stability results near the above equilibrium points can be presented in the following theorems.

Theorem 2. *The disease-free equilibrium $E_0 = (S_0,0,0,0)$ is locally asymptotically stable when $R_0 < 1$ and unstable for $R_0 > 1$.*

Proof. The characteristic equation of the Jacobian matrix of model (1) at the disease-free equilibrium can be written as

$$(\lambda+\mu)(\lambda-\beta_1 S_0+(\mu+\alpha_1+\delta-p_1)) \cdot (\lambda-\beta_2 S_0+(\mu+\alpha_2+\gamma-p_2))(\lambda+(\eta+\mu)) = 0. \quad (15)$$

So, if $R_0 < 1$, then according to (6), (15) has four negative real roots (eigenvalues). Hence, the DFE is locally asymptotically stable. Further, for $R_0 > 1$ (15) has at least one positive eigenvalue and then the DFE is a saddle point. □

Theorem 3. *The first disease-free equilibrium point $E_1 = (\overline{S},0,\overline{I}_2,0)$ of model (1) is locally asymptotically stable provided that*

$$T_2 > 1 > T_1. \quad (16)$$

Proof. The characteristic equation of the Jacobian matrix of model (1) at E_1 can be written as

$$\left[\overline{\lambda}-\beta_1\overline{S}+(\mu+\alpha_1+\delta-p_1)\right]\left[\overline{\lambda}+(\eta+\mu)\right] \cdot \left[\overline{\lambda}^2 - Tr_1\overline{\lambda}+D_1\right] = 0; \quad (17)$$

here $Tr_1 = -(\beta_2\overline{I}_2+\mu) < 0$ and $D_1 = \beta_2\overline{I}_2(\beta_2\overline{S}-(\gamma-p_1)) > 0$ due to condition (16). Hence both the eigenvalues $\overline{\lambda}_S$ and $\overline{\lambda}_{I_2}$, which describe the dynamics in the S-direction and I_2-direction, respectively, have negative real parts. Moreover, from (17), the eigenvalue in the I_1-direction can be written as

$$\overline{\lambda}_{I_1} = \beta_1\overline{S}-(\mu+\alpha_1+\delta-p_1) = \left(1-\frac{T_2}{T_1}\right)\beta_1\overline{S}. \quad (18)$$

Thus under the given condition (16), we have $\overline{\lambda}_{I_1} < 0$, while $\overline{\lambda}_R = -(\eta+\mu)$ is always negative. Hence, E_1 is locally asymptotically stable. □

Theorem 4. *The second disease-free equilibrium point $E_2 = (\widetilde{S},\widetilde{I}_1,0,\widetilde{R})$ of model (1) is locally asymptotically stable provided that*

$$T_1 > 1 > T_2\left(1+\widetilde{I}_1\right), \quad (19a)$$

$$\widetilde{P}_2 > \delta, \quad (19b)$$

where $\widetilde{P}_2$ is given in the proof.

Proof. The characteristic equation of the Jacobian matrix of model (1) at E_2 can be written as

$$\left[\widetilde{\lambda}-\beta_2\widetilde{S}+(\mu+\alpha_2+\gamma-p_2)\right]\left[\widetilde{\lambda}^3+A\widetilde{\lambda}^2+B\widetilde{\lambda}+C\right] = 0; \quad (20)$$

here

$$\begin{aligned} A &= \left(\widetilde{P}_1+\widetilde{P}_2\right)+(\eta+\mu), \\ B &= \left(\widetilde{P}_1+\widetilde{P}_2\right)(\eta+\mu)+\left(\widetilde{P}_1\widetilde{P}_2+\widetilde{P}_3\right), \\ C &= (\eta+\mu)\left(\widetilde{P}_1\widetilde{P}_2+\widetilde{P}_3\right)-\widetilde{P}_4 \end{aligned} \quad (21)$$

with

$$\begin{aligned} \widetilde{P}_1 &= \left(\frac{\beta_1\widetilde{I}_1}{1+\widetilde{I}_1}+\mu\right) > 0, \\ \widetilde{P}_2 &= \frac{\beta_1\widetilde{S}\widetilde{I}_1}{\left(1+\widetilde{I}_1\right)^2} > 0, \\ \widetilde{P}_3 &= \left(\frac{\beta_1\widetilde{I}_1}{1+\widetilde{I}_1}\left(\frac{\beta_1\widetilde{I}_1}{\left(1+\widetilde{I}_1\right)^2}+p_1\right)\right) > 0, \\ \widetilde{P}_4 &= \frac{\eta\delta\beta_1\widetilde{I}_1}{1+\widetilde{I}_1} > 0. \end{aligned} \quad (22)$$

Clearly, the eigenvalue $\widetilde{\lambda}_{I_2}$ in the I_2-direction can be written as

$$\begin{aligned} \widetilde{\lambda}_{I_2} &= \beta_2\widetilde{S}-(\mu+\alpha_2+\gamma-p_2) \\ &= \left(1-\frac{T_1}{T_2\left(1+\widetilde{I}_1\right)}\right)\beta_2\widetilde{S}, \end{aligned} \quad (23)$$

and thus $\widetilde{\lambda}_{I_2} < 0$ under the condition (19a). In addition from (20) we have $A > 0$ always, while C can be written as

$$\begin{aligned} C &= \widetilde{p}_5\frac{\widetilde{P}_2}{\delta}+\mu\widetilde{p}_5\frac{\widetilde{P}_2}{\delta}+\widetilde{p}_2\left(\beta_2 I_2^*(\eta+\mu)\right) \\ &+ \widetilde{p}_2\mu(\eta+\mu)+\widetilde{p}_3(\eta+\mu)+\widetilde{p}_4\left(\widetilde{p}_2+(\eta+\mu)\right) \\ &- \widetilde{p}_5. \end{aligned} \quad (24)$$

Hence, $C > 0$ provided that the sufficient condition (19b) holds. Further it is easy to verify that

$$AB-C = \left(\widetilde{P}_1+\widetilde{P}_2\right)\left(B+(\eta+\mu)^2\right)+\widetilde{P}_4 > 0. \quad (25)$$

Hence, due to the Routh-Hurwitz criterion the third-degree polynomial term in (20) has roots (eigenvalues) with negative real parts. Hence E_2 is locally asymptotically stable. □

Theorem 5. *The endemic equilibrium point $E_3 = (S^*,I_1^*,I_2^*,R^*)$ of model (1) is locally asymptotically stable provided that*

$$T_1 > T_2 > 1, \quad (26a)$$

$$\overline{\overline{Q}} > q_5 > \delta, \quad (26b)$$

where $\overline{\overline{Q}}$ and q_5 are given in the proof.

Proof. The characteristic equation of the Jacobian matrix of model (1) at E_3 can be written as

$$\lambda^{*4} + A_1\lambda^{*3} + B_1\lambda^{*2} + C_1\lambda^{*} + D_1 = 0. \tag{27}$$

Here

$$\begin{aligned} A_1 &= (q_1 + q_2) + (\eta + \mu) > 0, \\ B_1 &= (\eta + \mu)(q_1 + q_2) + q_1q_2 + q_3 + q_4 > 0, \\ C_1 &= (q_1q_2 + q_3)(\eta + \mu) + q_4(q_2 + (\eta + \mu)) - q_5, \\ D_1 &= q_4q_2(\eta + \mu) > 0 \end{aligned} \tag{28}$$

with

$$\begin{aligned} q_1 &= \frac{\beta_1 I_1^*}{1 + I_1^*} + \beta_2 I_2^* + \mu; \\ q_2 &= \frac{\beta_1 S^* I_1^*}{(1 + I_1^*)^2}; \\ q_3 &= \frac{\beta_1 I_1^*}{1 + I_1^*}\left(\frac{\beta_1 S^*}{(1 + I_1^*)^2} + p_1\right), \\ q_4 &= \beta_2 I_2^*\left[\beta_2 S^* - (\gamma - p_2)\right]; \\ q_5 &= \frac{\eta\delta\beta_1 I_1^*}{1 + I_1^*}. \end{aligned} \tag{29}$$

Obviously, $q_i > 0$, $i = 1, 2, 3, 5$, while q_4 is positive under condition (26a). Now, by using the values of q_i and the sufficient condition (26b), then straightforward computation gives

$$C_1 = \overline{\overline{Q}} + q_4(q_2 + (\eta + \mu)) - q_5 > 0, \tag{30}$$

and here $\overline{\overline{Q}} = (q_1q_2 + q_3)(\eta + \mu)$. Moreover we have

$$\begin{aligned} A_1B_1C_1 &= \left[Q + (q_1 + q_2)^2(\eta + \mu) + q_4q_1 \right. \\ &\quad \left. + \left(q_1q_2 + q_3 + (\eta + \mu)^2\right)(q_1 + q_2)\right]\left[Q - q_5\right], \\ C_1^2 + A_1^2D_1 &= (Q - q_5)^2 + \left[(q_1 + q_2) + (\eta + \mu)\right]^2 \\ &\quad \cdot (q_4q_2(\eta + \mu)), \end{aligned} \tag{31}$$

where $Q = \overline{\overline{Q}} + q_4(q_2 + (\eta + \mu))$. Therefore we obtain that

$$\begin{aligned} &A_1B_1C_1 - C_1^2 - A_1^2D_1 \\ &\quad = F\overline{\overline{Q}} + q_4\overline{\overline{Q}}(q_1 + q_2) + q_4q_1A_1(\eta + \mu)^2 \\ &\quad + q_4^2q_1(q_2 + (\eta + \mu)) \\ &\quad + q_4q_2(q_1 + q_2)(q_1q_2 + q_3) - Fq_5 + C_1q_5. \end{aligned} \tag{32}$$

Here $F = (q_1+q_2)^2(\eta+\mu)+(q_1q_2+q_3+(\eta+\mu)^2)(q_1+q_2)+q_4q_1$.

Hence, according to condition (26b) it is easy to verify that $A_1B_1C_1 - C_1^2 - A_1^2D_1 > 0$. Therefore, all the coefficients of (27) are positive and $A_1B_1C_1 - C_1^2 - A_1^2D_1 > 0$. Hence, due to the Routh-Hurwitz criterion all the eigenvalues ($\lambda_S^*, \lambda_{I_1}^*, \lambda_{I_2}^*$ and λ_R^*) of the Jacobian matrix near the endemic equilibrium point E_3 have negative real parts. Thus, the proof is complete. □

5. Global Stability Analysis

This section deals with the global stability of the equilibrium points of model (1) using Lyapunov methods with LaSalle's invariant principle. The obtained results are presented in the following theorems.

Theorem 6. *Assume that DFE $E_0 = (S_0, 0, 0, 0)$ of model (1) is locally asymptotically stable; then it is global asymptotically stable in Ω.*

Proof. Consider $\widehat{V} : \Omega \to \mathbb{R}$ that is defined by

$$\widehat{V}(S, I_1, I_2, R) = \sum_{i=1}^{2} I_i. \tag{33}$$

Computing the derivative of this positive semidefinite function with respect to time along the solution of model (1) and then simplifying the resulting terms give

$$\begin{aligned} \frac{d\widehat{V}}{dt} &= \left(\frac{\beta_1 S}{1 + I_1} - (\mu + \alpha_1 + \delta - p_1)\right) I_1 \\ &\quad + (\beta_2 S - (\mu + \alpha_2 + \gamma - p_2)) I_2. \end{aligned} \tag{34}$$

Since the solution of model (1) is bounded by $S_0 = \Lambda/\mu$ as $t \to \infty$,

$$\begin{aligned} \frac{d\widehat{V}}{dt} &\le (\beta_1 S_0 - (\mu + \alpha_1 + \delta - p_1)) I_1 \\ &\quad + (\beta_2 S_0 - (\mu + \alpha_2 + \gamma - p_2)) I_2 \\ &= (\mu + \alpha_1 + \delta - p_1)(T_1 - 1) I_1 \\ &\quad + (\mu + \alpha_2 + \gamma - p_2)(T_2 - 1) I_2. \end{aligned} \tag{35}$$

Since $T_i < 1$, $i = 1, 2$, due to the local stability condition of E_0 then $d\widehat{V}/dt < 0$. Also we have that $d\widehat{V}/dt = 0$ on the set $\{(S, I_1, I_2, R) \in \Omega : I_1 = I_2 = 0\}$, so $d\widehat{V}/dt$ is negative semidefinite and hence according to Lyapunov first theorem E_0 is globally stable point. Now, since on this set we have

$$\frac{dS}{dt} = \Lambda - \mu S + \eta R = 0, \tag{36}$$

if and only if $S = S_0$, $R = 0$, thus the largest invariant set contained in this set is reduced to the disease-free equilibrium point E_0. Hence according to LaSalle's invariant principle [18], E_0 is attractive point and hence it is globally asymptotically stable in Ω. □

Theorem 7. *Assume that the first disease-free equilibrium point $E_1 = (\overline{S}, 0, \overline{I}_2, 0)$ is locally asymptotically stable; then it is global asymptotically stable in Ω provided that*

$$I_2 > \overline{I}_2. \tag{37}$$

Proof. Consider that $L : \Omega \rightarrow \mathbb{R}$ that is defined by

$$L(S, I_1, I_2, R) = \frac{1}{2}\left[\left(S-\bar{S}\right)+I_1+\left(I_2-\bar{I}_2\right)+R\right]^2 + \frac{(2\mu+\alpha_1)}{\beta_1}I_1 + \frac{(2\mu+\alpha_2)}{\beta_2}\left(I_2-\bar{I}_2-\bar{I}_2\ln\frac{I_2}{\bar{I}_2}\right) + \frac{(2\mu+\alpha_1)}{2\delta}R^2. \tag{38}$$

Clearly L is continuous and positive definite function. Now by taking the derivative of L with respect to time along the solution of model (1), we get after simplifying the resulting terms that

$$\begin{aligned}\frac{dL}{dt} &= -\mu\left(S-\bar{S}\right)^2 - \mu R^2 - 2\mu R\left(S-\bar{S}\right) \\ &\quad - (2\mu+\alpha_1)\left(S-\bar{S}\right)I_1 \\ &\quad - (2\mu+\alpha_2)\left(S-\bar{S}\right)\left(I_2-\bar{I}_2\right) \\ &\quad - (2\mu+\alpha_1+\alpha_2)\left(I_2-\bar{I}_2\right)I_1 \\ &\quad - (2\mu+\alpha_1)I_1R - (\mu+\alpha_2)\left(I_2-\bar{I}_2\right)^2 \\ &\quad - (2\mu+\alpha_2)\left(I_2-\bar{I}_2\right)R - (\mu+\alpha_1)I_1^2,\end{aligned} \tag{39}$$

$$\begin{aligned}\frac{dL}{dt} &\le -\mu\left[\left(S-\bar{S}\right)+R\right]^2 - \mu\left[I_1+\left(I_2-\bar{I}_2\right)\right]^2 \\ &\quad - 2\mu\left(I_2-\bar{I}_2\right)R \\ &\quad + \frac{2\mu(\mu+\alpha_1+\delta-p_1)}{\beta_1}\left[\frac{T_1}{T_2}-1\right]I_1 \\ &\quad - \frac{\mu(\eta+\mu)}{\delta}R^2.\end{aligned}$$

Hence according to local stability condition (16) along with the sufficient condition (37) it obtains that dL/dt is negative definite function. Thus due to Lyapunov second theorem E_1 is global asymptotically stable in Ω. □

Theorem 8. *Assume that the second disease-free equilibrium point $E_2 = (\tilde{S}, \tilde{I}_1, 0, \tilde{R})$ of model (1) is locally asymptotically stable; then it is global asymptotically stable in Ω if*

$$q_{12}^2 < q_{11}q_{22}, \tag{40a}$$

$$q_{14}^2 < q_{11}q_{44}, \tag{40b}$$

$$q_{23}^2 < q_{22}q_{33}, \tag{40c}$$

$$q_{34}^2 < q_{33}q_{44}, \tag{40d}$$

where

$$\begin{aligned} q_{11} &= \mu; \\ q_{22} &= (\mu+\alpha_1) + \frac{\tilde{S}}{\left(1+\tilde{I}_1\right)\left(1+I_1\right)}; \\ q_{33} &= (\mu+\alpha_2); \\ q_{44} &= \frac{(2\mu+\alpha_1)(\eta+\mu)}{2\delta} + \mu; \\ q_{12} &= \frac{1}{1+I_1} - (2\mu+\alpha_1); \\ q_{14} &= 2\mu; \\ q_{23} &= (2\mu+\alpha_1+\alpha_2); \\ q_{34} &= (2\mu+\alpha_2). \end{aligned} \tag{41}$$

Proof. Consider the function $W : \Omega \rightarrow \mathbb{R}$ that is defined by

$$\begin{aligned} W(S, I_1, I_2, R) &= \frac{1}{2}\left[\left(S-\tilde{S}\right)+\left(I_1-\tilde{I}_1\right)+I_2+\left(R-\tilde{R}\right)\right]^2 \\ &\quad + \frac{1}{\beta_1}\left(I_1-\tilde{I}_1-\tilde{I}_1\ln\frac{I_1}{\tilde{I}_1}\right) + \frac{(2\mu+\alpha_2)}{\beta_2}I_2 \\ &\quad + \frac{(2\mu+\alpha_1)}{2\delta}\left(R-\tilde{R}\right)^2. \end{aligned} \tag{42}$$

Clearly W is continuous and positive definite function. Now by taking the derivative of W with respect to time along the solution of model (1), we get after simplifying the resulting terms that

$$\begin{aligned}\frac{dW}{dt} &= -q_{11}\left(S-\tilde{S}\right)^2 + q_{12}\left(S-\tilde{S}\right)\left(I_1-\tilde{I}_1\right) \\ &\quad - q_{14}\left(S-\tilde{S}\right)\left(R-\tilde{R}\right) - q_{22}\left(I_1-\tilde{I}_1\right)^2 \\ &\quad - q_{23}\left(I_1-\tilde{I}_1\right)I_2 - q_{33}I_2^2 - q_{34}\left(R-\tilde{R}\right)I_2 \\ &\quad - q_{44}\left(R-\tilde{R}\right)^2 \\ &\quad + \frac{2\mu(\mu+\alpha_2+\gamma-p_2)}{\beta_2}\left[\frac{T_2\left(1+\tilde{I}_1\right)}{T_1}-1\right]I_2.\end{aligned} \tag{43}$$

Now by using the given conditions (40a)–(40d) we get that

$$\begin{aligned}\frac{dW}{dt} &\le -\left[\sqrt{\frac{q_{11}}{2}}\left(S-\tilde{S}\right) - \sqrt{\frac{q_{22}}{2}}\left(I_1-\tilde{I}_1\right)\right]^2 \\ &\quad - \left[\sqrt{\frac{q_{22}}{2}}\left(I_1-\tilde{I}_1\right) + \sqrt{\frac{q_{33}}{2}}I_2\right]^2\end{aligned}$$

$$-\left[\sqrt{\frac{q_{11}}{2}}\left(S-\tilde{S}\right)+\sqrt{\frac{q_{44}}{2}}\left(R-\tilde{R}\right)\right]^2$$
$$-\left[\sqrt{\frac{q_{44}}{2}}\left(R-\tilde{R}\right)+\sqrt{\frac{q_{33}}{2}}I_2\right]^2$$
$$+\frac{2\mu\left(\mu+\alpha_2+\gamma-p_2\right)}{\beta_2}\left[\frac{T_2\left(1+\tilde{I}_1\right)}{T_1}-1\right]I_2. \quad (44)$$

Hence according to local stability condition (19a) it obtains that dW/dt is negative definite function. Thus due to Lyapunov second theorem E_1 is global asymptotically stable in Ω. □

Theorem 9. *Assume that the endemic equilibrium point $E_3 = (S^*, I_1^*, I_2^*, R^*)$ of model (1) is locally asymptotically stable; then it is global asymptotically stable in Ω if*

$$\alpha_{14}^2 < \alpha_{11}\alpha_{44}, \quad (45a)$$
$$\alpha_{12}^2 < \alpha_{11}\alpha_{22}, \quad (45b)$$
$$\alpha_{24}^2 < \alpha_{22}\alpha_{44}, \quad (45c)$$

where

$$\begin{aligned}\alpha_{11} &= \left(\frac{\beta_1 I_1}{1+I_1}+\beta_2 I_2+\mu\right),\\ \alpha_{22} &= \beta_1 S^*,\\ \alpha_{44} &= (\eta+\mu),\\ \alpha_{12} &= \frac{\beta_1}{1+I_1}-\left(\frac{\beta_1 S^*}{(1+I_1)(1+I_1^*)}+p_1\right),\\ \alpha_{24} &= \delta,\\ \alpha_{14} &= \eta.\end{aligned} \quad (46)$$

Proof. Consider the function $V:\Omega\to\mathbb{R}$ that is defined by

$$\begin{aligned}V(S,I_1,I_2,R) = {} & \frac{(S-S^*)^2}{2}+\left(I_1-I_1^*-I_1^*\ln\frac{I_1}{I_1^*}\right)\\ & +\frac{(2\mu+\alpha_2)}{\beta_2}\left(I_2-I_2^*-I_2^*\ln\frac{I_2}{I_2^*}\right)\\ & +\frac{(R-R^*)^2}{2}.\end{aligned} \quad (47)$$

Clearly the function V is continuous and positive definite function. By taking the derivative of V with respect to time along the solution of model (1), we get after simplifying the resulting terms that

$$\begin{aligned}\frac{dV}{dt} = {} & -\alpha_{11}(S-S^*)^2-\alpha_{22}(I_1-I_1^*)^2\\ & -\alpha_{44}(R-R^*)^2+\alpha_{12}(S-S^*)(I_1-I_1^*)\\ & +\alpha_{14}(R-R^*)(S-S^*)\\ & +\alpha_{24}(I_1-I_1^*)(R-R^*).\end{aligned} \quad (48)$$

Then by using the given conditions (45a)–(45c) we obtain that

$$\begin{aligned}\frac{dV}{dt} < {} & -\left[\sqrt{\alpha_{11}}(S-S^*)+\sqrt{\alpha_{22}}(I_1-I_1^*)\right]^2\\ & -\left[\sqrt{\alpha_{11}}(S-S^*)+\sqrt{\alpha_{44}}(R-R^*)\right]^2\\ & -\left[\sqrt{\alpha_{44}}(R-R^*)+\sqrt{\alpha_{22}}(I_1-I_1^*)\right]^2.\end{aligned} \quad (49)$$

Hence, dV/dt is negative semidefinite, and $dV/dt = 0$ on the set $\{(S,I_1,I_2,R)\in\Omega : S = S^*,\ I_1 = I_1^*,\ I_2 > 0,\ R = R^*\}$, so according to Lyapunov first theorem E_3 is globally stable point. Further, since on this set we have

$$\begin{aligned}\frac{dS}{dt} = {} & \Lambda-\left(\frac{\beta_1 I_1^*}{1+I_1^*}+\beta_2 I_2\right)S^*+(\gamma-p_2)I_2-\mu S^*\\ & -p_1 I_1^*+\eta R^* = 0,\end{aligned} \quad (50)$$

if and only if $I_2 = I_2^*$, then the largest compact invariant set contained in this set is reduced to the endemic equilibrium point E_3. Hence according to LaSalle's invariant principle [18], E_3 is attractive point and hence it is globally asymptotically stable in Ω. □

6. Bifurcation Analysis

In this section the local bifurcations near the equilibrium points of model (1) are investigated as shown in the following theorems with the help of Sotomayor theorem [20]. Note that model (1) can be rewritten in a vector form $dX/dt = f(X)$, where $X = (S,I_1,I_2,R)^T$ and $f = (f_1,f_2,f_3,f_4)^T$ with f_i, $i = 1,2,3,4$, are given in the right hand side of model (1). Moreover, straightforward computation gives that the general second derivative of the Jacobian matrix (14) can be written:

$$D^2 f(X,\beta)(U,U) = \begin{bmatrix}\dfrac{-2\beta_1\zeta_1\zeta_2}{(1+I_1)^2}-2\beta_2\zeta_1\zeta_3+\dfrac{2\beta_1 S(\zeta_2)^2}{(1+I_1)^3}\\ \dfrac{2\beta_1\zeta_1\zeta_2}{(1+I_1)^2}-\dfrac{2\beta_1 S(\zeta_2)^2}{(1+I_1)^3}\\ 2\beta_2\zeta_1\zeta_3\\ 0\end{bmatrix}, \quad (51)$$

where β is any bifurcation parameter and $U = (\zeta_1,\zeta_2,\zeta_3,\zeta_4)^T$ is any eigenvector.

Theorem 10. *Assume that $T_2 < 1$; then as T_1 passes through the value $T_1 = 1$, model (1) near the disease-free equilibrium E_0 has*

(1) *no saddle-node bifurcation;*

(2) *a transcritical bifurcation;*

(3) *no pitchfork bifurcation.*

Proof. Since $T_1 = \beta_1 S_0/(\mu + \alpha_1 + \delta - p_1) = 1$; then $\beta_1 = (\mu + \alpha_1 + \delta - p_1)/S_0 = \beta_1^*$. Now straightforward computation shows that the Jacobian matrix of model (1) at E_0 with $\beta_1 = \beta_1^*$ has zero eigenvalue ($\lambda_{I_1} = 0$) and can be written as follows:

$$J_0 = \begin{bmatrix} -\mu & -\beta_1^* S_0 - p_1 & -\beta_2 S_0 + (\gamma - p_2) & \eta \\ 0 & 0 & 0 & 0 \\ 0 & 0 & \beta_2 S_0 - (\mu + \alpha_2 + \gamma - p_2) & 0 \\ 0 & \delta & 0 & -(\eta + \mu) \end{bmatrix}. \tag{52}$$

Let $V = (v_1, v_2, v_3, v_4)^T$ be the eigenvector corresponding to $\lambda_{I_1} = 0$. Thus $J_0 V = 0$ gives

$$V = \begin{bmatrix} a v_2 \\ v_2 \\ 0 \\ b v_2 \end{bmatrix}, \tag{53}$$

where v_2 is any nonzero real number, $a = (\eta\delta/(\eta + \mu) - (\mu + \alpha_1 + \delta))/\mu$, and $b = \delta/(\eta + \mu)$.

Similarly, $W = (w_1, w_2, w_3, w_4)^T$ represents the eigenvector corresponding to eigenvalue $\lambda_{I_1} = 0$ of J_0^T. Hence $J_0^T W = 0$ gives that

$$W = \begin{bmatrix} 0 \\ w_2 \\ 0 \\ 0 \end{bmatrix}, \tag{54}$$

and here w_2 is any nonzero real number. Now, since

$$\frac{df}{d\beta_1} = f_{\beta_1}(X, \beta_1) = \left(\frac{-I_1 S}{1 + I_1}, \frac{I_1 S}{1 + I_1}, 0, 0 \right)^T, \tag{55}$$

thus $f_{\beta_1}(E_0, \beta_1^*) = (0, 0, 0, 0)^T$, which gives $W^T f_{\beta_1}(E_0, \beta_1^*) = 0$.

Thus, according to Sotomayor's theorem for local bifurcation, model (1) has no saddle-node bifurcation near DFE at $\beta_1 = \beta_1^*$.

Now since

$$Df_{\beta_1}(E_0, \beta_1^*) = \begin{bmatrix} 0 & -S_0 & 0 & 0 \\ 0 & S_0 & 0 & 0 \\ 0 & 0 & 0 & 0 \\ 0 & 0 & 0 & 0 \end{bmatrix}, \tag{56}$$

then, $W^T (Df_{\beta_1}(E_0, \beta_1^*) V) = S_0 v_2 w_2 \neq 0$. Now, by substituting E_0 and β_1^* in (51) we get

$$D^2 f(E_0, \beta_1^*) \cdot (V, V) = \begin{bmatrix} -2\beta_1^* v_1 v_2 - 2\beta_2 v_1 v_3 + 2\beta_1^* S_0 (v_2)^2 \\ 2\beta_1^* v_1 v_2 - 2\beta_1^* S_0 (v_2)^2 \\ 2\beta_2 v_1 v_3 \\ 0 \end{bmatrix}. \tag{57}$$

Therefore,

$$\begin{aligned} W^T &\left(D^2 f(E_0, \beta_1^*) \cdot (V, V) \right) \\ &= \left(2\beta_1^* v_1 v_2 - 2\beta_1^* S_0 (v_2)^2 \right) w_2 \\ &= \left[2\beta_1^* (v_2)^2 (a - S_0) \right] w_2 \neq 0. \end{aligned} \tag{58}$$

So, according to Sotomayor's theorem model (1) has a transcritical bifurcation at E_0 with parameter $\beta_1 = \beta_1^* \approx (T_1 = 1)$, while the pitchfork bifurcation cannot occur. □

Note that similar results as those of Theorem 10 are obtained at $T_2 = 1$ or $\beta_2 = \beta_2^* = (\mu + \alpha_2 + \gamma - p_2)/S_0$.

Theorem 11. *Assume that $T_1 = T_2$; then model (1) near the first disease-free equilibrium point $E_1 = (\overline{S}, 0, \overline{I}_2, 0)$ has*

(1) *no saddle-node bifurcation;*

(2) *a transcritical bifurcation;*

(3) *no pitchfork bifurcation.*

Proof. There are two cases; in the first case, it is assumed that $T_1 = T_2 = 1$; then straightforward computation shows that $J(E_1) = J(E_0)$; that is, $\overline{I}_2 = 0$. So (by Theorem 10) model (1) has no bifurcation and then the proof is complete.

Now in the second case it is assumed that $T_1 = T_2 \neq 1$ or equivalently $\beta_1 = \beta_1^{**} = \beta_2(\mu+\alpha_1+\delta-p_1)/(\mu+\alpha_2+\gamma-p_2)$. So straightforward computation shows that the Jacobian matrix of model (1) at E_1 with $\beta_1 = \beta_1^{**}$ has zero eigenvalue ($\lambda_{I_1} = 0$) and can be written as follows:

$$J_1 = \begin{bmatrix} -\beta_2 \overline{I}_2 - \mu & -\beta_1^{**} \overline{S} - p_1 & -\beta_2 \overline{S} + (\gamma - p_2) & \eta \\ 0 & 0 & 0 & 0 \\ \beta_2 \overline{I}_2 & 0 & \beta_2 \overline{S} - (\mu + \alpha_2 + \gamma - p_2) & 0 \\ 0 & \delta & 0 & -(\eta + \mu) \end{bmatrix}. \tag{59}$$

Let $X = (x_1, x_2, x_3, x_4)^T$ be the eigenvector corresponding to $\lambda_{I_1} = 0$, which satisfies $J_1 X = 0$, so we get

$$X = \begin{bmatrix} 0 \\ x_2 \\ a_1 x_2 \\ b x_2 \end{bmatrix}, \tag{60}$$

where x_2 is any nonzero real number, $a_1 = -[(\beta_1 \overline{S} + p_1) - \eta\delta/(\eta + \mu)]/\beta_1 \overline{S} - (\gamma - p_2)$, and $b = \delta/(\eta + \mu)$.

Similarly the eigenvector $Y = (y_1, y_2, y_3, y_4)^T$ that is corresponding to the eigenvalue $\lambda_{I_1} = 0$ of J_1^T satisfies $J_1^T Y = 0$, so we get

$$Y = \begin{bmatrix} 0 \\ y_2 \\ 0 \\ 0 \end{bmatrix}, \tag{61}$$

and here y_2 is any nonzero real number. Now, since

$$\frac{df}{d\beta_1} = f_{\beta_1}(X, \beta_1) = \left(\frac{-I_1 S}{1 + I_1}, \frac{I_1 S}{1 + I_1}, 0, 0 \right)^T \tag{62}$$

then by substituting the values of E_1 and β_1^{**} we obtain that $f_{\beta_1}(E_1, \beta_1^{**}) = (0, 0, 0, 0)^T$ and hence we get that $Y^T f_{\beta_1}(E_1, \beta_1^{**}) = 0$.

Thus according to Sotomayor's theorem for local bifurcation, model (1) has no saddle-node bifurcation near E_1 at $\beta_1 = \beta_1^{**}$. Now since

$$Df_{\beta_1}(E_1, \beta_1) = \begin{bmatrix} 0 & -\overline{S} & 0 & 0 \\ 0 & \overline{S} & 0 & 0 \\ 0 & 0 & 0 & 0 \\ 0 & 0 & 0 & 0 \end{bmatrix}, \tag{63}$$

then $Y^T (Df_{\beta_1}(E_1, \beta_1^{**}) X) = \overline{S} x_2 y_2 \neq 0$. Thus by substituting E_1 and β_1^{**} in (51) we get

$$D^2 f(E_1, \beta_1^{**}) \cdot (X, X) = \begin{bmatrix} -2\beta_1^{**} x_1 x_2 - 2\beta_2 x_1 x_3 + 2\beta_1^{**} \overline{S} (x_2)^2 \\ 2\beta_1^{**} x_1 x_2 - 2\beta_1^{**} \overline{S} (x_2)^2 \\ 2\beta_2 x_1 x_3 \\ 0 \end{bmatrix}. \tag{64}$$

Therefore, $[Y^T (D^2 f(E_1, \beta_1^{**}) \cdot (X, X))] = -2\beta_1^{**} \overline{S} (x_2)^2 y_2 \neq 0$.

So, model (1) has a transcritical bifurcation at E_1 with parameter $\beta_1 = \beta_1^{**} \approx (T_1 = T_2 = 1)$, while the pitchfork bifurcation cannot occur and hence the proof is complete. □

Theorem 12. *Assume that condition (19b) holds and let $T_1 = T_2(1 + \widetilde{I}_1)$; then model (1) near the second disease-free equilibrium point $E_2 = (\widetilde{S}, \widetilde{I}_1, 0, \widetilde{R})$ undergoes*

(1) *no saddle-node bifurcation;*

(2) *a transcritical bifurcation;*

(3) *no pitchfork bifurcation.*

Proof. From $T_1 = T_2(1 + \widetilde{I}_1)$ it is obtained that

$$\beta_2 = \beta_2^{**} = \frac{\beta_1 (\mu + \alpha_2 + \gamma - p_2)}{(\mu + \alpha_1 + \delta - p_1)(1 + \widetilde{I}_1)}, \tag{65}$$

and then straightforward computation shows that the Jacobian matrix of model (1) at E_2 with $\beta_2 = \beta_2^{**}$ has zero eigenvalue ($\lambda_{I_2} = 0$) and can be written as follows:

$$J_2 = \begin{bmatrix} \frac{-\beta_1 \widetilde{I}_1}{1 + \widetilde{I}_1} - \mu & \frac{-\beta_1 \widetilde{S}}{(1 + \widetilde{I}_1)^2} - p_1 & -\beta_2^{**} \widetilde{S} + (\gamma - p_2) & \eta \\ \frac{\beta_1 \widetilde{I}_1}{1 + \widetilde{I}_1} & \frac{-\beta_1 \widetilde{S} \widetilde{I}_1}{(1 + \widetilde{I}_1)^2} & 0 & 0 \\ 0 & 0 & 0 & 0 \\ 0 & \delta & 0 & -(\eta + \mu) \end{bmatrix}. \tag{66}$$

Further the eigenvector $L = (l_1, l_2, l_3, l_4)^T$ that is corresponding to $\lambda_{I_2} = 0$ satisfies $J_2 L = 0$, so we get

$$L = \begin{bmatrix} \frac{\widetilde{S}}{1 + \widetilde{I}_1} l_2 \\ l_2 \\ \frac{-E}{\beta_2^{**} \widetilde{S} - (\gamma - p_2)} l_2 \\ b l_2 \end{bmatrix}, \tag{67}$$

where l_2 is any nonzero real number and $E = (\beta_1 \widetilde{S}/(1 + \widetilde{I}_1)^2)(\widetilde{I}_1 + 1) + \mu\widetilde{S}/1 + \widetilde{I}_1 + p_1 - \eta b$.

Similarly the eigenvector $K = (k_1, k_2, k_3, k_4)^T$ that is corresponding to eigenvalue $\lambda_{I_2} = 0$ of J_2^T satisfies $J_2^T K = 0$, so we get

$$K = \begin{bmatrix} 0 \\ 0 \\ k_3 \\ 0 \end{bmatrix} \tag{68}$$

and here k_3 is any nonzero real number. Now, since $df/d\beta_2 = f_{\beta_2}(X, \beta_2) = (-I_2 S, I_2 S, 0, 0)^T$, therefore $f_{\beta_2}(E_2, \beta_2^{**}) = (0, 0, 0, 0)^T$, which yields $K^T f_{\beta_2}(E_2, \beta_2^{**}) = 0$. Consequently according to Sotomayor's theorem for local bifurcation, model (1) has no saddle-node bifurcation near E_2 at $\beta_2 = \beta_2^{**}$.

Now since

$$Df_{\beta_2}(E_2, \beta_2) = \begin{bmatrix} 0 & 0 & -\widetilde{S} & 0 \\ 0 & 0 & 0 & 0 \\ 0 & 0 & \widetilde{S} & 0 \\ 0 & 0 & 0 & 0 \end{bmatrix} \tag{69}$$

then $K^T(Df_{\beta_2}(E_2, \beta_2^{**})L) = \widetilde{S}l_3k_3 \neq 0$. Now, by substituting E_2 and β_2^{**} in (51) we get

$$D^2 f(E_2, \beta_2^{**}) \cdot (L, L)$$
$$= \begin{bmatrix} \dfrac{-\beta_1}{(1+\widetilde{I}_1)^2} l_1 l_2 - 2\beta_2^{**} l_1 l_3 - \dfrac{\beta_1 l_1 l_2}{(1+\widetilde{I}_1)^2} + \dfrac{2\beta_1 S(l_2)^2}{(1+\widetilde{I}_1)^3} \\ \dfrac{\beta_1 l_2}{(1+\widetilde{I}_1)^2} l_1 l_2 + \dfrac{\beta_1}{(1+\widetilde{I}_1)^2} l_1 l_2 - \dfrac{2\beta_1 S(l_2)^2}{(1+\widetilde{I}_1)^3} \\ 2\beta_2^{**} l_1 l_3 \\ 0 \end{bmatrix}. \quad (70)$$

Therefore, $[K^T(D^2 f(E_2, \beta_2^{**}) \cdot (L, L))] = 2\beta_2^{**} l_1 l_3 k_3 \neq 0$. Thus model (1) undergoes a transcritical bifurcation at E_2 with parameter $\beta_2 = \beta_2^{**} \approx (T_1 = T_2(1 + \widetilde{I}_1))$, while the pitchfork bifurcation cannot occur. □

Moreover, the following results are obtained too:

(1) Although $T_1 = T_2 \neq 1$ gives $\beta_2 = \grave{\beta}_2^{**}$, model (1) does not undergo any of the above types of bifurcation near the equilibrium point E_1 with parameter $\beta_2 = \grave{\beta}_2^{**}$.

(2) Although $T_1 = T_2(1 + \widetilde{I}_1)$ gives $\beta_1 = \grave{\beta}_1^{**}$, model (1) does not undergo any of the above types of bifurcation near the equilibrium point E_2 with parameter $\beta_1 = \grave{\beta}_1^{**}$.

(3) The determinant of the Jacobian matrix at E_3, say $J(E_3)$, cannot be zero and hence it has no real zero eigenvalue. So there is no bifurcation near E_3.

Keeping the above in view, in the following theorem we detect of the possibility of having Hopf bifurcation.

Theorem 13. *Assume that condition (26a) holds and let the following conditions be satisfied. Then model (1) undergoes Hopf bifurcation around the endemic equilibrium point when the parameter β_2 crosses a critical positive value $\grave{\beta}_2^{*}$,*

$$N_2 > 0, \qquad N_4 < 0, \quad (71a)$$

$$\sigma_1 > 2\sigma_2, \quad (71b)$$

$$\delta < q_5 < \{q_4 q_2, q_4 q_1\}, \quad (71c)$$

$$q_2 > 1, \quad (71d)$$

and here $\sigma_1 = q_1^2 + q_2^2 + (\eta + \mu)^2$ and $\sigma_2 = (q_1 + q_2)(\eta + \mu) + 2q_1 q_2 + 2q_3 + 2q_4$, while N_2 and N_4 are given in the proof.

Proof. It is well known that, in order for Hopf bifurcation in four-dimensional systems to occur, the following conditions should be satisfied [21, 22]:

(1) The characteristic equation given in (28) has two real and negative eigenvalues and two complex eigenvalues, say, $\lambda^*(\beta_2) = \tau_1(\beta_2) \pm \tau_2(\beta_2)$.

(2) $\tau_1(\grave{\beta}_2^{*}) = 0$.

(3) $(d/d\beta_2)\tau_1(\beta_2)|_{\beta_2 = \grave{\beta}_2^{*}} \neq 0$ (The transversality condition).

Accordingly the first two points are satisfied if and only if

$$\Delta_2 (\grave{\beta}_2^{*}) = A_1 B_1 C_1 - C_1^2 - A_1^2 D_1 = 0,$$
$$A_1^3 - 4\Delta_1 > 0, \quad (72)$$

while the third condition holds provided that

$$\frac{d}{d\beta_2}\tau_1(\beta_2)\bigg|_{\beta_2 = \grave{\beta}_2^{*}} = \frac{-(\Psi(\beta_2)\Theta(\beta_2) + \Phi(\beta_2)\Gamma(\beta_2))}{\Psi(\beta_2)^2 + \Phi(\beta_2)^2} \neq 0. \quad (73)$$

That means $\Psi(\beta_2)\Theta(\beta_2) + \Phi(\beta_2)\Gamma(\beta_2) \neq 0$.

Here

$$\Psi(\beta_2) = (4\tau_1^3 - 12\tau_1\tau_2^2) + 3A_1(\tau_1^2 - \tau_2^2) + 2B_1\tau_1 + C_1,$$
$$\Phi(\beta_2) = (12\tau_1^2\tau_2 - 4\tau_2^3) + 6A_1\tau_1\tau_2 + 2B_1\tau_2,$$
$$\Theta(\beta_2) = \dot{A}_1(\tau_1^3 - 3\tau_1\tau_2^2) + \dot{B}_1(\tau_1^2 - \tau_2^2) + \dot{C}_1\tau_1 + \dot{D}_1,$$
$$\Gamma(\beta_2) = \dot{A}_1(3\tau_1^2\tau_2 - \tau_2^3) + 2\dot{B}_1\tau_1\tau_2 + \dot{C}_1\tau_2. \quad (74)$$

Now, straightforward computation shows the condition

$$\Delta_2(\grave{\beta}_2^{*}) = 0$$
$$\text{gives that } N_1\beta_2^3 + N_2\beta_2^2 + N_3\beta_2 + N_4 = 0, \quad (75)$$

where

$$N_1 = [q_2(\eta + \mu)((\eta + \mu) + q_2) + (\eta + \mu)(\mu + \alpha_2)((\eta + \mu) + 2q_2 + (\mu + \alpha_2))] I_2^{*3} > 0,$$

$$N_2 = [(\eta + \mu)[q_2(\eta + \mu)(3q_6 + 2 + (\eta + \mu)) + q_3((\eta + \mu) + 2(\mu + \alpha_2) + 2q_2) + (\mu + \alpha_2) \cdot [q_6(2(\eta + \mu) + (\mu + \alpha_2)) + q_2((\eta + \mu) + 4q_6 + q_2)] + q_2^2(3q_6 + q_2) + (\mu + \alpha_2)(\eta + \mu)^2 - q_5] + (\mu + \alpha_2)[q_2^2(1 + q_2) + q_2(q_3 + q_6(\mu + \alpha_2))] - q_2 q_5] I_2^{*2},$$

$$N_3 = \left[(\eta+\mu)^2\left[q_2^3 + q_3(2q_6 + 2q_2 + (\eta+\mu)) + 3q_6q_2 - q_5 + (q_6+q_2)q_6(\mu+\alpha_2)\right] + (\eta+\mu) \cdot \left[q_6q_2^2(3q_6+2q_2) + (2q_6+q_2)\left[2q_2q_3 + q_2(\eta+\mu)^2\right] + (\mu+\alpha_2)\left[1+q_6(\eta+\mu)^2 + (2q_6+q_2)(q_6q_2+q_3-2q_6q_5) - q_2q_5\right]\right] + q_2(\mu + \alpha_2)\left[q_6q_2(1+q_2) + (q_6+q_2)q_3 + 1\right] - (q_2(2q_6 + q_2) + q_3)q_5\right]I_2^*,$$

$$N_4 = \left[(\eta+\mu)^2\left[q_6q_2\left(q_6^2+q_6q_2+q_2^2\right) + q_3q_6(q_6 + 2q_2) + (q_6+q_2)(\eta+\mu)(q_6q_2+q_3) + q_3q_2^2 - q_2\right] + (q_6+q_2)(\eta+\mu)\left[q_6^2q_2^2 + 2q_6q_2q_3 + q_3^2\right] + ((\eta+\mu)(q_3-q_6q_2) - q_6(\eta+\mu)(q_6+(\eta+\mu)) - q_2^2(\eta+\mu) - (q_6+q_2)(q_6q_2+q_3))q_5 - q_5^2\right] \tag{76}$$

with q_1, q_2, q_3, q_4, and q_5 given in (28) and $q_6 = q_1 - \beta_2 I_2^*$. Clearly, from condition (71a) there is unique positive root, say, $\beta_2 = \grave{\beta}_2^*$. Consequently by using $\Delta_2 = 0$ in the characteristic equation and then doing some algebraic computation we get four roots,

$$\lambda_{1,2}\left(\grave{\beta}_2^*\right) = \pm i\sqrt{\frac{C_1}{A_1}} = \tau_2\left(\grave{\beta}_2^*\right),$$
$$\lambda_{3,4} = \frac{1}{2}\left(-A_1 \pm \sqrt{A_1^2 - 4\frac{\Delta_1}{A_1}}\right). \tag{77}$$

Now, it is easy to verify that λ_3 and λ_4 are real and negative provided that (71b).

Further for $\beta_2 \in (\grave{\beta}_2^* - \epsilon, \grave{\beta}_2^* + \epsilon)$ the general form of complex eigenvalues can be written as

$$\lambda_1 = \tau_1(\beta_2) + i\tau_2(\beta_2),$$
$$\lambda_2 = \tau_1(\beta_2) - i\tau_2(\beta_2). \tag{78}$$

Substituting $\lambda(\beta_2) = \tau_1(\beta_2) + i\tau_2(\beta_2)$ into characteristic equation and after that calculating the derivative with respect to β_2 and then comparing the real and imaginary parts give that

$$\Psi(\beta_2)\dot{\tau}_1(\beta_2) - \Phi(\beta_2)\dot{\tau}_2(\beta_2) = -\Theta(\beta_2),$$
$$\Phi(\beta_2)\dot{\tau}_1(\beta_2) + \Psi(\beta_2)\dot{\tau}_2(\beta_2) = -\Gamma(\beta_2). \tag{79}$$

Moreover, by solving the above linear system for $\dot{\tau}_1(\beta_2)$ and $\dot{\tau}_2(\beta_2)$ then we get that

$$\Psi(\beta_2)\Theta(\beta_2) + \Phi(\beta_2)\Gamma(\beta_2) = q_1q_2\left[q_3(\mu+\alpha_2) + 2(\eta+\mu)^3 + (\mu+\alpha_2)(\eta+\mu)((\eta+\mu) + (3q_2+3q_1-1))\right] + q_2q_3\left[q_2(\mu+\alpha_2) + (\eta+\mu)((\mu+\alpha_2) + (2q_2+2q_1-1))\right] + q_1(\eta + \mu)(\mu+\alpha_2)\left[(\eta+\mu) + 2q_3 + (\eta+\mu)^2 + q_4\right] + (\eta + \mu)^3\left[q_2(\mu+\alpha_2) + q_2^2 + q_3 + q_4\right] + q_5(\mu+\alpha_2) \cdot \left[2q_2 + (\eta+\mu)\right] + q_1q_3(\eta+\mu)^2 + q_2(\eta+\mu)(\mu + \alpha_2)\left[q_2(q_2-1) + (\eta+\mu)(2q_2+2q_1-1)\right] + q_1(\mu+\alpha_2)\left[q_1q_2^2 + q_2^3 + q_2q_4 - q_5\right] + q_1(\eta+\mu) \cdot \left[2q_1q_2(1+q_2) + 2q_2^3 + q_2q_4 - q_5\right] + (q_2q_4 - q_5) \cdot \left(q_1q_2 + q_2^2\right) + (\eta+\mu)^2\left[3q_1q_2^2 + q_2^3 + q_1q_4 - q_5\right]. \tag{80}$$

Thus it is easy to verify that $\Psi(\beta_2)\Theta(\beta_2) + \Phi(\beta_2)\Gamma(\beta_2) \neq 0$ provided that (71c) and (71d). Thus, the proof is complete. □

7. Numerical Simulations

In this section, the global dynamics of model (1) is investigated numerically for different sets of initial values and different sets of parameters values. The objectives of such investigation are to determine the effect of varying the parameters values and confirm our obtained results. It is observed that, for the following biologically feasible set of hypothetical parameters values

$$\begin{aligned}
\Lambda &= 20,\\
\beta_1 &= 0.75,\\
\beta_2 &= 0.1,\\
\gamma &= 0.75,\\
p_1 &= 0.01,\\
p_2 &= 0.01\\
\mu &= 0.3,\\
\eta &= 0.5,\\
\alpha_1 &= 0.1,\\
\alpha_2 &= 0.6,\\
\delta &= 0.7,
\end{aligned} \tag{81}$$

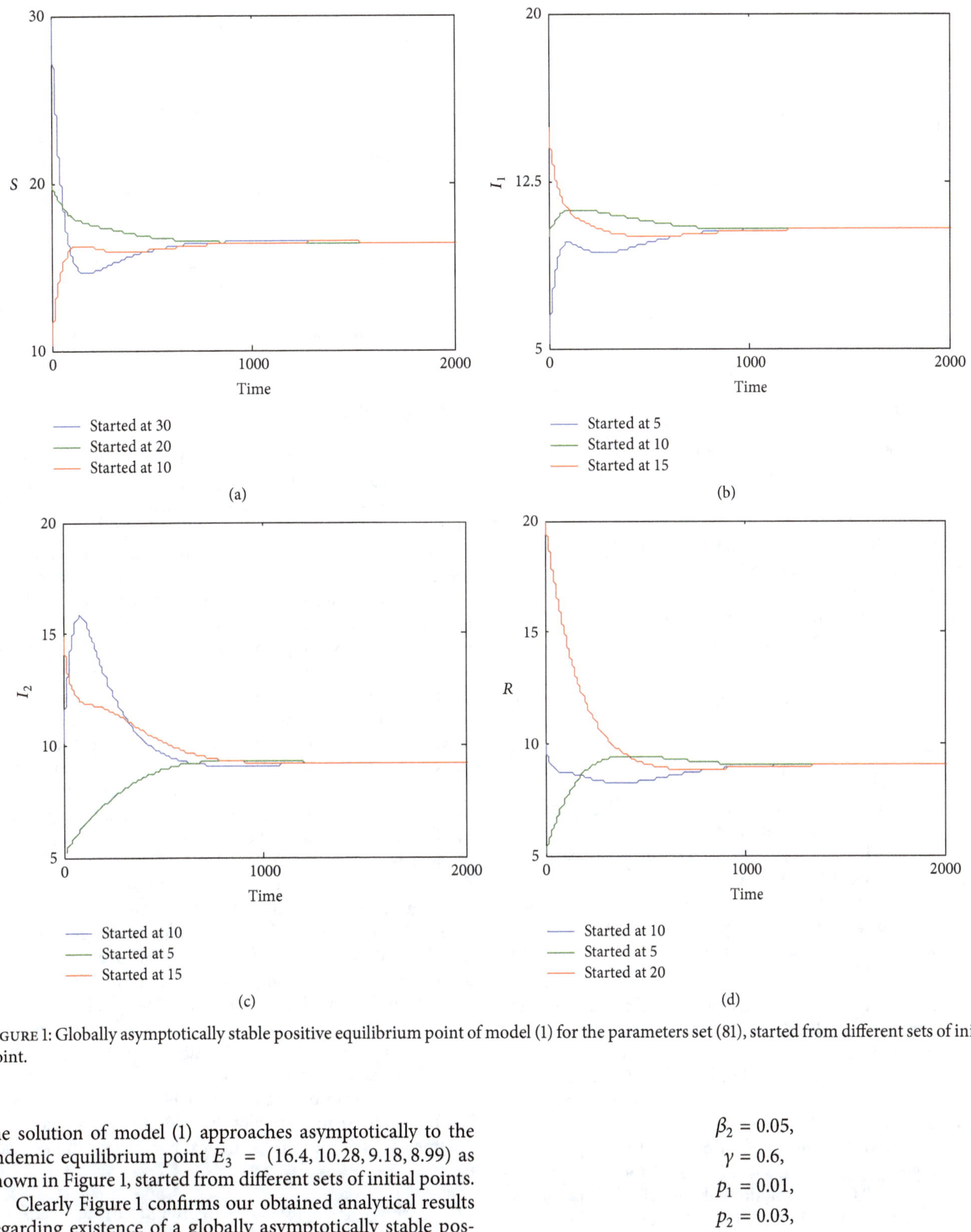

FIGURE 1: Globally asymptotically stable positive equilibrium point of model (1) for the parameters set (81), started from different sets of initial point.

the solution of model (1) approaches asymptotically to the endemic equilibrium point $E_3 = (16.4, 10.28, 9.18, 8.99)$ as shown in Figure 1, started from different sets of initial points.

Clearly Figure 1 confirms our obtained analytical results regarding existence of a globally asymptotically stable positive equilibrium point when the parameters values are satisfying $R_1 > R_2 > 1$. On the other hand, model (1) for the following set of hypothetical data approaches asymptotically to the DFE as shown in Figure 2:

$$\begin{aligned} \Lambda &= 20, \\ \beta_1 &= 0.05, \\ \beta_2 &= 0.05, \\ \gamma &= 0.6, \\ p_1 &= 0.01, \\ p_2 &= 0.03, \\ \eta &= 0.5, \\ \mu &= 0.9, \\ \alpha_1 &= 0.1, \\ \alpha_2 &= 0.1, \\ \delta &= 0.3. \end{aligned} \tag{82}$$

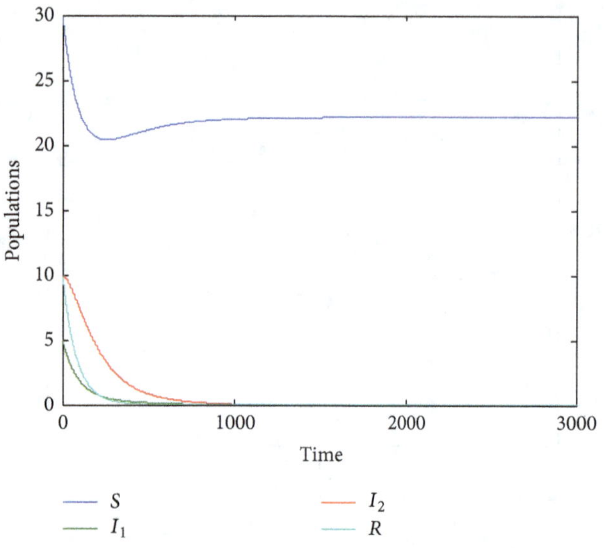

FIGURE 2: Time series of the solution of model (1) that approaches asymptotically to DFE for the data (82).

It is easy to verify that for the data (82) we have $R_0 = 0.86 < 1$, and the solution approaches to $E_0 = (22.22, 0, 0, 0)$.

Now in order to investigate the effect of varying one parameter value at a time on the dynamical behavior of model (1), the following results are observed.

(i) Varying of the parameters values $(\Lambda, \gamma, p_2, \eta, \alpha_1, \alpha_2, \delta)$ does not affect the dynamical behavior of model (1); that is, the system still approaches to coexistence equilibrium point.

(ii) For the data (81) with $\beta_1 \leq 0.015$, the solution of model (1) approaches asymptotically to $E_1 = (S, 0, I_2, 0)$ in the interior of positive quadrant of SI_2-plane with $R_1 < 1 < R_2$. However for $0.015 < \beta_1 < 0.066$ the solution of the model still approaches asymptotically to E_1 even when $1 < R_1 < R_2$. Finally when $\beta_1 \geq 0.066$ the solution of model (1) approaches to coexistence equilibrium point with $1 < R_2 < R_1$ as shown in Figure 3.

(iii) Similar results are obtained in case of varying the parameter β_2 keeping the rest of parameters values in (81) fixed. In fact for $\beta_2 \leq 0.024$, we have $R_2 < 1 < R_1$, and the solution of model (1) approaches asymptotically to $E_2 = (S, I_1, 0, R)$. However for $0.024 < \beta_2 < 0.06$, we have $1 < R_2 < R_1$, and it is observed that the solution of model (1) approaches asymptotically to $E_2 = (S, I_1, 0, R)$ too, while when $\beta_2 \geq 0.06$ the solution of model (1) approaches asymptotically to $E_3 = (S, I_1, I_2, R)$ as shown in Figure 4.

(iv) Now decreasing the parameter μ, keeping the rest of the parameters values in (81) fixed, gives similar dynamical behavior as that of varying β_2. Further it is observed that when $\mu > 0.9$, for which $R_2 < 1 < R_1$, the solution of model (1) approaches asymptotically to $E_2 = (S, I_1, 0, R)$; however when $0.57 < \mu < 0.9$, we have $1 < R_2 < R_1$, and the solution of model (1) still approaches to $E_2 = (S, I_1, 0, R)$. Finally, for $\mu \leq 0.57$ that satisfies $1 < R_2 < R_1$, the solution of model (1) approaches asymptotically to $E_3 = (S, I_1, I_2, R)$. Clearly this confirmed our obtained existence conditions (10) and (12) as well as stability conditions of these points.

(v) Finally, varying the parameter p_1, keeping the rest of the parameters values in (81) fixed, showed that, for $p_1 \geq 0.6$, that satisfies $1 < R_2 < R_1$, the solution of model (1) approaches asymptotically to $E_2 = (S, I_1, 0, R)$; however when $p_1 < 0.6$, which satisfies $1 < R_2 < R_1$ too, the solution of model (1) approaches asymptotically to $E_3 = (S, I_1, I_2, R)$ as shown in Figure 5.

8. Conclusion

In this paper, we proposed and analyzed an epidemic model involving vertical and horizontal transmission of infection with nonlinear incidence rate. It is assumed that the rates of infections p_1, p_2 are less than the recovery rates δ and γ, respectively. According to the diseases in model (1) the population is divided into four subclasses: susceptible individuals that are represented by $S(t)$, infected individuals for *SIRS*-type of disease that are represented by $I_1(t)$, infected individuals for *SIS*-type of disease that are represented by $I_2(t)$, and recovery individuals that are denoted by $R(t)$. The boundedness and invariant of the model are discussed. The basic reproduction number of the model and the associated threshold parameter values, namely, T_i, $i = 1, 2$, are determined. It is observed that if the basic reproduction number is less than unity then the diseases are eradicated from the model. The competitive exclusion principle occurred in model (1) such that only the second disease-free equilibrium point appeared in case of $R_2 < 1 < R_1$. However only the first disease-free equilibrium point appeared in case of $R_1 < 1 < R_2$. Finally the coexistence of both the diseases occurred in case of $1 < R_2 < R_1$ and the sufficient condition (12) holds. The dynamical behavior of model (1) has been investigated locally as well as globally using Routh-Hurwitz criterion and Lyapunov function, respectively. The local bifurcations of model (1) and the Hopf bifurcation around the endemic equilibrium point are studied. Finally to understand the effect of varying each parameter on the global dynamics of system (1) and to confirm our obtained analytical results, model (1) has been solved numerically and the following results are obtained for the set of hypothetical parameters values given by (81).

(1) Model (1) approaches asymptotically to a globally asymptotically stable point $E_3 = (16.4, 10.28, 9.18, 8.99)$.

(2) Varying one of the parameters values $(\Lambda, \gamma, p_2, \eta, \alpha_1, \alpha_2, \delta)$ at a time keeping other parameters fixed has no effect on the dynamical behavior of the model.

(3) As the infection rate of the first disease (β_1) decreases keeping other parameters fixed as in (81) the solution

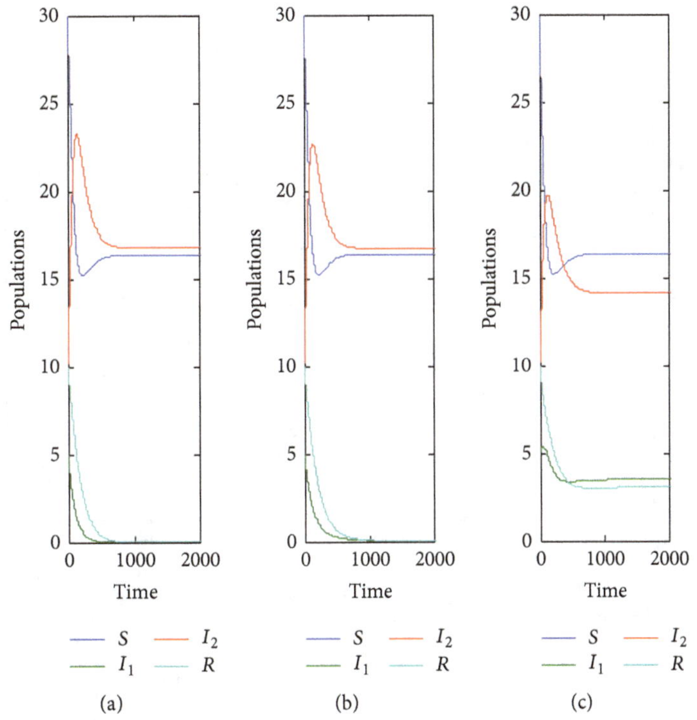

FIGURE 3: Time series of the solution of model (1) for the data given by (81). (a) For $\beta_1 = 0.015$ the model approaches to $E_1 = (16.4, 0, 16.75, 0)$. (b) For $\beta_1 = 0.06$ the model approaches to the same point. (c) For $\beta_1 = 0.3$ the model approaches to $E_3 = (16.4, 3.51, 14.1, 3.07)$.

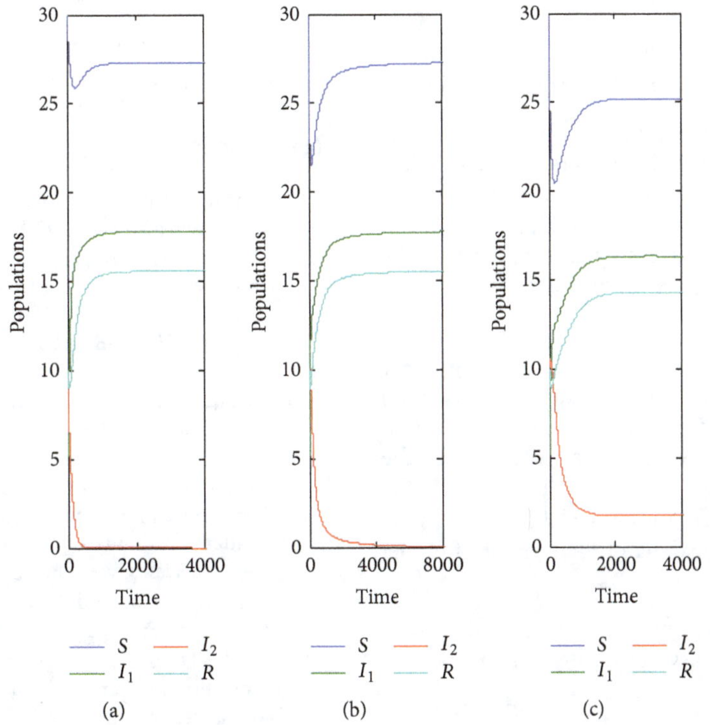

FIGURE 4: Time series of the solution of model (1) for the data given by (81). (a) For $\beta_2 = 0.024$ the model approaches to $E_2 = (27.33, 17.80, 0, 15.5)$. (b) For $\beta_2 = 0.059$ the model approaches to the same point. (c) For $\beta_2 = 0.065$ the model approaches to $E_3 = (25.2, 16.3, 1.76, 14.3)$.

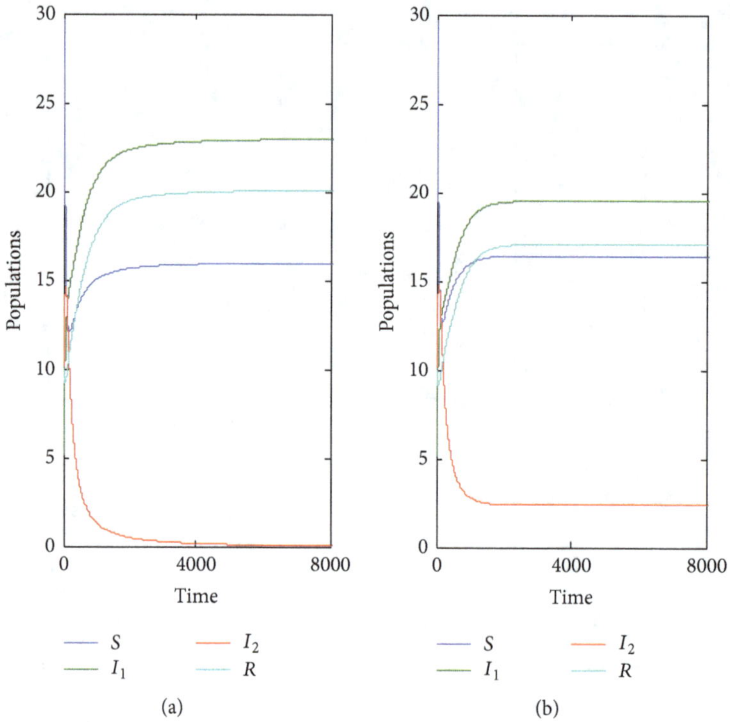

FIGURE 5: Time series of the solution of model (1) for the data given by (81). (a) For $p_1 = 0.6$ the model approaches to $E_2 = (15.9, 22.9, 0, 20.08)$. (b) For $p_1 = 0.5$ the model approaches to $E_3 = (16.4, 19.5, 2.40, 17.06)$.

of model (1) approaches asymptotically to the equilibrium point $E_1 = (16.4, 0, 16.75, 0)$. However in case of increasing this parameter the model is still globally asymptotically stable in the interior of $\mathbb{R}^4_+$.

(4) As the infection rate of the second disease (β_2) decreases keeping other parameters fixed as in (81) the solution of model (1) approaches asymptotically to the equilibrium point $E_2 = (27.33, 17.80, 0, 15.5)$. However in case of increasing this parameter the model is still globally asymptotically stable in the interior of $\mathbb{R}^4_+$.

(5) As the mortality rate (μ) increases keeping other parameters fixed as in (81) the solution of model (1) approaches asymptotically to the equilibrium point $E_2 = (13.8, 5.16, 0, 2.5)$ and when μ decreases the model is still globally asymptotically stable in the interior of $\mathbb{R}^4_+$. Further, it is observed that p_1 has the same effect as μ on the dynamical behavior of model (1).

(6) For the parameter set given in (82) the solution of model (1) approaches asymptotically to DFE.

(7) Finally, although for our selected parameters values model (1) does not undergo periodic dynamics, the model still has possibility to have periodic dynamics for other sets of parameters, especially Hopf bifurcation existing analytically.

Conflict of Interests

The authors declare that there is no conflict of interests regarding the publication of this paper.

References

[1] G. Hardin, "The competitive exclusion principle," *Science*, vol. 131, no. 3409, pp. 1292–1297, 1960.

[2] A. Iggidr, J.-C. Kamgang, G. Sallet, and J.-J. Tewa, "Global analysis of new malaria intrahost models with a competitive exclusion principle," *SIAM Journal on Applied Mathematics*, vol. 67, no. 1, pp. 260–278, 2006.

[3] V. Volterra, "Variations and fluctuations of the number of individuals in animal species living together," *Journal du Conseil International pour l'Exploration de la Mer*, vol. 3, no. 1, pp. 3–51, 1928.

[4] A. S. Ackleh and L. J. Allen, "Competitive exclusion in SIS and SIR epidemic models with total cross immunity and density-dependent host mortality," *Discrete and Continuous Dynamical Systems—Series B*, vol. 5, no. 2, pp. 175–188, 2005.

[5] F. Brauer, J. Wu, and P. van den Driessche, *Mathematical Epidemiology*, Springer, Berlin, Germany, 2008.

[6] P. van den Driessche and J. Watmough, "Reproduction numbers and sub-threshold endemic equilibria for compartmental models of disease transmission," *Mathematical Biosciences*, vol. 180, no. 1-2, pp. 29–48, 2002.

[7] J. M. Heffernan, R. J. Smith, and L. M. Wahl, "Perspectives on the basic reproductive ratio," *Journal of the Royal Society Interface*, vol. 2, no. 4, pp. 281–293, 2005.

[8] O. Diekmann, J. A. P. Heesterbeek, and J. A. J. Metz, "On the definition and the computation of the basic reproduction ratio R_0 in models for infectious diseases in heterogeneous populations in models for infectious diseases in heterogeneous populations," *Journal of Mathematical Biology*, vol. 28, no. 4, pp. 365–382, 1990.

[9] M. Martcheva, "A non-autonomous multi-strain SIS epidemic model," *Journal of Biological Dynamics*, vol. 3, no. 2-3, pp. 235–251, 2009.

[10] A. S. Ackleh and L. J. S. Allen, "Competitive exclusion and coexistence for pathogens in an epidemic model with variable population size," *Journal of Mathematical Biology*, vol. 47, no. 2, pp. 153–168, 2003.

[11] M. Lipsitch, S. Siller, and M. A. Nowak, "The evolution of virulence in pathogens with vertical and horizontal transmission," *Evolution*, vol. 50, no. 5, pp. 1729–1741, 1996.

[12] D. Bichara, A. Iggidr, and G. Sallet, "Global analysis of multi-strains SIS, SIR and MSIR epidemic models," *Journal of Applied Mathematics and Computing*, vol. 44, no. 1-2, pp. 273–292, 2014.

[13] M. Y. Li, H. L. Smith, and L. Wang, "Global dynamics an SEIR epidemic model with vertical transmission," *SIAM Journal on Applied Mathematics*, vol. 62, no. 1, pp. 58–69, 2001.

[14] W. M. Liu, H. W. Hethcote, and S. A. Levin, "Dynamical behavior of epidemiological models with nonlinear incidence rates," *Journal of Mathematical Biology*, vol. 25, no. 4, pp. 359–380, 1987.

[15] O. Adebimpe, A. A. Waheed, and B. Gbadamosi, "Modeling and analysis of an SEIRS epidemic model with saturated incidence," *Journal of Engineering Research and Application*, vol. 3, no. 5, pp. 1111–1116, 2013.

[16] R. K. Naji and A. N. Mustafa, "The dynamics of an eco-epidemiological model with nonlinear incidence rate," *Journal of Applied Mathematics*, vol. 2012, Article ID 852631, 24 pages, 2012.

[17] C. V. De-León, "On the global stability of SIS, SIR and SIRS epidemic models with standard incidence," *Chaos, Solitons & Fractals*, vol. 44, no. 12, pp. 1106–1110, 2011.

[18] J. P. LaSalle, "Stability theory for ordinary differential equations," *Journal of Differential Equations*, vol. 4, no. 1, pp. 57–65, 1968.

[19] E. Shim, *An epidemic model with immigration of infectives and vaccination [M.S. thesis]*, Department of Mathematics, Institute of Applied Mathematics, University of British Columbia, Vancouver, Canada, 2004.

[20] L. Perko, *Differential Equation and Dynamical Systems*, Springer, New York, NY, USA, 3rd edition, 2001.

[21] M. M. A. El-Sheikh and S. A. A. El-Marouf, "On stability and bifurcation of solutions of an SEIR epidemic model with vertical transmission," *International Journal of Mathematics and Mathematical Sciences*, vol. 2004, no. 56, pp. 2971–2987, 2004.

[22] X. Zhou and J. Cui, "Analysis of stability and bifurcation for an SEIV epidemic model with vaccination and nonlinear incidence rate," *Nonlinear Dynamics*, vol. 63, no. 4, pp. 639–653, 2011.

17

Bayesian Analysis for a Fractional Population Growth Model

Francisco J. Ariza-Hernandez, Jorge Sanchez-Ortiz, Martin P. Arciga-Alejandre, and Luis X. Vivas-Cruz

Facultad de Matemáticas, Universidad Autónoma de Guerrero, 39087 Chilpancingo, GRO, Mexico

Correspondence should be addressed to Jorge Sanchez-Ortiz; jsanchezmate@gmail.com

Academic Editor: Juan J. Trujillo

We implement the Bayesian statistical inversion theory to obtain the solution for an inverse problem of growth data, using a fractional population growth model. We estimate the parameters in the model and we make a comparison between this model and an exponential one, based on an approximation of Bayes factor. A simulation study is carried out to show the performance of the estimators and the Bayes factor. Finally, we present a real data example to illustrate the effectiveness of the method proposed here and the pertinence of using a fractional model.

1. Introduction

In recent years, models governed by fractional differential equations have been of great importance to describe anomalous dynamic, which is inherent in a wide range of phenomena (see [1–3]). This model under certain circumstances can offer a superior fit to experimental data [4]. It is well known that population growth can be modeled by ordinary differential equations and the exponential growth model is the simplest and most widely used of all. However, it can fail to describe an anomalous growth; that is, the population may have a slower or faster growth than exponential growth model. In order to fill this gap, we use a fractional population growth model.

On the order hand, the measurement procedure of growth data is under uncertainty or error data collection process. The statistical approach toward inverse problems from a Bayesian point of view has been very important for the development of uncertainty quantification. Textbooks such as [5–7] provide an extensive literature on the Bayesian statistical aspects in inverse problems. Donnet et al. [8] developed a Bayesian inference for the parameters of stochastic differential equations deduced from the standard deterministic growth function by adding random effects to the growth dynamic in a population of individuals over time. Capistrán et al. [9] also performed a Bayesian analysis of inverse problems in ordinary differential equations. Calvetti et al. [10] mentioned several articles on the development of Bayesian inverse problems in a broad range of applications, such as engineering, geophysics, life sciences, and economy. Dashti and Stuart [11] discussed the Bayesian approach to inverse problems in differential equations from mathematical and computational perspective and describe Markov Chain Monte Carlo and sequential Monte Carlo methods and measure-preserving reversible stochastic differential equations on infinite dimensional space.

In this paper, we implement the Bayesian statistical inversion theory to obtain a solution for an inverse problem of growth data using a fractional population growth model, defined in Section 2. In Section 3, we estimate the parameters of the fractional model proposed and we make a comparison between two different models, based on Bayes factor. In Section 4, we make a simulation study to show the performance of the estimators and the Bayes factor. Finally, we present a real data example to illustrate the effectiveness of the method proposed here and the pertinence of using a fractional model.

2. The Model

In this work, the proposed mathematical model is defined by two parts.

(1) A dynamical system is as follows:

$$\mathscr{D}^{\alpha} X(t) = aX(t) + f(t), \quad a > 0, \ \alpha \in (0,1]; \tag{1}$$

$$X(t_0) = X_0, \tag{2}$$

where $X(t)$ is the size of the population at the time t, X_0 is the initial size, a is the rate of change, and $f(t)$ is a harvesting function. Here $\mathscr{D}^{\alpha}$ is the ordinary derivative for $\alpha = 1$ and the Caputo fractional derivative for $\alpha \in (0,1)$, which is defined by

$$\mathscr{D}^{\alpha} X(t) = \frac{1}{\Gamma(1-\alpha)} \int_0^t \frac{X'(\tau)}{(t-\tau)^{\alpha}} d\tau, \tag{3}$$

where Γ is the Gamma function.

(2) An observation equation is as follows:

$$y_i = h(X(t_i)) + \varepsilon_i, \quad i = 1, \ldots, n, \tag{4}$$

where y_i correspond to the ith observed value under uncertainty from a solution of (1) at the discrete time $t_i \in [0,T]$, $i = 1,2,\ldots,n$; h is the observation function; and ε_i are measurement errors, which are considered as independent and identically distributed (i.i.d.) random variables from a normal distribution, with mean zero and constant variance σ^2, denoted by $\varepsilon_i \sim \mathscr{N}(0,\sigma^2)$ (in this paper, the symbol "∼" is used for "is distributed as").

Now, we find an explicit solution for (1) using the Laplace transform and its inverse, which are defined by

$$\widehat{X}(s) = \int_0^{\infty} e^{-st} X(t)\, dt,$$
$$X(t) = \frac{1}{2\pi i} \int_{s_0 - i\infty}^{s_0 + i\infty} e^{st} \widehat{X}(s)\, ds, \tag{5}$$

respectively. Applying the direct transform to (1), we obtain

$$\widehat{X}(s) = \frac{1}{s^{\alpha} - a} \left(s^{\alpha-1} X_0 + \widehat{f}(s)\right). \tag{6}$$

Then, using the following property

$$\left[t^{\beta-1} E_{\alpha,\beta}(at^{\alpha})\right]^{\frown}(s) = \frac{s^{\alpha-\beta}}{s^{\alpha} - a}, \tag{7}$$

where $E_{\alpha,\beta}$ is the Mittag-Leffler function,

$$E_{\alpha,\beta}(z) = \sum_{k=0}^{\infty} \frac{z^k}{\Gamma(\alpha k + \beta)}, \quad \alpha > 0, \ \beta > 0, \ z \in \mathbb{C}. \tag{8}$$

Finally, applying the inverse Laplace transform to (6) we arrive to

$$X(t) = X_0 E_{\alpha,1}(at^{\alpha}) + \int_0^t (t-\eta)^{\alpha-1} E_{\alpha,\alpha}\left(a(t-\eta)^{\alpha}\right) f(\eta)\, d\eta. \tag{9}$$

Note that if $\alpha = 1$ in the above equation we get the well known solution for (1), which is given by

$$X(t) = X_0 e^{at} + \int_0^t e^{a(t-\eta)} f(\eta)\, d\eta. \tag{10}$$

3. Bayesian Analysis

3.1. Bayesian Estimation. In recent years, the problem of parameter estimation in fractional models has been of great interest, and we use a Bayesian approach to estimate the parameters in a fractional population growth model. The Bayesian theory has proven to be efficient to solve fractional inverse problems (see [12, 13]). In this setting, the parameters are regarded as random variables, and they have prior distributions that reflect the knowledge about their true values before observing the data. From the perspective of Bayesian statistical inversion theory, the solution to an inverse problem is the posterior distribution of the quantity of interest given that all information available has been incorporated in the model [5].

Note that, for the model defined in (1) and (4), the parameter of interest is $\boldsymbol{\theta}' = (a, \alpha, \sigma^2, X_0)$. Here, we propose the following prior distributions:

$$a \sim \mathscr{G}(r_a, \lambda_a), \tag{11}$$

$$\alpha \sim \mathscr{U}(0,1), \tag{12}$$

$$\tau \sim \mathscr{G}(r_{\tau}, \lambda_{\tau}), \tag{13}$$

$$X_0 \sim \ln \mathscr{N}(\mu_0, \sigma_0^2), \tag{14}$$

where $\mathscr{G}(r,\lambda)$ denotes the Gamma distribution with shape parameter r and rate parameter λ, $\mathscr{U}$ is the continuous Uniform distribution on interval $(0,1)$, $\tau = 1/\sigma^2$ is called the precision parameter, and $\ln \mathscr{N}(\mu_0, \sigma_0^2)$ is a log-Normal distribution with parameters μ_0 and σ_0^2. The parameters involved in the prior distribution are called hyperparameters. The prior distributions were chosen according to our knowledge about the possible values of the parameters of interest. In models (1) and (4), a, X_0, and τ are positive and α take values in the interval $(0,1]$. Assuming prior independence of the parameters, we can write the joint prior distribution as

$$p(\boldsymbol{\theta} \mid \text{hyperparameters}) \propto p(a \mid r_a, \lambda_a)\, p(\alpha)\, p(\tau \mid r_{\tau}, \lambda_{\tau})\, p(X_0 \mid \mu_0, \sigma_0^2), \tag{15}$$

where $p(a \mid r_a, \lambda_a)$, $p(\alpha)$, $p(\tau \mid r_{\tau}, \lambda_{\tau})$, and $p(X_0 \mid \mu_0, \sigma_0^2)$ are defined in (11)–(14), respectively. Let $\mathbf{y}' = (y_1, y_2, \ldots, y_n)$ denote i.i.d. observed data at times $(t_1, t_2, \ldots, t_n)$ from the model defined by (1) and (4), and then the likelihood function is given by

$$L(\mathbf{y} \mid \boldsymbol{\theta}) = \tau^{n/2} (2\pi)^{-n/2} \exp\left\{-\frac{\tau}{2} \sum_{i=1}^{n} (y_i - h(X(t_i)))^2\right\}. \tag{16}$$

Thus, by Bayes' theorem the posterior distribution of the parameters of interest is given by

$$p(\boldsymbol{\theta} \mid \mathbf{y}) = \frac{L(\mathbf{y} \mid \boldsymbol{\theta})\, p(\boldsymbol{\theta})}{\int_{\Theta} L(\mathbf{y} \mid \boldsymbol{\theta})\, p(\boldsymbol{\theta})\, d\boldsymbol{\theta}}, \tag{17}$$

where Θ denotes the parameter space of $\boldsymbol{\theta}$. It is clear that $p(\boldsymbol{\theta} \mid \mathbf{y}) \propto L(\mathbf{y} \mid \boldsymbol{\theta})p(\boldsymbol{\theta})$, and then the posterior distribution can be expressed by

$$p(\boldsymbol{\theta} \mid \mathbf{y}) \propto \tau^{n/2} \exp\left\{-\frac{\tau}{2}\sum_{i=1}^{n}\left(y_i - X_0 E_{\alpha,1}(at_i^\alpha) - \int_0^{t_i}(t_i-\eta)^{\alpha-1}E_{\alpha,\alpha}\left(a(t_i-\eta)^\alpha\right)f(\eta)\,d\eta\right)^2\right\} \cdot a^{r_a-1}\tau^{r_\tau-1}X_0^{-1}\exp\left\{-\frac{(\ln X_0-\mu_0)^2}{2\sigma_0^2} - a\lambda_a - \tau\lambda_\tau\right\}. \quad (18)$$

Using a loss quadratic function, the Bayesian point estimation is the posterior mean of $\boldsymbol{\theta}$, which is given by

$$\widehat{\boldsymbol{\theta}}_B = E(\boldsymbol{\theta} \mid \mathbf{y}). \quad (19)$$

Note that the joint posterior distribution is analytically intractable and the marginal posterior distributions of the parameters are complicated; however, we can obtain samples from (18) using Markov Chain Monte Carlo (MCMC) techniques. The most common MCMC methods are the Gibbs sampling [14, 15] and the Metropolis-Hastings [16, 17]. Currently, many of the MCMC algorithms have been already implemented in computer programs, such as WinBUGS and JAGS [18] and Stan and t-walk [19]. All of these software packages provide programs for Bayesian modeling through posterior simulation given a specified model and data. The R [20] packages, such as R2WinBUGS [21], R2jags [22], rjags [23], and rstan [24], allow one to run WinBUGS, JAGS, and Stan from within R software, respectively. In this paper, we use JAGS within R to obtain samples from the marginal posterior distributions of interest.

JAGS (Just Another Gibbs Sampler) is a program for analysis of Bayesian models using MCMC methods, which was written in C++ by Martyn Plummer with three objectives: (a) to have a cross-platform engine for the BUGS language, (b) to be extensible, allowing users to write their own functions, distributions, and samplers, and (c) to be a platform for experimentation with ideas in Bayesian modeling (see [18]); moreover, it is a free package. To draw samples from the posterior distribution, JAGS use the Gibbs sampling and sometimes it is combined with or replaced by complex techniques such as Metropolis-Hastings algorithm, slice sampling [25], and Adaptive Rejection sampling [26]. In our work, JAGS used slice sampling, which is briefly described as follows.

Let $f(\theta)$ be univariate probability density function from which we want to obtain samples; for an initial value θ_0, $f(\theta_0) > 0$, the new value, θ_1, will be found by a three-step procedure:

(a) Sample a real value $y \sim \mathscr{U}(0, f(\theta_0))$ and we define a horizontal *slice*, $S = \{\theta : y < f(\theta)\}$.

(b) Find an interval, (l, r), around θ_0 that contains all, or much, of the slice.

(c) Sample the new value, θ_1, from the part of the slice within this interval.

For multivariate distributions, we can apply the single variable slice sampling for each θ_i, $i = 1, \ldots, p$, in turn. An example of JAGS implementation is given in Appendix.

3.2. Bayesian Model Comparison. In this section, our interest is to make a comparison between two different models. In one, the dynamic system changes in an ordinary way and in the other it changes anomalously, that is, with an entire derivative and with a fractional derivative, respectively. As soon as we obtain experimental data from a dynamical phenomenon, the uncertainty becomes present. Moreover, the mechanism that governs the evolution law is unknown at some level. So, the challenge here is to know which model best fits and best describes the data.

The Bayesian model selection in inverse problems framework has not been much investigated. In [8] it is suggested to validate the SDE approach via criteria based on the predictive posterior distribution. To compare a numerical and the theoretical posterior distribution, in [9] it was proposed to use Bayes factors, considering both of them as models for the data at hand. Other authors (see [27–29]) use Bayes factor to select the best model. We carried out a comparison of two models, defined below, from Bayesian point of view via Bayes factor, where the marginal likelihood is obtained using Gelfand and Dey's estimator.

Now, let us define the models that we are going to compare. The ordinary population growth model, $\mathscr{M}_1$, is given by the dynamical system

$$\frac{d}{dt}X(t) = aX(t), \quad a > 0; \quad (20)$$

$$X(t_0) = X_0, \quad (21)$$

and its corresponding observation equation, for n observations,

$$y_i^{(1)} = X_0 e^{at_i} + \varepsilon_i, \quad i = 1, \ldots, n. \quad (22)$$

The fractional population growth model, $\mathscr{M}_2$, is given by

$$\mathscr{D}^\alpha X(t) = aX(t), \quad a > 0, \ \alpha \in (0,1); \quad (23)$$

$$X(t_0) = X_0, \quad (24)$$

$$y_i^{(2)} = X_0 E_{\alpha,1}(at_i^\alpha) + \varepsilon_i, \quad i = 1, \ldots, n, \quad (25)$$

where $E_{\alpha,1}(\cdot)$ is defined in (8) and $\varepsilon_i \sim \mathscr{N}(0, \sigma^2)$. Note that for simplicity, in both models, the harvesting function $f(t)$ is assumed to be zero and the function h is taken as the identity function.

So, for model $\mathscr{M}_k$, $k = 1, 2$, the posterior distribution takes the form

$$p(\boldsymbol{\theta}_k \mid \mathbf{y}, \mathscr{M}_k) \propto L(\mathbf{y} \mid \boldsymbol{\theta}_k, \mathscr{M}_k)\, p(\boldsymbol{\theta}_k \mid \mathscr{M}_k), \quad (26)$$

where $L(\mathbf{y} \mid \boldsymbol{\theta}_k, \mathscr{M}_k)$ is the likelihood function of the data $\mathbf{y}' = (y_1, \ldots, y_n)$ under model $\mathscr{M}_k$ and $p(\boldsymbol{\theta}_k \mid \mathscr{M}_k)$ is the prior distribution of $\boldsymbol{\theta}_k$. By substituting these expressions for each model, we have

$$p(\boldsymbol{\theta}_1 \mid \mathbf{y}, \mathcal{M}_1) \propto \tau^{n/2+r_\tau-1} a^{r_a-1} X_0^{-1} \cdot \exp\left\{-\frac{\tau}{2}\sum_{i=1}^{n}\left(y_i - X_0 e^{at_i}\right)^2 - \frac{\left(\ln X_0 - \mu_0\right)^2}{2\sigma_0^2} - \lambda_a a - \lambda_\tau \tau\right\},$$

$$p(\boldsymbol{\theta}_2 \mid \mathbf{y}, \mathcal{M}_2) \propto \tau^{n/2+r_\tau-1} a^{r_a-1} X_0^{-1} \exp\left\{-\frac{\tau}{2} \cdot \sum_{i=1}^{n}\left(y_i - X_0 E_\alpha\left(at_i^\alpha\right)\right)^2 - \frac{\left(\ln X_0 - \mu_0\right)^2}{2\sigma_0^2} - \lambda_a a - \lambda_\tau \tau\right\}. \tag{27}$$

To compare these models, we need to calculate the marginal likelihood $m(\mathbf{y} \mid \mathcal{M}_k)$ for $k = 1, 2$, given by

$$m(\mathbf{y} \mid \mathcal{M}_k) = \int_{\Theta_k} L(\mathbf{y} \mid \boldsymbol{\theta}_k, \mathcal{M}_k)\, p(\boldsymbol{\theta}_k \mid \mathcal{M}_k)\, d\boldsymbol{\theta}_k, \tag{28}$$

and we choose the model which yields the largest marginal likelihood [30]. Note that the equation above is the normalizing constant of the posterior density in (26), so it can be written as

$$m(\mathbf{y} \mid \mathcal{M}_k) = \frac{L(\mathbf{y} \mid \boldsymbol{\theta}_k, \mathcal{M}_k)\, p(\boldsymbol{\theta}_k \mid \mathcal{M}_k)}{p(\boldsymbol{\theta}_k \mid \mathbf{y}, \mathcal{M}_k)}. \tag{29}$$

In general, $m(\mathbf{y} \mid \mathcal{M}_k)$ in (28) can not be represented in terms of elementary functions and obtaining a numerical approximation usually requires a huge computational effort. On the other hand, to calculate $m(\mathbf{y} \mid \mathcal{M}_k)$ using (29) is not an easy task because we have many MCMC draws falling in the tails of $p(\boldsymbol{\theta}_k \mid \mathbf{y}, \mathcal{M}_k)$ and this leads to a very unreliable estimate of $p(\boldsymbol{\theta}_k \mid \mathbf{y}, \mathcal{M}_k)$. Methods to estimate $m(\mathbf{y} \mid \mathcal{M}_k)$ via MCMC sampling are discussed in [30–32]. We propose the use of Gelfand and Dey's estimator [33] to obtain an approximation of $m(\mathbf{y} \mid \mathcal{M}_k)$, which is given by

$$\widehat{m}_{GD} = \left\{\frac{1}{G}\sum_{g=1}^{G}\left(\frac{\pi\left(\boldsymbol{\theta}_k^{(g)}\right)}{L\left(\mathbf{y} \mid \boldsymbol{\theta}_k^{(g)}, \mathcal{M}_k\right) p\left(\boldsymbol{\theta}_k^{(g)} \mid \mathcal{M}_k\right)}\right)\right\}^{-1}, \tag{30}$$

where $\{\boldsymbol{\theta}_k^{(1)}, \ldots, \boldsymbol{\theta}_k^{(G)}\}$ are draws from the posterior density $p(\boldsymbol{\theta}_k \mid \mathbf{y}, \mathcal{M}_k)$ obtained using MCMC method and $\pi(\cdot)$ is an importance sampling density. The stability of (30) depends on the importance density choice, which is recommended to be of thinner tails compared to the product of the prior and the likelihood. Natural choices of $\pi(\cdot)$ would be the multivariate normal or Student t densities with posterior sampling mean and covariance. Then, we compare the models $\mathcal{M}_1$ and $\mathcal{M}_2$ via Bayes factor:

$$\mathrm{BF}_{1,2} = \exp\left\{\ln\left[m(\mathbf{y} \mid \mathcal{M}_1)\right] - \ln\left[m(\mathbf{y} \mid \mathcal{M}_2)\right]\right\}, \tag{31}$$

where a value of $\mathrm{BF}_{1,2}$ greater than one indicates a preference for model $\mathcal{M}_1$ with respect to the model $\mathcal{M}_2$.

4. Simulation Study

We carried out a simulation study to assess the performance of the Bayesian estimators of the parameters of interest ($\boldsymbol{\theta}_k$, $k = 1, 2$) and the performance of the Bayes factor using Gelfand and Dey's estimator [32] for the marginal likelihood (29). For this task, we used R software [20] with R2jags package [22], under the following algorithm:

(1) We simulated $M = 1000$ synthetic data sets form a fractional population growth model using the true values of the parameters $\alpha = 0.5$, $a = 0.2$, $\sigma^2 = 30$, and $X_0 = 20$ for two sample sizes, $n = 10$ and $n = 30$, and two cases: X_0 known and X_0 unknown. Each data set was simulated using the model defined in (25).

(2) For each simulated sample, we obtained Bayesian estimators of the parameters both for model $\mathcal{M}_1$ and for model $\mathcal{M}_2$.

(a) We calculated the Bayes estimators as the sample mean from (18) using JAGS within R. We used two chains, each with 10000 iterations with a burn-in of 1000 iterations and a thinning rate of 9, so we kept a total of 2000 iterations to make inferences about the parameters of interest. The following prior distributions for the parameters in the models are assumed as

$$\begin{aligned} a &\sim \mathcal{G}(0.001, 0.001), \\ \alpha &\sim \mathcal{U}(0, 1), \\ \tau &\sim \mathcal{G}(0.001, 0.001), \end{aligned} \tag{32}$$

note that for the parameters a and τ we use diffuse Gamma distributions, for α we propose Uniform prior distribution in $(0, 1)$, and, for the case of X_0 unknown, a diffuse log-Normal distribution is used, $X_0 \sim \ln \mathcal{N}(0, 1\times10^3)$. These distributions are called *weakly informative* prior distributions, which contain some information about the parameters of interest but without affecting inferences by information external to the current data [34].

(b) We estimated 95% credible intervals for each parameter in each model; these intervals are based on the 0.025 quantile and the 0.975 quantile of the corresponding posterior sample.

(c) We calculate the marginal likelihood using Gelfand and Dey's estimator given in (30) using MCMC samples from the posterior density

$p(\boldsymbol{\theta}_k \mid \mathbf{y}, \mathcal{M}_k)$, $k = 1, 2$. The multivariate normal density with posterior sampling mean and covariance was used as importance sampling density.

(d) We obtain an estimation of the Bayes factor of $\mathcal{M}_2$ *versus* $\mathcal{M}_1$, denoted by $\widehat{\mathrm{BF}}_{2,1}$ using (31). If $\widehat{\mathrm{BF}}_{2,1} \geq 1$, then we prefer the model $\mathcal{M}_2$ with respect to the model $\mathcal{M}_1$.

(3) From these M samples, we obtained measures of the bias and mean-squared error of the estimators, as well as the coverage of the corresponding credible intervals.

Our simulations were carried out for two sample sizes ($n = 10$ and $n = 30$) and two cases: X_0 known and X_0 unknown. Tables 1 and 2 show the estimate bias, MSE, and coverage for the cases considered. We can observe that the parameters estimation of the model $\mathcal{M}_2$ was more accurate compared to the model $\mathcal{M}_1$, in terms of the estimated bias and MSE. Also, the estimates coverage was near to 100%, in the most cases considered, for the model $\mathcal{M}_2$. From our simulation, for the case X_0 known, the percentage of the estimated Bayes factors in favour of $\mathcal{M}_2$ was 98.4% with $n = 10$ observations and 100% for $n = 30$. For the case X_0 unknown, the percentage of the estimated Bayes factors in favour of $\mathcal{M}_2$ was 94.0% with $n = 10$ observations and 100% for $n = 30$.

5. Application

In this section, an example is given to illustrate the Bayesian estimation and selection model approach proposed in this paper. The data set used consists of 69 monthly measurements (in percentages) of the mobile web use in the world, from December 2008 to August 2014 (see Figure 2, dot points). We can find this data from the link http://stats.arepp-im.com/stats/stats_mobiwebsubstxtime.htm. At a first glance of data, one can think of an exponential growth model; however, as we will see, this model is unable to fit the data in satisfactory way. A generalization of this model is proposed, where we add a new parameter for the order of the derivative. The model defined in (23) is used to fit this data, and the parameters of interest to be estimated, based in available data, are α, a, and σ^2.

The prior distributions used were for $\alpha \sim \mathcal{U}(0,1)$, $a \sim \mathcal{G}(0.001, 0.001)$, and $1/\sigma^2 \sim \mathcal{G}(0.001, 0.001)$. For the model proposed, we used two chains, each with 500000 iterations, and the first 250000 were discarded, taking a thinning rate of 50. Thus, 10000 posterior samples were used to obtain the summary statistics about the parameters of interest. Standard convergence diagnostics were carried out. To mention a few, the value of $\widehat{R}$ for each parameter of interest was quite close to 1 for all models considered. Also, the Gelman-Rubin factor and Geweke diagnostics were calculated and showed evidence of convergence. Figure 1 shows the trace and estimated posterior distributions of the estimated parameters of interest.

```
library(R2jags)
library(coda)
library(lattice)
library(R2WinBUGS)
library(rjags)
```

Box 1

Posterior means are used as point estimates of the parameters of interest, for $\mathcal{M}_1 : \widehat{a} = 0.06$ and $\widehat{\sigma}^2 = 4.31$ and for $\mathcal{M}_2 : \widehat{\alpha} = 0.35$, $\widehat{a} = 0.33$, and $\widehat{\sigma}^2 = 0.52$. Figure 2 shows the fitted models considered. The Bayes factor was 64.25, giving very strong evidence in favour of the model $\mathcal{M}_2$.

6. Concluding Remarks

We estimated the parameters of a fractional population growth model using Bayesian framework. From the model proposed we can distinguish two different models, one with ordinary derivative, $\alpha = 1$, and another with fractional derivative, $\alpha \in (0, 1)$. An estimation of the Bayes factor was used to compare these two growth models. To measure the performance of the parameters estimation and Bayes factor a simulation study was conducted. As expected, the fractional population growth model outperforms the ordinary population growth model as is shown in Tables 1 and 2. In order to estimate the Bayes factor, Gelfand and Dey's estimator was used to approximate the marginal likelihood. It is known that this estimator could be unstable, even in the situation when the importance density is chosen as a multivariate normal density with mean and covariance obtained from MCMC sampling; however, in this work, it was not the case. A subsequent monitoring was carried on to check the stability. Our results give numerical evidence of good performance of the Bayes factor. Also, the inferences based on posterior simulation are relatively easy to implement in JAGS within the R software. Finally, we have shown a real data example where the fractional growth model had a better fit with respect to the ordinary model.

Appendix

JAGS Implementation

The proposed models were all implemented in JAGS using the R2jags package. R2jags was used to fit the models and to perform convergence diagnostics right within R. Here we use the data set of Section 5 to illustrate the implementation of model $\mathcal{M}_1$ in JAGS.

(1) Packages. Load the required R packages (see Box 1).

(2) Data. Read the data from the working directory (see Box 2).

(3) The Model. Write the model in BUGS code and save it as "movil.model.jags" in the working directory (see Box 3).

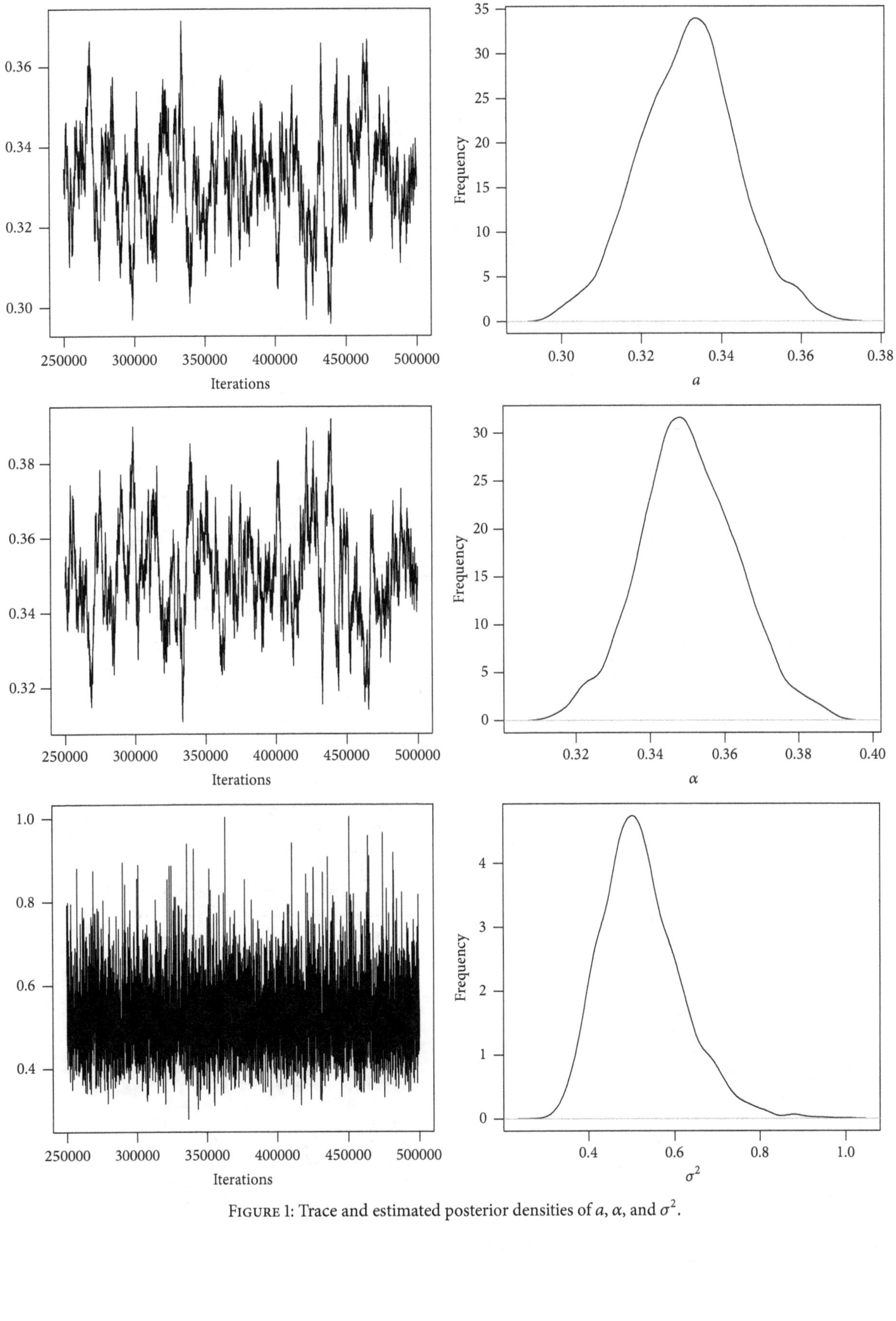

FIGURE 1: Trace and estimated posterior densities of a, α, and σ^2.

TABLE 1: Estimated Bias, MSE, and coverage, with X_0 known.

Parameter	$\mathscr{M}_1$ Bias	$\mathscr{M}_1$ MSE	$\mathscr{M}_1$ Cov (%)	$\mathscr{M}_2$ Bias	$\mathscr{M}_2$ MSE	$\mathscr{M}_2$ Cov (%)
			$n = 10$			
a	−0.105	0.011	0.0	0.008	0.002	97.4
α	—	—	—	0.013	0.018	98.3
σ^2	25.22	1218.77	82.7	10.51	480.56	97.0
			$n = 30$			
a	−0.135	0.018	0.0	−0.001	0.0004	92.3
α	—	—	—	0.008	0.0019	92.4
σ^2	77.36	6457.78	0.0	1.091	71.55	94.0

TABLE 2: Estimated Bias, MSE, and coverage, with X_0 unknown.

Parameter	$\mathscr{M}_1$ Bias	$\mathscr{M}_1$ MSE	$\mathscr{M}_1$ Cov (%)	$\mathscr{M}_2$ Bias	$\mathscr{M}_2$ MSE	$\mathscr{M}_2$ Cov (%)
			$n = 10$			
a	−0.100	0.010	0.0	0.028	0.003	100
α	—	—	—	−0.004	0.005	100
σ^2	10.68	606.67	93.0	10.24	610.42	95.0
X_0	5.82	40.76	60.2	0.68	9.61	100
			$n = 30$			
a	−0.151	0.023	0.0	−0.008	0.007	90.2
α	—	—	—	0.092	0.029	90.1
σ^2	5.97	131.63	90.1	1.60	79.42	95.3
X_0	9.44	90.63	0.0	1.44	14.69	92.0

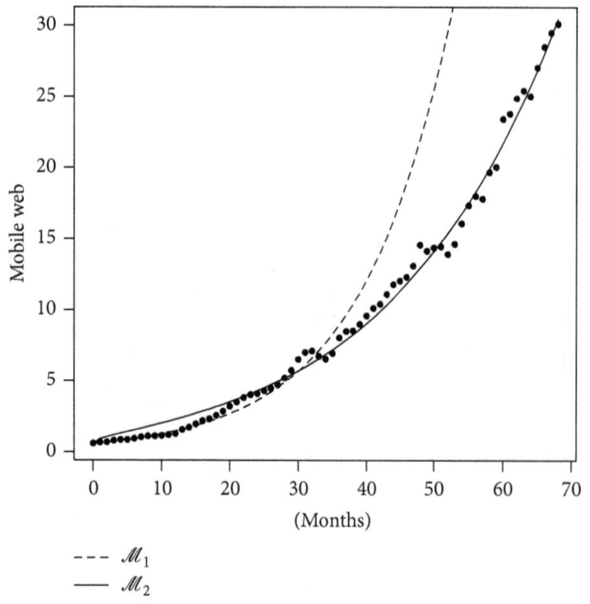

FIGURE 2: Mobile web (in percentage) from December 2008 to August 2014.

```
setwd("my directory")
movil<-read.table(file="movil.txt", sep=",")
N<-length(movil)
t<-1:N
movil.data <- list("movil","t","N")
```

Box 2

```
model{
for (i in 1:N){
movil[i] ~ dnorm(mu[i], tau)
mu[i] <- xo*exp(a*t[i])
}
xo <- 0.6
a ~ dgamma(0.001,0.001)
sigma2 <- 1/tau
tau ~ dgamma(0.001,0.001) # media =1 y var=1000
}", file="movil.model.jags", fill=T)
}
```

Box 3

```
movil.params<-c("a","sigma2")
```

Box 4

```
movil.inits<-function()
  {
       list("a"=c(0.1,0.5), "tau"=c(0.02,0.5))
  }
```

Box 5

```
set.seed(123)
fit.movil<-jags(data=movil.data, inits = movil.inits, parameters.to.save=
          movil.params, n.chains =2, n.iter = 500000, n.burnin=250000,
                     model.file="movil.model.jags")
    print(fit.movil)
```

Box 6

```
movil.mcmc<-as.mcmc(fit.movil)
densityplot(movil.mcmc)
autocorr.plot(movil.mcmc)
gelman.plot(movil.mcmc)
geweke.plot(movil.mcmc)
raftery.plot(movil.mcmc)
```

Box 7

(4) Parameters. Define the parameters of interest (see Box 4).

(5) Initial Values. Define the starting values for the MCMC runs (see Box 5).

(6) Fit. Fit the model in JAGS (see Box 6).

(7) Diagnostic. Convert the model output into an MCMC object in order to have access to several convergence diagnostics (see Box 7).

Competing Interests

The authors declare that there is no conflict of interests regarding the publication of this paper.

References

[1] C. A. Monje, Y. Chen, B. Vinagre, D. Xue, and V. Feliu, *Fractional-Order Systems and Controls: Fundamentals and Applications*, Advances in Industrial Control, Springer Science & Business Media, London, UK, 2010.

[2] L. Vázquez, J. J. Trujillo, and M. P. Velasco, "Fractional heat equation and the second law of thermodynamics," *Fractional Calculus and Applied Analysis*, vol. 14, no. 3, pp. 334–342, 2011.

[3] I. Area, H. Batarfi, J. Losada, J. J. Nieto, W. Shammakh, and Á. Torres, "On a fractional order Ebola epidemic model," *Advances in Difference Equations*, vol. 2015, article 278, 2015.

[4] B. Jin and W. Rundell, "A tutorial on inverse problems for anomalous diffusion processes," *Inverse Problems. An International Journal on the Theory and Practice of Inverse Problems, Inverse Methods and Computerized Inversion of Data*, vol. 31, no. 3, Article ID 035003, 2015.

[5] J. Kaipio and E. Somersalo, *Statistical and Computational Inverse Problems*, vol. 160 of *Applied Mathematical Sciences*, Springer, New York, NY, USA, 2006.

[6] H. T. Banks, S. Hu, and W. C. Thompson, *Modeling and Inverse Problems in the Presence of Uncertainty*, Monographs and Research Notes in Mathematics, CRC Press, Taylor & Francis, Boca Raton, Fla, USA, 2014.

[7] A. Tarantola, *Inverse Problem Theory and Methods for Model Parameter Estimation*, SIAM, 2005.

[8] S. Donnet, J.-L. Foulley, and A. Samson, "Bayesian analysis of growth curves using mixed models defined by stochastic differential equations," *Biometrics*, vol. 66, no. 3, pp. 733–741, 2010.

[9] M. A. Capistrán, J. A. Christen, and S. Donnet, "Bayesian analysis of ODEs: solver optimal accuracy and Bayes factors," *SIAM/ASA Journal on Uncertainty Quantification*, vol. 4, no. 1, pp. 829–849, 2016.

[10] D. Calvetti, J. P. Kaipio, and E. Somersalo, "Inverse problems in the Bayesian framework," *Inverse Problems*, vol. 30, no. 11, Article ID 110301, 2014.

[11] M. Dashti and A. M. Stuart, "The Bayesian approach to inverse problems," in *Handbook of Uncertainty Quantification*, Lecture Notes in Computer Science, Edited by R. Ghanem, D. Higdon and H. Owhadi, Springer, Berlin, Germany, 2016.

[12] W. Fan, X. Jiang, and H. Qi, "Parameter estimation for the generalized fractional element network Zener model based on the Bayesian method," *Physica A. Statistical Mechanics and its Applications*, vol. 427, pp. 40–49, 2015.

[13] W. Fan, X. Jiang, and S. Chen, "Parameter estimation for the fractional fractal diffusion model based on its numerical solution," *Computers & Mathematics with Applications*, vol. 71, no. 2, pp. 642–651, 2016.

[14] A. E. Gelfand and A. F. M. Smith, "Sampling-based approaches to calculating marginal densities," *Journal of the American Statistical Association*, vol. 85, no. 410, pp. 398–409, 1990.

[15] G. Casella and E. I. George, "Explaining the Gibbs sampler," *The American Statistician*, vol. 46, no. 3, pp. 167–174, 1992.

[16] N. Metropolis, A. W. Rosenbluth, M. N. Rosenbluth, A. H. Teller, and E. Teller, "Equation of state calculations by fast computing machines," *The Journal of Chemical Physics*, vol. 21, no. 6, pp. 1087–1092, 1953.

[17] S. Chib and E. Greenberg, "Understanding the metropolis-hastings algorithm," *The American Statistician*, vol. 49, no. 4, pp. 327–335, 1995.

[18] M. Plummer, JAGS: A program for analysis of bayesian graphical models using Gibbs sampling, 2012.

[19] J. Andrés Christen, C. Fox, D. Andrés Pérez-Ruiz, and M. Santana-Cibrian, "On optimal direction gibbs sampling," https://arxiv.org/abs/1205.4062.

[20] R Core Team, *R: A Language and Environment for Statistical Computing*, R Foundation for Statistical Computing, Vienna, Austria, 2014.

[21] S. Sturtz, U. Ligges, and A. Gelman, "R2WinBUGS: a package for running WinBUGS from R," *Journal of Statistical Software*, vol. 12, no. 3, pp. 1–16, 2005.

[22] Y.-S. Su and M. Yajima, R2jags: A Package for Running "JAGS' from R, R package version 0.05-03, 2015.

[23] M. Plummer, rjags: Bayesian Graphical Models using MCMC, R package version 3–15, 2015.

[24] Stan Development Team, Stan Modeling Language User's Guide and Reference Manual, Version 2.6.1., 2015.

[25] R. M. Neal, "Slice sampling," *The Annals of Statistics*, vol. 31, no. 3, pp. 705–767, 2003.

[26] W. R. Gilks and P. Wild, "Adaptive rejection sampling for gibbs sampling," *Applied Statistics*, vol. 41, no. 2, pp. 337–348, 1992.

[27] Y. Murakami, "Bayesian parameter inference and model selection by population annealing in systems biology," *PLOS ONE*, vol. 9, no. 8, Article ID e104057, 2014.

[28] X. Zhang, J. Cao, and R. J. Carroll, "On the selection of ordinary differential equation models with application to predator-prey dynamical models," *Biometrics*, vol. 71, no. 1, pp. 131–138, 2015.

[29] S. Hug, D. Schmidl, W. B. Li, M. B. Greiter, and F. J. Theis, "Uncertainty in biology," in *Bayesian Model Selection Methods and Their Application to Biological ODE Systems*, pp. 243–268, Springer International, Geneva, Switzerland, 2016.

[30] M.-H. Chen, Q.-M. Shao, and J. G. Ibrahim, *Monte Carlo Methods in Bayesian Computation*, Springer Science & Business Media, Berlin, Germany, 2012.

[31] R. E. Kass and A. E. Raftery, "Bayes factors," *Journal of the American Statistical Association*, vol. 90, no. 430, pp. 773–795, 1995.

[32] S. Chib, "Marginal likelihood from the Gibbs output," *Journal of the American Statistical Association*, vol. 90, no. 432, pp. 1313–1321, 1995.

[33] A. E. Gelfand and D. K. Dey, "Bayesian model choice: asymptotics and exact calculations," *Journal of the Royal Statistical Society. Series B. Methodological*, vol. 56, no. 3, pp. 501–514, 1994.

[34] A. Gelman, J. B. Carlin, H. S. Stern, D. B. Dunson, A. Vethari, and D. B. Rubin, *Bayesian Data Analysis*, Taylor &Francis, 3rd edition, 2014.

18

Study of Two-Sided Similarity Methods Using a Radiation "Switch on" Imploding Shock in a Magnetic Field

J. R. A. J. NiCastro

Systems Science Research Group, P.O. Box 4051 Walnut Creek CA., 94596, USA

Correspondence should be addressed to J. R. A. J. NiCastro; jrajnic@earthlink.net

Academic Editor: Syed Abdul Mohiuddine

This paper explores aspects of two-sided similarity modeling using cylindrical geometry for radiating shock waves embedded in a medium with a magnetic field. Two-sided similarity solution techniques may be used to link states influenced by long range near instantaneous fields that continually modify the pre- and postshock zones. Emergent radiation scaling relations are immediately available from consistent homologies. For both small angle and large angle measurements, an approximate analytic technique in the vicinity of luminous fronts together with the high symmetry implications delineated in Lemma provides direct access to the homology parameters. The parameters obtained using this process can augment the constraint relations and contribute to establishing relevant similarity homologies.

1. Introduction

In nonlinear materials, strong compression or shock phenomenon may result from an initially sufficiently strong impulse, mechanical or otherwise, propagating into the fluid. Similarity techniques have been used to study these problems involving radiation effects. Evidently there has been an interest in radiation driven cylindrical shocks in plasma systems. [1] Vishwakarma and Pandey have studied the propagation of shock waves under the action of monochromatic radiation in nonideal gas using a van deWaals model. [2] Nath and Sahu have studied propagation in an axisymmetric system involving a magnetic field and monochromatic radiation. Further studies have found solutions to spherical plasma systems. [3] NiCastro, in an early work using similarity analysis, studied the possibility of initiating nuclear effects in an electromagnetically driven quasispherical convergent geometry dominated by thermal radiation. [4] Singh has found a solution for the self-similar flow behind a radiative spherical shock wave. [5] Singh and Vishwakarma have studied propagation of spherical waves with radiation heat flux in a dusty gas, a mixture of gas, and small particles and include the effect of an exponential density gradient. [6] D. Mihalas and B. W. Mihalas have prepared a summary bibliography of work on radiation hydrodynamics that spans a history of work and deals with a diversity of problems. [7] Nath, Dutta, and Pathak have obtained an exact solution for propagating shocks including magnetic field, radiative heat flux, and gravitating effects.

When one departs from purely fluid mechanics variables to include radiation (photons and neutrons) and electromagnetic field effects, pre- and postshock regions are coupled by these fields which have long range near instantaneous effects. Both regions will inevitably be modified by these fields during the shock development. Similarity techniques applied to the pre- and postshock regions, and two-sided similarity seems to suggest itself in approaching these problems. More generally, if problems require that the domain be divided by contact surfaces into "n" regions with "p" regions in a constant state, then "n-p" similarity models may be warranted. Some form of conservation equations would be expected to be preserved across the "n-1" contact surfaces.

Similarity equations are based on a symmetrization of the original physics equations and as such imply a strong underlying assumption about the nature of the process. There is no a priori justification for these symmetry assumptions. Assumptions must be justified through an iterative process involving experimental observation and if necessary homology redesign. If the symmetry properties discussed in Lemma 1 are valid, an analytic process, that is an extension of a perspective of [8] Auluck and Tandon, can be used to address this issue using "emergent radiation observables" in the vicinity of a luminous front. The objective of this paper is to develop this process.

Additionally, radiation phenomenon while not initially dominant may become so as the shock onset goes through its development phase and becomes a developed shock front. The initial conditions are masked by what is essentially a radiation switch on (RSO) phenomenon. A radiative "pre-cursor" and a "postcursor" detach the radiating shock from information (initial conditions) about its formation. Indeed it may easily be the case that many different initial conditions lead to the same or indistinguishably similar RSO results! While it is convenient to use assigned values on boundaries to solve numerical problems, the results are academic if the boundary values do not evolve reflecting the energy released in the pre- and postshock zones (fissions and fusion events or even exothermic chemical explosives). Also it would be suspect to define the solution by initial conditions when the dominant phenomenology determining the homology does not develop until the shock is formed. Consider the example case of a tamper plate (contact surface) moving into a latent explosive material. A compression wave may build into a shock and a detonation front is formed. The fluid and thermodynamic parameters at the plate and the movement of the tamper must reflect the evolutionary release of energy occurring behind the detonation front.

These comments suggest in many cases the shock may have to be considered as a standalone event. Complex problems may require dividing the solution domain into more than two regions partitioned by appropriate contact surfaces. In these cases models should be built from the conservation laws and shock jump conditions meshed with measurements and observations. Radiation observables at luminous fronts may play an important role in this process.

Historically the analysis of various aspects of the nature of a shock front has undergone considerable discussion in the literature. It has been suggested by [9] Germain that, to some degree, there appears to have been two avenues of shock wave studies.

In the first case, the continuum physics equations are used to describe the asymptotic character of the shock transition which is represented as a discontinuity across which the conservation laws must apply. Rankine and Hugoniot (R-H) produced the classic result formulating the shock jump conditions from the conservation laws in the shock moving frame which are summarized in [10] Friedrich and Courant. Implosions studied by [11] Guderly and explosions studied by [12] Taylor have used similarity techniques in the strong shock limit. Similarity solutions in the strong shock limit of a magneto hydrodynamic shock have also received some attention by [13] Cole and Greifinger and [14] Greenspan. It would appear that these early works and others might have been motivated by the study of large weapons systems. The "Strong Shock" model assumption, acceptable in modeling strong explosions, belies the issue of coupling the two regions using any two-sided similarity methods.

In the second case, attention is concentrated on the shock transition. Important are the effects of radiation and the dissipative mean free paths, among others, on the character of the transition. Numerous research papers have developed these issues. Some work by [15] Heaslet and Baldwin and [16] Clarke suggests that radiation may smooth the shock structure under certain flow conditions. [17] Mitchner and Vinokur suggest the presence of a magnetic field might to some extent countervail this smoothing effects. [18] Imshennik has discussed shock wave structure in the case the radiation energy and pressure are significant and may exceed the kinetic pressure and energy. These conditions are likely to prevail in stars. One might expect they could also prevail in the early stages of a fusion explosion.

2. Problem

The predicate for using two-sided similarity techniques is that the pre- and postshock regions are tied together by some form of radiation exchange. In this paper, we develop the discussion with thermal radiation but neutron transport and electromagnetic effects are also issues. For example, a neutron "switch on" wave is likely to be accompanied by significant thermal radiation. A neutron "switch on" wave may be characterized as a wave initiated in a medium that has the potentiality of a self-supporting process of fission or fusion or both leading to the emission of neutron radiation. When the wave steepens into a shock or compression wave and either the number of MFP or the "ρx"product of the fluid reaches a critical value, the process may become self-sustaining or may simply be a dissipating pulse. The neutron and thermal radiation fronts are likely to coexist and require new physical considerations. These adjunct processes distinguish these problems from purely mechanical shock problems.

In this paper we use imploding radiating shock in a magnetic field as a vehicle to study some aspects of two-sided similarity problem formulations. The same ideas used in this model can be used in more complex problems, if the solution space is divided by "(n-1)" contact surfaces into "n" zones and "p" zones are a constant states.

Assume a cylindrical column of gas is permeated by a background magnetic field. The cylinder implodes. Conventional explosives would be one of several means of accomplishing this. In laboratory plasmas, external pulsed fields initiate snowplow and shock wave processes. A compression wave is initiated and may build into an isothermal shock. The gas is ionized in front of and behind the shock front by the radiation cooling of the shock zone. The resulting high conductivity can lead to what is referred to as "frozen in magnetic field." The fluid, radiation, and magnetic field sensitively depend on the homologies and the initial values of the dependent variables. It is of importance to exploit the high symmetry characteristics, to be discussed in Lemma 1, of a presumed similarity solution to obtain measurements that would validate or guide the values of homology parameters.

3. Basic Model Equations

The descriptive equations can be tailored depending on the specific engineering problem. In this discussion we will use basic conservation laws. In the text by [19] Sutton and

Sherman, these equations are sometimes referred to as the MHD approximation. It is frequently useful to gain insight into the nature of the physics with a less detailed model, especially if there are measurements available to guide the model development. In cylindrical coordinates the equations take the form

$$r\partial_t \rho + \partial_r (r\rho v) = 0 \quad (1)$$

$$\rho (\partial_t v + v\partial_r v) + \partial_r P - jxB = 0 \quad (2)$$

$$\rho (\partial_t E + v\partial_r E) = -P\left(\partial_r v + \frac{v}{r}\right) - \left(\partial_r F + \frac{F}{r}\right) \quad (3)$$

$$\frac{-1}{r\partial_r (r\partial_r B)} = \mu\sigma\left[-\partial_t B - B\left(\partial_r v + \frac{v}{r}\right) - v\partial_r B\right] \quad (4)$$

Equations (1)-(4) are the continuity, momentum, and energy and magnetic diffusion, respectively. The symbols ρ, v, P, B, j, F, and E are the density, velocity, kinetic pressure, magnetic field, current, radiation flux, and internal energy. In general, the radiation flux may be of any kind, photons or neutrons, for example. Also, the pressure may generally include hydrodynamic, magnetic, and radiation contributions. We have excluded Ohmic heating, assuming the conductivity will be sufficiently large in the region of interest to justify this. It is unlikely this will be the case in all phases of an experiment or phenomenon under study. If thermal gradients are large, then (4) needs to be reassessed.

Generally, P=P(ρ,S). Since F and $\Delta F \neq 0$, $\Delta S \neq 0$. We will use the ideal gas equation of state functional form together with an internal energy proportional to the temperature. So $P=R\rho T$ and $E=c_v T$ serve in this analysis. This is called the polytropic model. We view these equations and the specific heat ratio as an heuristic. If S=constant, the pressure would be proportional to the density to a constant power γ, the specific heat ratio. Also, the ideal gas form of the constitutive equation of state does not always represent the best modeling of high temperature processes.

In some analysis, different values of γ are used to capture a range of behavior due to different degrees of ionization, collision frequencies, and magnetic field strength. In a "frozen in field" B/ρ is constant. One might be inclined to discuss a $\gamma_\perp$ and $\gamma_\parallel$ for the fluid due to the field in a collisionless or weakly collisional gas, as might occur in some laboratory or an astrophysical problem. This might be a fruitful modeling approach in such cases, but in this analysis, we are assuming a collision dominated fluid. Additionally, we will not consider contributions from the thermal conductivity or viscosity. It is important to model the detailed constitutive properties of the fluid, if one does a complete computer calculation.

These basic model equations are usually augmented by Faraday and Ampere equation (5) and generalized Ohm's Law. In this example the magnetic field is axial, the bulk fluid velocity is radial, and the induction electric field and consequent currents are azimuthal. The displacement field term in Ampere equation is ignored.

$$\nabla XE = -\partial_t B \nabla XB = \mu jj = \sigma (E + vXB) \quad (5)$$

Because of the geometry of the B field and its gradient, jxB in (2) can be represented as in

$$j \times B = \left(\nabla \times \frac{B}{\mu}\right) \times B = -\frac{\partial_r B^2}{(2\mu)} \quad (6)$$

We note that the diffusion equation (4) for the magnetic field is derived by using the generalized Ohm's Law in Ampere's equation (5), then by taking the curl of (5).While the high conductivity approximation is acceptable in a high temperature zone, ignoring gradients in the conductivity is suspect if there are strong thermal gradients. In this case (4) would have to be reassessed. Faraday's equation is necessary if the relative gradient in conductivity cannot be ignored and if flux slippage is something to be modeled. Ignoring the tensor character of the conductivity, some researchers have used a constant value or the Spitzer Harm (SH) power law approximation to capture aspects of the temperature dependence of conductivity. $\sigma_c=\chi T^\zeta$; $\chi \approx 1.5x10^{-2}/\ln \Omega$; $\zeta=3/2$. Note that depending on the structure of the plasma, ln $\Lambda \approx 6$-23. This covers a significant range effect.

Within a given problem it is unlikely that just one radiation functional structure will represent the entire domain of the problem. This fact was noted by F.C. Auluck and J.N. Tandon, hereafter referred to as AT, and their perspective will form the basis of the approximate calculations to follow. The Flux "F," (3), is radiation energy transfer internal to the plasma and is likely due to many different processes. The mean free path is of significance, for example, in determining whether the plasma is optically thick, thin, or something intermediate. There are a number of recognized expressions for emergent radiation loss mechanisms that would constitute the "external observables" of the experiment or phenomenon depending on the properties of the plasma.

By "external observables" the signature a measuring instrument external to the phenomenon would record is meant. The measurements might be a superposition of the functions in (7). The "external observables" might have to be decomposed into individual contributions from different sources before using (45). This might require deconvolving the frequency distribution of the measured radiation signature.

Equations (7) are a listing of some these radiation processes along with their dependence on thermodynamic and fluid variables. It is likely that a measurement would contain a superposition of the various sources indicated in (7). In order to condense the presentation, modeling details are condensed in the scaling coefficients. The values of the coefficients depend on the particular model calculation and plasma composition. For example, for Bremsstrahlung, for a pure deuterium plasma ρ^2 would be proportional to n_d^2 and for a mixed deuterium-tritium plasma it would be proportional to $n_d n_t$. For Cyclotron radiation ρ is proportional to the electron number density. In both cases T is proportional to T_e the electron temperature. For a Black body, T is the equilibrium temperature.

Bremsstrahlung: $P = P_{brs}\rho^2\sqrt{T} watts/cm^3$

Cyclotron: $P = P_{cys}\rho T B^2 watts/cm^3$

$$\begin{aligned} \text{Black Body: } & P \\ & = P_{bbs} 5.75\text{x}10^{-5} T^4 ergscm^{-2} sec^{-1} \left(K^0\right)^{-4} \\ \text{Radio Frequency: } & P = P_{rfs} E \times H = P_{rfs} v \times \frac{B^2}{\mu_0} wattscm^{-2} \end{aligned} \tag{7}$$

RF emissions, which are the result of the bulk compression (or movement) of any "embedded" axial magnetic field and the azimuthal induction field at the shock wave interface, are included. RF emission is present in a complete treatment of the energy equation and might serve as a useful diagnostic signature.

Black Body radiation flux diffusion is characterized by the gradient of the radiative energy and is a model considered in many papers. It is not the focus in this paper, but we will indicate the transformation constraints for that radiative model. The frequency averaged mass absorption or Rosseland opacity is κ. The mean free path, μ=1/$\rho\kappa$, is frequently modeled as a product of a power of the density times a separate power of the temperature. This is not done but recognize it allows modeling flexibility and will change the homology representation of the fluid's opacity. It may be possible to model the pre- and postshock material with different mean free path functions, when using the more general representation of the mean free path, as long as the homologies cancel across the shock. The quantity ac/4 is the Stephan Boltzmann constant σ_{R}. This is summarized in

$$\begin{aligned} U^R &= aT^4; \\ F &= -\left(\frac{4\sigma_R}{3\langle\kappa\rangle\rho}\right)\nabla T^4; \quad \sigma_R = \frac{ac}{4} \end{aligned} \tag{8}$$

$$\begin{aligned} r^{-1}\partial_r rF & \\ &= -\left(\frac{4\sigma_R}{3\langle\kappa\rangle}\right)\left[\rho^{-1}\partial_{rr}T^4 + \left((r\rho)^{-1} - \rho^{-2}\partial_r\rho\right)\partial_r T^4\right] \end{aligned} \tag{9}$$

It is unlikely that the electron and ion kinetic temperatures and the radiation field are in thermal equilibrium in most cases. A shape factor equal to the ratio of the systems dimensions to the radiation MFP has been used in modeling radiation diffusion to compensate for this property. In addition to Bremsstrahlung, a significant contribution of the radiation in the presence of a magnetic field would likely come from Cyclotron effects. Unless one is analyzing extreme radiation environments, radiation pressure and energy density can usually be ignored in the energy and momentum equations. In those cases a more complete description of the radiative process is required.

4. Solution

4.1. The Overall Approach. To use (45), a solution (model) on both sides of the vicinity of the shock front needs to be obtained. This calculation is an example of the iterative process of calculation and measurement to help define relevant homologies consistent with measured radiation spectral structure. The solution involves solving or constructing a preshock state and joining the solution across the shock front to the postshock state using the shock jump conditions. We choose to illustrate the technique using analytic approximation but a detailed computer calculation is equally acceptable. However the neighborhood of the luminous front is the only region of interest to perform this analysis. The homologies of (17a), (17b1), (17b2), (17c), (17d), (17e), and (17f) reveal important aspects of the flow, if narrow angle observations about a luminous front, shock front, or contact surface can be made. The approximate analytic or computer solutions about the shock front can be used when the experimental observations about a line of constant η require more than a very narrow subtended observation angle.

To obtain the analytic approximation, this paper extends the suggestion and perspective of AT. In their particular calculation, they work under the assumption that the details of the radiation field are difficult to postulate a priori. This is well suited to the process that involves incorporating experimental data of the luminous front to determine the character of the radiation instead of stipulating it as an assumption. This analysis asserts that if the radiation loss is not a dominant mechanism, the homology should not determine the dynamics. Their analysis centers on the jump equations. We expand their point of view by developing simple approximate two-sided similarity representations in the vicinity of the shock front. The complex dependence on homology parameters, which is frequently lost with detailed discreet computer calculations, can be retained. Then as an operational alternative to postulating the homology of the radiation, the homologies of the external emergent radiation functions can be evaluated and compared with observation ((44a)-(46)). There are at least three approaches to consider.

(1) Assume a strong shock condition where the shock assumes limiting values and the preshock state is not strongly influential in the process. This eliminates the need for structuring the preshock state explicitly but has limited applicability. It is well known that there are substantial photon and neutron precursors that strongly influence the preshock regions in nuclear weapons effects into which the blast wave eventually propagates. Nonetheless the strong blast approach produces useful insights into the mechanical nature of the blast waves and has been successfully utilized in a number of past studies.

(2) The relevant equation can be solved with boundary point conditions that are physically reasonable, such as bounded physical variable values at a well-defined contact surface (piston) for the functions and if necessary their first derivatives. Assigning initial values at select points to obtain solutions is sometimes a necessity but it is in reality at best a convenience since discreet contact surfaces seldom exist in real situations for all the dependent variables. The solutions are more useful if the initial conditions are tied to experimental measurements. There are various techniques for measuring the plasma properties, as described by [20] Griem. Temperature and densities are among them.

(3) The preshock state being strongly influenced by near instantaneous long range field effects can be postulated (calculated), be a prepared laboratory experimental state,

or be the observed result of a natural process. The state can be mathematically structured enabling wide modeling applicability. The strong blast assumption is a limiting case.

5. The Transformation

Introduce the homology representation for the dependent and independent variables equations (10a)-(10b). One of the first applications of this was by [21] Boltzman in his treatment of the heat diffusion problem. Before WWII, [22] Weizacker studied similarity solutions presumable as part of the fission implosion program. While (10c) is a composite variable, the quantities, ρ_λ, T_λ, B_λ, F_λ, r_λ, and t_λ are, respectively, the scale density, temperature and magnetic field and radiation flux, scale length, and time.

$$\rho = r^k \rho_0 \varrho(\eta), \qquad v = \left(\frac{r}{t}\right) \upsilon(\eta), \tag{10a}$$

$$T = \left(\frac{r^m}{t^n}\right) T_0 T(\eta), \qquad B = \left(\frac{r^w}{t^l}\right) B_0 B(\eta), \qquad F = \left(\frac{r^a}{t^b}\right) F_0 F(\eta) \tag{10b}$$

$$\frac{B(r,t)}{\rho(r,t)} = \left(\frac{r^{w-k}}{t^l}\right)\left(\frac{B_0 B(\eta)}{\rho_0 \varrho(\eta)}\right) \tag{10c}$$

$$\eta = \frac{r}{t^\delta}, \qquad \partial_r \eta = \left(\frac{\eta}{r}\right), \qquad \partial_t \eta = -\frac{\delta \eta}{t}, \tag{10d}$$

$$\rho_0 = \frac{\rho_\lambda}{r_\lambda^k}; \qquad T_0 = T_\lambda \left(\frac{t_\lambda^n}{r_\lambda^m}\right); \qquad B_0 = \left(\frac{t_\lambda^l}{r_\lambda^w}\right) B_\lambda; \qquad F_0 = \left(\frac{t_\lambda^b}{r_\lambda^a}\right) F_\lambda \tag{11}$$

The functional structures of (10a)-(11) are not unique and have been used by many authors. The similarity transformation, while a simple mathematical process, implies a strong underlying symmetry assumption about the dependent variables. Equations (12a) are the similarity structures and the scaling relations of the composite emergent radiation functions in (7). The terms in brackets are the respective homologies.

$$\begin{aligned}
&\text{Bremsstrahlung:} \quad P = P_{brs}\left[\frac{r^{(2k+m/2)}}{t^{(n/2)}}\right] \varrho^2(\eta)\sqrt{T(\eta)} watts/cm^3 \\
&\text{Cyclotron:} \quad P = P_{cys}\left[\frac{r^{(k+m+2w)}}{t^{(n+2l)}}\right] \varrho(\eta)\, T(\eta)\, B(\eta)^2\, watts/cm^3 \\
&\text{Black Body:} \quad P = P_{bbs}\left[\frac{r^{4m}}{t^{4n}}\right] 5.75\text{x}10^{-5} T(\eta)^4\, ergscm^{-2}sec^{-1}\left(K^0\right)^{-4} \\
&\text{Radio Frequency} \quad P = P_{rfs} E\times H = \overline{P_{rfs}}\left[r^{(1+2w)}/t^{(1+2l)}\right] v(\eta) \times \frac{B(\eta)^2}{\varsigma_0} wattscm^{-2}
\end{aligned} \tag{12a}$$

The approximate analysis in subsequent sections provides a representation for the dependent invariants in (12a), (12b), and (12c). Equations 12(a) are the radiation observables, among others, that can be used to test the similarity hypothesis.

Lemma 1. *If the problem is self-similar and if any set of functions such as $B(\eta)$,$T(\eta)$,$\upsilon(\eta)$, and $\rho(\eta)$ are dependent invariants, then any function Ψ, (12b), of dependent invariants is a dependent invariant and a constant for each η. The physical function's explicit time and space dependence is determined solely by the composite homology.*

$$\Psi = \Psi\left(\rho(\eta)^{\theta_1}, B(\eta)^{\theta_2}, T(\eta)^{\theta_3}, \upsilon(\eta)^{\theta_4}\right); \quad \forall \theta_1,\theta_2,\theta_3,\theta_4 \tag{12b}$$

This would apply to (12a), the quantities in brackets in (12a) being the composite homology. Furthermore Γ, (12c), is a three-parameter family of functions of physical variables that is invariant for every consistent solution of (13)-(16). This follows since there are four-theta parameters and one δ constraint.

$$\Gamma = \left[\frac{r^{k\theta_1+w\theta_2+m\theta_3+\theta_4}}{t^{\epsilon\theta_1+l\theta_2+n\theta_3+\theta_4}}\right]\Psi;$$

$$\Psi = \rho(\eta)^{\theta_1} B(\eta)^{\theta_2} T(\eta)^{\theta_3} \upsilon(\eta)^{\theta_4}$$

$$\eta^Q = \frac{r^{k\theta_1+w\theta_2+m\theta_3+\theta_4}}{t^{\epsilon\theta_1+l\theta_2+n\theta_3+\theta_4}} \quad Q = k\theta_1 + w\theta_2 + m\theta_3 + \theta_4$$

$$\delta = \frac{(\epsilon\theta_1 + l\theta_2 + n\theta_3 + \theta_4)}{(k\theta_1 + w\theta_2 + m\theta_3 + \theta_4)} \tag{12c}$$

There are strong rationals for physical symmetry transformation such as the Lorentz transformation. There is no a priori physical law that requires sets of physical variables to be mapped (evolve) by these automorphism symmetries. Collapsing space and time into a single independent variable η has induced these symmetries. The symmetry is a modeling assumption. If the similarity hypothesis is true for the particular problem under study, it makes possible comparisons between calculated emergent radiation characteristics and small angle measurements about a line of constant η to test the similarity hypothesis. This includes the trajectories of contact surfaces, shocks and plasma sheaths, and density time profiles. No differential equations have to be solved at this point to accomplish these comparisons, if Lemma 1 is found to apply. If large angle measurements need to be made in the vicinity of the shock line, an approximation procedure based on an extension of the AT concept could be used. This is developed further on.

Comment on Lemma 1. It is likely that in real physical problems many cases would involve a weakly broken automorphism symmetry. The measured homology parameters might be close too but not precisely the calculated values from (17a), (17b1), (17b2), (17c), (17d), (17e), and (17f). This might suggest consideration of a calculation that is a perturbation approach to the similarity equations. Depending upon measurements, this could involve a perturbation to the shock path or any of the dependent invariants. The perturbation would likely break the strong internal symmetries of (12b) and (12c).

Using the transformation equations (10a), (10b), (10c), and (10d), a symmetrization of (1)-(4) is being modeled. Furthermore in reducing the PDEs to ODEs we are not modeling the full physical equations subject to boundary values that might lead to a unique solution in a strict mathematical sense. Furthermore, the process predicate (switch on waves) has detached the shock from initial condition of its formation. The notion of constructing a solution to the ODEs that satisfies some arbitrary initial value problem becomes at best moot unless there is a well-defined contact surface or line of constant η on which there has been measured values or observations.

With these caveats in mind, the PDEs (1)-(4) are transformed into the ODEs (13)-(16). We note that for explosion "t""" increases and implosions "t" decreases but it always remains positive. Since the fundamental equations are time translation invariant one could introduce a new time variable $\tau=T_s-t$, $t < T_s$ for implosions where T_s is the implosion time.

$$(v-\delta)\eta\partial_\eta\varrho + (k+2)\varrho v + \eta\varrho\partial_\eta v = 0 \tag{13}$$

$$\varrho\left[(v-\delta)\eta\partial_\eta v + v(v-1)\right] + \eta^{q_1}M_0\left[(m+k)\varrho T + \eta\partial_\eta\varrho T\right] + \eta^{q_2}M_1\left[2wB^2 + \eta\partial_\eta B^2\right] = 0 \tag{14}$$

$$\left[E_0\varrho\left((mv-n)T + (v-\delta)\eta\partial_\eta T\right) + E_1 T\varrho\left(2v + \eta\partial_\eta v\right)\right] + \eta^{q_3}F_0\left((a+1)F + \eta\partial_\eta F\right) = 0 \tag{15}$$

$$D_0^{-1}\eta^{(-q_4)}\left[w^2B + (2w+1)\eta\partial_\eta B + \eta^2\partial_{\eta_\eta}^{\ 2}B\right] = (v-\delta)\cdot\eta\partial_\eta B + \left((w+2)v - l + \eta\partial_\eta v\right)B \tag{16}$$

where $E_0 = \rho_0 c_v T_0$; $E_1 = R\rho_0 T_0$; $M_0 = RT_0$; $M_1 = B_0^2/2\mu\rho_0$; $D_0 = \mu\sigma$

The continuity equation does not generate a constraint. The constraints from the momentum, energy, and magnetic diffusion are indicated below, respectively. Equation (17b2) is an alternate and more restrictive form than (17b1). Equation (17f) replaces (17c) in the case one uses (8) as a radiation model.

$$\eta^{(q_1)} = \frac{r^{m-2}}{t^{n-2}}; \quad (q_1) = (m-2); \qquad \delta = \frac{(n-2)}{(m-2)} \tag{17a}$$

$$\eta^{(q_2)} = \frac{r^{2w-2-k}}{t^{2l-2}}; \quad (q_2) = 2w-2-k; \qquad \delta = \frac{(l-1)}{(w-1-k/2)} \tag{17b1}$$

$$\eta^{q_1} = \eta^{q_2}; \quad q_1 \equiv q_2; \qquad k+m = 2w, \qquad n = 2l \tag{17b2}$$

$$\eta^{(q_3)} = \frac{r^{a-(1+k+m)}}{t^{b-n-1}}; \quad q_3 = a-(1+k+m); \qquad \delta = \frac{(b-n-1)}{(a-(1+k+m))} \tag{17c}$$

$$\eta^{(q_4)} = \frac{r^{(\zeta m)+2}}{t^{(\zeta n)+1}}; \quad q_4 = (\zeta m)+2; \qquad \delta = \frac{((\zeta n)+1)}{((\zeta m)+2)} \tag{17d}$$

$$\eta^{(q_5)} = \frac{r^{w-k}}{t^l}; \quad q_5 = w-k; \qquad \delta = \frac{l}{(w-k)} \tag{17e}$$

$$\eta^{(q_6)} = \frac{r^{3m-2k-2}}{t^{3n-1}}; \quad q_6 = 3m-2k-2; \qquad \delta = \frac{(3n-1)}{(3m-2k-2)} \tag{17f}$$

Note that ζ=3/2 and ζ=0 correspond, respectively, to SH and a constant conductivity. Equations (13)-(16) are the transformed continuity, momentum, energy, and field diffusion,

TABLE 1

	k	δ	v_0	w	l	m	n	ς	k	δ	v_0	w	l	m	n	ς
NO RADIAL FLOW	0	0.8	0	1.156	1.125	2.313	2.250	1.5	0	0.5	0	2.250	1.625	4.5	3.25	0
RADIAL FLOW	-1	0.8	0.4	0.656	1.125	2.313	2.250	1.5	-1	0.5	0.25	1.750	1.625	4.5	3.25	0

respectively. Equations (17a), (17b1), (17b2), (17c), (17d), (17e), and (17f) are the constraints. Since we must match similarity structures across the shock front along a line of constant η further constraints may be imposed on the homologies.

Corollary 2. *For the particular class of transformations characterized by equations (10a), (10b), (10c), and (10d), the origin is a singular point. By Lemma 1, it will be the case that dependent invariant functions are multivalued at an intersection point of η lines. Hence they are not uniquely defined at this point. For this reason, the solution space should exclude a small region centered about the origin (r=0,t=0). Depending on the circumstances, it may be possible to generate similarity transformations that do not have this problem.*

6. Homology Analysis

It is necessary to find a consistent set of parameters (n, m, δ, ζ, l,w,k) to satisfy the conditions of equations (17a), (17b1), (17b2), (17c), (17d), (17e), and (17f). There is no unique way of doing this. The goal is to obtain (13)-(16) solely a function of η. Powers of η to the qth power will necessarily appear in these equations. In obtaining an homology solution to transform the system, it tacitly assumes the governing constraint properties are valid throughout the entire domain. Plasmas are unlikely to be that consistent. This situation is not reflected in global similarity solution applications. This study is focused on locally approximate solutions in the vicinity of luminous contact surfaces. We consider several cases.

In both Cases 1(a) and 1(b), the conductivity is assumed to be very large so explicit solution of the entire diffusion equation can be ignored but still characterized by the conductivity homology. The parameters ζ,k and n are free.

Case 1(a). For ζ=3/2, σ corresponds to the SH law. The parameter "m" is determined from (17a) and (17d). Using (17a) determines δ. From (17b2), use k+m=2w and n=2l which makes the kinetic and magnetic pressure terms completely homogeneous in η, i.e., $\eta^{q_1} = \eta^{q_2}$; $q_1=q_2$. Flow chart (18) follows:

$$m = \frac{(5n-2)}{4} \longrightarrow \delta = \frac{(n-2)}{(m-2)} \longrightarrow l = \frac{n}{2} \longrightarrow w = \frac{(k+m)}{2} \tag{18}$$

Case 1(b). For ζ =0, σ is a constant. Then from (17d), necessarily δ=1/2. Equation (17a) defines "m" and as before use (17b2) for l and w. Flow chart (19) follows

$$\delta = \frac{1}{2} \longrightarrow m = 2(n-1) \longrightarrow l = \frac{n}{2} \longrightarrow w = \frac{(k+m)}{2} \tag{19}$$

Since the radiation is not a priori specified, for any values of k, m, n, and δ, a relationship exists between a and b from (17c). The flowcharts are not unique and the parameter space has many solutions. The solutions generated by the transformation of equation (10a), (10b), (10c), and (10d) are a symmetry approximation and are used to build an approximate model. Some solutions using (18) and (19) are indicated in Table 1.

Table 1 contains values for k=0 and -1. Because of (28) this corresponds to a zero and nonzero preshock velocity. In some modeling cases a nonzero preshock velocity might be useful. In explosion geometries, near instantaneous preshock radiative heating will begin an expansion before the arrival of the shock front. In implosion geometries as radiation streams into the preshock region there may be a preshock radially inward or outward flow or both in different parts of the region, depending on the nature of the energy transfer process. Photon and neutron mean free paths (MFP) and resultant energy deposition and possibly energy release will influence this.

Case 2. Homology solutions that include an expression of the radiation flux according to (8) and (9) in the energy equation can be constructed. Additionally, the magnetic diffusion can be used for both constant and SH conductivity approximations. Other sequences built from (17a), (17b1), (17b2), (17c), (17d), (17e), and (17f) are possible. The calculation sequence, in flowchart (20), produces consistent homology solutions for this combination. Table 2 summarizes some of these.

For the radiation flux to conform to the T^4 law of (8), set a=4m-k-1 and b=4n in (17c) to obtain (17f). It is useful to note that models express the mean free path as a product of different powers of the density and temperature may be constructed differently in the pre- and postregions as long as the homologies cancel across the shock. Choose "n" and k as independent parameters. Express "m" as a function of n and k using (17a) and (17f). Use (17a) from the momentum equation to express δ as a function of n and m. The diffusion equation, which is part of the solution set, places no constraint on B but the momentum equation does. Thus use (17b2) to establish w and l. Then use relation (17d) from the diffusion equation to express ζ as a function of n, δ, and m. Find solution sets for ζ =0 and 1.5 by searching k, n parameter space. Some parameters satisfying (20) are indicated in Table 2.

$$m = \frac{[n(4-2k)+4k+2]}{5} \longrightarrow \delta = \frac{(n-2)}{(m-2)} \longrightarrow l = \frac{n}{2} \longrightarrow w = \frac{(k+m)}{2} \longrightarrow \zeta = \frac{(2\delta-1)}{(n-m\delta)} \tag{20}$$

These calculation sequences (18), (19), and (20) are obviously not unique. Solutions were obtained by putting these relations

TABLE 2

K	N	M	δ	L	W	ς
0.75	2.5	2.25	2	1.25	1.5	1.5
0.75	3	2.5	2	1.5	1.63	1.5
-3	2.5	3	0.5	1.25	0	0
-3	3	4	0.5	1.5	0.5	0

in a spreadsheet. Many solutions that are available in a spreadsheet analysis make sense and many more do not. Part of the modeling process is to select appropriate homologies to model the problem. The fact that homology solutions to transform the system of equations can be found does not imply that they usefully represent a physically meaningful problem. Observational validation has been the known historical requirement of scaling models.

7. The Preshock Model

In subsequent sections the consequences of extrapolating the basic observation of AT are examined. The spectral character of the radiation flux is left unspecified. The expectation is that measurements of the emergent radiation, (12a), compared with their pre- and postshock similarity functional forms and homologies, (44a)-(46), would provide input on the characteristics of the plasma.

In modeling the core, if there is a strong developing thermal gradient, there would be some density variation and fluid movement to compensate for the pressure imbalance, the result of the energy streaming into the core. Furthermore experimental systems are open-ended and the cylinder being finite would incur end losses. These are difficult nuances to accommodate in a similarity representation. The process used to model the core, (21)-(25), is indicated below. In building core models, the implications of Corollary 2 are applicable. Since the problem involves a standalone RSO system, initial conditions are still an issue. If we measure dependent variable values at (r_o, t_o) then the values at this point are the initial condition at $\eta_o = r_{o/}/t_o^{\delta}$ for developing the solution to the shock front $\eta_s = r_s/t_s^{\delta}$.

(1) Density Profile. When the preshock state is strongly influenced by near instantaneous long range field effects, the density can be parametrically postulated, obtained by observation of a naturally occurring event or structured as a prepared experimental state. When building a density model structure, for any the above reasons, it may be required to cancel homologies involving powers of the space variable. This poses no problem. A function of η can be chosen to always produce bounded values of the real physical variable as indicated.

$$\rho = r^k \rho_0 \eta^{-k} g(\eta);$$
$$\varrho(\eta) = \eta^{-k} g(\eta)$$
$$\eta \partial_\eta \ln \varrho(\eta) = -k + \eta \partial_\eta \ln g(\eta) = H(\eta)$$
$$\text{if } g(\eta) = \exp A\eta;$$
$$H(\eta) = [A\eta - k] \tag{21}$$

This example will address the case of A=0 and H=-k. Consistent with this, the physical density can be a constant (k=0) or spatially uniform and increasing in time (k≠0). Exponential density profiles have received some recent study and, for A≠0, such exponential profiles can be addressed.

(2) Obtain a Velocity Profile from the Continuity Equation Using the Modeled Density Profile in (21). Since A=0, the velocity is determined by the density homology and the dimensionless shock speed. Equation (22) is a linear ODE that can be put in standard form and solved for arbitrary functions H(η) and in particular for constant H.

$$\upsilon(H(\eta) + k + 2) + \eta \partial_\eta \upsilon = \delta H(\eta)$$
$$\upsilon(\eta) = \left(\frac{\eta}{\eta_r}\right)^{-\Phi} \left[\upsilon_0 + \int \left(\frac{\eta}{\eta_r}\right)^{\Phi} \left(\frac{\delta H}{\eta}\right) d\eta\right]; \tag{22}$$
$$\Phi = (k + 2 + H)$$

H=-k in this model and the physical velocity should trend to zero as you approach the origin.

$$\upsilon = \upsilon_0 \left(\frac{\eta}{\eta_r}\right)^{-2} - \frac{k\delta}{2}; \quad \upsilon_0 = 0;$$
$$v(r,t) = -\frac{kr\delta}{2t} \tag{23}$$

(3) The Magnetic Field Is Obtained from an Infinite or Limiting Large Conductivity Model. In the limit, the inverse diffusion parameter D_0 goes to infinity. A second-order equation would become a first-order equation. If it is of interest to model the evolution of the field capture in the core with initially small finite conductivity, the complete diffusion equation (16) may be used in this region. First substitute (23) into the limiting large conductivity approximate form of the similarity diffusion equation (16). Then in the case of a constant radial velocity field, (24) results in (25). The solution is a monomial.

$$\partial_\eta \ln B(\eta) = -\left[\frac{((w + 2)k\delta - 2l)}{\delta(k + 2)}\right]\eta^{-1} \tag{24}$$

$$B(\eta) = B(\eta_o)\left(\frac{\eta}{\eta_o}\right)^{-\beta}; \quad \beta = \frac{[(w + 2)k\delta + 2l]}{\delta(k + 2)} \tag{25}$$

(4) With a Modeling Structure for the Density, Velocity, and Magnetic Field, the Momentum Equation (26) Is in Standard Format and May Be Integrated to Determine the Temperature Field

$$T(\eta) = \exp^{-\int_{\eta_o}^{\eta}((m+k+H)/\eta)d\eta}\left[T(\eta_0) + \int_{\eta_o}^{\eta}\exp^{\int_{\eta_o}^{\eta}((m+k+H)/\eta)d\eta}Q(\eta)\,d\eta\right]$$

$$Q(\eta) = \left[\left(-\frac{M_1\varrho^{-1}}{M_0}\right)\left(2\mathrm{wB}^2 + \eta\partial_\eta B^2\right)\eta^{q_2-q_1-1} - M_0^{-1}\left((v-\delta)\,\eta\partial_\eta v - v + v^2\right)\eta^{-(q_1+1)}\right] \tag{26}$$

Using the model values for the velocity and density equations (21)-(23) in (26) results in (31) and can be expressed as a two-term polynomial.

(5) The Energy Equation Can Be Used to Establish a Consistent Heuristic Functional Form for the Radiation Flux in the Core. There is no specification of the spectral form and homology. The radiation is treated as a moderate to weak loss mechanism. The radiation flux equation can be expressed from the energy equation in standard form.

$$F(\eta) = \left(\frac{\eta}{\eta_0}\right)^{-(a+1)}\left[F(\eta_0) + \int\left(\frac{\eta}{\eta_0}\right)^{(a+1)} Q_E(\eta)\,d\eta\right]$$
$$Q_E = -\frac{\eta^{-q_3-1}\left[E_0\varrho\left((mv-n)\,T + (v-\delta)\,\eta\partial_\eta T(\eta)\right) + E_1 T\varrho\left(\eta\partial_\eta v + 2v\right)\right]}{F_0} \tag{27}$$

The calculation, driven by the density profile, is consistent. The homology of a weak radiation loss mechanism does not drive the dynamics. The values of the variables in the preshock model at $\eta = \eta_o$ are the values at observation or measurement points and the preshock values to use in the jump equations are the functions evaluated at $\eta = \eta_s$. Solution of the jump equation at the shock front, along a line of constant η, provides the initial value to perform the postshock integration. Without observation there are free parameters in the calculation. The solutions (constructed model) are summarized in (28)-(31). The equations, particular to the assumption of the preshock model, can use any of the parameter sets from Table 1.

$$v = -\frac{k\delta}{2} \tag{28}$$

$$\varrho = \varrho_o\left(\frac{\eta}{\eta_o}\right)^{-k}; \tag{29}$$

$$B(\eta) = B(\eta_o)\left(\frac{\eta}{\eta_o}\right)^{-\beta}, \quad \beta = \frac{((w+2)\,v - l)}{(v-\delta)} \tag{30}$$

$$T = \left(\frac{\eta}{\eta_o}\right)^{-m}\left[T_0 + \int_{\eta_o}^{\eta}\left(\frac{\eta}{\eta_o}\right)^{m} Q(\eta)\,d\eta\right]$$
$$Q(\eta) = \varrho_0^{-1}B(\eta_0)^2\left(\frac{\eta}{\eta_0}\right)^{-(k+2\beta)}\left[2(\beta-w)\frac{M_1}{M_0}\eta^{q_2-q_1-1}\right] - \frac{\eta^{-q_1-1}}{M_0}\left(k\frac{\delta}{2}\left(1+k\frac{\delta}{2}\right)\right) \tag{31}$$

Although the process seems somewhat unorthodox, it is in reality not conceptually different from a strong blast model which assumes limiting values. The long range near instantaneous fields will condition the preshock fluid before the arrival of the mechanical shock. Building an appropriate preshock model, by any of the processes indicated above, acknowledges that reality. The solution set is complete and displays the dependence explicitly on the homology numbers. The expense has been to leave unspecified the spectral structure of the internal core radiation flux although the radiation flux can be calculated as a loss. More to the point, the dynamics are not being determined by the radiation homology which in this approximation is not considered a dominant mechanism. The process is sufficiently simple that many computer experiments can be conducted in a spreadsheet varying the structure of the density profile and admissible homologies resulting in corresponding structures for the dependent invariants. Using preshock model equations (28)-(31) and postshock model equations (41a), (41b), and (41c)-(42), the homology parameters can be compared with experimental data using (44a)-(48).

In the infinite conductivity case, D_0→ ∞, the fluid in the core is entrained by the field lines. The fluid is necessarily compressed and both the kinetic and magnetic pressures increase. The increase in postshock density will also occur, among other reasons, due to geometric factors. If the conductivity in the core remains finite, D_0 ≈ O(1), the gas can slip away from the fields and the shock can continue to trap some flux. The field will slip through as well and will build behind the shock front where the conductivity is highest. The full magnetic diffusion equation needs to be considered. With a finite conductivity, in particular a tensor conductivity, the analysis quickly becomes more complicated. Tensor conductivity may be an important issue in studying emergent radiation characteristics such as cyclotron and synchrotron signatures. The models considered are spatially uniform both with and without a temporally increasing density. It is apparent from (21) that other structures, such as a density gradient, simulating a snowplough model profile can be built. This process differs substantially from a foil implosion which is simply compressing field where the density ρ_0 interior to the foil is negligible. The strong shock approximation might be relevant in that type of problem. At some point the shock might eventually dissipate or bounce but the similarity analysis is limited because of Corollary 2.

8. Shock Jump Conditions

The shock jump conditions arise from the conservation equations across the sharp transition and are given in (32b)-(32d). U is the shock velocity. To find a solution across the jump equations it is presumed that the preshock values and shock speed are known, supplemented by a caloric and state equations; the postshock values are then determined from the jump equations. When limiting strong shock waves are assumed, the necessity of structuring a more detailed preshock model is preempted. Equations (32a) summarize this for a polytropic gas. The compression is specified in this limit by the specific heat ratio.

$$\mu^2 = \frac{(\gamma - 1)}{(\gamma + 1)}$$
$$\frac{\rho_b}{\rho_f} = \frac{1}{\mu^2} = \varphi \tag{32a}$$
$$v_b = U\left(1 - \frac{1}{\varphi}\right) = U\left(1 - \mu^2\right)$$

With two-sided similarity this generally is not necessarily the case. In an RSO standalone event, the data classically discussed that determines the transition is not immediately available. In a very detailed computer calculation the value of the long range fields in this region and their effects could be calculated as the shock structure forms. This does not preempt the need to make measurements. In lieu of this, observations are required in the core to execute the similarity calculation.

In this two-sided similarity analysis we have built a preshock model to use with (32b) through (32d). The preshock variables, (29)-(31), evaluated at the shock line are the values that are the necessary conditions. The shock moves on a line of constant η, r=.$\eta_s t^{\delta_i}$; $\partial_t r = \delta\eta_s t^{\delta-1} = \delta r/t$. Now U=r$\delta$/t. In a standalone event, these values are based on observations at $\eta = \eta_o$. Observations or measurements have to be made at an interior point. If observations are not available, there are free modeling parameters. The quantity "m" is the mass flow through the shock.

$$\left[\rho\left(v - U\right)\right]_f^b = 0; \qquad m = \left|\rho\left(v - U\right)\right| \tag{32b}$$

$$\left[\rho\left(v - U\right)^2 + P + \frac{B^2}{2\mu}\right]_f^b = 0 \tag{32c}$$

$$\left[\frac{v^2}{2} + E + \frac{P}{\rho} + \frac{B^2}{\mu\rho} + \frac{F}{m}\right]_f^b = 0 \tag{32d}$$

In formulating (32d), the transition is expressed in terms of a change in total enthalpy. The rate of work done by Lorentz force on the bulk fluid, (jXB·v) has been included. The fluid is assumed infinitely conducting at the jump interface. These approximations have been used by many authors and relaxing these conditions to include finite conductivity, ion slippage, and nontangential fields leads to considerably more complicated equations. The post- and preshock model variables are matched across the shock front on the shock line of constant $\eta = \eta_s$. To form the similarity jump equations (33a)-(33c), substitute the homology functions of equations (10a), (10b), (10c), and (10d) and use the constraints (17a), (17b2), and (17c) in the jump equations (32b)-(32d).

$$\left[\varrho\left(\upsilon - \delta\right)\right]_f^b = 0; \qquad \{m\} = \left|\varrho\left(\upsilon - \delta\right)\right|; \quad m = \frac{\rho_0 r^{k+1}\{m\}}{t} \tag{33a}$$

$$\left[\varrho\left(\upsilon - \delta\right)^2 + M_0\eta_s^{q_1}\varrho T + \eta_s^{q_2}M_1B^2\right]_f^b = 0 \tag{33b}$$

$$\left[\frac{\eta_s^{-q_1}\upsilon^2}{2} + \left(\frac{E_0}{\rho_0} + M_0\right)T + \frac{2M_1B^2}{\varrho} + \frac{FF_0\eta_s^{q_3}}{\left(\{m\}\rho_0\right)}\right]_f^b = 0 \tag{33c}$$

There are terms, $\eta_s{}^q$, that appear in two-sided similarity. These terms have to be included and are the result of (17a), (17b1), (17b2), (17c), (17d), (17e), and (17f) and the two-sided similarity process. Since the evaluation of the jump is along a line of constant η=η_s these terms become multiplicative constants. The preshock and postshock radiation models can be studied as quite different structures. In particular this would apply to the mean free path dependence on the temperature and density. This is possible because we matched the homology in both regions. The value of η_s is a property of the shock trajectory and must be determined either from a measurement, (48), or considered a free modeling parameter. It is a key parameter in building models.

The velocity is expressed as a function of the compression from the continuity equation. The quotient of the magnetic field divided by the density is constant. Substituting the expressions into the momentum equation allows a solution for the compression ratio. The continuity jump, the isothermal limit, frozen in field modeling assumptions, and the momentum are indicated in (34)-(37).

$$\text{Continuity;}\quad v_b = \left(\frac{1}{\varphi}\right)\left(v_f - \delta\right) + \delta \quad \varphi = \frac{\varrho_b}{\varrho_f} \tag{34}$$

$$\text{Isothermal assumption;}\quad T_b = T_f \tag{35}$$

$$\text{Frozen field assumption;}\quad B_b = B_f\varphi \tag{36}$$

$$\text{Momentum Jump}\quad \varphi = \frac{\left[-\left(P_f^k + P_f^m\right) + \sqrt{\left(P_f^k + P_f^m\right)^2 + 4\mathrm{KP}_f^m}\right]}{2P_f^m};\quad P_f^m > 0 \tag{37}$$

where $K(\eta) = \varrho_f(v_f - \delta)^2 P_f^k = M_0\eta_s^{(q_1)}\varrho_f T_f(\eta);\ P_f^m = M_1\eta_s^{(q_2)}B_f^2(\eta)$

The compression φ is determined from the cubic polynomial derived from the momentum equation. So φ =1 is a root and a degenerate case corresponding to a smooth density transition or no shock transition. The remainder is a quadratic and only the positive branch is acceptable. For an isothermal approximation the temperature is assumed continuous. While the radiation flux is usually considered continuous, it need not be unless it is assumed proportional to the gradient in the temperature. The postshock radiation flux amplitude follows from (33c). Because of the assumptions of (35) and (36), the jump in flux is only a function of the change in kinetic energy and magnetic pressure. It is easy to see from (38) the conditions under which the flux would be continuous.

$$F_b = F_f - \left[\frac{F_0\eta_s^{(q_3)}}{\rho_0\varrho_f\left(v_f - \delta\right)}\right]^{-1} \cdot \left[\eta_s^{-q_1}\left(\frac{1}{2v_b^2} - \frac{1}{2v_f^2}\right) + 2M_1\left(\frac{B_f^2}{\varrho_f}\right)(\varphi - 1)\right] \tag{38}$$

The jump equations provide v_b, ρ_b, B_b, T_b, and F_b. The values so obtained can be used to continue the model into the postshock region. If the preshock gas is available for measurement, then there would be an experimental imperative to fashion a preshock model to reflect those measurements. Thinking in terms of preshock and postshock fission or fusion events triggered by the RSO shock, it might be important to build a pre- and postshock model that likely would not be the same.

9. Postshock Model

The postshock calculations continue to use the augmented AT process in dealing with the energy equation. The solution proceeds from the shock front using the initial values derived from the jump equations (34)-(37). The solutions are approximations in the vicinity of the shock front, are not intended to be global in nature and to be used for wide angle observations of the luminous fronts. Without calibrating observation of the similarity model we cannot have rest assurance of the validity of the similarity hypothesis.

The continuity equation (22) was written in standard form. The coefficients for the velocity and density are established from the continuity equation and jump values. For this example, the calculation proceeds by truncating an expansion for the density to one term. Close to the shock line, the density model, either postulated or obtained by measurements, will drive the calculation. So ψ and $\rho_b=\varphi\rho_f$ and δ, the dimensionless shock speed, are central parameters. The compression φ is obtained from (37) and involves the preshock state. The velocity is a solution to (22) with A=0 and H=-ψ

$$\begin{aligned} &\varrho \approx a_1\eta^{-\psi};\quad \psi = (k+1);\\ &\Phi = k + 2 + H;\quad H = -\psi;\\ &v = b_0 + b_1\eta^{-1} \end{aligned} \tag{39}$$

The postshock value initial condition is $v = v_b$. With negative powers of η, it is expected that dependent variables would fall off as "r" increases. This need not be the case if the plasma is being driven by a mechanical piston, explosives, or JXB current sheath. The values for the magnetic field follow by using the solution for the velocity in the high conductivity diffusion equation (40) and using the jump condition.

$$\ln B = -\int\left[\frac{\left[-l + (w+2)\,v + \eta\partial_\eta v\right]}{\eta\,[v - \delta]}\right]d\eta \tag{40}$$

If you depart from the shock line into a region where D_0 cannot be ignored then the diffusion equation must be solved in its entirety. The momentum equation and the energy equation are entirely the same as equations (14) and (15). Note that if ψ=k and b_{-1} = 0 we recover the preshock solution for the density and velocity. Substituting the forms (39) into (13) through (16) and remembering D_0 is large yields (41c) and (42). Both the temperature and flux are ordinary differential equations and can be put in standard form. The temperature is obtained from the momentum equation (26). The flux follows from the energy equation (27). Equations (41c) and (42) are integrable for this model choice of homology parameters.

$$\varrho(\eta) = \frac{a_1}{\eta^{(k+1)}}; \quad a_1 = \varrho_b \eta_s^{(k+1)} \tag{41a}$$

$$v(\eta) = -\delta(k+1) + \left(\frac{\eta_s}{\eta}\right)(\delta(k+1) + v_b) \tag{41b}$$

$$B(\eta) = B_b \exp - \Omega(\eta, \eta_s);$$

$$\Omega(\eta, \eta_s) = \int_{\eta_s}^{\eta} \frac{[(w+2)b_0 - l]\eta + (w+1)b_1}{\eta[(b_0 - \delta)\eta + b_1]} d\eta \tag{41c}$$

$$T(\eta) = \left(\frac{\eta}{\eta_s}\right)^{-(m-1)} \left(T_s + \int_{\eta_s}^{\eta} \left(\frac{\eta}{\eta_s}\right)^{(m-1)} Q(\eta)\, d\eta\right)$$

$$Q(\eta) = -\left(\frac{M_1}{M_0}\right)\left(\frac{\eta}{\eta_s}\right)^{(k+1)}\left(\frac{B_b^2}{\rho_b}\right) \cdot \eta^{q_2 - q_1 - 1}\left[2w - 2\eta\partial_\eta(\Omega)\right]\exp - 2\Omega - \frac{\eta^{-q_1-1}\left[b_0^2 - b_0 + (b_0 b_1 + b_1(\delta - 1))\eta^{-1}\right]}{M_0} \tag{42}$$

The predicate for this entire pre- to postshock approximation modeling process is that a presumed spectral structure for the radiation flux does not drive the process and radiation is a moderate loss mechanism. The changes in the energy are primarily due to PdV work but the process is not isentropic; 0<dF/PdV<<1. Furthermore the density structure function obtained through trial approximation or observation drives the calculation. If the conductivity at the shock front is not limiting large, then Joule heating should be added to the energy equation. Note that since homology solutions exist for the full system, as indicated in Tables 1 and 2, numerical calculations can be accomplished with both radiation and magnetic diffusion. Computer solution would compromise some of the transparency of the homology dependence. However (45)-(48) can still be used to evaluate the homology hypothesis.

If radiation is a dominant process, it would require including radiation pressure in the momentum equation and radiation energy in the energy equation. The approximation process could be reversed. The temperature would be determined by the energy equation and would drive the density and flow velocity through the momentum and continuity equations. This is not likely to be the case in laboratory plasmas.

Had the polytropic and isentropic model been strictly assumed, from the equation of state, the temperature can be expressed explicitly as a function of the density. The continuity equation is a linear ODE and can be integrated, so the velocity can be expressed explicitly as a function of the density. By substitution, the magnetic field is then similarly a function of the density. Although cumbersome, the momentum equation becomes a single equation for the density. Solving that equation determines all the dependent variables consistently. This approximation approach was not taken.

10. Power Radiated

Corollary 3. *By Lemma 1, any integral between two lines of constant η is a constant. Hence narrow angle and wide angle observations have the same homology if taken between lines of constant η. Let the luminous front under observation (shock) and the power function be described generically by*

$$\begin{aligned} r_s &= \eta_s t_s^{\delta}; \\ P &= \overline{P} r^{\alpha} t^{-\beta} P(\eta) \end{aligned} \tag{43}$$

As a result of Lemma 1, Corollary 3 applies to observational integral equation (44a). The narrow angle and wide angle observations have the same time evolution. Also the use of a computer calculation, instead of the above analytic approximation for the dependent invariants, does not change the time evolution in this case. The amplitude changes. For this reason it might be sufficient to compare the normalized outputs of the time evolution of each of the emergent radiation function. The power radiated from a small interval $d\eta$ about a line of constant η at any time t and from a finite interval are given respectively by (44a). In this formulation, the shock thickness or spatial observation interval is collapsing.

$$P = \overline{P}\Delta\theta t^{(\delta(\alpha+2)-\beta)}\left[P(\eta)\eta^{(\alpha+1)}\right] d\eta; \tag{44a}$$

$$P = \overline{P}\Delta\theta t^{(\delta(\alpha+2)-\beta)}\left[\int_{\eta_1}^{\eta_2} P(\eta)\eta^{(\alpha+1)} d\eta\right] \tag{44b}$$

Corollary 4. *From Corollary 3, the time evolution of the emergent power during convergence depends only on the homology and the geometry since the quantities in brackets, (44a), are a constant in time. Hence the quotient of the logarithm of the measured emergent radiation power at observation times t_1 and t_2 divided by the logarithm of the quotient of the observation times is equal to the homology. The number is unique for each type of radiation in (12a), (12b), and (12c). Using the generic formula (43) and (44a), this can be expressed in*

$$\delta(\alpha+2) - \beta = \frac{\ln(P(t_1)/P(t_2))_{Observation}}{\ln(t_1/(t_2))} \tag{45}$$

This holds whether the process consists of observations or is an analytic or computer model. An estimate of the error in the analytic model may be obtained by comparing the difference of the ratio of the observed and analytic or computed power ratios as per

$$\epsilon_r = \left|1 - \frac{\ln(P(t_1)/P(t_2))_{calculated}}{\ln(P(t_1)/P(t_2))_{Observation}}\right| \tag{46}$$

There is another perspective, depending on the nature of the observing instrument, in which a fixed spatial interval Δ_r about the converging front needs to be kept constant. Corollary 3 does not apply to (47). In this case, as convergence occurs, more lines of constant η are added to the field of view and the integral *continues to change. This means the model functions for the dependent variables have to be taken explicitly into account.*

$$P\left(r_j,t\right) = \Delta\theta\overline{P}\int_{r_j-\Delta_r/2}^{(r_j+\Delta_r/2)} r^{\alpha}t^{-\beta}P\left(\eta\right)rdr;$$
$$P\left(r_j,t_j\right) = \overline{P}\Delta\theta t_j^{\delta(\alpha+2)-\beta}\int_{((r_j-\Delta_r/2)/t_j^{\delta})}^{((r_j+\Delta_r/2)/t_j^{\delta})} P\left(\eta\right)\eta^{\alpha+1}d\eta \qquad (47)$$

The pre- and postshock functions to be used for the emergent radiation function in (12a), (12b), and (12c) to evaluate (44a)-(47) are, respectively, available in (28)-(31) and (41a), (41b), and (41c)-(42). For narrow angle observation the integral on each side of the front can be replaced by $d\eta=|\eta^+-\eta^-|$. This interval should be just sufficient to capture the luminous front, since the approximate calculation of the dependent invariants is only intended in the vicinity of the shock front.

$$S = \left\{r_j,t_j\right\};$$
$$\langle\delta\rangle = \left\langle\frac{\ln\left(r_j/r_k\right)}{\ln\left(t_j/t_k\right)}\right\rangle; \qquad (48)$$
$$\langle\eta_s\rangle = \left\langle\frac{r_j}{t_j^{\langle\delta\rangle}}\right\rangle$$

From a streak photograph, the shock line parameters and deviations can be established from a measurement set "S" as indicated in (48). The brackets indicate averages over "S." The transformation, (10d), depends on the structure of the shock line. There will be a statistical deviation in δ and η_s and if they are too large might negate the similarity hypothesis or suggest consideration of a perturbation correction as indicated in the comment to Lemma 1.

11. Concluding Remarks

(1) Two-sided similarity models are built to bridge the gap between conservation equations and physical observables in regions coupled by long range and near instantaneous fields. The bulk emergent radiation functions, (12a), (12b), and (12c), are the observables measured closely around a line of constant η (narrow angle) or a small bundle of lines of constant η (wide angle). The homology of the emergent radiation functions alone determines their time evolution in a narrow angle measurement To accommodate wide angle measurement, the AT concept was extended and was used to obtain approximate characterizations of the thermodynamic parameters of the plasma around shock fronts, contact surfaces, or plasma sheaths. Detailed calculation change the absolute amplitude; they do not change the scaling relations. The analytic approximation explicitly displays the dependence on the homology parameters. Complex dependence of the solutions on the similarity transformation parameters can be obscured with individual discreet numerical calculation.

Measurements must play an important role in building and testing scaling models.

(2) The transformation introduced by Boltzmann in the classic treatment of the heat problem involved one dependent variable that satisfied transformable boundary conditions. The automorphism (10a)-(12a), (12b), and (12c) deals with four variables and results in a strong symmetry relationship between them as per Lemma 1. Symmetries such as time translation and space translation invariance are well studied but similarity automorphisms are in a distinctly different class of their own. The techniques impose a strong symmetry on the physics and hence are part of the solution hypothesis. The implication is that arbitrary composite functions of dependent invariants, (12b), are also invariant along composite space time lines. In addition, (12c) indicate that for every solution to (13)-(16) there exists an internal symmetry of a three-parameter family of physical invariant function sets. This powerful symmetry may or may not capture the nature of the physics. If the phenomenon that drives the homology is a dominant physical process then the transformation would likely capture the physics. As per the comment on Lemma 1, it is suggested that when measurement does not precisely conform to homology parameters, as is likely to be the case, the similarity equations might be the subject of a perturbation analysis and the perturbation would break the strong internal symmetries.

(3) "Switch on waves" are disconnected from initial values by virtue of the long range fields developed after the switch on process occurs. This is especially the case if the radiation field is a "switch on" neutron radiation field. We have discussed Radiation "switch on" in the context of isolated radiation loss mechanisms. Neutron switch on waves are inevitably accompanied by strong thermal radiation environments and when the "switch on" is triggered, the shock is essentially a standalone event. It might seem that the process resembles a classical detonation wave. This would be misleading since there do not exist nearly spontaneous long range influences in classical detonation wave theory.

The preshock and postshock solution can be matched across a shock front (or contact discontinuity) as long as the homologies in both regions cancel or can be absorbed. This provides a form of solution "continuation" of the analytic model. In a standalone event, the shock jump conditions along with the determination of the dimensionless shock speed are insufficient to determine the problem. More data is needed to build models of standalone events. If there are no experiments or measurements then there are free parameters and the solutions must be tested by some other means. In the example calculation, while δ is determined as part of the homology calculation, the shock line $\eta=\eta_s$ would have to be determined by an observation, (48). Also in the preshock model there is an observation point and hence a line of constant $\eta_o= r_o/ t_o^{\delta}$ where observations of the dependent variables are taken. The functions calculated from η_o to η_s determine the preshock values. Along with δ, this determines the postshock values by continuation across the jump equations. These solutions (models) can be used in (44a)-(48) for comparisons with measurements of the emergent radiation functions and measurement of the shock trajectory. In a comparison with the similarity homology structure, (17a), (17b1), (17b2), (17c), (17d), (17e), and (17f) may provide corroborative homology design data.

Disclosure

Dr. J. NiCastro* received his Ph.D. degree from the Case Institute of Technology and was formerly associated with Lawrence Radiation laboratory, Science Applications Incorporated, Boeing Corporation, and Sandia Laboratory.

Conflicts of Interest

*The author declares that there are no conflicts of interest regarding the publication of this paper.

References

[1] J. P. Vishwakarma and V. Kumar Pandey, "Self-Similar Flow under the Action of Monochromatic Radiation Behind a Cylindrical MHD Shock in a Non-Ideal Gas," *Journal of Applied Mathematics*, vol. 2, no. 2, pp. 28–33, 2012.

[2] G. Nath and P. K. Sahu, "Similarity solution for the flow behind a cylindrical shock wave in a rotational axisymmetric gas with magnetic field and monochromatic radiation," *Ain Shams Engineering Journal*, 2016.

[3] J. R. NiCastro, "Similarity Analysis of the Radiative Gas Dynamic Equations with Spherical Symmetry," *Physics of Fluids*, vol. 13, no. 8, pp. 2000–2006, 1970.

[4] S. Singh, "Self Similar flow behind a radiative spherical shock wave," *Advances in Applied Science Research*, vol. 5, no. 5, pp. 94–98, 2014.

[5] K. Singh and J. Vishwakarma, "Propagation of Spherical Shock Waves in a Dusty Gas with Radiation Heat Flux," *Journal of Theoretical and Applied Mechanics*, vol. 45, no. 4, pp. 801–817, 2007.

[6] D. Mihalas and B. W. Mihalas, *Foundations of Radiation Hydrodynamics*, Dover Publications, New York, 2nd edition, 1999.

[7] G. Nath, M. Dutta, and R. P. Pathak, "An exact Solution for the propagation of Shock Waves in Self Gravitating Perfect Gas in the Presence of Magnetic Field and Radiative Heat Flux," *AMSE JOURNALS_AMSE IIETA publication-2017 Series; Modeling B*, vol. 86, no. 4, pp. 907–927, 2017.

[8] F. Auluck and J. Tandon, "Radiation effects on the propagation of plane perpendicular Magnetogasdynamic shock in a plasma," *Proceedings of the National Academy of Sciences India*, no. A, 1966.

[9] P. Germain, "Shock waves and shock-wave structure in magneto-fluid dynamics," *Reviews of Modern Physics*, vol. 32, pp. 951–958, 1960.

[10] K. O. Friedrichs and R. Courant, *Supersonic Flow and Shock Waves*, Springer-Verlag, New York, NY, USA, 1976.

[11] G. Guderley, *"Starke kugelige und zylindrische Verdichtungsstosse in der Nahe des Kugelmittelpunktes bzw. Der Zylinderachse., Luftfahrtforschung", band 19, Lfg.9*, 1942.

[12] G. Taylor, "The formation of a blast wave by a very intense explosion," *Proceedings of the Royal Society*, vol. 201, 1950.

[13] J. D. Cole and C. Greifinger, "Similarity solution for cylindrical magnetohydrodynamic blast waves," *Physics of Fluids*, vol. 5, pp. 1597–1607, 1962.

[14] H. P. Greenspan, "Similarity solution for a cylindrical shock-magnetic field interaction," *Physics of Fluids*, vol. 5, pp. 255–259, 1962.

[15] M. A. Heaslet and B. S. Baldwin, "Predictions of the structure of radiation-resisted shock waves," *Physics of Fluids*, vol. 6, pp. 781–791, 1963.

[16] J. F. Clarke, "Radiation-resisted shock waves," *Physics of Fluids*, vol. 5, no. 11, pp. 1347–1361, 1962.

[17] M. Mitchner and M. Vinokur, "Radiation smoothing of shocks with and without a magnetic field," *Physics of Fluids*, vol. 6, no. 12, pp. 1682–1692, 1963.

[18] V. S. Imshennik, "Shock Wave Structure in a Dense High Temperature Plasma," *Soviet Physics—JETP*, vol. 15, pp. 167–174, 1962.

[19] Sutton and Sherman, *Engineering Magnetohydrodynamics*, Mcgraw Hill, 1965.

[20] H. Greim, *Plasma Spectroscopy*, McGraw Hill, 1964.

[21] L. Boltzmann, "Zur integration des diffusiongleichung bei variabeln diffusions coefficienten," *Annals of Physics*, vol. 53, p. 959, 1894.

[22] C. F. Weizsäcker, "Genäherte Darstellung starker instationärer Stoßwellen durch Homologie-Lösungen," *Zeitschrift für Naturforschung A*, vol. 9, no. 4, 1954.

Permissions

All chapters in this book were first published in JAM, by Hindawi Publishing Corporation; hereby published with permission under the Creative Commons Attribution License or equivalent. Every chapter published in this book has been scrutinized by our experts. Their significance has been extensively debated. The topics covered herein carry significant findings which will fuel the growth of the discipline. They may even be implemented as practical applications or may be referred to as a beginning point for another development.

The contributors of this book come from diverse backgrounds, making this book a truly international effort. This book will bring forth new frontiers with its revolutionizing research information and detailed analysis of the nascent developments around the world.

We would like to thank all the contributing authors for lending their expertise to make the book truly unique. They have played a crucial role in the development of this book. Without their invaluable contributions this book wouldn't have been possible. They have made vital efforts to compile up to date information on the varied aspects of this subject to make this book a valuable addition to the collection of many professionals and students.

This book was conceptualized with the vision of imparting up-to-date information and advanced data in this field. To ensure the same, a matchless editorial board was set up. Every individual on the board went through rigorous rounds of assessment to prove their worth. After which they invested a large part of their time researching and compiling the most relevant data for our readers.

The editorial board has been involved in producing this book since its inception. They have spent rigorous hours researching and exploring the diverse topics which have resulted in the successful publishing of this book. They have passed on their knowledge of decades through this book. To expedite this challenging task, the publisher supported the team at every step. A small team of assistant editors was also appointed to further simplify the editing procedure and attain best results for the readers.

Apart from the editorial board, the designing team has also invested a significant amount of their time in understanding the subject and creating the most relevant covers. They scrutinized every image to scout for the most suitable representation of the subject and create an appropriate cover for the book.

The publishing team has been an ardent support to the editorial, designing and production team. Their endless efforts to recruit the best for this project, has resulted in the accomplishment of this book. They are a veteran in the field of academics and their pool of knowledge is as vast as their experience in printing. Their expertise and guidance has proved useful at every step. Their uncompromising quality standards have made this book an exceptional effort. Their encouragement from time to time has been an inspiration for everyone.

The publisher and the editorial board hope that this book will prove to be a valuable piece of knowledge for researchers, students, practitioners and scholars across the globe.

List of Contributors

Maslina Darus
School of Mathematical Sciences, Faculty of Science and Technology, Universiti Kebangsaan Malaysia, Bangi 43600 Selangor Darul Ehsan, Malaysia

Hameed Ur Rehman
School of Mathematical Sciences, Faculty of Science and Technology, Universiti Kebangsaan Malaysia, Bangi 43600 Selangor Darul Ehsan, Malaysia
Center for Language and Foundation Studies, A' Sharqiyah University, Ibra, Oman

Jamal Salah
College of Applied and Health Sciences, A' Sharqiyah University, Ibra, Oman

Jawad Raza, Azizah Mohd Rohni and Zurni Omar
School of Quantitative Sciences, Universiti Utara Malaysia, 06010 Sintok, Kedah, Malaysia

Limin Zhou, Xinxin Niu and Jing Yuan
Information Security Center, Beijing University of Posts and Telecommunications, Beijing 100876, China

El-Sayed M. E. Mostafa, Abdallah W. Aboutahoun and Fatma F. S. Omar
Department of Mathematics and Computer Science, Faculty of Science, Alexandria University, Moharam Bek, Alexandria 21511, Egypt

Mohamed Ben Said
Mathematical Modeling and Control, Department of Mathematics, Faculty of Sciences and Techniques of Tangier, Abdelmalek Essaadi University, Tangier, Morocco

Lahcen Azrar
Research Center STIS, Department of Applied Mathematics and Informatics, ENSET, Mohammed V University, Rabat, Morocco
Department of Mechanical Engineering, Faculty of Engineering, King Abdulaziz University, Jeddah, Saudi Arabia

Driss Sarsri
LTI, Ecole Nationale des Sciences Appliquées de Tanger, Abdelmalek Essaadi University, Tangier, Morocco

Nissaya Chuathong
Faculty of Science Environment and Energy, King Mongkut's University of Technology North Bangkok (Rayong Campus), Rayong 21120,Thailand

Sayan Kaennakham
School of Mathematics, Institute of Science, Suranaree University of Technology, Nakhon Ratchasima 30000,Thailand
Centre of Excellence in Mathematics, Bangkok 10400, Thailand

Mohammad Alkasassbeh and Zurni Omar
Department of Mathematics, School of Quantitative Sciences, College of Arts and Sciences, Universiti Utara Malaysia, Kedah, Malaysia

J. C. Ndogmo
Department of Mathematics and Applied Mathematics, University of Venda, P/B X5050, Thohoyandou, Limpopo 0950, South Africa

C. Garcia-Martinez
Departamento de Ciencias Básicas, Universidad Autóma Metropolitana Unidad Azcapotzalco, Av. San Pablo 180, Col. Reynosa Tamaulipas, C.P. 02200 Ciudad de México, Mexico

F. R. McMorris
Department of Applied Mathematics, Illinois Institute of Technology, Chicago, IL 60616, USA
Department of Mathematics, University of Louisville, Louisville, KY 40292, USA

R. C. Powers
Department of Mathematics, University of Louisville, Louisville, KY 40292, USA

O. Ortega
Department of Mathematics, Harold Washington College, Chicago, IL 60601, USA

Alibay Iskakbayev
Department of Mechanics, Al-Farabi Kazakh National University, 71 Al-Farabi, Almaty 050040, Kazakhstan

Bagdat Teltayev
Kazakhstan Highway Research Institute, 2A Nurpeisov, Almaty 050061, Kazakhstan

Sergei Alexandrov
Institute for Problems in Mechanics, Russian Academy of Sciences, 101 Prospect Vernadskogo, Moscow 117526, Russia

Gamal M. Abdel-Rahman Rashed
Department of Mathematics, Faculty of Science, Benha University, Benha 13518, Egypt

Mohammad Alkasassbeh and Zurni Omar
Department of Mathematics, School of Quantitative Sciences, College of Arts and Sciences, Universiti Utara Malaysia, Sintok, Malaysia

Ni Hua
Faculty of Science, Jiangsu University, 301 Xuefu Road, Zhenjiang, Jiangsu 212013, China

Dirk Hennig
Erich-Steinfurth-Str. 10/11, 16227 Eberswalde, Germany

D. V. Hieu and D. T. Hung
Thai Nguyen University of Technology, Thai Nguyen, Vietnam

N. Q. Hai
Hanoi Architectural University, Ha Noi, Vietnam

Raid Kamel Naji and Reem Mudar Hussien
Department of Mathematics, College of Science, University of Baghdad, Baghdad, Iraq

Francisco J. Ariza-Hernandez, Jorge Sanchez-Ortiz, Martin P. Arciga-Alejandre and Luis X. Vivas-Cruz
Facultad de Matemáticas, Universidad Autónoma de Guerrero, 39087 Chilpancingo, GRO, Mexico

J. R. A. J. NiCastro
Systems Science Research Group, Walnut Creek CA., 94596, USA

Index

www.ingramcontent.com/pod-product-compliance
Lightning Source LLC
LaVergne TN
LVHW060942220926
848485LV00015B/1376